DATE DUE

Topology and Geometry of Manifolds

Proceedings of Symposia in
PURE MATHEMATICS

Volume 71

Topology and Geometry of Manifolds

2001 Georgia International Topology Conference
May 21 – June 2, 2001
University of Georgia
Athens, Georgia

Gordana Matić
Clint McCrory
Editors

American Mathematical Society
Providence, Rhode Island

PROCEEDINGS OF THE 2001 GEORGIA INTERNATIONAL TOPOLOGY CONFERENCE, HELD AT THE UNIVERSITY OF GEORGIA, ATHENS, GEORGIA, MAY 21–JUNE 2, 2001

with support from the National Science Foundation,
Grant DMS-0110085

Any opinions, findings, and conclusions or recommendations expressed in this material are those of the authors and do not necessarily reflect the views of the National Science Foundation.

2000 *Mathematics Subject Classification.* Primary 57-06, 57M50, 53D10, 57R30, 53D05, 57M27.

Library of Congress Cataloging-in-Publication Data

Georgia International Topology Conference (2001 : University of Georgia)
Topology and geometry of manifolds : 2001 Georgia International Topology Conference, May 21–June 2, 2001, University of Georgia, Athens, Georgia / Gordana Matic, Clint McCrory, editors.
p. cm. – (Proceedings of symposia in pure mathematics, ISSN 0082-0717 ; v. 71)
Includes bibliographical references.
ISBN 0-8218-3507-6 (acid-free paper)
1. Manifolds (Mathematics)–Congresses. 2. Low-dimensional topology–Congresses. 3. Differential topology–Congresses. I. Matic, Gordana, 1957– II. McCrory, Clint. III. Title. IV. Series.

QA613.G48 2001
514–dc22 2003062859

∞ The paper used in this book is acid-free and falls within the guidelines
established to ensure permanence and durability.
Visit the AMS home page at http://www.ams.org/

10 9 8 7 6 5 4 3 2 1 08 07 06 05 04 03

Contents

Preface

The 2001 Georgia International Topology Conference was held at the University of Georgia in Athens, Georgia, from May 21 to June 2. There were forty talks and four problem sessions over this two-week period. This conference has been held every eight years since 1961, for the purpose of bringing together mathematicians to discuss the most exciting new developments in topology. While the invited talks always represent a wide variety of topics, a large number of talks usually focus on a common theme. The choice of speakers for the 2001 conference reflects the current interest in low dimensional topology, symplectic and contact topology, and foliations.

In the last decade gauge theoretic methods in low dimensional topology, in particular Seiberg-Witten theory, were developed, and relationships between gauge theoretic invariants and symplectic and contact topology and geometry were established, as well as deep relations between foliations and contact topology. Analysis of symplectic and contact structures and holomorphic curves emerged as tools to study and construct new invariants for both three and four dimensional manifolds. There were also significant advances in our understanding of foliations and laminations of 3-manifolds. Other rich sources of invariants for 3-manifolds which emerged over the last decade are Vassiliev invariants and topological quantum field theory, including Rozansky-Witten theory.

The two main goals of the Georgia Topology Conference are to give wide exposure to new and important results and to encourage interaction among researchers in different stages of their careers. Towards this end, we encourage speakers to aim their talks at a broad audience of topologists. Some, unfortunately not all, of the expository talks given at the 2001 conference are represented in this volume. Several of the papers are surveys, some with the goal of introducing students and newcomers to the subjects presented, and some with the goal of summarizing the results of years of research. Several papers present new research that was the subject of talks at the conference.

This volume also contains two extensive problem lists. There is a history of problem lists in low dimensional topology and the Georgia Topology Conference. Rob Kirby's famous 1977 problem list was updated and extended for the proceedings of the 1993 Georgia conference, and it is still an important reference. David Gabai's list of problems in foliations and laminations, published in the same 1993 proceedings, is complemented here by *Problems in foliations and laminations of 3-manifolds*, a new list assembled by Danny Calegari, who led a problem session on the topic during the conference.

Contact topology, a new topic of interest in low dimensional topology which emerged in the last decade, was present in the Kirby problem list, but only in a very modest way. John Etnyre and Lenhard Ng's *Problems in low dimensional contact topology* is a new list assembled for these proceedings. It is a product of two problem sessions held at the conference, one on 3-dimensional contact topology led by Emmanuel Giroux, and another led by John Etnyre which focused on Legendrian knots and contact homology.

Both problem lists have been assembled with essential help from many mathematicians, some of whom were present at the conference and some of whom gave input afterwards.

During the conference there were several talks organized for the purpose of introducing graduate students to topics of current interest which were subjects of invited talks at the conference. One of these lectures is presented in extended form in this volume. John Etnyre gave an introductory lecture on contact topology, and the notes in this volume will be a welcome resource for students entering the subject. While there are several sources of introductory material for the more geometric aspects of symplectic and contact geometry, the more topological approach has been a subject of very few expository articles. Etynre's lecture notes are followed by several papers containing expositions of results in the field.

It is our wish to express special thanks to Rob Kirby, who has been a supporter of the Georgia Topology Conference for more than thirty years, and whose participation in the early stages of planing for the 2001 conference was invaluable. The list of principal speakers for the 2001 conference was made with the help of a scientific organizing committee consisting of Mladen Bestvina, Simon Donaldson, Yasha Eliashberg, Steve Ferry, Dave Gabai, Helmut Hofer, Rob Kirby, and Cliff Taubes. As is the tradition at Georgia conferences, there were thirty invited principal speakers over two weeks, and the remaining slots were filled by people whose names were suggested to the scientific committee during the conference.

We would like to thank the authors and the referees for the work they have done in preparing this volume for publication. Many thanks also go to Julie McEver and Gail Suggs, who provided invaluable logistical support for the conference, and to Suzie Radosic Pagnut, who did the final LaTeX formatting.

Funding for the conference was provided by the National Science Foundation (grant DMS-0110085) and the University of Georgia—the Vice President for Academic Affairs, the Georgia Research Foundation, the Franklin College of Arts and Sciences, and the Department of Mathematics. We are grateful for their support.

G. M.
C. M.

Athens, Georgia
May 2003

Proceedings of Symposia in Pure Mathematics
Volume **71**, 2003

Rozansky-Witten theory

Justin Roberts

ABSTRACT. Rozansky and Witten proposed in 1996 a family of new three-dimensional topological quantum field theories, indexed by compact (or asymptotically flat) hyperkähler manifolds. As a byproduct they proved that hyperkähler manifolds also give rise to Vassiliev weight systems. These may be thought of as invariants of hyperkähler manifolds, so the theory is of interest to geometers as well as to quantum topologists. This paper surveys the geometrical construction of the weight systems, how they may be integrated into the framework of Lie algebra weight systems (joint work with Simon Willerton), their applications, and an approach to a rigorous construction of the TQFTs (joint work with Justin Sawon and Simon Willerton).

1. Introduction

The 1996 paper by Rozansky and Witten [**RzW**] opened up a very interesting new area of geometry and topology. They wrote down a physical path integral, a sigma model involving integration over the space of all maps from a closed oriented 3-manifold M^3 into a hyperkähler manifold X^{4n}, which should give rise to a numerical invariant $Z_X(M) \in \mathbb{C}$, at least in the case when X is compact or asymptotically flat. We can view this as a construction of a family of invariants $Z_X(-)$ of 3-manifolds, or of a family of invariants $Z_{(-)}(M)$ of hyperkähler manifolds. Part of the appeal of the theory is the marriage of such different worlds.

Subsequent work by Kontsevich, Kapranov, Hitchin and Sawon [**Ko2, Ka, Sa, HS**] has developed the theory from the geometer's point of view. In particular, Kapranov showed that the hyperkähler condition is unnecessarily strong, and that the theory works for holomorphic symplectic manifolds. This is the point of view we take here.

There are three natural contexts for Rozansky-Witten theory, each with its own motivations:

1. Geometry. Perhaps the most concrete application of Rozansky-Witten theory is to study hyperkähler manifolds. Compact hyperkähler manifolds seem quite

2000 *Mathematics Subject Classification.* Primary 53C26, 57R20; Secondary 57M27, 57R56.

The research described here was partially supported by an EPSRC Advanced Fellowship and NSF grant DMS-0103922. I would like to thank Gordana Matic and the UGA topologists for hosting such an excellent conference.

rare, though as yet there is no guess at a classification theorem. The Rozansky-Witten invariants amount to various complicated computations involving the curvature tensor of the manifold, and as such can be related to characteristic numbers. By combining the Rozansky-Witten invariants with the wheeling theorem of Bar-Natan, Le and Thurston [**Thu**], Hitchin and Sawon obtained a formula for the L^2-norm of the curvature in terms of a topological invariant, which is described in section 9. One might hope to extract (perhaps using the TQFT structure) further identities of this type, and obtain more constraints on the topology of compact hyperkähler manifolds.

A more extensive discussion of the geometrical point of view and its potential applications, together with a list of problems, appears in [**RS**]; the present paper will concentrate on the other two viewpoints.

2. Vassiliev theory. A compact holomorphic symplectic manifold (X^{4n}, ω), together with a holomorphic vector bundle $E \to X$, defines a Vassiliev weight system $w_{X,E} : \mathcal{A} \to \mathbb{C}$ on the usual algebra $\mathcal{A}$ of Jacobi diagrams with an external circle. This should be compared with the more familiar construction: a metric (for example, semisimple) Lie algebra $\mathfrak{g}$, together with a representation V, defines a weight system $w_{\mathfrak{g},V} : \mathcal{A} \to \mathbb{C}$.

In sections 2-4 I will describe the differential-geometric construction of the weight systems, following Kapranov. In sections 5-8 I describe how Simon Willerton and I have integrated these two rather different-looking constructions, using the derived category of coherent sheaves on X.

Our original motivation in studying Rozansky-Witten theory was in fact to try to understand better the algebra $\mathcal{A}$ and its relatives. All our intuition about such diagram algebras seems to be derived from the invariant theory of Lie algebras, but as Vogel [**Vo**] has shown, there is more to them than that. In Rozansky-Witten theory, diagrams behave more like cohomology classes, and this seems a valuable alternative point of view. Whether we will gain true insight remains to be seen.

3. TQFT. A compact holomorphic symplectic manifold X^{4n} defines a 3-dimensional topological quantum field theory Z_X, a functor from the category of 3-dimensional cobordisms to the category of finite-dimensional complex vector spaces. Again, there is a more familiar example: the Chern-Simons theory of Witten, Reshetikhin and Turaev [**Wi, RT**], in which a semisimple Lie algebra $\mathfrak{g}$ and a root of unity $q = e^{2\pi i/r}$ define a TQFT $Z_{\mathfrak{g},q}$. The work of Rozansky and Witten does not actually give a rigorous construction of such a theory, but Justin Sawon, Simon Willerton and I are currently working on providing one: this is described in sections 10 and 11.

This theory is a really non-trivial new kind of TQFT in three dimensions, and exhibits important differences from Chern-Simons theory, most notably that it is not semi-simple, and therefore does not satisfy the usual kinds of gluing and splitting axioms. Though many generalisations of the original axioms of TQFT have been investigated since their introduction twelve years ago, there were until now no compelling examples to make such study worthwhile.

The TQFT also appears to be related to quantum cohomology, quantum K-theory, and mirror symmetry. Perhaps the techniques of 3-*dimensional* TQFT will prove useful in these areas, and this theory will finally provide a link between the mysterious world of quantum and Vassiliev invariants and the more concrete geometric world of Gromov-Witten-type invariants, which seems to include almost all the other modern topological invariants.

2. Weight systems

The *Kontsevich integral* is an invariant of framed oriented knots in S^3 which takes values in (the completion of) a certain rational, graded, algebra $\mathcal{A}$. This algebra is defined as the rational span of vertex-oriented trivalent *Jacobi diagrams* which have a preferred oriented circle, modulo the vertex-antisymmetry and IHX/STU relations. The relations are like the skein relations in knot theory: they relate diagrams which differ only locally. The pictures below shows an example Jacobi diagram and the three relations. In planar pictures such as these, vertex orientations are taken to be anticlockwise. We grade diagrams by their total number of vertices; this is *twice* the conventional grading.

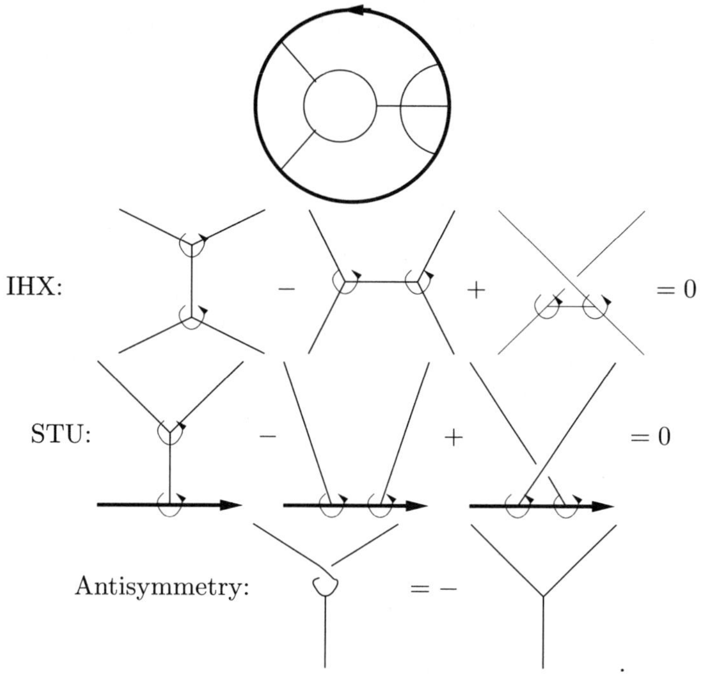

A *weight system* is a linear map $\mathcal{A} \to \mathbb{Q}$, or perhaps to some other finite-dimensional vector space. Such maps enable us to study the structure of $\mathcal{A}$, and by composing with the Kontsevich integral, to construct more manageable scalar-valued knot invariants, the Vassiliev invariants. This is described by Bar-Natan [**BN**].

The standard examples of weight systems, for a long time believed to be essentially the only examples, are those coming from *metric Lie algebras*. Let $\mathfrak{g}$ be a finite-dimensional Lie algebra with a non-degenerate invariant symmetric bilinear form b, and let V be a finite-dimensional representation of $\mathfrak{g}$. These structures are encoded by linear maps $\mathfrak{g} \otimes \mathfrak{g} \to \mathfrak{g}$, $\mathfrak{g} \otimes \mathfrak{g} \to \mathbb{Q}$, and $\mathfrak{g} \otimes V \to V$ respectively. Using the metric to identify $\mathfrak{g} \cong \mathfrak{g}^*$ enables us to transform these into equivalent but more suitable structure tensors: a totally skew version $f \in \Lambda^3 \mathfrak{g}^*$ of the bracket; the Casimir $c \in S^2\mathfrak{g}$; and the action $a_V \in V^* \otimes V \otimes \mathfrak{g}^*$.

A Jacobi diagram of the kind spanning $\mathcal{A}$ defines a procedure for contracting these tensors together and obtaining simply a rational number. To do this, associate a copy of f with each "internal" vertex, a copy of a_V with each "external" vertex (those on the preferred circle), and a copy of c with each edge of the graph; now contract the $\mathfrak{g}$–$\mathfrak{g}^*$ and V–V^* pairs throughout. Note that the orientation conventions correspond: we associate skew 3-tensors to skew vertices, and symmetric 2-tensors to unoriented edges. The Jacobi identity and the identity expressing that V is a representation of $\mathfrak{g}$ correspond to the IHX and STU relations, showing that this evaluation descends to the algebra $\mathcal{A}$ and gives a weight system $w_{\mathfrak{g},V} : \mathcal{A} \to \mathbb{Q}$.

3. Chern-Weil theory on a complex manifold

An attractive way to introduce Rozansky-Witten invariants is as a generalisation of the usual Chern-Weil theory. Suppose $E \to X$ is a smooth complex vector bundle on a smooth manifold X. Picking any smooth connection on E defines a curvature form $F \in \Omega^2(X; \mathrm{End}(E))$, from which we seek to produce topological invariants of E, quantities independent of the choice of connection. It is well-known that the forms

$$\mathrm{ch}_d(E) = \frac{1}{d!} \operatorname{tr} \left\{ \left(\frac{-F}{2\pi i} \right)^d \right\}$$

are closed, and that varying the connection alters them by coboundaries, so that their de Rham cohomology classes in $H^*(X; \mathbb{C})$ are indeed invariants, namely the components of the Chern character of E. The theory of characteristic classes shows that these are the only functions of the curvature one need consider; in any case, there are no other sensible ways to combine an endomorphism-valued 2-form with itself and end up with a complex-valued differential form on X. A schematic picture of the form representing $\mathrm{ch}_d(E)$ (in the case $d = 8$) is shown below. Each arrow represents a copy of F, the blob denoting the "2-form-ness" and the arrow shows that F, as a section of an endomorphism bundle has one "input" and one "output". Eight disjoint arrows would represent the wedge product $F^8 \in \Omega^{16}(X; \mathrm{End}(E)^{\otimes 8})$; the circle illustrates the contraction of consecutive inputs and outputs which computes the trace of the product of the endomorphisms.

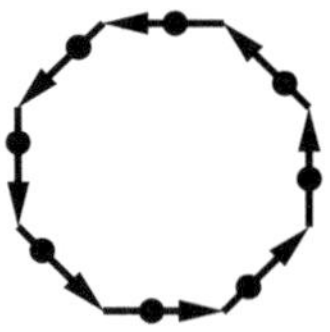

If we look at a bundle with a structure group other than $GL(n, \mathbb{C})$ then there may be more subtle algebraic operations we can use; in fact, the ring of invariant polynomials on the Lie algebra parametrises the different (complex-valued) characteristic classes for G-bundles. For example, in the real oriented case with structure group $SO(2n)$, the ring is generated by the traces of even powers, corresponding to the Pontrjagin classes, together with the Pfaffian, corresponding to the Euler class.

The basis of the Rozansky-Witten refinement (due, in this form, to Kapranov) occurs when E is a *holomorphic* bundle on a *complex manifold* X. In this case there is a preferred class of connections, namely those compatible with the holomorphic structure and with some smooth hermitian metric on E. The curvature form F_E of such a connection lies in $\Omega^{1,1}(\mathrm{End}(E))$, but can also be thought of as an element $R_E \in \Omega^{0,1}(T^* \otimes \mathrm{End}(E))$, where T^* denotes the holomorphic cotangent bundle of X. Let R_T denote a corresponding curvature form $R_T \in \Omega^{0,1}(T^* \otimes \mathrm{End}(T))$ for the *holomorphic tangent bundle* T. Each of these can be pictured as a trivalent vertex with two input legs, one output leg, and a vertex carrying the "1-form-ness"; larger trivalent graphs can then be used to index the ways in which they can be combined tensorially, yielding a richer range of possibilities than in the basic Chern-Weil case.

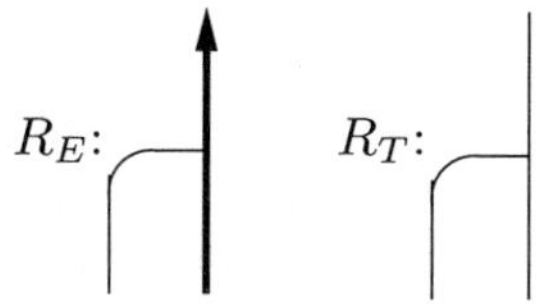

In particular, the following three pictures denote elements of $\Omega^{0,2}(T^* \otimes T^* \otimes \mathrm{End}(E))$ obtained by forming the exterior products $R_T \wedge R_E, R_E \wedge R_E, R_E \wedge R_E$ and then contracting tensorially according to the graph. The basic lemma is that the sum of these three elements is a $\bar\partial$-coboundary. Cohomologically, it will become the Jacobi or IHX identity.

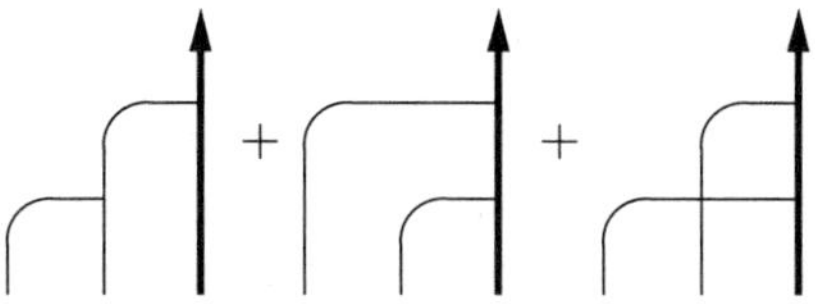

$= \bar\partial(\ldots).$

A similar lemma shows that the element R_T is symmetric in its two inputs, up to a coboundary. Kapranov shows how this identity (which is really just the Bianchi identity on $T^* \otimes E^* \otimes E$, expanded using the Leibniz identity), defines the structure of an L_∞*-algebra* on the Dolbeault complex of forms with values in T. We will not use this result here, preferring to work at the level of cohomology.

4. Rozansky-Witten weight systems

Now let (X^{4n}, ω) be a *holomorphic symplectic manifold*, that is a complex manifold of *real* dimension $4n$, with $\omega \in \Omega^0(\Lambda^2 T^*)$ a holomorphic non-degenerate skew 2-form on X. Non-degeneracy of ω gives a holomorphic identification $T \cong T^*$, and we can use this to convert the curvature form R_T and symplectic form ω itself into alternative versions

$$f_T \in \Omega^{0,1}(T^* \otimes T^* \otimes T^*), \qquad \omega^{-1} \in \Omega^0(\Lambda^2 T).$$

The 1-form f_T turns out to be totally symmetric in its three tensor indices.

Suppose we take a v-vertex trivalent graph with a preferred oriented circle. By associating f_T to its internal vertices, ω^{-1} to its edges, R_E to its external vertices and wedging/contracting accordingly, we obtain an element of $\Omega^{0,v}(X)$. This element is $\bar\partial$-closed, and varying the connections on E, T alters it by a coboundary, giving a well-defined cohomology class. To avoid a sign ambiguity here we need to orient the edges of the graph (because ω^{-1} is skew) and order its vertices (because the vertices carry 1-forms, which anticommute). Remarkably, such an orientation of a graph is naturally equivalent to a vertex-orientation of the kind required in defining $\mathcal{A}$. The IHX and STU identities are satisfied because of the Bianchi identity, and the result is that we have a weight system

$$w_{X,E} : \mathcal{A}^v \to H^{0,v}_{\bar{\partial}}(X).$$

If X^{4n} is *compact* then we can make a weight system $\mathcal{A}^{2n} \to \mathbb{C}$ by associating to a graph Γ the integral

$$b_\Gamma(X, E) = \int_X w_{X,E}(\Gamma) \wedge \omega^n.$$

These are the original Rozansky-Witten invariants studied by Sawon [**Sa**]. Now irreducible compact holomorphic symplectic manifolds seem quite rare: Kähler ones are hyperkähler, by a theorem of Yau, and the known examples comprise two infinite families (the Hilbert schemes of points on the 4-torus or $K3$ surface) and two exceptional examples due to O'Grady; even non-Kähler ones are relatively few.

However, the construction of the cohomology-valued $w_{X,E}$ described above does not require X to be compact. Although in general we may know little about the target Dolbeault cohomology groups, these cases may turn out to be the most interesting for geometers, simply because there are so many examples of (non-compact) moduli spaces arising in geometry and physics which are naturally hyperkähler, and therefore holomorphic symplectic.

5. Weight systems revisited

What I will now explain is how to bring together the two extremely different constructions of weight systems we have seen. The basic ingredient is a reformulation of the construction of section 2 in a way that generalises to metric Lie algebras *in categories other than the category of vector spaces.* We will see subsequently that a holomorphic symplectic manifold gives rise to such a category and Lie algebra.

Suppose $\mathcal{C}$ is a *symmetric $\mathbb{C}$-linear tensor category.* This is a category whose morphism sets are complex vector spaces, and which is equipped with an associative tensor operation $\otimes : \mathcal{C} \times \mathcal{C} \to \mathcal{C}$, commutative in the sense that there are isomorphisms $\tau_{A,B} : A \otimes B \to B \otimes A$, for all pairs of objects A, B, satisfying $\tau_{A,B}\tau_{B,A} = \text{id}$. The most obvious example other than the usual category of complex vector spaces is the category of *super-* (or $\mathbb{Z}_2$*-graded*) vector spaces, in which the commutativity isomorphism incorporates a sign to make odd elements anticommute with one another. It is the fact that this commutativity isomorphism cannot be ignored that led Vogel and Vaintrob [**Vo, Va**] to the picture I am about to describe: for Lie superalgebras, the standard construction explained earlier just doesn't work.

An object L in $\mathcal{C}$ is a *Lie algebra* if it is equipped with a morphism $L \otimes L \to L$ satisfying the Jacobi identity, interpreted as a linear relation between three morphisms in $\text{Mor}(L^{\otimes 3}, L)$. Pictorially this is described by

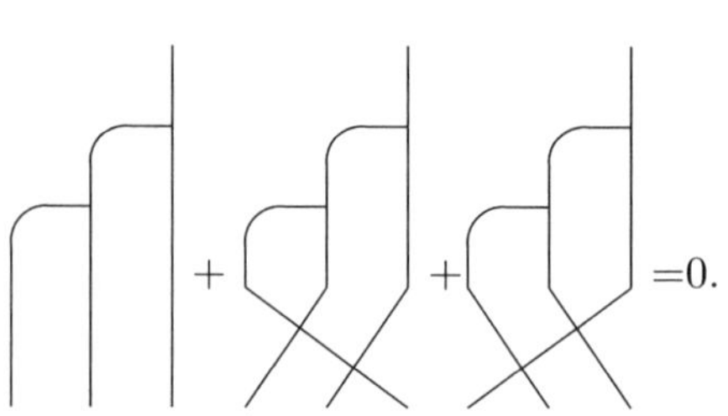

where the pictures are read from bottom to top, the trivalent vertex represents the bracket, and the quadrivalent crossing represents the commutativity morphism $\tau_{L,L}$. Such planar pictures are very useful for describing compositions in a tensor category, and are very common in the TQFT literature. See Bakalov and Kirillov [**BK**], for example.

Let 1 be the unit object for the tensor product in $\mathcal{C}$. A *non-degenerate metric* on a Lie algebra is a pair of morphisms $L \otimes L \to 1, 1 \to L \otimes L$, pictured using a "cap" and "cup", which satisfy a certain "S-bend" relation. A *left module* M over L is an object with a morphism $L \otimes M \to M$, pictured (just like the form R_E in section 3) as a trivalent vertex with a special oriented line marking the legs corresponding to M.

A pair (L, M) determines a weight system

$$w_{L,M} : \mathcal{A} \to \mathrm{Mor}_{\mathcal{C}}(1, 1).$$

To compute it, a diagram in $\mathcal{A}$ must first be drawn in the plane in a Morse position with respect to the vertical coordinate. Slicing it into horizontal bands, each containing a single vertex, cup or cap, defines a way to compose the structural morphisms above, starting and ending at the trivial object 1. (An attempt at Morsifying the example diagram from section 1 is shown below.) The result turns out to be independent of the planar picture used (remember that the diagrams in $\mathcal{A}$ are simply abstract trivalent graphs) and satisfies the correct orientation and IHX relations.

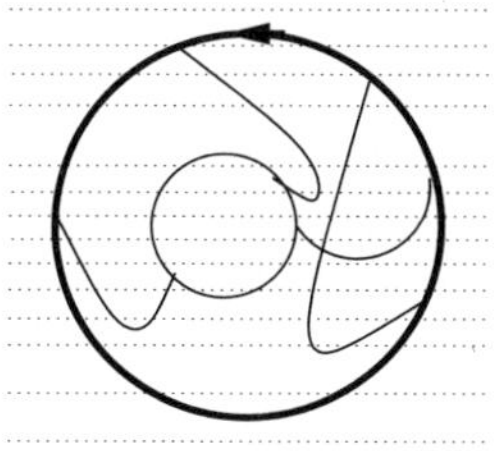

6. Derived categories

The use of derived categories is necessary for the algebraic unification we are proposing, and is even more important for the construction of the TQFT. This section contains a brief description of their structure, concentrating on the fact that *morphism sets in a derived category are cohomology groups.* Because of this, the derived world is the natural place to work if one wants to compose lots of cohomology classes. For details, see Thomas [**Tho**] and Gelfand and Manin [**GM**].

Let $\mathcal{C}$ be an abelian category, for example the category of modules over some ring R. Thus, $\mathcal{C}$ has direct sums, kernels and cokernels, and it makes sense to consider the category of (bounded) chain complexes $\mathrm{Ch}(\mathcal{C})$. The most straightforward notion of equivalence for chain complexes is *chain homotopy equivalence*, and it is straightforward to pass to the corresponding homotopy category $K(\mathcal{C})$. But this is not really the correct notion: better is to use the *quasi-isomorphisms*, the chain maps which induce isomorphisms on homology, to generate the equivalence. (Whitehead's theorem in topology, that a map between simply-connected CW-complexes inducing isomorphisms on homology is a homotopy-equivalence, is a helpful justification here.) Now homotopy-equivalences are always quasi-isomorphisms, but the converse is not true. For example, a quasi-isomorphism from a complex of free modules to a complex of torsion modules has no inverse, and this also shows that quasi-isomorphism isn't an equivalence relation. By forcibly inverting the quasi-isomorphisms in $K(\mathcal{C})$, we obtain a quotient category $D(\mathcal{C})$ — the derived category of $\mathcal{C}$.

It is characterised by the universal property that any functor defined on $\mathrm{Ch}(\mathcal{C})$ which takes quasi-isomorphisms to isomorphisms (the most obvious example being the homology-group functors $h^i : \mathrm{Ch}(\mathcal{C}) \to \mathcal{C}$) factors through $D(\mathcal{C})$. The objects of the derived category are the same as those of $\mathrm{Ch}(\mathcal{C})$. In particular, objects of the original category $\mathcal{C}$ may be identified with chain complexes whose only non-zero term lies in degree zero.

Under suitable conditions, a functor $F : \mathcal{A} \to \mathcal{B}$ between abelian categories may also be derived to a functor $D(F) : D(\mathcal{A}) \to D(\mathcal{B})$. The *classical derived functors* associated to F are just the composites of the homology functors h^i with $D(F)$. For example, if $\mathcal{C}$ is the category of R-modules and $\mathcal{A}$ the category of abelian groups, then the "R-invariant part" functor $\Gamma = \mathrm{Hom}_R(1, -) : \mathcal{C} \to \mathcal{A}$ has a derived functor denoted $\mathbf{R}\Gamma$, whose classical derived functors $H^i = h^i\mathbf{R}\Gamma$ are the standard cohomology functors for R-modules, and are computable using resolutions. (In classical homological algebra there is an inessential distinction between applying such a functor to an object of $\mathcal{A}$, computing e.g. the cohomology of a module, and to an object of $D(\mathcal{A})$, computing e.g. the hypercohomology of a complex of modules.) A second example is that the bifunctor $\mathrm{Hom}_R(-,-) : \mathcal{C}^{\mathrm{op}} \times \mathcal{C} \to \mathcal{A}$ has a derived functor $D(\mathcal{C})^{\mathrm{op}} \times D(\mathcal{C}) \to D(\mathcal{A})$ denoted $\mathbf{R}\mathrm{Hom}$ or Ext, whose classical derived functors are the bifunctors $\mathrm{Ext}^i_R(-,-)$.

The morphisms in $D(\mathcal{C})$ are therefore (hyper)cohomology groups, in the form of Ext-groups. In particular, for objects A, B of $\mathcal{C}$,

$$\mathrm{Mor}_{D(\mathcal{C})}(A, B) = \mathrm{Ext}^0_{\mathcal{C}}(A, B) = \mathrm{Hom}_{\mathcal{C}}(A, B)$$

which shows that $\mathcal{C}$ is embedded in $D(\mathcal{C})$ as a full subcategory. More generally, we have

$$\mathrm{Mor}_{D(\mathcal{C})}(A, B[i]) = \mathrm{Ext}^i_{\mathcal{C}}(A, B),$$

where $[i]$ denotes the functor which shifts the gradings of complexes by $-i$. Under this identification, the Yoneda product of cohomology classes

$$\mathrm{Ext}^i(A, B) \otimes \mathrm{Ext}^j(B, C) \to \mathrm{Ext}^{i+j}(A, C)$$

becomes composition of morphisms, shifted so as to make sense. The construction of weight systems is, in effect, a complicated Yoneda product, which is far better expressed as a composition of morphisms.

7. Rozansky-Witten weight systems

Let X be a holomorphic symplectic manifold. We know that we can construct weight systems $w_{X,E}$ out of holomorphic vector bundles E on X. In order to fit these into the framework of section 5, we want to view these bundles as modules over a Lie algebra in some category.

Ignore the symplectic structure for now and think of X simply as a complex manifold. Let $\mathcal{O}_X$ be its *structure sheaf*, the sheaf of germs of holomorphic functions on X. It is a sheaf of *rings*, and the sheaves of germs of sections of vector bundles are examples of sheaves of $\mathcal{O}_X$*-modules*, in fact *locally-free* ones. If we consider the more general *coherent sheaves of* $\mathcal{O}_X$*-modules*, those which are locally quotients of finite-rank locally-free sheaves, then we obtain an abelian category. It is then a natural step to pass to the derived category, which we denote by $D(X)$.

The objects of $D(X)$, then, are bounded chain complexes of coherent sheaves, though thinking of them as complexes of holomorphic vector bundles won't do too much harm. The morphism sets are cohomology groups: in particular, for a pair of sheaves E, F (viewed as complexes concentrated in degree zero, as described earlier) we have

$$\mathrm{Mor}_{D(X)}(E, F[i]) = \mathrm{Ext}^i(E, F),$$

a convenient description of the usual sheaf cohomology Ext-groups. This language subsumes the earlier differential-geometric language: in fact, when E is a holomorphic vector bundle, we can identify the Dolbeault cohomology groups, via sheaf cohomology groups, as morphism sets:

$$H^{0,i}_{\bar\partial}(X; E) = H^i(E) = \mathrm{Ext}^i(\mathcal{O}_X, E) = \mathrm{Mor}_{D(X)}(\mathcal{O}_X, E[i]).$$

Notice that $\mathcal{O}_X$ is the *unit object* in the tensor category $D(X)$, and that the last term in this sequence may be thought of as a kind of "space of invariants" of the object $E[i]$, by analogy with the situation in a module category.

In order to construct a Lie algebra in this category we need a sheaf-theoretic version of the curvature form of a bundle. Suppose E is a holomorphic vector bundle on X. Its *Atiyah class* $\alpha_E \in \mathrm{Ext}^1(E, E \otimes T^*)$ is the extension class (obstruction to splitting) arising from the exact sequence

$$0 \to E \otimes T^* \to JE \to E \to 0,$$

where JE denotes the bundle of 1-jets of sections of E. (See [**A**] for the details.) It corresponds to the cohomology classes of the curvature forms R_E and F_E used in section 3 under the isomorphisms

$$\mathrm{Ext}^1(E, E \otimes T^*) \cong H^1(E^* \otimes E \otimes T^*) \cong H^{0,1}_{\bar\partial}(E^* \otimes E \otimes T^*) \cong H^{1,1}_{\bar\partial}(\mathrm{End}(E)).$$

For our purposes it is better to make the identification

$$\mathrm{Ext}^1(E, E \otimes T^*) \cong \mathrm{Ext}^1(E \otimes T, E) \cong \mathrm{Mor}_{D(X)}(E \otimes T, E[1]),$$

because this begins to look like the structural map of a right module. Applying the shift functor $[-1]$ to each side produces a morphism

$$\alpha_E : E \otimes T[-1] \to E.$$

Repeating this for the tangent bundle T, with a shift of $[-2]$ gives

$$\alpha_T : T[-1] \otimes T[-1] \to T[-1].$$

This is in fact a Lie bracket: it is *skew* rather than symmetric, because of the parity shift, and the earlier Bianchi identity becomes the Jacobi identity. It is possible to extend the definition of the Atiyah class α_E to arbitary coherent sheaves and even complexes of them, so that we have the following theorem.

THEOREM. *If X is a complex manifold then the object $T[-1]$ is a Lie algebra in the category $D(X)$, and all other objects in $D(X)$ are modules over it.*

Note that this is true for any complex manifold. So what about the symplectic structure? By viewing ω first as as a morphism $\mathcal{O}_X \to \Lambda^2 T^*$ and then dualising and shifting judiciously, we construct a pair of morphisms

$$\omega : T[-1] \otimes T[-1] \to \mathcal{O}_X[-2], \qquad \omega^{-1} : \mathcal{O}_X \to T[-1] \otimes T[-1][2].$$

These are actually *symmetric* rather than skew, again as a result of the interaction of the shift with the notion of commutativity.

Because these morphisms have non-trivial "degrees", namely the outstanding shifts $[\pm 2]$, we have to alter the category $D(X)$ in order to be able to interpret them correctly as defining a non-degenerate metric. Define the *graded derived category* $\tilde{D}(X)$ to have the same set of objects as $D(X)$, but with the space of morphisms $A \to B$ being the graded vector space $\mathrm{Ext}^*(A, B)$, instead of just $\mathrm{Ext}^0(A, B)$. Composition of morphisms in $\tilde{D}(X)$ is graded bilinear.) After this "deformation", we can view ω as a genuine morphism $T[-1] \otimes T[-1] \to \mathcal{O}_X$, and hence as a metric.

This deformation seems pleasing rather than puzzling if one compares it with the deformation quantization of the category of representations of a semisimple Lie algebra arising from the Knizhnik-Zamolodchikov equation, which is explained further in section 10. In summary:

THEOREM. *If X is a holomorphic symplectic manifold then $T[-1]$ is a metric Lie algebra in the graded derived category $\tilde{D}(X)$, and all other objects are modules over it.*

From this theorem and the methods of section 5 we get weight systems $w_{X,E} : \mathcal{A} \to H^*(\mathcal{O}_X)$ for each object E of $D(X)$.

8. Wheels and wheeling

Let $\mathfrak{g}$ be a complex semisimple Lie algebra, $U(\mathfrak{g})$ its universal enveloping algebra, and $S(\mathfrak{g})$ its symmetric algebra. The *Poincaré-Birkhoff-Witt theorem* says that the map $PBW : S(\mathfrak{g}) \to U(\mathfrak{g})$, given by including $S(\mathfrak{g})$ into the tensor algebra $T(\mathfrak{g})$ and then projecting, is an isomorphism of (filtered) vector spaces. It is equivariant, so gives an isomorphism $S(\mathfrak{g})^{\mathfrak{g}} \cong U(\mathfrak{g})^{\mathfrak{g}} = Z(\mathfrak{g})$.

Now $S(\mathfrak{g})$, $U(\mathfrak{g})$ are algebras, and though they aren't isomorphic as algebras (one is commutative and one non-commutative!), their invariant parts are isomorphic as algebras, though not by the restriction of the PBW map. The isomorphism is called the *Duflo isomorphism*, and is built from the function

$$j^{\frac{1}{2}}(x) = \det{}^{\frac{1}{2}} \left\{ \frac{\sinh(\mathrm{ad}\, x/2)}{(\mathrm{ad}\, x/2)} \right\}$$

on the Lie algebra $\mathfrak{g}$. One can define a map $S(\mathfrak{g}) \to S(\mathfrak{g})$ by viewing this function as lying in the completion of $S\mathfrak{g}^*$ and applying the contraction $S(\mathfrak{g})^* \otimes S(\mathfrak{g}) \to S(\mathfrak{g})$, which is a kind of convolution or "cap product". The Duflo theorem is that the composite

$$S(\mathfrak{g})^{\mathfrak{g}} \xrightarrow{j^{\frac{1}{2}} \cap} S(\mathfrak{g})^{\mathfrak{g}} \xrightarrow{PBW} U(\mathfrak{g})^{\mathfrak{g}}$$

is an isomorphism.

Bar-Natan, Le and Thurston [**Thu**] discovered a purely diagrammatic version of this statement. Let $\mathcal{B}$ be the space spanned by Jacobi diagrams with both trivalent and unitrivalent vertices, subject to the usual orientation convention and relations. It is a bigraded rational commutative algebra under disjoint union: write $\mathcal{B}^{v,l}$ for the part with v internal trivalent vertices, and l legs.

This algebra $\mathcal{B}$ plays the role of $S(\mathfrak{g})^{\mathfrak{g}}$; the original algebra $\mathcal{A}$, with connect-sum of diagrams as its product, plays the role of $U(\mathfrak{g})^{\mathfrak{g}}$. The analogue of the PBW map is $\chi \colon \mathcal{B} \to \mathcal{A}$, defined by sending an l-legged diagram to the average of the $l!$

diagrams obtained by attaching its legs in all possible orders to an oriented circle. This is an isomorphism of rational graded vector spaces [**BN**], but not of algebras.

The analogue of the function $j^{\frac{1}{2}}$ is the special *wheels element*

$$\Omega = \exp\left\{\sum_{i=1}^{\infty} b_{2i} w_{2i}\right\},$$

living in the completion of $\mathcal{B}$. Here, w_{2i} is a *wheel* diagram (a circular hub with $2i$ legs) and the b_{2i} are versions of Bernoulli numbers such that replacing w_{2i} by the function $\operatorname{tr}(\operatorname{ad} x)^{2i}$ recovers $j^{\frac{1}{2}}(x)$. *Contraction* (we will again write a cap product) with a diagram $C \in \mathcal{B}$ is the operation $C\cap : \mathcal{B} \to \mathcal{B}$ given by summing over all attachments of legs of C to legs of the target diagram. This extends to an action of the completion of $\mathcal{B}$. The *wheeling theorem* is then that

$$\mathcal{B} \xrightarrow{\Omega\cap} \mathcal{B} \xrightarrow{\chi} \mathcal{A}$$

is an isomorphism of algebras.

A metric Lie algebra $\mathfrak{g}$ defines weight systems $\mathcal{B} \to S(\mathfrak{g})^{\mathfrak{g}}$ and $\mathcal{A} \to U(\mathfrak{g})^{\mathfrak{g}}$, setting up a commuting diagram which intertwines the wheeling and Duflo isomorphisms (see [**BGRT**] for the details). (It seems peculiar here that the Duflo isomorphism itself is true for *all* finite-dimensional Lie algebras, not just the metric ones.)

This whole picture has an analogue for holomorphic symplectic manifolds. We construct in [**RW**] a pair of objects S, U in $\tilde{D}(X)$ which are the symmetric and universal enveloping algebras of the Lie algebra $T[-1]$, in the sense that they are associative algebras with appropriate universal properties. The description of S is straightforward, because one can take tensor powers and form symmetrisers inside $\tilde{D}(X)$. Constructing U is not so straightforward, because $\tilde{D}(X)$ is not abelian and the usual quotient construction therefore doesn't make sense. As one might expect there is also a PBW isomorphism $S \cong U$ which does not respect their structure as associative algebras.

The "invariant parts" of these two algebras are the algebras $\operatorname{Mor}(\mathcal{O}_X, S)$ and $\operatorname{Mor}(\mathcal{O}_X, U)$, which are just the total cohomology groups of these objects and are finite-dimensional graded complex algebras. They appear in Kontsevich's remarkable paper [**Ko1**] under the names HT^* (cohomology of polyvector fields) and HH^* (Hochschild cohomology). To be precise, we have

$$\operatorname{Mor}(\mathcal{O}_X, S) = \bigoplus_n HT^n(X) = \bigoplus_n \bigoplus_{i+j=n} H^i(\Lambda^j T)$$

$$\operatorname{Mor}(\mathcal{O}_X, U) = \bigoplus_n HH^n(X) = \bigoplus_n \operatorname{Ext}^n_{X\times X}(\mathcal{O}_\Delta, \mathcal{O}_\Delta)$$

where $\mathcal{O}_\Delta$ is the structure sheaf of the diagonal in $X \times X$. The algebra structures here are the natural ones: wedge product and Yoneda product, respectively. The usual Dolbeault cohomology $H^*(\Lambda^* T^*)$ acts by contraction on HT^*, and Kontsevich proved that cap product with the *root-A-hat-class*, the characteristic class corresponding to the power series

$$\left\{\frac{\sinh(x/2)}{x/2}\right\}^{\frac{1}{2}}$$

gives a Duflo-style isomorphism of algebras

$$HT^* \xrightarrow{\hat{A}^{\frac{1}{2}}\cap} HT^* \xrightarrow{PBW} HH^*.$$

A holomorphic symplectic X supplies weight system maps $\mathcal{B} \to HT^*$ and $\mathcal{A} \to HH^*$ intertwining the wheeling and Kontsevich isomorphisms. (As in the Duflo case though, Kontsevich's isomorphism is true for any complex manifold, and not just for holomorphic symplectic ones.)

9. The Hitchin-Sawon theorem

THEOREM. *Let X^{4n} be a compact hyperkähler manifold. Then there is an identity relating the root-A-hat-genus of X, its volume and the L^2-norm $\|R\|$ of its Riemann curvature tensor:*

$$\hat{A}^{\frac{1}{2}}[X] = \frac{1}{(192\pi^2 n)^n} \frac{\|R\|^{2n}}{\mathrm{vol}(X)^{n-1}}.$$

This theorem, proved in [**HS**], is a striking identity between a topological and a geometric invariant of X. Its proof is a consequence of the wheeling theorem, and shows the potential of Rozansky-Witten theory in geometry.

The basic idea of the proof is worth explaining. We use the Rozansky-Witten weight systems in the form $w_X : \mathcal{B}^{v,l} \to H^{l,v}_{\bar{\partial}}(X)$. There are two numerical evaluations one can consider: if $\Gamma \in \mathcal{B}^{2n,2n}$ is a diagram with $2n$ internal vertices and $2n$ legs then we can define

$$c_X(\Gamma) = \int_X w_X(\Gamma),$$

whereas if $\Gamma \in \mathcal{B}^{2n,0}$ is a "closed" trivalent diagram with $2n$ vertices then we can define

$$b_X(\Gamma) = \int_X w_X(\Gamma) \wedge \omega^n.$$

The key observation is that $c_X(\Gamma) = b_X(\mathrm{cl}(\Gamma))$ (up to normalisations and signs) when $\mathrm{cl}(\Gamma)$ is the *closure* of Γ, the element obtained by summing over all ways of pairing up the legs of Γ. (Remember that we are always really working with linear combinations of diagrams.) This is a straightforward consequence of the shuffle product formula for the top power ω^n of a symplectic form evaluated on a list of $2n$ vectors.

If Γ is chosen to be a *polywheel* diagram, that is, a disjoint union of wheels with $2n$ legs in total, then $w_X(\Gamma)$ turns out to be a product of Chern character terms with total degree $2n$, and so $c_X(\Gamma)$ is a linear combination of Chern numbers. In particular, for the special wheels element Ω we get the root-A-hat polynomial $w_X(\Omega) = \hat{A}^{\frac{1}{2}}(TX)$, and hence $c_X(\Omega) = \hat{A}^{\frac{1}{2}}[X]$.

On the other hand, the closure $\mathrm{cl}(\Omega)$ may be computed using Jacobi diagram techniques worked out by Bar-Natan, Le and Thurston [**Thu**], and it turns out to be $\exp(\Theta/24)$, where Θ is the two-vertex theta-graph. A standard result in hyperkähler geometry expresses this graph in terms of the norm of the curvature and gives the result.

Huybrechts used the identity to prove a finiteness theorem for hyperkähler manifolds [**H**]. It is to be hoped that further identities of a similar nature will emerge from the TQFT structure of RW theory, and help in the programme of classification of hyperkähler manifolds.

10. Link invariants

Let us turn now to the topological invariants arising from Rozansky-Witten theory. Recall that a semisimple Lie algebra $\mathfrak{g}$, together with a representation V, defines a weight system $w_{\mathfrak{g},V} : \mathcal{A} \to \mathbb{C}$, and that composing this with the Kontsevich integral should give us an invariant of framed knots in S^3. Because the Kontsevich integral actually lies in the completion of $\mathcal{A}$, we must first introduce a variable $\hbar$ of degree two and complete the weight system into a graded map

$$w_{\mathfrak{g},V} : \hat{\mathcal{A}} \to \mathbb{C}[[\hbar]],$$

from which we obtain a $\mathbb{C}[[\hbar]]$-valued invariant $Z_{\mathfrak{g},V}$ of framed knots in S^3.

There is an alternative formulation of this invariant using the framework of ribbon categories, invented by Turaev [**Tu**] to handle the invariants such as the Jones polynomial which arise from quantum groups. A *ribbon category* is a braided tensor category with a compatible notion of duality. It gives rise to representations of the category of framed "coloured" tangles: one assigns an object of the category (a "colour") to each string of a tangle, and then composes the elementary structural morphisms in the category (braidings and dualities) according to a Morse-theoretic slicing of the tangle into crossings, cups and caps. In particular, one gets invariants of framed coloured links, with values in the endomorphisms of the unit object of the category.

The *Knizhnik-Zamolodchikov equation* builds an interesting ribbon category from a semisimple Lie algebra. We start with the usual symmetric tensor category of modules over $\mathfrak{g}$, tensor it with $\mathbb{C}[[\hbar]]$, and then use the monodromy of the KZ equation to introduce a new, braided tensor (and in fact ribbon) structure. The resulting category has non-trivial associator morphisms (it is sometimes called a *quasitensor category*), but turns out to be equivalent to the strictly associative category of representations of the quantum group $U_\hbar(\mathfrak{g})$; see Bakalov and Kirillov [**BK**], for example. Applying Turaev's machinery to this category gives back the $\mathbb{C}[[\hbar]]$-valued invariant $Z_{\mathfrak{g},V}$ of framed knots.

It turns out that there is an analogous picture for a holomorphic symplectic manifold X. As we have seen, a vector bundle or sheaf E on X gives a weight system $w_{X,E} : \hat{\mathcal{A}} \to H^*(\mathcal{O}_X)$, and therefore a $H^*(\mathcal{O}_X)$-valued Vassiliev invariant $Z_{X,E}$ of framed knots.

THEOREM. *The graded derived category $\tilde{D}(X)$ of a holomorphic symplectic manifold X may be given a ribbon structure so that the associated invariant of a knot coloured by $E \in D(X)$ is $Z_{X,E}$.*

The starting point of this construction is the usual derived category $D(X)$, which is a symmetric tensor category. The analogue of tensoring with $\mathbb{C}[[\hbar]]$ is its replacement by the graded version $\tilde{D}(X)$, in which the shift [2] plays the role of $\hbar$. The [2] is attached to the symplectic form ω, just as in the Lie algebra case, $\hbar$ may be thought of as attached to the metric. We then use the Kontsevich integral, which underlies the KZ equation, to define the ribbon structure $\tilde{D}(X)$.

By construction, the tensor structure on $\tilde{D}(X)$ is not strictly associative. We don't know whether it is possible to make a gauge transformation, as Drinfeld does in the case of quantum groups [**D**], to a form which is strictly associative and has a local, but more complicated, braiding. Drinfeld's transformation is defined purely algebraically, and it is not clear how to derive it from geometry in the way we would like. Equally remarkably, the $\mathbb{C}[[\hbar]]$-valued invariants coming from Lie algebras turn out to be the expansions, on setting $q = e^{\hbar}$, of Laurent polynomials in q. Is there some analogous hidden structure to the Rozansky-Witten link invariants?

11. TQFT

The basis of Rozansky and Witten's work is the path integral which defines the partition function for a closed oriented 3-manifold M:

$$Z_X(M) = \int_{\mathrm{Map}(M,X)} e^{iS} d\phi.$$

The integral is over all smooth maps $\phi : M \to X$, and the action S is an expression involving a Riemannian metric on M and the hyperkähler metric on X. It turns out to be independent of the metric on M, meaning that the associated quantum field theory, too, is *topological*. The link invariants described above may also be formulated using the path integral, by insertion of suitable "Wilson loop" observables.

We would like to have a rigorous construction of this TQFT, with which we can calculate and explore. That is, we would like to build a TQFT which shares the predicted properties of the genuine physical theory; a "Reshetikhin-Turaev invariant" for our "Witten invariant". This is underway in joint work with Justin Sawon and Simon Willerton.

To do it we have to work combinatorially. We will build the TQFT by specifying its values on elementary pieces of surfaces and 3-manifolds, calculate for larger manifolds by means of decompositions into such pieces, and prove that our specified elementary data satisfies the coherence relations which ensure that different decompositions of a 3-manifold compute the same invariant. This is not, in all honesty, a very satisfactory method, but there's little reasonable alternative.

A standard TQFT, such as Witten's Chern-Simons theory [**Wi**], is a functor from the category of (2+1)-dimensional cobordisms to the category of vector spaces. It is typically a *tensor functor*, taking disjoint unions to tensor product; and a *unitary* one, taking surfaces to hermitian vector spaces. There is a way to enhance such a TQFT into an *extended* or $(1+1+1)$-*dimensional* theory, in which we also assign a *category* to each closed 1-manifold. In the usual theories this category is *semisimple*, with finitely many simple objects, and is promptly replaced by a set of *colours* corresponding to these objects. In Chern-Simons theory, the category for a single circle is the (truncated) representation category of a quantum group at a root of unity. Its simple objects correspond to the irreps of the classical group which lie inside a certain compact Weyl alcove.

The Rozansky-Witten TQFT is somewhat different, and requires a different formalism, which has been studied by Freed, Segal, Tillmann and Khovanov [**Ti, Kh**], amongst others. In this formalism we view a $(1+1+1)$-dimensional TQFT as a functor from the 2-category of 3-cobordisms with corners, to the 2-category of linear categories. Thus, a 1-manifold is sent to a category; a "vertical" 2-dimensional cobordism between 1-manifolds defines a functor between these categories; and a "horizontal" 3-dimensional cobordism between two surfaces which have common lower and upper 1-manifolds defines a natural transformation of functors. There's no good reason here for imposing a tensor product axiom, unitarity, or semisimplicity of the categories associated to 1-manifolds.

Indeed, our theory has none of these. Its basic feature is that a 1-manifold consisting of k circles gets sent to the graded derived category $\tilde{D}(X^k)$: clearly $\tilde{D}(X)$ is not semisimple and it is *not* true that $\tilde{D}(X \times X) \cong \tilde{D}(X) \times \tilde{D}(X)$.

To construct the functor associated to a surface, we break it using Morse theory into elementary pieces, and write down the basic functors associated to the different types of handles. For example, a 2-handle gives the pushforward functor $\tilde{D}(X) \to \tilde{D}(\mathrm{pt})$; a 0-handle, the functor $\tilde{D}(\mathrm{pt}) \to \tilde{D}(X)$ which sends $\mathbb{C}$ to $\mathcal{O}_X$; and an index 1-handle which joins two circles, the functor $\tilde{D}(X \times X) \to \tilde{D}(X)$ obtained by taking derived tensor product with the structure sheaf of the diagonal, then pushing forward once.

To prove independence of the decomposition we have to use the ribbon structure on $\tilde{D}(X)$, and some extra properties resembling modularity which arise from diagrammatic identities and the Kontsevich integral. These are enough to check that the Moore-Seiberg equations are satisfied. To define the invariants for elementary 3-manifolds is more complicated, and uses Walker's framework [**Wa**].

It is possible to recover a traditional $(2+1)$-dimensional TQFT functor which sends surfaces to vector spaces instead of functors from our axioms. The empty 1-manifold is sent to $\tilde{D}(\mathrm{pt})$, which is essentially the category of graded vector spaces. A closed surface defines a functor $\tilde{D}(\mathrm{pt}) \to \tilde{D}(\mathrm{pt})$, which when applied to the generating object $\mathbb{C}$ in $\tilde{D}(\mathrm{pt})$ outputs the desired graded vector space.

Using the basic data specified above, we can compute that the graded vector space associated to a closed surface Σ_g is isomorphic (non-canonically) to the cohomology $H^*((\Lambda^*T)^{\otimes g})$, as postulated by Rozansky and Witten. An action of the mapping class group on this space emerges from the Moore-Seiberg equations. The superdimensions of these spaces are given for the sphere and torus by the Todd genus and Euler characteristic of X, respectively, and they vanish for $g \geq 2$. The space associated to S^2, which is naturally an associative algebra in any TQFT, is in fact the ring $H^*(\mathcal{O}_X)$, which for an irreducible hyperkähler X is the truncated polynomial ring $\mathbb{C}[\bar{\omega}]/(\bar{\omega}^{n+1})$. It is not semisimple, which shows that the TQFT cannot be unitary.

This TQFT has a close relationship with the Le-Murakami-Ohtsuki invariant [**LMO**], which is a kind of extension of the Kontsevich integral to 3-manifolds, with values in the algebra of trivalent Jacobi diagrams. In principle, the TQFT invariant of a closed 3-manifold should equal the LMO invariant, evaluated using a weight system coming from X; the TQFT itself should be the evaluation of the universal TQFT constructed by Murakami and Ohtsuki [**MO**], which underlies the LMO invariant. However, their TQFT is somewhat badly behaved axiomatically, and there are subtle differences in normalisation which make it unsatisfactory to try to use this approach as a definition. The $(1+1+1)$-dimensional framework seems to work so well when we are building the specific theory associated to X that it is better to use it as a foundation, and to make connections with the universal theory in retrospect.

12. Future directions

In [**RS**] we give an extensive discussion of the potential applications of RW invariants to the geometry of hyperkähler (and certain other types of) manifolds, and conclude with a problem list. Further discussion of the interaction with Vassiliev theory, the theory of quantum invariants and TQFT will appear in [**RSW**]. But it is worth mentioning a few problems here, just to give the flavour of possible future research.

QUESTION. *What is the meaning of the* $\hat{A}^{\frac{1}{2}}$ *genus?*

This genus appears in Kontsevich's theorem and the Hitchin-Sawon theorem, as we have seen. But it also appears in physics, in the context of cohomological D-brane charge. It seems natural that these two occurrences are related, but how and why? (With impeccable timing, physicists are also now adopting derived categories of coherent sheaves as basic structures in string theory!) A straightforward question is whether the root-A-hat genus, like the honest A-hat genus, has interesting integrality properties, and can be interpreted for manifolds with some appropriate geometric structure as the index of a natural elliptic operator.

QUESTION. *Are the Rozansky-Witten weight systems new, and can their associated Vassiliev invariants detect orientation of knots?*

Vogel showed that the primary examples of Vassiliev weight systems, those coming from complex semisimple Lie algebras and superalgebras, do not span the whole space of weight systems. But we do not yet know whether the Rozansky-Witten weight systems lie outside their span or not.

It's easy to prove that the Vassiliev invariants coming from Lie algebra weight systems are unable to distinguish knots from their reverses. It is in fact thought likely that no Vassiliev invariants can separate knots from their reverses, and there is an alternative purely diagrammatic statement of this conjecture. But the proof that Lie algebra weight systems fail doesn't work for Rozansky-Witten invariants, so there is still potential here.

QUESTION. *Is there a geometric quantization approach to the vector spaces* $Z(\Sigma_g)$*?*

These vector spaces, as we have seen, can be constructed from a combinatorial approach to the TQFT. This is a reasonable but crude approach: much better would be to give a direct geometric (and completely functorial) construction. In Witten's Chern-Simons theory with gauge group G, there is an approach via geometric quantization of the moduli space of flat G-connections on a surface. Is there an analogue in Rozansky-Witten theory? A straightforward guess is that it might be possible to define a virtual structure sheaf of the moduli space of holomorphic maps from a closed Riemann surface Σ_g to X, and take its cohomology as the graded vector space. We would then require a "projectively flat connection", a system of coherent isomorphisms between these sheaves, as the complex structure on Σ_g varies. Going a level deeper, how does one construct the category $\tilde{D}(X)$ functorially from a given circle and manifold X? In the Chern-Simons case, one uses the representation category of the group of loops from the circle to G. What is the analogue? We would require surfaces with boundary to generate functors between such categories by means of a Fourier-Mukai transform operation.

References

[A] M. F. Atiyah. *Complex analytic connections in fibre bundles*, Trans. Amer. Math. Soc. 85 (1957), 181-207.

[BK] B. Bakalov, A. Kirillov. *Lectures on tensor categories and modular functors* University Lecture Series 21, AMS (2001).

[BN] D. Bar-Natan. *On the Vassiliev knot invariants*, Topology 34 (1995), 423-472.

[BGRT] D. Bar-Natan, S. Garoufalidis, L. Rozansky, D. P. Thurston. *Wheels, wheeling, and the Kontsevich integral of the unknot*, Israel J. Math. 119 (2000), 217-237.

[D] V. G. Drinfel'd. *Quasi-Hopf algebras*, Leningrad J. Math. 1 (1990) 1419-1457.

[GM] S. I. Gelfand, Yu. I. Manin. *Homological algebra*, Springer (1999).

[HS] N. Hitchin, J. Sawon. *Curvature and characteristic numbers of hyper-Kähler manifolds*, Duke Math. J. 106 (2001), no. 3, 599-615.

[H] D. Huybrechts. *Finiteness results for hyperkähler manifolds*, arXiv preprint `math.AG/0109024`

[Ka] M. Kapranov. *Rozansky-Witten invariants via Atiyah classes*, Compositio Math. 115 (1999), 71-113.

[Kh] M. Khovanov. *A functor-valued invariant of tangles*, arXiv preprint `math.QA/0103190`

[Ko1] M. Kontsevich. *Deformation quantization of Poisson manifolds I*, arXiv preprint `q-alg/9709040`.

[Ko2] M. Kontsevich. *Rozansky-Witten invariants via formal geometry*, Compositio Math. 115 (1999), 115-127.

[LMO] T. Q. T. Le, J. Murakami, T. Ohtsuki. *On a universal perturbative invariant of 3-manifolds*, Topology 37 (1998), 539-574.

[MO] J. Murakami, T. Ohtsuki. *Topological quantum field theory for the universal quantum invariant*, Comm. Math. Phys. 188 (1997), 501-520.

[RT] N. Yu. Reshetikhin, V. G. Turaev. *Invariants of 3-manifolds via link polynomials and quantum groups*, Invent. Math. 103 (1991), 547-597.

[RS] J. D. Roberts, J. Sawon. *Generalisations of Rozansky-Witten invariants*, in preparation.

[RSW] J. D. Roberts, J. Sawon, S. Willerton. *On the Rozansky-Witten TQFT*, in preparation.

[RW] J. D. Roberts, S. Willerton. *On the Rozansky-Witten weight systems*, in preparation.

[RzW] L. Rozansky, E. Witten. *Hyper-Kähler geometry and invariants of 3-manifolds*, Selecta Math. **3** (1997) 401–458.

[Sa] J. Sawon. *Rozansky-Witten invariants of hyperkähler manifolds*, PhD Thesis, University of Cambridge (1999).

[Tho] R. P. Thomas. *Derived categories for the working mathematician*, arXiv preprint `math.AG/0001045`.

[Thu] D. P. Thurston. *Wheeling: a diagrammatic analogue of the Duflo isomorphism*, PhD thesis, UC Berkeley (2000), arXiv preprint `math.QA/0006083`.

[Ti] U. Tillmann. *S-Structures for k-linear categories and the definition of a modular functor*, arXiv preprint `math.GT/9802089`

[Tu] V. G. Turaev. *Quantum invariants of knots and 3-manifolds*, de Gruyter Studies in Mathematics, 18 (1994).

[Va] A. Vaintrob. *Vassiliev knot invariants and Lie S-algebras.*, Math. Res. Lett. 1 (1994), 579-595.

[Vo] P. Vogel. *Algebraic structures on modules of diagrams*, preprint (1995).

[Wa] K. Walker. *On Witten's 3-manifold invariants*, preprint (1991).

[Wi] E. Witten. *Quantum field theory and the Jones polynomial*, Comm. Math. Phys. 121 (1989), 351-399.

Department of Mathematics, UC San Diego, 9500 Gilman Drive, La Jolla CA 92093
E-mail address: `justin@math.ucsd.edu`

Proceedings of Symposia in Pure Mathematics
Volume **71**, 2003

Some recent applications of the barycenter method in geometry

Christopher Connell and Benson Farb †

CONTENTS

1. The Entropy Rigidity Conjecture

In this paper we describe some recent applications of the barycenter method in geometry. This method was first used by Douady-Earle and later greatly extended by Besson-Courtois-Gallot in their solution of a number of long-standing problems, in particular in their proof of entropy rigidity for closed, negatively curved locally symmetric manifolds. Since there are already a number of surveys describing this work (see [**BCG2, BCG3, Pa, Ga**]), we will concentrate here only on advances that have occurred after these surveys appeared. While most of this paper is a report on results appearing in other papers, some of the material here is new (e.g. Proposition 12 and the examples in Section 6).

1.1. The main conjecture. The *volume entropy* of a closed Riemannian n-manifold (M, g), denoted by $h(g)$, is defined to be

$$h(g) = \lim_{R\to\infty} \frac{1}{R} \log(\mathrm{Vol}(B(x, R)))$$

where $B(x, R)$ is the ball of radius R around a fixed point x in the universal cover X. The number $h(g)$ is independent of the choice of x, and equals the topological entropy of the geodesic flow on (M, g) when the curvature $K(g)$ satisfies $K(g) \leq 0$

†Both authors are supported in part by the NSF. This paper grew out of talks given by the first and second authors at the Cornell and Georgia Topology conferences, respectively.

(see [**Ma**]). Note that while neither the volume $\mathrm{Vol}(M,g)$ nor the entropy $h(g)$ is invariant under scaling the metric g, the *normalized entropy*

$$\mathrm{ent}(g) = h(g)^n \, \mathrm{Vol}(M,g)$$

is scale invariant.

Now let M be a closed n-manifold that admits a locally symmetric Riemannian metric g_{loc} of nonpositive sectional curvature. When (M, g_{loc}) is not locally isometric to a product then g_{loc} is unique up to homothety (i.e. multiplying the metric by a number). When M is locally a product, one may show using Lagrange multipliers that there is a unique (up to homothety) locally symmetric metric on M which minimizes $\mathrm{ent}(g)$; see [**CF1**]. Henceforth we will abuse notation and denote this metric by g_{loc}, and call it *the* locally symmetric metric.

The *Entropy Rigidity Conjecture*, stated in various forms by Katok, Gromov, and Besson-Courtois-Gallot (see, e.g., [**BCG2**], Open Question 5), has two components. The purely metric component posits that for most M, the metric g_{loc} minimizes the functional $\mathrm{ent}(g)$ over the space of all Riemannian metrics on M, and in fact *uniquely* minimizes $\mathrm{ent}(g)$ (up to homothety). This property would characterize the locally symmetric metric by essentially a single number. The topological component of the conjecture is an extension of this statement from manifolds to maps.

CONJECTURE 1 (Entropy Rigidity Conjecture). *Let M be a closed manifold which admits a locally symmetric Riemannian metric g_{loc} with nonpositive sectional curvature. Assume that (M, g_{loc}) has no local factors isometric to* **R**. *Let (N, g) be any closed Riemannian manifold, and let $f : N \longrightarrow M$ be any continuous map. Then*

$$\mathrm{ent}(N,g) \geq |\mathrm{deg} f| \, \mathrm{ent}(M, g_{loc}) \tag{1}$$

with equality if and only if f is homothetic to a Riemannian covering.

Remarks:

(1) The case when f is a homeomorphism, or even the identity map, gives that g_{loc} uniquely minimizes $\mathrm{ent}(g)$.
(2) Easy examples show that the restrictions to M with no local **R** factors is necessary.
(3) Conjecture 1 easily implies Mostow Rigidity; the argument is the same as in the case when M is negatively curved, which is given in [**BCG2**].

Conjecture 1 may be extended to a variety of other contexts, for example to finite volume metrics, Finsler metrics, foliations with locally symmetric leaves, contact flows, and magnetic flows. To be more precise, the table below shows some natural choices for the volume and entropy in various metric settings. In the table we have adopted the following conventions:

- θ represents a contact 1-form
- h_m is metric entropy, i.e. the measure theoretic entropy for the Liouville measure
- h_{top} represents the topological entropy of the geodesic flow
- h_{vol} is the volume growth entropy
- $\dim_{\mathcal{H}}(\Lambda(\Gamma))$ is the Hausdorff dimension of the limit set $\Lambda(\Gamma)$ of a discrete group Γ of hyperbolic isometries

Setting	**Volume** Vol(M)	**entropy** $h(d)$
Riemannian metrics on closed M	Riem. volume	h_{vol}
Finite vol. metrics on noncompact M	Riem. volume	$\inf\{s : \int_Y e^{-sd(p,y)} dg(y) < \infty\}$
Contact flows	$\int_M \theta \wedge (d\theta)^{n-1}$	h_{top}
Finsler metrics	Finsler Volume	h_{top}
Magnetic field flows	Euler-Lagrange Volume	h_{top}
Foliations on M	$d\operatorname{Vol}_{g_L} \times d\operatorname{Vol}(L)$: leafwise volume $\times$ transverse volume	$\left(\frac{1}{\operatorname{Vol}(M)} \int_M h(g_L)^n d\operatorname{Vol}\right)^{1/n}$
Geometrically finite M	Volume of core	$\dim_{\mathcal{H}}(\Lambda(\Gamma))$

FIGURE 1. Some natural choices for the volume and entropy in different settings. The functional being minimized is ent $:= h(d)^n \operatorname{Vol}(M)$.

- $M = Y/\Gamma$

These extensions are discussed in greater depth in Sections 4 and 5.

1.2. Progress. The main evidence for Conjecture 1 is given by the following theorem.

THEOREM 2 (Besson-Courtois-Gallot [**BCG1**]). *Conjecture 1 is true when (M, g_{loc}) has* **R***-rank one, i.e. when (M, g_{loc}) is negatively curved.*

This result and its proof have a number of corollaries, including solutions to long-standing problems on geodesic flows, asymptotic harmonicity, and Gromov's Minvol invariant; these are described in [**BCG2**].

In higher rank very little is known. The following was announced in [**BCG2**] and later in [**BCG3**], and was proved in [**CF1**] (in the finite volume case as well).

THEOREM 3. *Conjecture 1 is true when (M, g_{loc}) is locally (but not necessarily globally) isometric to a product of negatively curved locally symmetric spaces.*

It seems that a significantly new idea is needed to prove Conjecture 1. See, however, §1.4 below.

1.3. The Besson-Courtois-Gallot map. Building on earlier work of Douady-Earle, Besson-Courtois-Gallot constructed a remarkable map F in every homotopy class of maps $f : N \to M$ from an arbitrary closed manifold N to any closed manifold M of negative sectional curvature. This map is a kind of replacement for, and has several advantages over, the harmonic map in the homotopy class of f; in particular, it can often be differentiated explicitly. The map F is a key ingredient in the proof of Theorem 2.

Now suppose that, in addition to being negatively curved, M is also locally symmetric with metric g_{loc}. In this case Besson-Courtois-Gallot were able to give a precise estimate on the Jacobian $|\operatorname{Jac} F|$, proving Theorem 2. To state things precisely, it will be useful to fix a parameter $s > h(g)$, where g is the metric on N and $h(g)$ is the volume growth entropy of g. When (M, g_{loc}) has $\mathbf{R}$-rank one, that is when it is negatively curved, Besson-Courtois-Gallot proved that, given any continuous $f : N \longrightarrow M$, there exists a smooth map $F_s : N \longrightarrow M$ homotopic to f with the following two important properties:

Universal Jacobian bound: For all $s > h(g)$ and all $y \in N$ we have:

$$|\operatorname{Jac} F_s(y)| \leq \left(\frac{s}{h(g_{loc})}\right)^n \tag{2}$$

Infinitesimal rigidity: There is equality in equation (2) at the point $y \in N$ if and only if $D_y F_s$ is a homothety.

Before explaining how to construct such an F_s (which we do in §2), let us deduce some consequences of its existence.

The existence of any F_s satisfying the two properties above implies Conjecture 1 by an elementary degree argument, as follows. Since for $s > h(g)$, the map F_s is a C^1 map, we may simply compute:

$$\begin{aligned}
|\deg(f)| \operatorname{Vol}(M) &= |\deg(F_s)| \operatorname{Vol}(M) \\
&= \left|\int_N F_s^* dg_{loc}\right| \\
&\leq \int_N |\operatorname{Jac} F_s| \, dg \\
&\leq \left(\tfrac{s}{h(g_{loc})}\right)^n \operatorname{Vol}(N)
\end{aligned}$$

Letting $s \longrightarrow h(g)$ gives the inequality in Conjecture 1. In the case when equality is achieved, after scaling the metric g by the constant $\frac{h(g)}{h(g_{loc})}$, we have $h(g) = h(g_{loc})$ and $\operatorname{Vol}(N) = |\deg(f)| \operatorname{Vol}(M)$.

1.4. The degree theorem. It is useful for us to restate the inequality (1) of the Entropy Rigidity Conjecture as

$$\deg(f) \leq C \frac{\operatorname{Vol}(N)}{\operatorname{Vol}(M)} \tag{3}$$

where $C = \left(\frac{h(g)}{h(g_0)}\right)^n$. In particular, proving the inequality (3) with this precise value of C gives that the locally symmetric metric g_{loc} minimizes $\operatorname{ent}(g)$ among all metrics g on N.

When the inequality (3) holds for *some* (universal) C, perhaps bigger than $\left(\frac{h(g)}{h(g_0)}\right)^n$, then the degree argument given in §1.3 gives a universal bound on the degree of any map of any manifold into N. This can be viewed as the topological part of Conjecture 1. When N is negatively curved, Gromov proved this fundamental relationship between degree and volume in his paper [**Gr**] (see his Volume Comparison Theorem).

We note that by *universal* we mean that C depends only on the dimension of the manifolds M, N and on their smallest Ricci curvatures. Scaling the metrics shows that these dependencies are necessary.

When the sectional curvatures of N are not necessarily negative but only non-positive, the situation is more complicated. Since the n-dimensional torus has flat metrics and also has self-maps of arbitrary degree, one needs at least to assume that N has no local $\mathbf{R}$ factors in order that (3) be true. Using the barycenter technique (see below), we proved the following in [**CF2**].

THEOREM 4 (The Degree Theorem). *Let M be a closed, locally symmetric n-manifold with nonpositive sectional curvatures. Assume that M has no local direct factors locally isometric to $\mathbf{R}, \mathbf{H}^2$, or $\mathrm{SL}_3(\mathbf{R})/\mathrm{SO}_3(\mathbf{R})$. Then for any closed Riemannian manifold N and any continuous map $f : N \longrightarrow M$,*

$$\deg(f) \leq C\frac{\mathrm{Vol}(N)}{\mathrm{Vol}(M)}$$

where $C > 0$ depends only on n and the smallest Ricci curvatures of N and M.

While the "no local $\mathbf{R}$ factors" assumption is necessary, we believe that Theorem 4 should hold in all other cases. The case when M is locally modelled on $\mathrm{SL}_n(\mathbf{R})/SO_n(\mathbf{R}), n \geq 2$ also follows from work of Savage [**Sa**]; see §1.5 below.

Theorem 4 follows immediately from the degree argument given above (in §1.3) together with the following theorem, which is the main theorem of [**CF2**], and which we believe is of independent interest.

THEOREM 5 (Universal Jacobian bound). *Any continuous map $f : N \to M$ between closed n-manifolds, with M nonpositively curved and locally symmetric (barring the exceptions of Theorem 4), is homotopic to a C^1 map F with*

$$|\operatorname{Jac} F| \leq C$$

for some constant C depending only on n and on the smallest Ricci curvatures of M and N.

The map F is the extension to nonpositively curved locally symmetric manifolds of the Besson-Courtois-Gallot map; we discuss this below.

As we will discuss in §3, the universal bound for $|\operatorname{Jac} F|$ is obtained by reducing the problem to a minimization problem over a space of measures on a certain Lie group. The Lie group appears because the natural boundary attached to symmetric spaces may be described *algebraically.* For general nonpositively curved manifolds no such description is available. But the following still seems possible.

CONJECTURE 6 (Degree Conjecture in nonpositive curvature). *Let M be a closed n-manifold with nonpositive sectional curvature, negative Ricci curvature, and no local $\mathbf{R}$ factors. Then for any closed Riemannian manifold N and any continuous map $f : N \longrightarrow M$,*

$$\deg(f) \leq C\frac{\mathrm{Vol}(N)}{\mathrm{Vol}(M)}$$

where $C > 0$ depends only on n and the smallest Ricci curvatures of N and M.

The evidence for Conjecture 6 is that it is true when M has negative sectional curvature (Gromov [**Gr**]), and (almost always) when M is locally symmetric (by

Theorem 4, since for locally symmetric manifolds of nonpositive sectional curvature, no local **R** factors implies negative Ricci curvature). The "negative Ricci curvature" hypothesis in Conjecture 6 is necessary, as the following example (see [**BGS**]) shows.

EXAMPLE 7. Let S be a negatively curved surface with one boundary component C. While keeping S nonpositively curved we may smooth a neighborhood of C into a cylinder which is metrically $[0, \epsilon) \times S^1$, where C is identified to $\{0\} \times S^1$. Call the resulting surface T.

Now take two copies X_1 and X_2 of the manifold $T \times S^1$ and glue X_1 to X_2 by identifying $[0, \epsilon/3] \times S^1 \times S^1 \subset X_1$ and $[0, \epsilon/3] \times S^1 \times S^1 \subset X_2$ by the map $(t, x, y) \mapsto (\epsilon/3 - t, y, x)$. The resulting 3-manifold X has no flats, but has some zero curvature at every point and even has an open subset of zero curvature, namely the subset $Z \subset X$ formed from the union of the images of $[0, \epsilon) \times S^1$ in each X_i.

We note that the two overlapping pieces, $(X_1 \setminus Z) \cup (\epsilon/2, \epsilon) \times S^1 \times S^1$ and $(X_1 \setminus Z) \cup (\epsilon/2, \epsilon) \times S^1 \times S^1$, each carry an S^1 action that is coherent with a torus action on Z which acts the same in each factor. Therefore, the whole of X carries an *F-structure* in the sense of Cheeger and Gromov [**CG**]. This implies via [**PaPe, CG**] that $\text{Minvol}(X) = 0$ (see below for the definition of $\text{Minvol}(X)$); in particular, the conclusion of Conjecture 6 cannot hold for X, even when f is the identity map.

Note that X has no local direct factors locally isomorphic to **R**, but that X does have some points with some direction of zero Ricci curvature.

1.5. A related conjecture.

Gromov has made the following:

CONJECTURE 8 (Positivity of Gromov norm). *The Gromov norm of a closed, nonpositively curved, locally symmetric manifold M with no local* **R** *factors is positive.*

Conjecture 8 was proven by Savage [**Sa**] when M is locally isometric to $\text{SL}_n(\mathbf{R})/SO_n(\mathbf{R})$. While we do not see how positivity of Gromov norm for M directly implies the inequality (3), Savage's proof does imply this inequality. The key point is that Savage proves that any simplex in M can be "straightened" to a simplex with universally bounded volume. This should hold for all symmetric spaces of noncompact type, but as far as we know this is still an open question.

1.6. Some consequences of the Degree Theorem.

We end this section by recalling some consequences, given in [**CF2**], of Theorem 4.

The Minvol invariant. One of the basic invariants associated to a smooth manifold M is its *minimal volume*:

$$\text{Minvol}(M) := \inf_g \{\text{Vol}(M, g) : |K(g)| \leq 1\}$$

where g ranges over all smooth, complete metrics on M and $K(g)$ denotes the sectional curvature of g. The basic questions about $\text{Minvol}(M)$ are: for which M is $\text{Minvol}(M) > 0$? when is $\text{Minvol}(M)$ realized by some metric g?

When a closed, nonpositively curved manifold M has a local direct factor locally isometric to **R**, then M has some finite cover M' with S^1 as a direct factor. This in turn implies that M' has positive degree self-maps and that $\text{Minvol}(M') = 0$. On the other hand, in the statement of Theorem 4 we may take f to be the identity map for an M which admits a locally symmetric metric and allow only the metric g on M to vary. Theorem 4 then shows $\text{Minvol}(M) > 0$ for locally symmetric M, barring the possible exceptional cases listed in Theorem 4. However, positivity was already known more generally, by the following result of Gromov (see also [**Sa**] for the case of M locally isometric to the symmetric space for $\text{SL}(n, \mathbf{R})$).

COROLLARY 9 (Positivity of Minvol). *Let M be a closed, locally symmetric manifold with nonpositive curvature and no local $\mathbf{R}$ factors. Then* $\text{Minvol}(M) > 0$.

When M is real hyperbolic, Besson-Courtois-Gallot [**BCG1**] proved that $\text{Minvol}(M)$ is *uniquely* realized by the locally symmetric metric.

PROBLEM 10. *Compute* $\text{Minvol}(M)$ *for all closed M which admit a locally symmetric metric of nonpositive curvature, with no local $\mathbf{R}$ factors. Is* $\text{Minvol}(M)$ *always realized by the locally symmetric metric? Is it realized uniquely?*

Self maps and the co-Hopf property. As $\deg(f^n) = \deg(f)^n$, an immediate corollary of Theorem 4 is the following.

COROLLARY 11 (Self maps). *Let M be a closed, locally symmetric manifold as in Theorem 4. Then M admits no self-maps of degree > 1. In particular, $\pi_1(M)$ is* co-Hopfian*: every injective endomorphism of $\pi_1(M)$ is surjective.*

Note that Corollary 11 may also be deduced from Margulis' Superrigidity theorem (for higher rank M). The co-Hopf property for lattices was first proved by Prasad [**Pr**].

2. The Douady-Earle-Besson-Courtois-Gallot map

2.1. The construction. In this section we describe the canonical map F discussed above. This construction is due to Douady-Earle for the hyperbolic plane, was extended by Besson-Courtois-Gallot [**BCG4**] to negatively curved targets, and was extended to symmetric spaces of noncompact type in [**CF1, CF2**]. In this section we will extend this construction further to nonpositively curved targets with negative Ricci curvature. For background on nonpositively curved manifolds and symmetric spaces, see for example [**BGS, Eb**].

As above, let M, N be closed, Riemannian n-manifolds with M nonpositively curved, and let $f : N \longrightarrow M$ be any continuous map. Denote by Y (resp. X) the universal cover of N (resp. M). Denote by ∂X the visual boundary of X; that is, the set of equivalence classes of geodesic rays in X, endowed with the cone topology. Hence $X \cup \partial X$ is a compactification of X which is homeomorphic to a closed ball; see, e.g. [**BGS**].

Idea of the main construction. Let ϕ denote the lift to universal covers of f with basepoint $p \in Y$ (resp. $f(p) \in X$), i.e. $\phi = \widetilde{f} : Y \longrightarrow X$. We first construct a map $\widetilde{F} : Y \longrightarrow X$ by "averaging" ϕ as follows: first embed Y into a space of measures on Y, then push forward each measure via ϕ, then smooth out the measure onto ∂X by convolving with a canonical measure on ∂X, and finally take the "barycenter" of the resulting measure. An essential feature of each of these steps is that they are *canonical*, i.e. they are equivariant with respect to the actions of fundamental groups on each of the spaces involved. It follows from this that $\widetilde{F}$ descends to a map $F : N \longrightarrow M$.

Now to define $\widetilde{F}$ precisely. Actually, it is useful to fix a parameter $s > h(g)$ for which we define a map $\widetilde{F}_s$. We denote the Riemannian metric and corresponding volume form on universal cover Y by g and dg respectively.

Let $\mathcal{M}(Z)$ denote the space of probability measures on the set Z. Following the method of [**BCG1**], we define a map $\widetilde{F}_s : Y \longrightarrow X$ as a composition

$$\begin{array}{ccccc} \mathcal{M}(Y) & \xrightarrow{\phi_*} & \mathcal{M}(X) & \xrightarrow{\circledast\nu_z} & \mathcal{M}(\partial X) \\ \uparrow{\scriptstyle \mu_x^s} & & \curvearrowright & & \downarrow{\scriptstyle \text{bar}} \\ Y & - - - - - & \overset{\widetilde{F}_s}{-} & - - - - \rightarrow & X \end{array}$$

where the individual maps are defined as follows:

- The inclusion $Y \longrightarrow \mathcal{M}(Y)$, denoted $y \mapsto \mu_y^s$, is given by localizing the Riemannian volume form dg on Y and normalizing it to be a probability measure; that is, μ_y^s is the probability measure on Y in the Lebesgue class with density given by
$$\frac{d\mu_y^s}{dg}(z) = \frac{e^{-sd(y,z)}}{\displaystyle\int_Y e^{-sd(y,z)}dg}$$
Note that each $\mu_y^s, y \in Y$ is well-defined by the choice of s.
- The map ϕ_* is the push forward of measures.
- The symbol $\circledast\nu_z$ indicates the operation of convolution with the *Patterson-Sullivan measures* $\{\nu_x\}_{x\in X}$ corresponding to $\pi_1(M) < \mathrm{Isom}(X)$ (see §2.2). In other words, the resulting measure σ_y^s is defined on a Borel set $U \subset \partial X$ by
$$\sigma_y^s(U) = \int_X \nu_z(U)d(\phi_*\mu_y^s)(z)$$
Since $\|\nu_z\| = 1$, we have
$$\|\sigma_y^s\| = \|\mu_y^s\| = 1.$$
- The map bar is the *barycenter* of the measure σ_y^s, which is defined as the unique minimum of a certain functional on X, depending on the measure (see §2.3). The map bar is not always defined; when it is defined its domain of definition often must be restricted.

We now describe the last two maps in more detail. For full details in the current context we refer the reader to [**CF2**].

2.2. Patterson-Sullivan measures. Let Γ be a discrete group of isometries of a connected, simply-connected, complete, nonpositively curved manifold X with no Euclidean direct factors. Fix a basepoint $p \in X$.

Generalizing work of Patterson and Sullivan, a number of authors, including Coornaert, Margulis (unpublished), Albuquerque, and Knieper constructed a remarkable family $\{\nu_x\}$ of probability measures on ∂X. These measures, called *Patterson-Sullivan measures*, are meant to encode the density of an orbit $\Gamma \cdot p$ at infinity as viewed from $y \in Y$, giving the Patterson-Sullivan measure ν_y on $\partial_\infty Y$. See [**Al**] and [**Kn**] for the construction of Patterson-Sullivan measures in nonpositive curvature.

The key properties of Patterson-Sullivan measures ν_x are:

(1) **No atoms:** Each ν_x has no atoms.
(2) **Equivariance:** $\nu_{\gamma x} = \gamma_* \nu_x$ for all $\gamma \in \Gamma$.
(3) **Explicit Radon-Nikodym derivative:** For all $x, y \in X$, the measure ν_y is absolutely continuous with respect to ν_x. In fact the Radon-Nikodym derivative is given explicitly by:

$$\frac{d\nu_x}{d\nu_y}(\xi) = e^{h(g)B(x,y,\xi)} \tag{4}$$

where $B(x, y, \xi)$ is the *Busemann function* on X. For points $x, y \in X$ and $\xi \in \partial X$, the function $B : X \times X \times \partial X \to \mathbf{R}$ is defined by

$$B(x, y, \xi) = \lim_{t \to \infty} d_X(y, \gamma_\xi(t)) - t$$

where γ_ξ is the unique geodesic ray with $\gamma(0) = x$ and $\gamma(\infty) = \xi$.
(4) **Support**(Knieper [**Kn**], Albuquerque [**Al**]): Each ν_x is supported on a specific Γ-orbit in $\partial_\infty X$.

The importance of property (3) is that it makes possible a reasonably explicit computation of the Jacobian of the map F_s; see below.

Property (4) is the most difficult of the properties to prove (see [**Al**] for the locally symmetric case and [**Kn**] for the geometric rank one case). Its importance is that when X is a symmetric space of noncompact type, the support of each ν_x lies in a specific copy of the *Furstenberg boundary* $\partial_F X$, and thus may be identified algebraically. This allows one to convert a geometry problem into an algebra problem; see §3.1 below.

2.3. The barycenter functional. For a probability measure on ∂X, one can try to define its *barycenter*, or *center of mass*, as follows. One first defines a map

$$\mathcal{B} : X \times \mathcal{M}(\partial X) \longrightarrow \mathbf{R}$$

via

$$\mathcal{B}(x, \lambda) = \int_{\partial_\infty X} B(x, \theta) d\lambda(\theta)$$

where B is the Busemann function on X, based at a fixed basepoint. Since X has nonpositive sectional curvatures, the distance function d on X is convex, and hence so is the Busemann function B. It follows that for any fixed measure $\lambda \in \mathcal{M}(\partial X)$, the function $\mathcal{B}(\cdot, \lambda)$ is a convex function on X. While $\mathcal{B}(\cdot, \lambda)$ is not in general strictly convex (for example when X is Euclidean, it is in many important cases.

PROPOSITION 12 (strict convexity). *Suppose that X has nonpositive sectional curvature, negative Ricci curvature uniformly bounded away from 0, and no local* $\mathbf{R}$ *factors. Let $\lambda \in \mathcal{M}(\partial X)$ be a fixed measure. Suppose that either*

- *X is a symmetric space of noncompact type and* $\mathrm{supp}(\lambda)$ *is the Furstenberg boundary of X, or*
- *X has geometric rank one and supp(λ) $= \partial X$.*

Then the function $\mathcal{B}(\cdot, \lambda)$ is strictly convex on X.

We note that in both cases of Proposition 12 the hypotheses are satisfied when λ is Patterson-Sullivan measure.

PROOF. The case when X is a symmetric space of noncompact type is Proposition 3.1 of [**CF2**]. Hence we assume that X has geometric rank one and negative

Ricci curvature (say bounded by $-a^2$). We also fix a basepoint p for the Busemann function B on X.

As the Busemann B function is convex, the functional

$$\mathcal{B}(y,\lambda) := \int_{\partial X} B(y,p,\xi)d\lambda(\xi)$$

is convex, being an integral of convex functions. We must prove that $\mathcal{B}(\cdot,\lambda)$ is strictly convex on X.

To see this, we first note that while the Hessian of $B(\cdot,p,\xi)$ is only semi-definite for fixed ξ, after taking the integral it becomes positive definite provided that $D_v d_y B$ does not have a common 0 direction $w \in S_y Y$ for each $v \in S_y Y$ which lies in the support of λ. Since the support of λ is all of ∂X, we must show that the tensors $D_v d_y B$ do not have a common zero w.

Now we have the decomposition

$$(D_v d_y B)^* = 0 \oplus U(v)$$

where $U(v)$ is the second fundamental form of the horosphere through v. In particular the Ricatti equations imply that $U(v)$ is a semi-definite tensor on the subspace of $T_y Y$ orthogonal to v. Therefore w must lie in $< v > \oplus \ker U(v)$ for all $v \in S_y Y$, or for all $v \neq \pm w$ we have $w- < w,v > v \in \ker U(v)$.

Consider such a $w \neq v$. Let $J(t)$ be the unique Jacobi field $J(t)$ along the geodesic $v(t)$ with $v'(0) = v$ satisfying $J(0) = w$ and $J(t)$ bounded for all $t > 0$. Then J also satisfies $J'(0) = 0$, that is $J(t)$ is parallel along v at $t = 0$. In particular the Ricatti equation then implies that $R(v,w,v,w) = 0$. Since this is true for all $v \in S_y Y$ with $v \neq w$ then $R_w := R(w,\cdot,w,\cdot)$ is the zero tensor, but we assumed that the Ricci curvature $\operatorname{tr} R_w$ at w was nonzero, and we are done. □

Note that Example 7 on page 23 shows that the negativity condition on the Ricci curvature is necessary.

Suppose that X satisfies the hypotheses of Proposition 12. Since we also know that the strictly convex function $\mathcal{B}(y,\lambda)$ goes to infinity as x goes to infinity we can define:

$$\text{bar} : \mathcal{M}(\partial_\infty X) \longrightarrow X$$
$$\text{bar}(\lambda) := \text{unique minimum of } \mathcal{B}(\cdot,\lambda)$$

We can now define $\widetilde{F} : Y \longrightarrow X$ by

$$\widetilde{F}(x) := \text{bar}(\widetilde{f}_*(\mu_x^s) \circledast \nu_x)$$

as described above. From the equivariance properties described above, it follows that $\widetilde{F}$ is equivariant with respect to the homomorphism $f_* : \pi_1(N) \longrightarrow \pi_1(M)$, so that $\widetilde{F}$ descends to a map $F : N \longrightarrow M$.

3. Bounding the Jacobian

The power of the map F comes from the fact that one can often obtain explicit estimates on its Jacobian. As we saw in §1.3, such estimates are the key to all applications of the barycenter method. In this section, we sketch how estimates on $|\operatorname{Jac}(F)|$ have been obtained in some cases.

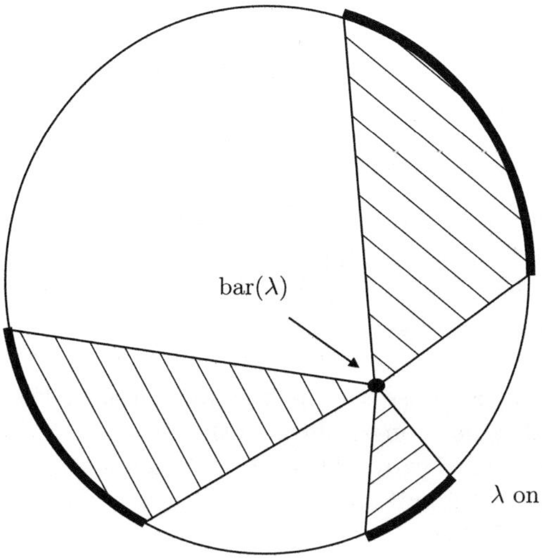

FIGURE 2. The barycenter of a measure.

3.1. Existence of some bound. Let $\sigma_y^s = \widetilde{f}_*(\mu_x^s) \circledast \nu_x$. Now F_s is defined by the implicit vector equation:

$$\int_{\partial_\infty X} dB_{(F_s(y),\theta)}(\cdot) d\sigma_y^s(\theta) = 0 \tag{5}$$

Differentiating (5) and applying the Implicit Function Theorem then gives a formula for $|\operatorname{Jac} F_s|$. An application of Hölder's inequality and some further estimates then give:

$$|\operatorname{Jac} F_s| \leq \left(\frac{s}{\sqrt{n}}\right)^n \frac{\det\left(\int_{\partial_\infty X} dB^2_{(F_s(y),\theta)}(\cdot) d\sigma_y^s(\theta)\right)^{1/2}}{\det\left(\int_{\partial_\infty X} DdB_{(F_s(y),\theta)}(\cdot,\cdot) d\sigma_y^s(\theta)\right)} \tag{6}$$

Note that all of this works as long as F_s is well-defined.

We wish to bound the right hand side of (6). In order to do this we first convert the problem to a Lie groups problem. This is where we use crucially the assumption that M is locally symmetric. An easy argument allows us to assume that the universal cover of M is irreducible, which we now do.

Recall Property (4) of Patterson-Sullivan measures, which states that in the locally symmetric case, the support of each Patterson-Sullivan is contained in a specific Γ-orbit which may also be identified with the *Furstenberg boundary* corresponding to the Lie group $G = \operatorname{Isom}(\widetilde{M})$ of isometries of the symmetric space $\widetilde{M}$. From this one can deduce the following:

Key fact: The expression $\partial_\infty X$ in (6) can be replaced by the maximal compact subgroup $K \subset G = \operatorname{Isom}(X)$.

To further simplify the right hand side of (6), we combine :

- Eigenvalue estimates on DdB, and
- The fact that, for M_i positive semidefinite, $\det(\sum M_i)$ is a nondecreasing homogeneous polynomial in eigenvalues of the M_i.

to obtain

$$|\operatorname{Jac} F| \leq C \frac{\left(\det \displaystyle\int_K O_\theta \begin{pmatrix} 1 & 0 \\ 0 & 0 \end{pmatrix} O_\theta^* \, d\sigma_y(\theta)\right)^{\frac{1}{2}}}{\det \displaystyle\int_K O_\theta \begin{pmatrix} 0 & 0 \\ 0 & I \end{pmatrix} O_\theta^* \, d\sigma_y(\theta)} \tag{7}$$

where I is the identity matrix of size $n - \operatorname{rank}(X)$, the matrix O_θ is an element of K, and the constant C depends only on n and on the Ricci curvatures of M, N. We remark that the eigenvalues of DdB are determined purely by algebraic data attached to the Lie algebra of G.

The strategy now is to bound the right hand side of (7) *over all probability measures* on K. In some sense this is the weakness of the barycenter method: without knowing additional constraints on the measure σ_y, one does not always obtain a bound as needed to prove Theorem 4, much less Conjecture 1, at least for certain locally symmetric manifolds; such examples which show that the method can fail are given in §6 below.

Some bounds on the right hand side of (7) are possible in most cases, however; the main technique is called "eigenvalue matching" in [**CF2**]. The idea is that for each small eigenvalue in the denominator, one tries to find two comparably small eigenvalues in the numerator. The hard part is to make "comparably small" independent of the measure. This involves a detailed analysis of action of the maximal compact subgroup K on subspaces of Lie algebra of $G = \operatorname{Isom}(X)$ (a simple Lie group). For details see [**CF2**].

The first basic fact used here is that the kernel of the operator in the numerator of the right hand side of (7) contains the cokernel of the operator in the denominator. The second main ingredient is the following proposition, which gives a K-invariant subspace perpendicular to K-orbit of a flat, and of twice the dimension of the flat (=rank(X)). This fact is precisely where the hypothesis that $G \neq \mathrm{SL}(3, \mathbf{R})$ is used.

PROPOSITION 13 (Eigenspace matching). *There is a constant C, depending only on $\dim X$ such that for any subspace $V \subset T_x X$ with $\dim V \leq \operatorname{rank}(X)$ there is a subspace $V' \subset V^\perp$ of dimension $2 \cdot \dim V$ satisfying*

$$\measuredangle(O_\theta \cdot V', A^\perp) \leq C \measuredangle(O_\theta \cdot V, A)$$

for all $O_\theta \in K$.

3.2. The perfect bound in rank one. When M has real-rank one, i.e. when M is negatively curved, Besson-Courtois-Gallot are able to obtain the infinitesimal rigidity described on page 22 (and hence a proof of the entropy rigidity conjecture) because of the following remarkable phenomenon. It is possible to write the operator in the denominator of the right hand side of the inequality (7) explicitly in terms of the operator in the numerator and the complex structure on the symmetric space; see [**BCG2**] for details. For example, for the case when M is real-hyperbolic, the denominator operator is the identity minus the numerator. They are then able to solve the minimization problem explicitly by using (quite nontrivial) linear algebra.

4. Negative results in various settings

In this section we describe in greater detail some negative results in the settings given in the table on page 21

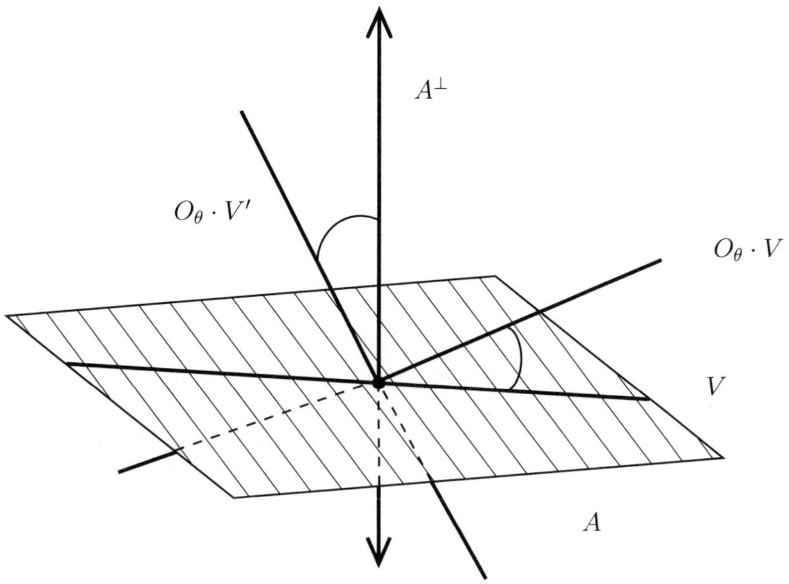

FIGURE 3. Eigenvalue matching.

4.1. Finsler metrics. Given a manifold, M, if we equip each tangent space T_xM with a norm F_x which depends smoothly on x then this gives rise to a Finsler metric by taking the distance between two points x and y to be the infimum of lengths of curves connecting x to y where length is measured using the norm on their tangent vectors. For our purposes we shall assume that the unit ball in each T_xM given by the norm is strictly convex with C^2 boundary. This is a common assumption since it allows us to form a positive definite inner product, g_u, for each $u \in T_xM$ whose matrix in local coordinates $(x_i, \dot{x}_i)$ for TM is given by entries

$$\frac{\partial^2 F^2}{\partial \dot{x}_i \partial \dot{x}_j}(u)$$

As expected, when F comes from a Riemannian inner product then the g_u all equal the inner product on T_xM. The notion of a geodesic flow makes sense for F as it does for the Riemannian case, and therefore we have a dynamical notion of topological entropy.

Let $B_x^F(R)$ be the ball of radius R in T_xM in the norm F. For any Riemannian metric g on TM we can similarly define $B_x^g(R)$. Then the volume form for F is defined at each point x to be,

$$dF(x) := \frac{\mathrm{Vol}_g\, B_x^g(1)}{\mathrm{Vol}_g\, B_x^F(1)} dg(x),$$

which is independent of the choice of inner product g. In particular, the quantities $\mathrm{Vol}_F(M)$ and $h_{\mathrm{Vol}}(F)$ make sense.

For deformations of a single Finsler metric, we can keep dF fixed by simply keeping the $\mathrm{Vol}_g\, B_x^F(1)$ fixed for each x relative to a fixed inner product g. Keeping the topological entropy constant as well is more subtle. P. Vérovic ([**Ve**]) has shown for any volume-preserving deformation F_t for $t \in (-\epsilon, \epsilon)$ of a compact hyperbolic manifold through Finsler metrics where F_0 is a hyperbolic Riemannian metric,

$$\left.\frac{d}{dt}\right|_{t=0} \mathrm{h}_{\mathrm{top}}(F_t) = 0.$$

For surfaces he shows the same result for deformations which preserve the Liouville volume of each unit tangent sphere. Moreover, there exist one-parameter families F_t where topological entropy can be held constant; in particular the rigidity part of Theorem 2 does not hold when the class of metrics is extended to Finsler metrics.

In higher rank things get worse: Vérovic also proved in [**Ve**] that for compact higher rank symmetric spaces (M, g_0), there exists a Finsler metric F_0 (invariant under all local isometries) with $\mathrm{Vol}_{F_0}(M) = \mathrm{Vol}_{g_0}(M)$ and $h_{\mathrm{Vol}}(F_0) < h_{\mathrm{Vol}}(g_0)$; that is, the locally symmetric metric on a higher rank manifold does *not* minimize normalized entropy among Finsler metrics.

Any invariant Finsler metric is determined by its values on a single maximal flat A. In fact, if $\mathfrak{a}$ is the corresponding maximal abelian subalgebra and Λ^+ is a set of positive roots α each with multiplicity m_α then F_0 is given by

$$F_0(H) := \sum_{\alpha \in \Lambda^+_{\mathfrak{a}}} m_\alpha |\alpha(H)|$$

for each $H \in \mathfrak{a}$.

In [**BCG3**] Besson, Courtois and Gallot asked the following intriguing question.

QUESTION 14. *Does Vérovic's Finsler metric F_0 minimize normalized entropy among all Finsler metrics?*

4.2. Magnetic field flows. Let (M, g) be a Riemannian manifold, let $\pi : TM \longrightarrow M$ be natural projection from the tangent bundle TM of M. The metric g determines an isomorphism of TM with the cotangent bundle; let ω_0 be the symplectic 2-form on TM formed by pulling back the canonical symplectic form on T^*M by this isomorphism. Now consider any closed 2-form Ω on M. Then for any $\lambda \geq 0$ it is easy to verify that

$$\omega_\lambda = \omega_0 + \lambda \pi^* \Omega$$

is also a symplectic form on TM.

Consider the usual Hamiltonian

$$H_x(v) := \frac{1}{2} g_x(v, v)$$

and define the λ-*magnetic flow* of the pair (g, Ω), denoted ϕ_λ, to be the Hamiltonian flow of H with respect to ω_λ. Consider the bundle map $Y : TM \longrightarrow TM$ given implicitly by

$$\lambda \Omega_x(u, v) = g_x(Y_x u, v)$$

Then for $v \in TM$ the curve $t \mapsto \phi^t_\lambda(v) = (\gamma(t), \dot\gamma(t))$ is characterized by the fact that

$$\nabla_{\dot\gamma(t)} \dot\gamma(t) = Y_{\gamma(t)} \dot\gamma(t),$$

where ∇ is the connection for the metric g; in particular ϕ^t_0 is just the geodesic flow for g.

Now we consider the case where M admits an Anosov geodesic flow. For instance we could take M to be any negatively curved compact manifold. In [**PaPa1**] and [**PaPa2**], G. and M. Paternain show that for $\Omega \neq 0$ the function $\mathrm{h}_{\mathrm{top}}(\phi_\lambda)$ is strictly decreasing for $\lambda \geq 0$ in the (nonempty) interval containing $\lambda = 0$ for which ϕ_λ is an Anosov flow. K. Burns and G. P. Paternain further show that $\mathrm{h}_{\mathrm{top}}(\phi_\lambda)$ is decreasing between "Anosov intervals" (see [**BP**]). In this paper the authors also give an example of a higher genus surface M, with a certain choice of Ω, where $\mathrm{h}_{\mathrm{top}}(\phi_\lambda)$ increases for λ in between disjoint intervals where ϕ_λ is Anosov. Therefore, while metrics corresponding to local maxima and minima for $\mathrm{h}_{\mathrm{top}}(\phi_\lambda)$ may be special, no standard entropy rigidity exists for such flows.

4.3. Metric entropy. Lastly, we mention the work of L. Flaminio in [**Fl**]. The *metric entropy* is the measure-theoretic entropy of the Liouville measure on the unit tangent bundle of M. The measure-theoretic entropy of the Bowen-Margulis measure is the same as the topological entropy, which in turn is the same as the volume entropy. Hence for hyperbolic metrics, for instance, the metric entropy and volume entropy coincide. In general the topological entropy is always greater than or equal to the metric entropy. Flaminio shows that the metric entropy for volume preserving deformations of a hyperbolic metric on a 3-manifold M does not have a local maximum at the hyperbolic metric. This is in contrast to the two dimensional case where A. Katok [**Ka**] showed that the metric entropy does have a local maximum there.

5. Positive results in various settings

In this section we describe some known cases where a locally symmetric metric structure minimizes $h(g)$ for a suitable normalization.

5.1. Finsler metrics. Recall the definition of a Finsler metric given above. We call a Finsler metric F *reversible* if it satisfies $F(-v) = F(v)$ for all $v \in TM$. Moreover we define an *eccentricity factor* $N(F)$ for a Finsler metric F on a manifold X to be

$$N(F) := \max_{x\in X} \max_{u\in S_x^F(1)} \max_{v\in S_x^{g_u}(1)} \frac{F_x(v)^n \operatorname{Vol}_{g_u}(B_x^F(1))}{\operatorname{Vol}_{g_u}(B_x^{g_u}(1))}$$

where $S_x^F(1)$ and $S_x^{g_u}(1)$ are the unit spheres in the norms F and g_u respectively. For this setting we redefine the volume entropy functional ent to be

$$\operatorname{ent}(F) := N(F)h(F)^n \operatorname{Vol}_F(X)$$

where n is the dimension of X.

To each direction in T_xX we may ascribe a Riemannian metric, and consider the corresponding curvature tensor. These are the flag curvatures (see [**BaCS**]). With this notation, J. Boland and F. Newberger [**BN**] proved the following theorem.

THEOREM 15 (Finsler entropy rigidity). *Let (X, g_0) be a compact, n-dimensional, locally symmetric Riemannian manifold of negative curvature ($n \geq 3$) and (Y, F) a compact, reversible, Finsler manifold of negative flag curvature that is homotopy equivalent to (X, g_0). Then*

$$1 \leq \frac{\operatorname{ent}(F)}{\operatorname{ent}(g_0)}$$

with equality if and only if F is homothetic to g_0

Notice that if F is Riemannian then $N(F) = N(g_0) = 1$ and so this reduces to the usual entropy rigidity theorem for the case of maps homotopic to the identity.

5.2. Foliations. Let N and M be compact topological manifolds supporting continuous foliations $\mathcal{F}_N$ and $\mathcal{F}_M$ by leaves which are smooth Riemannian manifolds, and such that the metrics on the leaves vary continuously in the transverse direction. We suppose that the leaves of $\mathcal{F}_M$ are locally isometric to n-dimensional symmetric spaces of negative curvature, $n \geq 3$; by continuity of the metrics these are all locally homothetic to a fixed symmetric space $(\widetilde{X}, g_o)$.

For any leaf (L, g_L), we may define

$$\overline{h}(g_L) = \limsup_{R \longrightarrow \infty} \frac{\log \operatorname{Vol}(B(x,R))}{R}$$

where $B(x,R)$ is the ball of radius R in the universal cover $\widetilde{L}$ of L. Similarly we can define $\underline{h}(g_L)$ using the $\liminf_{R\longrightarrow\infty}$. These numbers do not depend on the choice of x. We then define the *volume growth entropy* $h(g_L)$ to be

$$h(g_L) = \inf\left\{ s > 0 \,\middle|\, \int_0^\infty e^{-st} \operatorname{Vol} S(x,t)dt < \infty \right\}$$

where $S(x,t)$ is the sphere of radius t about x in the universal cover $\widetilde{L}$ of L. This quantity is independent of $x \in L$. Hence we can define a function $f : N \longrightarrow [0, \infty]$ by letting $f(x)$ be $h(g_L)$ for the leaf L containing x. Of course f is constant on each leaf.

The function f is also measurable. This follows from the fact that the transverse continuity of the leafwise metrics implies that for each R, the function

$$x \mapsto \int_0^R e^{-st} \operatorname{Vol} S(x,t)dt$$

is continuous on N. On $(M, \mathcal{F}_M)$ the entropy is constant and we denote it by $h(g_o)$. From the definition of $h(g_L)$ it follows almost immediately that

$$\underline{h}(g_L) \leq h(g_L) \leq \overline{h}(g_L)$$

For the foliation $(N, \mathcal{F}_N)$ we assume that the leaves (L, g_L) are strictly negatively curved, and satisfy the stronger condition that for all $x, y \in L$ there is a $\delta < h(g_L)$ such that

$$\limsup_{R \longrightarrow \infty} \frac{\operatorname{Vol} S(x,R)}{\operatorname{Vol} S(y,R)} \leq Ce^{\delta d(x,y)} \tag{8}$$

where $S(x,R)$ is the sphere of radius R in L. Such a leaf will be called a *Patterson-Sullivan manifold.*

Finally, assume we have a leaf-preserving homeomorphism

$$f : (N, \mathcal{F}_N) \to (M, \mathcal{F}_M)$$

which is leafwise C^1 with transversally continuous leafwise derivatives, but not necessarily transversally differentiable. Equip the foliation $(N, \mathcal{F}_N)$ with any choice of finite, transverse, holonomy quasi-invariant measure ν (see Hurder [**Hu**] or Zimmer [**Zi**] for the definition and existence). Holonomy quasi-invariance simply means that the push forward of ν under any holonomy map is in the same measure class as ν. This measure ν provides us with a global finite measure μ_N on N which is locally a product of ν with the Riemannian volumes dvol_L of the leaves L. In this setting J. Boland and the first author ([**BC**]) showed the following.

THEOREM 16 (Foliated entropy rigidity, I). *Let $(N, \mathcal{F}_N)$ be a continuous foliation of the compact manifold N such that ν-almost every leaf is a Patterson-Sullivan manifold. Suppose that $f : (N, \mathcal{F}_N) \longrightarrow (M, \mathcal{F}_M)$ is a foliation-preserving homeomorphism, leafwise C^1 with transversally continuous leafwise derivatives, and that $f_*\nu$-almost every leaf of $(M, \mathcal{F}_M)$ is a rank one locally symmetric space. Then there exists a finite measure μ_M on M which is locally the product of* dvol_o *with a transverse quasi-invariant measure ν_o such that*

$$\int_M h(g_o)^n d\mu_M \leq \int_N h(g_L)^n d\mu_N$$

and equality holds if and only if ν-almost every leaf (L, g_L) is homothetic to its image $(f(L), g_o)$.

When the foliation admits a holonomy-invariant measure ν then we may take $\nu_o = f_*\nu$. When ν is just holonomy quasi-invariant however, then ν_o is the push forward of ν under the natural map F defined below.

When the foliation $(N, \mathcal{F}_N)$ is ergodic with respect to ν, then the entropy function $h(g_L) = h(g)$ is constant on N, and we get the following (see [**BC**]).

COROLLARY 17 (Foliated entropy rigidity, II). *Under the same assumptions as in the main theorem, if $(N, \mathcal{F}_N)$ is ergodic, then $h(g_o)^n$* $\mathrm{Vol}(M, \mu_M)$ $\leq h(g)^n \,\mathrm{Vol}(N, \mu_N)$ *with equality if and only if ν-almost every leaf (L, g_L) is homothetic to $(f(L), g_o)$.*

Remark. If $(N, \mathcal{F}_N)$ and $(M, \mathcal{F}_M)$ are foliations such that almost every leaf is compact or simply connected, then the requirement that the homeomorphism f be leafwise C^1 can be dropped. In particular if the foliations have just one leaf and $\dim N \neq 3, 4$, any homotopy equivalence induces a homeomorphism between N and M (see [**FJ**]). Therefore, when $\dim N \neq 3, 4$, Corollary 17 recovers Theorem 2.

Of course one can also ask for foliated versions of Conjecture 1 as well.

5.3. Finite volume manifolds. For the case of finite volume manifolds J. Boland, J. Souto and the second author showed the following analogue of the Real Schwarz Lemma.

THEOREM 18 (Volume Theorem [**BCS**]). *Let (M, g) and (M_o, g_o) be two oriented complete finite volume Riemannian manifolds of the same dimension $n \geq 3$ and suppose that*

$$Ric_g \geq -(n-1)g, \quad \text{and} \quad -a \leq K_{g_o} \leq -1.$$

Then for all proper continuous maps $f : M \longrightarrow M_o$,

$$\mathrm{Vol}(M, g) \geq |\deg(f)| \,\mathrm{Vol}(M_o, g_o),$$

and equality holds if and only if M and M_o are hyperbolic and f is proper homotopic to a Riemannian covering.

When M and M_o are compact, Theorem 18 follows from a real Schwarz lemma proved by Besson, Courtois, and Gallot in [**BCG5**]. As in the case of all noncompact situations, to apply the method of [**BCG1**] and [**BCG5**], the fundamental difficulty is proving the properness of the natural map.

Theorem 18 implies the following.

COROLLARY 19. *Under the hypotheses of Theorem 18,*

$$\mathrm{Minvol}(M) \geq |\deg(f)| \,\mathrm{Vol}(M_o).$$

Now restrict f to be degree 1, and consider the case when $\mathrm{Minvol}(M) = \mathrm{Vol}(M_o)$. Bessières proved in [**Be1**] that if there is a degree 1 map $f : M \longrightarrow M_o$ from the compact n-dimensional M to the compact n-dimensional real hyperbolic manifold M_o such that $\mathrm{Minvol}(M) = \mathrm{Vol}(M_o)$, then M and M_o are diffeomorphic. He also gave examples in [**Be2**] of a finite volume manifold M with the same simplicial volume as a hyperbolic manifold M_o such that there is a degree 1 map from M to M_o and $\mathrm{Minvol}(M) \leq \mathrm{Vol}(M_o)$ but M and M_o are not even homeomorphic. By the above corollary, $\mathrm{Minvol}(M) = \mathrm{Vol}(M_o)$. Despite such examples, our next result ([**BCS**]) shows that the pointed Lipschitz limit of some subsequence of any sequence of metrics whose volumes achieve $\mathrm{Minvol}(M)$ is isometric to M_o.

THEOREM 20 (Hyperbolic **Minvol** rigidity, finite volume case). *Let M and M_o be oriented finite volume manifolds of the same dimension $n \geq 3$, M_o real hyperbolic, and $f : M \longrightarrow M_o$ a continuous, proper, degree 1 map. If* $\mathrm{Minvol}(M) = \mathrm{Vol}(M_o)$*, then for any sequence of metrics g_i realizing the minimal volume of M, there are $p_i \in M$ and a subsequence g_{i_j} such that (M, p_{i_j}, g_{i_j}) converges in the pointed Lipschitz topology to a Riemannian manifold isometric to M_o.*

In particular, the topology of the limit manifold changes to that of M_o.

The last main result of [**BCS**] is the finite volume version of the entropy rigidity result found in [**BCG1**]. We say that a Riemannian manifold (M, g) has *bounded local geometry* if the sectional curvatures and the injectivity radius of the universal cover $\widetilde{M}$ are bounded above and below respectively.

THEOREM 21 (Rank one entropy rigidity, finite volume case). *Let (M, g) and (M_o, g_o) be two complete, oriented, finite volume manifolds of the same dimension $n \geq 3$ with respective volume growth entropies $h(g)$ and $h(g_o)$. Suppose (M_o, g_o) is a rank one locally symmetric manifold, and that (M, g) has bounded local geometry. If $f : M \longrightarrow M_o$ is a proper continuous map, then*

$$h(g)^n \,\mathrm{Vol}(M, g) \geq |\deg(f)| h(g_o)^n \,\mathrm{Vol}(M_o, g_o)$$

and equality holds if and only if f is proper homotopic to a Riemannian covering.

As in [**BCG2**], this gives a quick proof of the Mostow rigidity theorem for finite volume negatively curved locally symmetric manifolds.

5.4. Quasifuchsian representations. The barycenter method has recently been applied by Besson-Courtois-Gallot [**BCG5**] to convex cocompact (infinite covolume) representations, generalizing Bowen's rigidity theorem about quasifuchsian groups. We now describe one of the several results in [**BCG5**] along these lines.

Let X be a negatively curved manifold. A faithful representation $\rho : \Gamma \longrightarrow \mathrm{Isom}(X)$ is *convex cocompact* if $\rho(\Gamma)$ acts cocompactly on the convex hull of its limit set $\Lambda(\rho(\Gamma))$. Attached to each ρ is a number, namely the Hausdorff dimension $\dim_{\mathcal{H}}(\rho(\Lambda(\Gamma)))$. Amazingly, this single number can be used to characterize the *totally geodesic* representations, i.e. those representations leaving invariant a totally geodesic submanifold. The following theorem was originally proved by Pansu, Bourdon, and Yue, generalizing an earlier theorem of Bowen. Besson-Courtois-Gallot gave another proof in [**BCG5**] using the barycenter method.

THEOREM 22 (Quasifuchsian rigidity). *Let Γ be the fundamental group of a closed, hyperbolic n-manifold, and let X be a connected, simply-connected manifold with sectional curvature ≤ -1. Then*

$$\dim_{\mathcal{H}}(\rho(\Lambda(\Gamma))) \geq n-1$$

with equality if and only if ρ is totally geodesic.

There is also a version for complex hyperbolic manifolds (see [**BCG5**]). The idea of the proof is to define the barycenter map as we discussed above, and to bound the Jacobian of this map; Theorem 22 then follows reasonably quickly. The main difficulty is that the dimensions of the domain and target spaces are different. To repair this one notes that ρ gives a quasi-isometric embedding between negatively curved spaces, hence a homeomorphic embedding of one boundary at infinity into another. The general technique can then be carried out once one finds the right notion of volume.

5.5. Alexandroff Spaces. In this section we report on recent work of P. Storm, who extended some of the results of [**BCG1**] to certain Alexandroff space domains within the same bilipschitz class as a fixed hyperbolic manifold. All of the results can be found in [**St**].

In brief, an Alexandroff space X with curvature bounded below by -1 is any complete, locally compact metric space of finite Hausdorff dimension, such that every point $x \in X$ has a neighborhood containing a geodesic triangle such that the comparison triangle in $\mathbf{H}^2$ with the same side lengths has the property that the distance form any vertex of the triangle to any point on the opposite side is shorter than the corresponding distance on the triangle in X. Storm's main result in [**St**] is the following.

THEOREM 23 (Alexandroff domain). *Let (X, d) be an Alexandroff space with curvature bounded below by -1, and Y a closed, hyperbolic n-manifold, $n \geq 3$. If there is a bilipschitz homeomorphism between X and Y, then*

$$\mathrm{Vol}(X) \geq \mathrm{Vol}(Y).$$

The proof of this theorem is at first glance unrelated to the method of Besson, Courtois, and Gallot as outlined in the opening sections of this paper. Nevertheless, it is actually of the same vein. The main technique is to extend the idea of *Spherical Volume* as presented in [**BCG1**] to this new setting. Instead of going through the (perhaps insurmountable) difficulty of forming a barycenter map from X to Y, one can in a sense stop "half-way" and attempt to do the analysis in $\mathcal{M}(X)$ and $\mathcal{M}(Y)$. Since the embedding via measures is in essence an L^1 approach it is preferable to modify this to an L^2 approach which leaves us working in a Hilbert space.

More precisely, one embeds $\widetilde{X}$ into $L^2(\widetilde{X})$ in a π_1-equivariant way via the map

$$\Phi_s(x) := e^{s/2d(x,\cdot)}$$

for any $s > h(X)$, and then projects this to the unit Hilbert sphere $S^\infty(\widetilde{X})$. We recall that this is the embedding achieved by using the square root of the Radon-Nikodym derivatives of the un-normalized σ^s_y, which are in L^2 by definition of the volume growth entropy $h(X)$. If $f : \widetilde{Y} \longrightarrow \widetilde{X}$ represents a bilipschitz π_1-equivariant map given by the hypotheses of the theorem, then the map

$$\widetilde{I} : L_2(\widetilde{X}) \longrightarrow L_2(\widetilde{Y})$$

given by

$$\mathcal{I}(g) := (g \circ f)\,\mathrm{Jac}(f)$$

is a π_1-equivariant isometry. Hence it restricts to $\mathcal{I} : S^\infty \widetilde{X} \longrightarrow S^\infty \widetilde{Y}$. The composition

$$\widetilde{F} := \mathcal{I} \circ \Phi_s \circ f : \widetilde{Y} \longrightarrow S^\infty \widetilde{Y}$$

descends to a map $F : Y \longrightarrow S^\infty(Y)$ with image in the positive orthant.

For such maps there is a notion of volume, and without defining it, we simply point out that spherical volume is the infimum of the volume of such maps. In particular

$$\mathrm{Vol}(F) \geq \mathrm{Spherical\,Vol}(Y) = \left(\frac{(n-1)^2}{4n}\right)^{n/2} \mathrm{Vol}(Y)$$

The last equality was proven by Besson-Courtois-Gallot in [**BCG1**].

Since Alexandroff spaces are in a sense Riemannian manifolds off of a measure zero set, and Y is a Riemannian manifold, Storm can use standard integration theory. Of course we now come to the difficult (and omitted) part of the proof of Theorem 23, which is to show

$$\mathrm{Vol}(F) \leq \left(\frac{s^2}{4n}\right)^{n/2} \mathrm{Vol}(X)$$

It then follows that

$$\mathrm{Vol}(X) \geq \left(\frac{n-1}{s}\right)^{n} \mathrm{Vol}(Y)$$

The final step is to apply a result of Burago-Gromov-Perel'man which states that for Alexandroff spaces with curvature bounded below by -1, $h(X) \leq n-1$; Theorem 23 then follows.

The main applications of Theorem 23 are to two important classes of Alexandroff spaces which arise naturally in the study of hyperbolic manifolds: doublings of convex cores and cone manifolds.

5.5.1. *Convex Cores.* The *(metric) double* DC_M of the convex core C_M of a convex cocompact manifold M is the result of identifying the boundaries of two copies of C_M and then extending the induced metric on each copy of C_M to the whole. While as a topological manifold DC_M can always be smoothed (e.g. by taking the double a neighborhood of C_M), the point is that any Riemannian metric on the resulting smooth manifold cannot agree with the metric on each copy of C_M as a subspace, unless these had totally geodesic boundaries to begin with. On the other hand, DC_M naturally carries the structure of an Alexandroff space with Alexandroff curvature bounded below by -1 and Theorem 23 applies.

Let $\mathrm{CC}(N)$ be the space of complete convex cocompact hyperbolic manifolds diffeomorphic to the *interior* of smooth compact n-manifold N. In analogy with the volume entropy results discussed in this paper, one wants to know that the topological invariant

$$\mathcal{V}(N) := \inf_{M \in \mathrm{CC}(N)} \{\text{volume of the convex core } C_M \text{ of } M\}$$

is minimized for certain special hyperbolic manifolds.

It is consequence of Thurston's work together with Mostow Rigidity that a 3-manifold N is acylindrical if and only if there exists unique $M_0 \in \mathrm{CC}(N)$ such that ∂C_{M_0} is totally geodesic (see [**Th**]). Combining this with Theorem 23, Storm obtains the following.

COROLLARY 24. *Let N be an acylindrical compact irreducible 3-manifold such that $CC(N)$ is nonempty. Then there exists a unique $M_0 \in CC(N)$ such that $\mathcal{V}(N) = Vol(C_{M_0})$. Moreover, ∂C_{M_0} is totally geodesic.*

In fact he shows that any C_{M_0} with totally geodesic boundary is the unique minimizer of $\mathrm{Vol}(C_N)$ among hyperbolic N homotopy equivalent to M_0. (Note we are using that M_0 is tame). We should also mention that Bonahon had previously shown in [**Bon**] that M_0 is a strict local minimum of $\mathrm{Vol}(C_N)$.

REMARK. Storm is able to also obtain results giving exact relations between the Gromov norm of such DC_N and covers in terms of $\mathcal{V}(N)$.

5.5.2. *Application to Cone Manifolds.* Theorem 23 may also be applied to cone-manifolds with all cone angles $\leq 2\pi$.

DEFINITION ([**CHK**], page 53). An *n-dimensional cone-manifold* is a topological manifold, M, which admits a triangulation giving it the structure of a PL manifold and M is equipped with a complete path metric such that the restriction of the metric to each simplex is isometric to a geodesic simplex of constant curvature K. The singular locus Σ consists of the points with no neighborhood isometric to a ball in a Riemannian manifold.

It follows that

- Σ is a union of totally geodesic closed simplices of dimension $n-2$.
- At each point of Σ in an open $(n-2)$-simplex, there is a *cone angle* which is the sum of dihedral angles of n-simplices containing the point.

Notice that cone-manifolds whose singular locus has vertices are allowed.

LEMMA 25 ([**BGP**], page 7). *If all cone angles of n-dimensional cone-manifold M are $\leq 2\pi$, and $K \geq -1$, then M is an Alexandroff space with curvature bounded below by -1.*

An n-dimensional cone-manifold clearly has Hausdorff dimension n. Therefore Theorem 23 implies the following.

COROLLARY 26. *Let M be an 3-dimensional cone-manifold with all cone angles $\leq 2\pi$ and $K \geq -1$. Let M_0 be a closed hyperbolic 3-manifold. If M and M_0 are homeomorphic then*

$$Vol(M) \geq Vol(M_0)$$

All of these corollaries have only slightly weaker generalizations to any dimension greater than 2. We refer the reader to [**St**]. There are other results whose methods follow along these lines; in particular we would like to point out the work relating to Einstein metrics, the Gromov norm, and simplicial volume for manifolds and covers carried out by A. Sambusetti (e.g. [**Sam1, Sam2**]). However, these recent results have already been well surveyed in [**BCG5**].

6. Cautionary examples

Our proof in [**CF2**] of the Degree Theorem (Theorem 4) can be viewed as a step towards the Entropy Rigidity Conjecture (Conjecture 1), or at least the inequality of that conjecture; to do this one "only" needs to prove the inequality (3) on page 22 with the lowest possible C, namely $C = \left(\frac{h(g)}{h(g_0)}\right)^n$.

While the value of C which comes out of the proof (see [**CF2**]) of inequality (3) can be explicitly computed, finding the best C seems much harder. In fact, it soon became clear to us that the barycenter method applied without *a priori* constraints on the types of measures considered is not sufficiently precise to obtain the rigidity aspect of the theory, or even to prove that the locally symmetric metric minimizes entropy. In this section we describe some explicit examples which demonstrate some of the problems.

The Jacobian estimates on the Besson-Courtois-Gallot map F are obtained by bounding the right hand side of (7), that is, bounding

$$\frac{\left(\det \int_K O_\theta \begin{pmatrix} 1 & 0 \\ 0 & 0 \end{pmatrix} O_\theta^* \, d\sigma_y(\theta)\right)^{\frac{1}{2}}}{\det \int_K O_\theta \begin{pmatrix} 0 & 0 \\ 0 & I \end{pmatrix} O_\theta^* \, d\sigma_y(\theta)}$$

independently of the measure. In other words, no special property of the measure $d\sigma_y$ is used. This is also the case in the proof of Entropy Rigidity in rank one (see [**BCG2**]).

We recall that in a symmetric space of noncompact type, the Furstenberg boundary is the space of Hausdorff equivalence classes of (maximal) Weyl chambers; this space can be identified with G/P, where P is a minimal parabolic subgroup of G. This in turn is naturally homeomorphic to $K/(K \cap M)$ where K is the maximal compact subgroup of G and M is the centralizer of $\mathfrak{a}$ in G. Since K is isomorphic to K_p, isotropy subgroup of $p \in X$, and M is isomorphic to M_p, the centralizer of a Weyl chamber in $\mathcal{F}_p$, we may interpret the Furstenberg boundary geometrically. Namely it is naturally identified to any G-orbit of a regular point in ∂X. We will denote by $\partial_F X$ the particular orbit of the point at infinity corresponding to the normalized barycenter b^+ of positive root vectors. By the above, $\partial_F X$ projects to the unit tangent sphere $S_p X$ as the orbit of b^+ by K_p.

Hence if we write $J(\mu)$ for the square of the right hand side of the expression (6) with σ_y^s replaced by μ, then we would like for it to be uniquely maximized when μ is the projection to $\partial_F X$ of the Haar measure on K. Here we only care about maximizing over the space of $M_p \cap K_p$-invariant μ supported on K_p.

6.1. An example in $\mathbf{H}^2 \times \mathbf{H}^2$. In the case of $X = \mathbf{H}^2 \times \mathbf{H}^2$ the Furstenberg boundary $\partial_F X$ is simply the torus $S^1 \times S^1$. We will parameterize this by (e^{it_1}, e^{it_2}). Let M be a compact quotient of X and let $f : M \longrightarrow M$ a continuous map. We first consider the case $f = \mathrm{Id}$ and the simple two-parameter family of probability measures $\mu_{a,b}$ for $a, b > 0$ on $\partial_F X$ given by

$$d\mu_{a,b} = \frac{2 + b + 2\,a\cos(\frac{t_1}{2})^4 + b\,\cos(\,t_2\,)}{(8 + 3\,a + 4\,b)\;\pi^2} d\mu_0$$

where μ_0 the unit Haar measure on the torus. Note that $\mu_{0,0}$ is Haar measure. For this family, by integrating over the torus we can compute the expression for

$$J_0(\mu_{a,b}) := J(\mu_{a,b})/J(\mathrm{Haar})$$

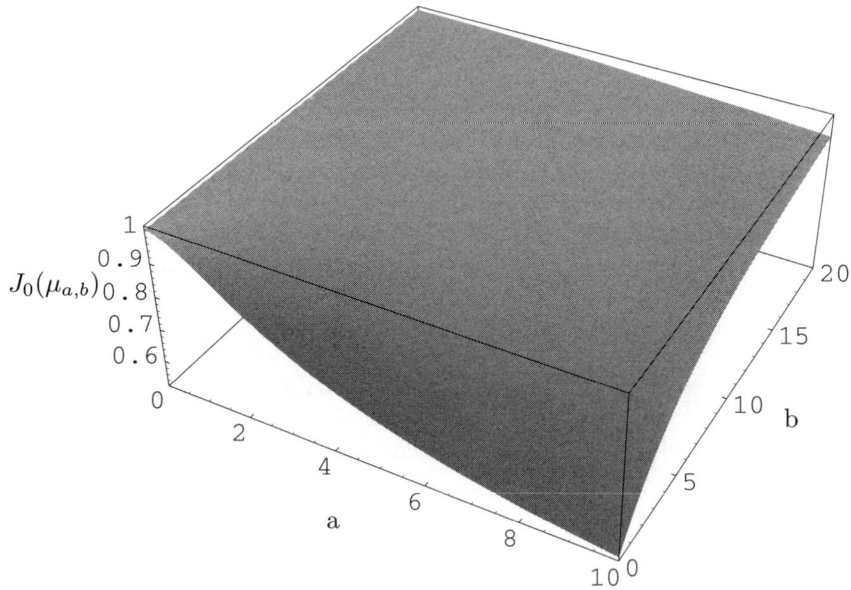

FIGURE 4. Graph of $J(\mu_{a,b})/J(\text{Haar})$ for several values of a and b.

to be

$$J_0(\mu_{a,b}) = \frac{4\left(5\,a^2 + 24\,a\,(2+b) + 16\,(2+b)^2\right)^2}{(8+3\,a+4\,b)^2\,(5\,a+8\,(2+b))\,(7\,a+8\,(2+b))}.$$

Note that the measures $\mu_{0,b}$ are distinct, but that $J(\mu_{0,b})/J(\mu_{0,0}) = 1$ identically. In particular the Haar measure does not uniquely minimize this quantity. This may not be that surprising in light of the fact that $\mathbf{H}^2$ does not have a unique minimizer either, nor is there a entropy rigidity theorem for X, at least not for reducible lattices.

6.2. An example in $\mathrm{SL}_3(\mathbf{R})/\mathrm{SO}_3(\mathbf{R})$. Now we examine the case of $X = \mathrm{SL}(3,\mathbf{R})/\mathrm{SO}(3)$. Suppose for a single flat $\mathcal{F}_x$ and a sequence of $y_i \in \mathcal{F}_x$, the measures $\sigma^s_{y_i}$ tend to the sum of Dirac measures

$$\mu = \frac{1}{6}\sum_{i=1^6} \delta_{b_i^+(\infty)}$$

where the b_i^+ are all the images of b^+ under the Weyl group in $\mathcal{F}_x$. Hence there is one atom for each Weyl chamber at x and they are symmetrically placed making x the barycenter of μ. Then we claim that if $N = M$ and $\widetilde{f}$ induces the identity map on $\partial\mathcal{F}_x$ then $\operatorname{Jac} F_s(y_i) \longrightarrow \infty$.

The hypotheses that $N = M$ and that $\widetilde{f}$ is identity on $\partial\mathcal{F}_x$, which contains the support of μ, implies that $\operatorname{Jac} F_s(y_i)$ is identical in the limit $i \longrightarrow \infty$ to the estimate on the right-hand side of (6) with σ^s_y replaced by μ. Therefore it remains to show that this right-hand side is unbounded.

First note that the sum

$$\lim_{i\longrightarrow\infty}\sum_{j=1}^{6} dB^2_{(F_s(y_i),b_j^+(\infty))}$$

has only a 3-dimensional kernel, while

$$\lim_{i\longrightarrow\infty}\sum_{j=1}^{6} DdB_{(F_s(y_i),b_j^+(\infty))}$$

has a 2-dimensional kernel. Furthermore the numerator and denominator

$$Q_1 := \int_{\partial_F X} dB^2_{(F_s(y_i),\theta)} d\sigma^s_{y_i} \quad\text{and}\quad Q_2 := \int_{\partial_F X} DdB_{(F_s(y_i),\theta)} d\sigma^s_{y_i}$$

degenerate homogeneously. In particular, the quantity $\det(Q_1)/\det(Q_2)^2$ is unbounded. This can be easily verified explicitly in the case of a sum of eight Dirac measures for which both integrals are nonsingular degenerating to the sum of the six Dirac measures given above.

6.3. An example in $\mathrm{SL}_4(\mathbf{R})/\mathrm{SO}_4(\mathbf{R})$. In the case when $X = \mathrm{SL}(4,\mathbf{R})/\mathrm{SO}(4)$, Theorem 4 holds and the quantity (6) is bounded independently of σ^s_y. Nevertheless, here we present a smooth (with respect to Haar) probability measure μ on $\partial_F X$ such that $J_0(\mu)$ is strictly larger than 1. In fact, μ will be very close to the projection of Haar measure μ_0 only differing by the addition of a few very small and sharp "thorns" and removal of a few "drillings." What we mean by this will become evident from the final construction.

Recall that X may be identified with the space P of positive definite symmetric matrices in $\mathrm{SL}_4(\mathbf{R})$. After translating from the point $p \in P$ to $\mathrm{Id} \in P$, the tangent space T_pP may be identified with the space $\mathfrak{p}$ of traceless symmetric matrices in $\mathfrak{sl}(4,\mathbf{R})$. The linear space $\mathfrak{p}$ is in turn naturally identified with P via the exponential map.

The standard symmetric metric on X is the one induced from the killing form. For all $X, Y \in \mathfrak{p} \equiv T_pP$ at the point $p \in P$ this inner product has the form

$$\langle X, Y\rangle_p = 8\cdot \mathrm{Tr}(p^{-1}Xp^{-1}Y),$$

which at $p = \mathrm{Id}$ is simply the restriction to $\mathfrak{p}$ of 8 times the standard Euclidean inner product on $\mathbf{R}^{4^2}$.

Now let E_{ij} be the 4×4 matrix with 1 in the entry i, j and 0 elsewhere. Then we span the diagonal of $\mathfrak{p}$ with the three basis elements $e_1 = \frac{1}{4\sqrt{2}}(E_{11}+E_{22}-E_{33}-E_{44})$, $e_2 = \frac{1}{4\sqrt{2}}(E_{11}-E_{22}+E_{33}-E_{44})$ and $e_3 = \frac{1}{4\sqrt{2}}(E_{11}-E_{22}-E_{33}+E_{44})$. We complete this to a basis of $\mathfrak{p}$ by adding the six elements given by $\frac{1}{4}(E_{ij}+E_{ji})$ for $1 \leq i < j \leq 4$ which we denote $e_4, \ldots, e_9$. With respect to this basis, $\langle \sum x_i e_i, \sum y_j e_j\rangle_{\mathrm{Id}} = \sum_{i=1}^n x_i y_i$.

Our procedure is to consider second derivatives of the Jacobian

$$f(t) := J_0\left(\frac{\mu_0 + t\mu}{\|\mu_0 + t\mu\|}\right)$$

with respect to a parameter t where μ is an arbitrary signed measure. In fact, if we can find a measure μ for which the second derivative at zero is positive then we will show how one can obtain such an example where μ is a smooth (with respect to μ_0) positive probability measure.

One can verify directly that, for general X,

$$J(\mu_0) = \left(\frac{s^2/n}{(\mathrm{Tr}(DdB)/n)^2}\right)^n.$$

Since X is symmetric, $\mathrm{Tr}(DdB)$ is constant and by the Ricatti equation it equals $\mathrm{Tr}(\sqrt{-R(b^+,\cdot,b^+,\cdot)}) = \sqrt{\frac{5}{2}}$ where R is the curvature tensor and b^+ is the normalized algebraic barycenter of a choice of positive roots. One choice of b^+ in our coordinates works out to be $(\frac{2}{\sqrt{5}}, \frac{1}{\sqrt{5}}, 0, 0, 0, 0, 0, 0, 0)$. Therefore in $\mathrm{SL}(4, \mathbf{R})$ we have

$$J(\mu_0) = \left(\frac{18s^2}{5}\right)^9.$$

Hence, after clearing the denominator of the measures, we can rewrite

$$f(t) = \left(\frac{18}{5}\|\mu_0 + t\mu\|\right)^9 \frac{\det\left(\int_{\partial_F X} dB^2_{(F_s(y),\theta)} d[\mu_0 + t\mu](\theta)\right)}{\det\left(\int_{\partial_F X} DdB_{(F_s(y),\theta)}(\cdot,\cdot) d[\mu_0 + t\mu](\theta)\right)^2}$$

Taking the derivative of the log we have

$$\partial_t f(t) = f(t)\partial_t \log f(t)$$

Hence the second derivative is

$$\partial_t\partial_t f(t) = f(t)\left[(\partial_t \log f(t))^2 + \partial_t\partial_t \log f(t)\right]$$

For an invertible matrix M we have the operator identities

$$\log\det M = \mathrm{Tr}\log M \quad \text{and} \quad \partial_t \mathrm{Tr}\log M = \mathrm{Tr}\left((\partial_t M)M^{-1}\right)$$

Applying these identities we directly obtain

$$\begin{aligned}\partial_t \log f(t) &= 9\frac{\|\mu\|}{\|\mu_0 + t\mu\|} + \\ &\mathrm{Tr}\left(\int_{\partial_F X} dB^2_{(F_s(y),\theta)} d\mu(\theta)\right)\left(\int_{\partial_F X} dB^2_{(F_s(y),\theta)} d[\mu_0 + t\mu](\theta)\right)^{-1} - \\ &2\,\mathrm{Tr}\left(\int_{\partial_F X} DdB_{(F_s(y),\theta)} d\mu(\theta)\right)\left(\int_{\partial_F X} DdB_{(F_s(y),\theta)} d[\mu_0 + t\mu](\theta)\right)^{-1}\end{aligned}$$

Similarly,

$$\begin{aligned}\partial_t\partial_t \log f(t) &= -9\left(\frac{\|\mu\|}{\|\mu_0 + t\mu\|}\right)^2 - \\ &\mathrm{Tr}\left(\left(\int_{\partial_F X} dB^2_{(F_s(y),\theta)} d\mu(\theta)\right)\left(\int_{\partial_F X} dB^2_{(F_s(y),\theta)} d[\mu_0 + t\mu](\theta)\right)^{-1}\right)^2 + \\ &2\,\mathrm{Tr}\left(\left(\int_{\partial_F X} DdB_{(F_s(y),\theta)} d\mu(\theta)\right)\left(\int_{\partial_F X} DdB_{(F_s(y),\theta)} d[\mu_0 + t\mu](\theta)\right)^{-1}\right)^2\end{aligned}$$

Since $f(0) = 1$ and

$$\int_{\partial_F X} dB^2_{(F_s(y),\theta)} d\mu_0(\theta) = \frac{1}{9}\,\mathrm{Id} \quad \text{and} \quad \int_{\partial_F X} DdB_{(F_s(y),\theta)} d\mu_0(\theta) = \frac{\sqrt{5}}{9\sqrt{2}}\,\mathrm{Id}$$

we see that $\partial_t|_{t=0}f(t) = 0$ and

$$\partial_t\partial_t|_{t=0}f(t) = -9^2\operatorname{Tr}\left(\int_{\partial_F X} dB^2_{(F_s(y),\theta)}d\mu(\theta)\right)^2 - 9\left(\int_{\partial_F X} d\mu\right)^2 + 9^2\frac{4}{5}\operatorname{Tr}\left(\int_{\partial_F X} DdB_{(F_s(y),\theta)}d\mu(\theta)\right)^2.$$

We denote $F_s(y)$ by p, and without loss of generality we may assume $p = \mathrm{Id} \in P$. Consequently, $K = \mathrm{SO}(4)$ is the stabilizer of the point p. We also have the natural representation of K in $\mathrm{SO}(\mathfrak{p})$ induced by the action of K on P by conjugation.

We may lift a measure μ on $\partial_F X$ to K by averaging over orbits of $M \cap K$, the stabilizer in K of b^+ under conjugation. This lifted measure, which we again denote by μ, has push-forward equal to μ on $\partial_F X$. For finite atomic measures we may equally as well choose, for each atom in $\partial_F X$, one atom of the corresponding $M \cap K$ coset in K. In this way we may therefore treat the integration over K. It is convenient to parameterize K by $\mathbf{R}^6 \equiv \mathfrak{s}$ the subalgebra of skew-symmetric matrices in $\mathfrak{sl}(4, \mathbf{R})$. For $\theta \in \mathbf{R}^6$ let $\mathrm{O}(\theta) \in \mathrm{SO}(\mathfrak{p})$ be the corresponding element. Let A be the matrix (with respect to the e_i basis) representing the Kronecker product of b^+ with itself. Let B be the matrix representing $\sqrt{-R(b^+, \cdot, b^+, \cdot)}$. In the e_i basis, B is the diagonal matrix with diagonal

$$\left(0, 0, 0, \frac{1}{2\sqrt{10}}, \frac{1}{2\sqrt{10}}, \frac{1}{2\sqrt{10}}, \frac{1}{\sqrt{10}}, \frac{1}{\sqrt{10}}, \frac{3}{2\sqrt{10}}\right).$$

Relative to the e_i basis on $T_{F_s(y)}X$ we may write

$$dB^2_{(F_s(y),\theta)} = O(\theta)AO(\theta)^*$$

and

$$DdB_{(F_s(y),\theta)} = O(\theta)BO(\theta)^*$$

With these notations, the expression above for $\partial_t\partial_t|_{t=0}f(t)$ becomes

$$\int_K\int_K q(\sigma,\tau)d\mu(\sigma)d\mu(\tau)$$

where

$$q(\sigma,\tau) = -9^2\operatorname{Tr}\left(O(\sigma)AO(\sigma)^*O(\tau)AO(\tau)^*\right) - 9 + \frac{4\cdot 9^2}{5}\operatorname{Tr}\left(O(\sigma)BO(\sigma)^*O(\tau)BO(\tau)^*\right).$$

As expected, for $\sigma \in M \cap K$, $O(\sigma)$ commutes with A and B and so q descends to a map on $\partial_F X$. In fact, the elements in the representation of $M \cap K$ are shown after a bit of calculation to be identity on $\mathfrak{a}$ and diagonal with resepect to the chosen basis.

The above formula shows that we may treat $\partial_t\partial_t|_{t=0}f(t)$ as a symmetric 2-form $\Omega(\cdot,\cdot)$ on the space of signed measures. In fact, the function q is equivariant, so $q(\sigma,\tau)$ can be written as $g(\sigma^{-1}\tau) := q(1, \sigma^{-1}\tau)$ and one can easily check that $g(\sigma) = g(\sigma^{-1})$.

Three representative graphs of $g(\sigma)$ over two-parameter subspaces spanned by (respectively) v_1 and v_2, v_3 and v_4, and v_5 and v_6, are shown in Figures 5,6 and 7. Here for each $i = 1, \ldots, 6$, $v_i \in \mathbf{R}^6$ is the basis element representing s_i in (9) below. We see that the function g is mostly negative but there are smaller positive regions as well.

Now we can take μ_1 to be the sum of the atomic measures at the 55 points of K, at most one in each $M \cap K$ coset, corresponding to the following rational parameters in $\mathbf{R}^6$. Namely the points are of the form $K_i \cdot b^+$ where b^+ is the barycenter of a set of positive roots (note that any two choices will produce equivalent measures differing by a push forward by an isometry in the Weyl group) and that K_i is the element of $K = SO(4)$ acting on b^+ by conjugation corresponding to the element

$$\exp \begin{pmatrix} 0 & s1 & s4 & s6 \\ -s1 & 0 & s2 & s5 \\ -s4 & -s2 & 0 & s3 \\ -s6 & -s5 & -s3 & 0 \end{pmatrix} \in SO(4). \tag{9}$$

Where we are using the previously mentioned choice for b^+ which has coordinates $(2/\sqrt{5}, 1/\sqrt{5}, 0)$ restricted to $\mathfrak{a}$. The corresponding values of the $s_1, \ldots, s_6$ for K_i may be read across the i-th row of the table in Figure 8. Each of the 55 atoms is weighted by the one of the three vectors of 55 weights each listed in Figure 9.

For μ_1 chosen to be any of these three weighted sums of atomic measures on $\partial_F X$ we use the explicit form of the representation for K to show that the barycenter is p in each case, i.e. the vector sum in T_pX is 0. Evaluating the resulting sums, yields that $\Omega(\mu_1, \mu_1)$ is greater than $2.29526, 1.63584$, or 2.02786 corresponding to the three sets of weights chosen above.

These three values were proven to be lower bounds using interval arithmetic in the *Mathematica* computer algebra environment for all of the computations. (The particular implementation used is able to guarantee lower and upper bounds for the actual value of a given computation.)

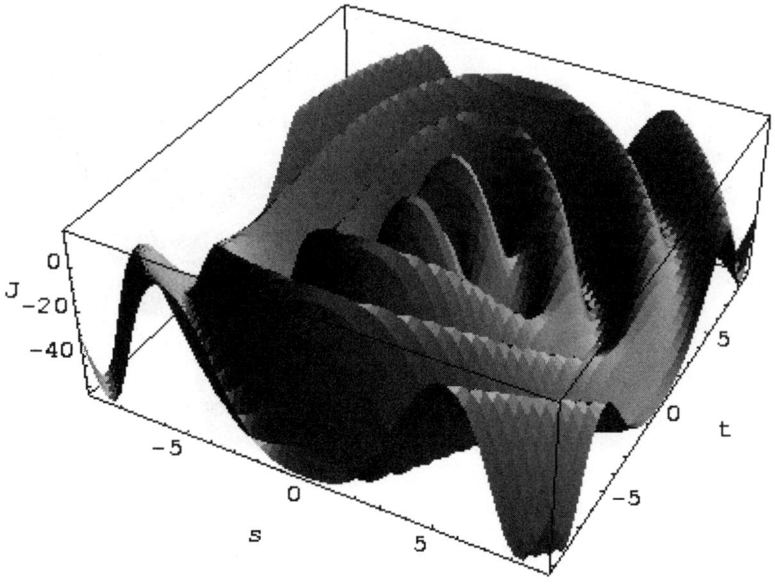

FIGURE 5. Graph of $g(sv_1 + tv_2)$

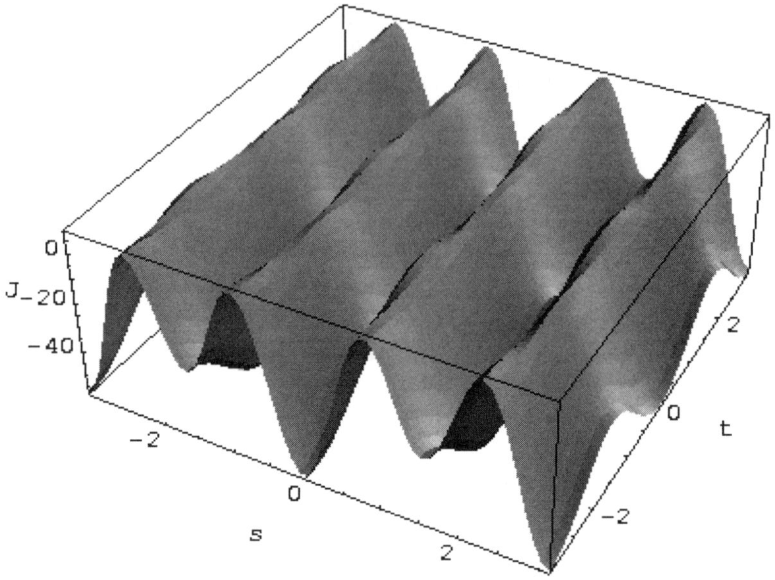

FIGURE 6. Graph of $g(sv_3 + tv_4)$

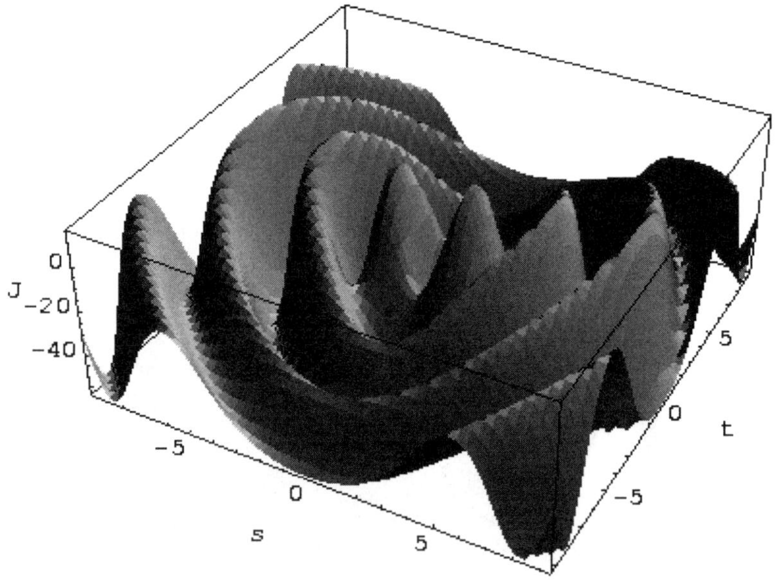

FIGURE 7. Graph of $g(sv_5 + tv_6)$

The only remaining problem is that μ_1 is not a positive measure since each of the above systems of weights has negative values. Nevertheless, we can fix this by adding elements in the kernel of Ω. For any $\epsilon > 0$ we may take a smooth (with respect to μ_0) and symmetric (keeping the barycenter 0) approximation μ_1^ϵ to μ_1 such that

$$|\Omega(\mu_1, \mu_1) - \Omega(\mu_1^\epsilon, \mu_1^\epsilon)| < \epsilon$$

Then if $c(\epsilon) = \min\{\inf \frac{d\mu_1^\epsilon}{\mu_0}, 0\}$, the measure

$$\mu_2 = \frac{-c(\epsilon)\mu_0 + \mu_1^\epsilon}{|c(\epsilon)| + \|\mu_1^\epsilon\|}$$

(−2.1763	−1.1507	−2.04369	0.44192	0.348006	−0.0873793)
(1.10122	−1.30303	−1.07559	1.69668	−0.537171	−0.790495)
(−2.66081	−2.51729	0.874064	0.0170786	2.76331	0.918491)
(3.06666	−1.20383	3.10876	−1.8274	1.33181	1.36971)
(2.14346	2.46489	0.233902	−2.2138	−1.34613	−0.589322)
(2.27427	2.23082	2.87105	0.855587	−0.33015	−0.120273)
(2.39027	0.231281	1.93738	3.00424	2.76855	2.45438)
(2.01231	1.06647	2.80138	1.14019	−2.46108	2.83835)
(−2.48367	1.81689	0.446606	1.91056	2.00405	−0.735379)
(1.31393	2.82133	2.2746	1.55063	−1.49752	−0.199991)
(3.02592	−1.82225	−0.293302	−0.0626386	−2.88422	−1.13504)
(0.835977	2.01248	−2.54401	0.866367	0.15547	2.31572)
(3.08126	2.19107	2.85046	−2.73643	−2.06439	−0.215145)
(−1.60506	−2.41617	−1.19739	1.37582	3.03405	0.925416)
(−1.08172	0.0564745	0.185755	−2.15354	−1.3391	−1.95008)
(2.49137	−1.02443	−1.93669	0.325143	−0.805692	−0.198555)
(−1.87635	1.27567	−0.514556	−0.603716	−2.95356	−1.65078)
(−2.05109	−1.32914	1.38543	0.114991	−1.94354	0.887035)
(−0.674438	−3.08308	1.0123	−0.10102	−2.47693	2.0086)
(1.66252	−2.21819	2.60135	−1.45814	−0.67338	1.12196)
(1.33611	0.407789	2.98277	−1.41592	1.14807	−1.08302)
(1.89226	3.05482	2.90424	1.94358	0.694209	−0.973808)
(0.437082	1.88506	2.8235	2.26881	−0.227581	3.01805)
(−1.98061	1.3454	0.31266	1.3346	1.83436	−2.91815)
(2.11814	−2.21478	1.99318	1.63935	−2.17152	2.00983)
(−3.04067	1.72613	−1.93417	−3.07534	−0.593287	−0.441659)
(0.770342	−1.81881	−0.275197	0.431129	−2.14367	−1.69527)
(−1.43618	2.22732	0.685263	0.111728	−0.128947	2.00388)
(1.70871	−0.815082	1.01946	−2.77706	0.738644	0.316677)
(0.918539	−1.36159	−0.46878	0.250422	−1.62977	2.22166)
(1.90247	−1.07237	1.78702	−1.35106	0.904547	−2.51869)
(0.0816078	−0.436794	−2.92231	0.511173	−2.93104	0.700914)
(−1.48943	−1.81534	−0.808906	0.336383	0.913519	1.00958)
(1.41415	−1.44362	−1.75929	−2.38244	−0.0976769	−0.52368)
(−0.520171	1.83152	1.25689	−2.31421	1.71687	1.20862)
(−1.96631	1.26418	1.49759	−2.44414	−2.17686	−2.57833)
(−0.154573	2.51279	1.77364	0.226881	2.0735	−1.63838)
(−2.7821	−1.4711	0.6912	−2.39754	0.457165	2.19418)
(−1.93022	−1.08747	2.34186	1.36679	−0.505503	0.8455)
(1.16658	−3.03898	1.1385	0.148052	0.201847	2.68094)
(−1.84852	0.776858	1.5698	−0.687528	−0.780427	−0.72635)
(1.21031	−2.35802	1.66997	−1.4704	−2.38844	−1.41061)
(0.458594	2.75866	−1.58872	0.364191	−2.1775	−1.22843)
(0.386299	0.261576	−0.174403	1.76511	−2.95714	0.722224)
(−1.46748	−2.15334	−1.38535	−1.73184	2.45454	1.7146)
(0.545927	−2.51541	−2.35701	0.0434106	−0.207223	2.03679)
(0.325985	0.426342	−1.7601	−1.46899	−0.638112	−1.48682)
(0.995193	1.41102	2.67788	−0.110336	0.810741	−2.45279)
(1.00377	−1.09859	−0.9455	2.42064	1.69081	0.328404)
(1.65017	1.79446	0.906236	−2.8566	−1.2842	2.89926)
(−2.56134	−0.141348	−2.6657	1.22666	1.21836	−1.79612)
(−0.519297	2.95723	1.68207	1.45581	1.81155	2.26842)
(−2.46329	−0.587199	−0.384539	2.98937	−1.01251	2.22599)
(1.10689	−1.94668	1.22285	1.941	−0.7505	−1.70434)
(0.642599	−1.05925	−1.2264	0.210594	2.56583	−2.40472).

FIGURE 8. Values of $s_1, \ldots, s_6$ for the atoms comprising μ_1

$$
\begin{aligned}
&\{0.0903801, 0.0969632, 0.145474, -0.0591319, -0.0214708, 0.0919131, -0.000241965, 0.0241533,\\
&-0.0886868, 0.104168, 0.0392277, 0.0888558, 0.0227218, 0.00893166, -0.212648, 0.120529,\\
&0.0324734, -0.0546992, 0.111557, -0.159074, -0.130784, 0.0239644, 0.191044, -0.212465,\\
&-0.057288, -0.130236, -0.0515242, 0.043965, 0.0421474, 0.0869512, 0.0922342, 0.186051,\\
&0.169673, 0.365987, 0.0227764, -0.0812135, -0.0426228, 0.371221, -0.126781, -0.377116,\\
&-0.172414, -0.00152986, -0.0305896, -0.00327764, -0.0494426, 0.130664, -0.0487419, 0.244147,\\
&0.0402412, -0.0658492, 0.044671, -0.153327, -0.0747899, 0.0137243, 0.155563\}\\
&\{0.0237331, 0.199178, 0.123541, 0.112583, -0.203652, -0.137257, 0.11308, 0.118312,\\
&0.088158, -0.0285321, -0.00992443, 0.106434, -0.0561753, 0.118035, -0.101555, 0.10806,\\
&-0.0693862, 0.0678379, 0.277438, -0.100565, -0.135501, 0.105018, 0.113, 0.0720556,\\
&-0.022502, 0.116922, -0.0346328, 0.0573696, -0.0847348, -0.257755, 0.1629, -0.119937,\\
&-0.0503877, -0.0524023, 0.0808267, 0.0327606, 0.133811, -0.177716, 0.0564386, 0.0494853,\\
&-0.075546, 0.215228, -0.286515, -0.222407, -0.397166, 0.081443, 0.202924, -0.0419609,\\
&-0.111385, -0.0710704, 0.0650887, 0.0620108, -0.150127, 0.031396, 0.0245637\}\\
&\{0.112757, 0.172212, 0.0460243, -0.0769192, 0.151564, -0.0015713, 0.280719, -0.194652,\\
&0.0117092, -0.000721688, -0.177536, 0.0200862, -0.0896234, -0.0198031, -0.0127717, -0.00823162,\\
&0.0344373, -0.19736, 0.137134, 0.0857473, -0.00273806, -0.00872684, -0.105853, -0.157082,\\
&0.0635683, 0.106352, -0.13377, -0.304801, -0.0583263, 0.220548, 0.20856, 0.0743173,\\
&-0.184458, 0.0193141, -0.000287821, -0.0463311, -0.223555, -0.284636, -0.0380433, 0.00918501,\\
&0.286329, 0.131752, 0.00101416, -0.0225734, -0.0120999, 0.136267, -0.0511013, 0.122597,\\
&0.0157355, 0.183215, 0.198623, -0.144791, 0.0955358, 0.0789536, 0.157696\}
\end{aligned}
$$

FIGURE 9. The three possible sets of weights for the atoms in μ_1.

is a positive probability measure with barycenter p. Observe that

$$\int_K O(\sigma)AO(\sigma)^* d\mu_0(\sigma) = \frac{1}{9}\,\mathrm{Id}$$

and that

$$\int_K O(\sigma)BO(\sigma)^* d\mu_0(\sigma) = \frac{\sqrt{5}}{9\sqrt{2}}\,\mathrm{Id}$$

It follows from integrating inside the traces of the function $g(\sigma^{-1}\tau)$ that $\Omega(\mu_0, \mu) = 0$ for any measure μ. Therefore,

$$
\begin{aligned}
\Omega(\mu_2, \mu_2) &= \frac{1}{(|c(\epsilon)| + \|\mu_1^\epsilon\|)^2}\Omega(\mu_1^\epsilon, \mu_1^\epsilon)\\
&\geq \frac{1}{(|c(\epsilon)| + \|\mu_1^\epsilon\|)^2}(1.6 - \epsilon)
\end{aligned}
$$

which is positive for $\epsilon < 1.6$.

The consequence of this is that $f(t)$ is larger than 1 for some sufficiently small t, and $\mu = \mu_2$. Hence there is no iso-derivative type inequality for the Jacobian among (positive) probability measures which is sharp at μ_0.

In actuality the Jacobian at $\frac{\mu_0 + t\mu_2}{\|\mu_0 + t\mu_2\|}$ stays less than 1.01, and probably much less, because $\frac{1}{(|c(\epsilon)| + \|\mu_1^\epsilon\|)^2}$ decays at least quadratically in ϵ as $\epsilon \longrightarrow 0$ for any choice of μ_1^ϵ.

References

[Al] P. Albuquerque, *Patterson-Sullivan theory in higher rank symmetric spaces,* Geom. Funct. Anal. (GAFA), Vol. **9** (1999), 1-28.

[BGS] W. Ballmann, M. Gromov and V. Schroeder, *Manifolds of nonpositive curvature,* Progress in Mathematics, Vol. **61**, Birkaüser, 1985.

[BaCS] D. Bao, S.S. Chern and Z. Shen, *An introduction to Riemann-Finsler geometry,* Graduate Texts in Mathematics, 200. Springer-Verlag, New York, 2000.

[Be1] L. Bessières, *Un theoreme de rigidite differentielle,* Comment. Math. Helv. **73** (1998), No.3, 443-479.

[Be2] ———, *Sur le volume minimal des varits ouvertes,* Ann. Inst. Fourier (Grenoble) **50** (2000), no. 3, 965–980.

[BCG1] G. Besson, G. Courtois and S. Gallot, *Entropies et rigidités des espaces localement symétriques de courbure strictement négative,* Geom. Funct. Anal. **5** (1995), no. 5, 731–799.

[BCG2] ———, *Minimal entropy and Mostow's rigidity theorems,* Ergodic Theory Dynam. Sys. **16** (1996), no. 4, 623–649.

[BCG3] ———, *Volumes and rigidities of Riemannian manifolds,* Lecture Notes, Korean Advanced Institute for Sciences and Technlogy, July 1998.

[BCG4] ———, *A real Schwarz lemma and some applications.* Rend. Mat. Appl., VII. Ser. **18**, No.2, 381-410 (1998).

[BCG5] ———, *Lemme de Schwarz rel et applications gomtriques,* Acta Math. **183** (1999), no. 2, 145–169.

[BC] J. Boland and C. Connell, *Entropy rigidity for foliations of compact spaces,* To appear in Israel J. Math.

[BCS] J. Boland, C. Connell and J. Souto, *Minimal entropy rigidity for finite volume manifolds,* Preprint (1999).

[BN] J. Boland and F. Newberger, *Minimal entropy rigidity for Finsler manifolds of negative flag curvature.* Ergodic Theory Dynam. Systems **21** (2001), no. 1, 13–23.

[Bon] F. Bonahon *A Schläfli-type formula for convex cores of hyperbolic 3-manifolds*, J. Differential Geom. **50** (1998), 25-58.

[BGP] Y. Burago, M. Gromov and G.A. Perel′man, *A. D. Aleksandrov spaces with curvatures bounded below,* Uspekhi Mat. Nauk **47** (1992), no. 2(284), 3–51, 222; translation in Russian Math. Surveys **47** (1992), no. 2, 1–58.

[BP] K. Burns and G.P. Paternain, *Anosov magnetic flows, critical values and topological entropy,* Preprint (2001).

[CMT] R. Canary, Y. Minsky and E. Taylor, *Spectral theory, Hausdorff dimension and the topology of hyperbolic 3-manifolds*, Journal of Geometric Analysis **9** (1999), 17-40.

[CG] J. Cheeger and M. Gromov, *Chopping Riemannian manifolds,* Differential geometry, 85–94, Pitman Monogr. Surveys Pure Appl. Math. **52**, Longman Sci. Tech., Harlow, 1991.

[CF1] C. Connell and B. Farb, *Minimal entropy rigidity for lattices in products of rank one symmetric spaces,*, Comm. in Analysis and Geometry, to appear.

[CF2] ———, *The Degree Theorem in higher rank,* November 2000 preprint.

[CHK] D. Cooper, C.D. Hodgson and S.P. Kerckhoff, *Three-dimensional orbifolds and cone-manifolds,* With a postface by Sadayoshi Kojima. MSJ Memoirs **5**, Mathematical Society of Japan, Tokyo, 2000.

[Eb] P. Eberlein, *Geometry of nonpositively curved manifolds*, Chicago Lectures in Math. Univ. of Chicago Press, 1996.

[FJ] F.T. Farrell and L.E. Jones,*Topological rigidity for compact non-positively curved manifolds,* Differential geometry: Riemannian geometry (Los Angeles, CA, 1990), 229–274, Proc. Sympos. Pure Math. **54**, Part 3, Amer. Math. Soc., Providence, RI, 1993.

[Fl] L. Flaminio, *Local entropy rigidity for hyperbolic manifolds,* Comm. Anal. Geom. **3** (1995), no. 3-4, 555–596.

[Ga] S. Gallot, Curvature-decreasing maps are volume-decreasing (on joint work with G. Besson and G. Courtois), Proc. ICM, Vol. II (Berlin, 1998), Doc. Math. 1998, 339–348.

[Gr] M. Gromov, *Volume and bounded cohomology,* Inst. Hautes Études Sci. Publ. Math. **56** (1983), 5–99.

[Gr2] ———, *Filling Riemannian manifolds,* J. Diff. Geom. **18** (1983), 1–147.

[Hu] S. Hurder, *Coarse geometry of foliations. Geometric study of foliations,* 35–96, World Sci. Publishing, River Edge, NJ, 1994.

[Ka] A. Katok, *Entropy and closed geodesics,* Ergodic Theory Dynam. Systems **2** (1982), no. 3-4, 339–365.

[Kn] G. Knieper, *On the asymptotic geometry of nonpositively curved manifolds,* Geom. Funct. Anal. (GAFA) **7** (1997), 755–782.

[Ma] A. Manning, *Topological entropy for geodesic flows,* Annals of Math. **110** (1979), 567–573.

[Pa] P. Pansu, *Volume, courbure et entropie (d'après G. Besson, G. Courtois et S. Gallot),* Séminaire Bourbaki, Vol. 1996-97, Astérisque No. **245** (1997), Exp. No. 823, 3, 83–103.

[PaPa1] G.P. Paternain and M. Paternain, *Anosov geodesic flows and twisted symplectic structures,* International Conference on Dynamical Systems (Montevideo, 1995), Pitman Res. Notes Math. Ser., **362** (1996), 132–145.

[PaPa2] ———, *First derivative of topological entropy for Anosov geodesic flows in the presence of magnetic fields,* Nonlinearity **10** (1997), no. 1, 121–131.

[PaPe] G.P. Paternain and J. Petean, *Minimal entropy and collapsing with curvature bounded from below,* Preprint (2001).

[Pr] G. Prasad, *Discrete subgroups isomorphic to lattices in semisimple Lie groups,* Amer. J. Math. **98** (1976), no. 1, 241–261.

[Sam1] A. Sambusetti, *Einstein manifolds, volume rigidity and Seiberg-Witten theory,* Sémin. Théor. Spectr. Géom. Année 1998–1999, Univ. Grenoble I, Saint-Martin-d'Héres **17** (1999), 163–184.

[Sam2] ———, *Minimal entropy and simplicial volume,* Manuscripta Math. **99** (1999), no. 4, 541–560.

[Sa] R. Savage, *The space of positive definite matrices and Gromov's invariant,* Trans. Amer. Math. Soc. **274** (1982), no. 1, 239–263.

[St] P. Storm, *Minimal volume Alexandrov spaces,* Preprint (2001).

[Th] ———, *Hyperbolic geometry and 3-manifolds*, in *Low-dimensional topology* (Bangor, 1979), London Math. Soc. Lecture Note Ser., **48**, Cambridge Univ. Press, Cambridge-New York (1982), 9–25.

[Ve] P. Verovic, *Problème de l'entropie minimale pour les métriques de Finsler,* Erg. Theo. and Dynam. Syst., **19** (1999), no. 6, 1637–1654.

[Zi] R. Zimmer, *Ergodic theory, semisimple Lie groups, and foliations by manifolds of negative curvature,* Inst. Hautes tudes Sci. Publ. Math. **55** (1982), 37–62.

Department of Mathematics, University of Chicago, Chicago, Il 60637
E-mail address: `cconnell@math.uchicago.edu`
E-mail address: `farb@math.uchicago.edu`

Proceedings of Symposia in Pure Mathematics
Volume **71**, 2003

The Lawrence-Krammer Representation

Stephen Bigelow

ABSTRACT. The Lawrence-Krammer representation of the braid groups recently came to prominence when it was shown to be faithful by myself and Krammer. It is an action of the braid group on a certain homology module $H_2(\tilde{C})$ over the ring of Laurent polynomials in q and t. In this paper we describe some surfaces in $\tilde{C}$ representing elements of homology. We use these to give a new proof that $H_2(\tilde{C})$ is a free module. We also show that the $(n-2,2)$ representation of the Temperley-Lieb algebra is the image of a map to relative homology at $t = -q^{-1}$, clarifying work of Lawrence.

1. Introduction

The Lawrence-Krammer representation is the action of the braid group B_n on a certain homology module $H_2(\tilde{C})$ over the ring $\Lambda = \mathbf{Z}[q^{\pm 1}, t^{\pm 1}]$. It was introduced by Lawrence [**Law90**], except that she worked over $\mathbf{C}$ instead of Λ. It recently came to prominence when it was shown to be faithful for $n = 4$ [**Kra00**], and then for all n [**Big01**], [**Kra02**]. Since then, a number of papers have appeared which closely examine certain aspects of the representation. See [**PP01**], [**Son02**], and [**Bud02**]. We now continue in this tradition, with the specific goal of understanding the connection with the Temperley-Lieb algebra.

This paper was partly motivated by an attempt to clarify two points from [**Big01**]. First, the pairing between a "noodle" and a "fork" involved an algebraic intersection number between two non-compact surfaces in $\tilde{C}$. Such a thing is not necessarily well-defined, so I gave an indirect proof of the existence of a certain closed surface corresponding to a fork. I now have an explicit description of this surface, which will be given in Section 3.4. In fact it is possible to define the pairing without reference to this surface by using results from [**Kaw96**, Appendix E]. (Thanks to Won Taek Song, whose paper [**Son02**] drew my attention to this.)

Second, to compute matrices for the representation, I tensored $H_2(\tilde{C})$ with a field containing Λ. The resulting vector space contains $H_2(\tilde{C})$, but strictly speaking, the action of B_n on this vector space should not be called the Lawrence-Krammer representation. In Section 5 we give a new proof that $H_2(\tilde{C})$ is a free Λ-module. This is originally due to Paoluzzi and Paris [**PP01**], but our proof uses an explicit

---

2000 *Mathematics Subject Classification.* Primary 20F36; Secondary 20C08.
Key words and phrases. Braid group, Temperley-Lieb algebra, linear representations.
Research supported by the Australian Research Council.

description of surfaces representing elements of a free basis of $H_2(\tilde{C})$. The correct matrices for the Lawrence-Krammer representation could be computed from this, but we do not do so since they are quite complicated.

By addressing these two issues from [**Big01**] we also shed new light on the work of Lawrence [**Law90**]. There, the Lawrence-Krammer representation was used to give a topological interpretation to a representation of the Hecke algebra. The idea is to specialise t to the value $-q^{-1}$, at which point the Lawrence-Krammer representation becomes reducible and the desired representation appears as a quotient. Somewhat complicated methods are used in [**Law90**] to define the required quotient. In Section 6 we show that it is simply the image of a map to a certain relative homology module.

Throughout this paper n is a positive integer, D is the unit disk centred at the origin in the complex plane, $-1 < p_1 < \cdots < p_n < 1$ are real numbers, and $D_n = D \setminus \{p_1, \ldots, p_n\}$ is the n-times punctured disk. The braid group B_n is the mapping class group of D_n. We also assume familiarity with the presentation of B_n using Artin generators σ_i, and the interpretation of B_n as the fundamental group of a configuration space in the plane.

This research was supported by the Australian Research Council. Thanks to Gordana Matic and the other organisers for such an inspiring and entertaining conference.

2. The Lawrence-Krammer representation

We briefly review the definition of the Lawrence-Krammer representation as given in [**Big01**]. Let C be the space of unordered pairs of distinct points in D_n. Let $c_0 = \{d_1, d_2\}$ be a basepoint in C, where d_1 and d_2 are distinct points on the boundary of the disk.

We define a homomorphism

$$\Phi\colon \pi_1(C, c_0) \to \langle q \rangle \oplus \langle t \rangle$$

as follows. Suppose $\alpha\colon I \to C$ is a closed loop in C representing an element of $\pi_1(C, c_0)$. By ignoring the puncture points we can consider α as a loop in the space of unordered pairs of points in the disk, and hence as a braid in B_2. Let j be the exponent of this braid in the Artin generator σ_1. Similarly, the map

$$s \mapsto \{p_1, \ldots, p_n\} \cup \alpha(s)$$

determines a braid in B_{n+2}. Let j' be the the exponent sum of this braid in the Artin generators of B_{n+2}. Note that j and j' have the same parity. Let $i = \frac{1}{2}(j' - j)$. We define

$$\Phi(\alpha) = q^i t^j.$$

Let $\tilde{C}$ be the connected covering space of C such that $\pi_1(\tilde{C}) = \ker(\Phi)$. Fix a choice of $\tilde{c}_0$ in the fibre over c_0. The homology group $H_2(\tilde{C})$ admits a Λ-module structure, where q and t act by covering transformations.

Suppose f is a homeomorphism from D_n to itself, representing an element of B_n. Let f_* be the induced map from $\pi_1(C, c_0)$ to itself. I claim that

$$\Phi \circ f_* = \Phi.$$

To see this, think of $\pi_1(C, c_0)$ as the subgroup of B_{n+2} consisting of braids whose first n strands are straight. Then f_* acts on $\pi_1(C, c_0)$ by conjugation with a braid on the first n strands. This preserves the image under Φ, since Φ maps to an abelian group.

It follows that f has a unique lift

$$\tilde{f}\colon (\tilde{C}, \tilde{c}_0) \to (\tilde{C}, \tilde{c}_0),$$

and that the induced map

$$\tilde{f}_*\colon H_2(\tilde{C}) \to H_2(\tilde{C})$$

is a Λ-module automorphism. We define the Lawrence-Krammer representation to be this action of B_n on $H_2(\tilde{C})$.

We now introduce some relative homology modules. For $\epsilon > 0$, let ν_ϵ be the set of points $\{x, y\} \in C$ such that either x and y are within distance ϵ of each other, or at least one of them is within distance ϵ of a puncture point. Let $\tilde{\nu}_\epsilon$ be the preimage of ν_ϵ in $\tilde{C}$. The relative homology modules $H_2(\tilde{C}, \tilde{\nu}_\epsilon)$ are nested by inclusion. Let

$$H_2(\tilde{C}, \tilde{\nu}) = \lim_{\epsilon \to 0} H_2(\tilde{C}, \tilde{\nu}_\epsilon),$$

and

$$H_2(\tilde{C}, \partial\tilde{C} \cup \tilde{\nu}) = \lim_{\epsilon \to 0} H_2(\tilde{C}, \partial\tilde{C} \cup \tilde{\nu}_\epsilon).$$

The braid group B_n acts on these, and on $H_2(\tilde{C}, \partial\tilde{C})$, by Λ-module automorphisms.

3. Some surfaces

In this section we describe some immersed surfaces in $\tilde{C}$. These will be used to represent elements of homology or relative homology. Each immersed surface will be specified by giving a map from a surface to C and showing that it can be lifted to $\tilde{C}$. We will not specify a choice of lift, and we will not pay much attention to issues of orientation. Thus the resulting element of homology or relative homology will only be defined up to multiplication by a unit in Λ. This will be sufficient for our purposes.

3.1. Squares and triangles. We describe some properly embedded surfaces representing elements of the second homology of $\tilde{C}$ relative to $\partial\tilde{C}$, $\tilde{\nu}$ and $\partial\tilde{C} \cup \tilde{\nu}$. These will be represented by one or two embedded edges in the disk.

Suppose $\alpha_1, \alpha_2\colon I \to D$ are disjoint embeddings of the interior of I into D_n, and map the endpoints of I to puncture points (not necessarily injectively). Let f be the map from the interior of $I \times I$ to C given by $f(x, y) = \{\alpha_1(x), \alpha_2(y)\}$. A lift of f to $\tilde{C}$ will represent an element of $H_2(\tilde{C}, \tilde{\nu})$. Similarly, we can define an element of $H_2(\tilde{C}, \partial\tilde{C})$ corresponding to a pair of disjoint edges in D_n with endpoints on ∂D. Finally, we can define an element of $H_2(\tilde{C}, \partial\tilde{C} \cup \tilde{\nu})$ corresponding to a pair of edges which have a mixture of endpoints on ∂D and on puncture points. For all of these examples, call the resulting element of relative homology the *square* corresponding to the edges α_1 and α_2.

Now suppose $\alpha\colon I \to D$ is an embedding of the interior of I into D_n, and maps the endpoints of I to the puncture points. Define a map

$$f\colon \{(x, y) \in I \times I : 0 < x < y < 1\} \to C$$

by $f(x, y) = \{f(x), f(y)\}$. A lift of f to $\tilde{C}$ will represent an element of $H_2(\tilde{C}, \tilde{\nu})$. Similarly, we can obtain an element of $H_2(\tilde{C}, \partial\tilde{C} \cup \tilde{\nu})$ if we allow one or both endpoints of the edge to be on ∂D. For all of these examples, call the resulting element of relative homology the *triangle* corresponding to α.

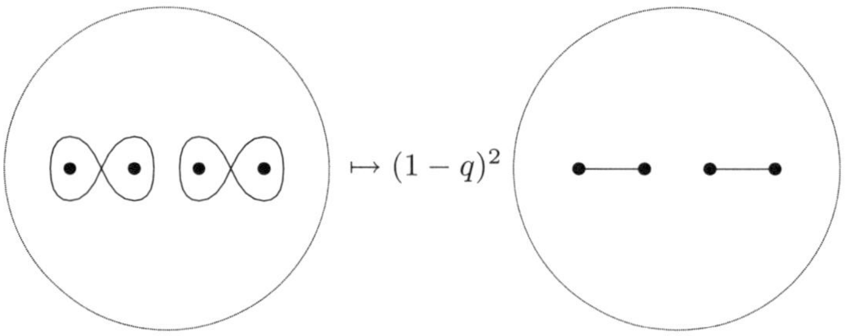

FIGURE 1. A torus and a square.

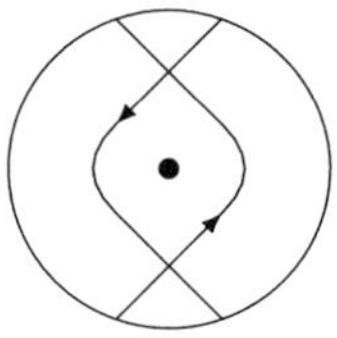

FIGURE 2. The disk B and edges $\alpha_1(I)$ and $\alpha_2(I)$.

3.2. A genus one surface. Suppose $\alpha_1, \alpha_2 \colon S^1 \to D_n$ are disjoint figure-eights, each going around two puncture points, as in Figure 1. Define a map f from the torus $S^1 \times S^1$ to C by $f(x,y) = \{\alpha_1(x), \alpha_2(y)\}$. Both the meridian and longitude of the torus are mapped into the kernel of Φ, so f lifts to a map

$$\tilde{f} \colon S^1 \times S^1 \to \tilde{C}.$$

We obtain an element $[\tilde{f}]$ of $H_2(\tilde{C})$.

It will be useful to know the image of $[\tilde{f}]$ in $H_2(\tilde{C}, \tilde{\nu})$. I claim this is $(1-q)^2$ times the square as indicated in Figure 1. To see this, homotope each figure-eight α_i so as to map the points $\pm 1 \in S^1$ into respective ϵ-neighbourhoods of the two puncture points enclosed by α_i. Then f maps the four lines $\{\pm 1\} \times S^1$ and $S^1 \times \{\pm 1\}$ into ν_ϵ. These lines cut the torus into four squares. The restriction of $\tilde{f}$ to each of these squares represents an element of $H_2(\tilde{C}, \tilde{\nu})$. The claim now follows from a careful comparison of the lifts and orientations of these squares, which is left to the reader.

3.3. A genus two surface. Suppose $\alpha_1, \alpha_2 \colon S^1 \to D_n$ are figure-eights such that α_1 passes around p_i and p_j, α_2 passes around p_j and p_k, and α_1 intersects α_2 twice, as in Figure 3. Let $B \subset D$ be a disk centred at p_j, containing the two points of intersection, and meeting each of α_1 and α_2 in a single edge. For $i = 1, 2$, let $I_i \subset S^1$ be the interval $\alpha_i^{-1}(B)$. For convenience, assume $I_1 = I_2$, and identify both with $I = [0,1]$, oriented as shown in Figure 2.

Let T be the closure of

$$(S^1 \times S^1) \setminus (I_1 \times I_2).$$

Note that for $(x,y) \in T$ we have $\alpha_1(x) \neq \alpha_2(y)$. We can therefore define $f \colon T \to C$ by $f(x,y) = \{\alpha_1(x), \alpha_2(y)\}$.

Let f_0 be the restriction of f to $\partial T = \partial(I \times I)$. For $s \in I$, let f_s be the composition of f_0 with an anticlockwise rotation of B by an angle of $s\pi$ about the centre p_j. We can assume B has a rotational symmetry so that $f_1(x,y) = f_0(y,x)$

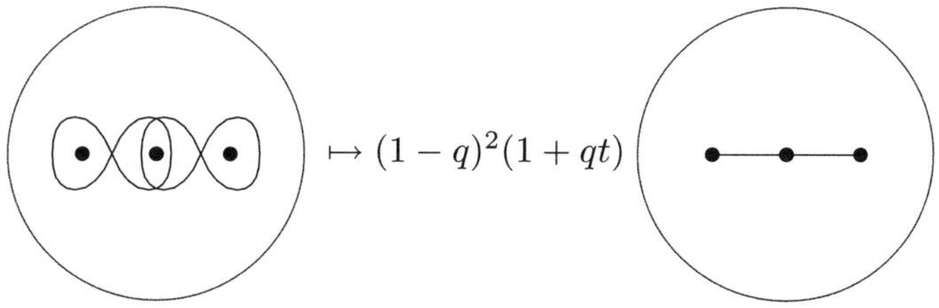

FIGURE 3. A genus two surface and a square.

for all $(x, y) \in \partial(I \times I)$. Thus f_0 and f_1 represent the same loops, but with opposite orientations.

We now build a closed genus two surface Σ_2 by gluing together $\partial T \times I$ and two copies of T as follows. First glue $\partial T \times I$ to T by $((x, y), 0) \sim (x, y)$. Then glue $\partial T \times I$ to a second copy T' of T by $((x, y), 1) \sim (y, x)$. Let Σ_2 be the surface so obtained. Let $g\colon \Sigma_2 \to C$ be given by $g|_T = g|_{T'} = f$ and $g|_{(\partial T \times \{s\})} = f_s$.

The fundamental group of Σ_2 is generated by the meridian and longitude of T and T'. Each of these is mapped by g into the kernel of Φ, so g lifts to a map $\tilde{g}\colon \Sigma_2 \to \tilde{C}$. We obtain an element $[\tilde{g}]$ of $H_2(\tilde{C})$.

We now compute the image of $[\tilde{g}]$ in $H_2(\tilde{C}, \tilde{\nu})$. For any $\epsilon > 0$ we can assume that the disk B has radius less than ϵ. Then g maps $\partial T \times I$ into ν_ϵ, so $\tilde{g}|_T$ represents an element of $H_2(\tilde{C}, \tilde{\nu}_\epsilon)$. This element is $(1-q)^2$ times a square, by a similar argument to the one given for the genus one surface. It remains to figure out how $\tilde{g}|_{T'}$ is related to $\tilde{g}|_T$.

Consider the path $\delta\colon I \to \Sigma_2$ given by

$$\delta(s) = ((0, 0), s) \in \partial T \times I.$$

This goes from $(0, 0) \in T$ to $(0, 0) \in T'$. Now $g \circ \delta$ is a loop in C in which the pair of points switch places by an anticlockwise rotation through an angle of π around p_j. Thus $\Phi(g \circ \delta) = qt$. It follows that $\tilde{g}|_{T'} = qt \circ \tilde{g}|_T$.

Also note that T' and T inherit the same orientation from Σ_2. This is because the ends of the annulus $\partial T \times I$ were attached with opposite orientations. We conclude that the image of $[\tilde{g}]$ in $H_2(\tilde{C}, \tilde{\nu})$ is $(1-q)^2(1+qt)$ times a square, as shown in Figure 3.

3.4. A genus three surface. Suppose $\alpha_1, \alpha_2\colon S^1 \to D_n$ are figure-eights, both passing around p_i and p_j, and intersecting transversely at four points, as in Figure 4. We construct a map from a genus three surface Σ_3 into $\tilde{C}$ by a slight modification of the procedure used above for Σ_2. This time the surface T will be a torus with two disks removed, one for each of p_i and p_j. Two annuli are then needed to glue T to another copy T' of T. We obtain a genus three surface Σ_3 and a map $g\colon \Sigma_3 \to C$.

We now show that g lifts to a map $\tilde{g}\colon \Sigma_3 \to \tilde{C}$. The image of the longitude, meridian, and both boundary components of T all lie in the kernel of Φ. Thus $g|_T$ lifts to $\tilde{g}|_T$. This lift can be extended to the annuli $\partial T \times I$. Finally, define $\tilde{g}|_{T'}$ to be the covering transformation qt applied to $\tilde{g}|_T$.

We now compute the image of $[\tilde{g}]$ in $H_2(\tilde{C}, \tilde{\nu})$. This is $(1-q)^2(1+qt)$ times a square, by a similar argument to the one given for the genus two surface. This square corresponds to a parallel pair of edges from p_i to p_j. These edges can be homotoped

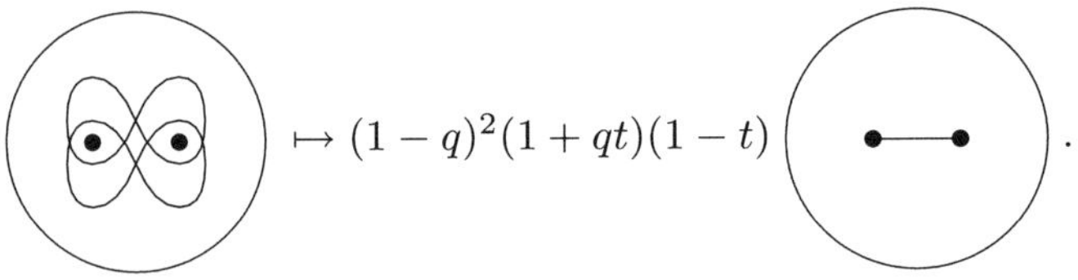

FIGURE 4. A genus three surface and a triangle.

so as to lie within ϵ of each other. Then a diagonal of the square in $\tilde{C}$ will lie in $\tilde{\nu}_\epsilon$. This diagonal cuts the square into two triangles, which have opposite orientations and differ by the covering transformation t. Thus $[\tilde{g}]$ represents $(1-q)^2(1+qt)(1-t)$ times a triangle, as shown in Figure 4.

4. An intersection pairing

It is not immediately obvious that the surfaces described in Section 3 represent non-trivial elements of homology or relative homology. In Sections 5 and 6 we will need to prove even stronger results concerning the linear independence of various sets of such elements. Our main tool will be the following intersection pairing.

For $x \in H_2(\tilde{C})$ and $y \in H_2(\tilde{C}, \partial\tilde{C} \cup \tilde{\nu})$ let $(x \cdot y) \in \mathbf{Z}$ denote the standard intersection number. We define an intersection pairing

$$\langle \cdot, \cdot \rangle \colon H_2(\tilde{C}) \times H_2(\tilde{C}, \partial\tilde{C} \cup \tilde{\nu}) \to \Lambda$$

by

$$\langle x, y \rangle = \sum_{i,j \in \mathbf{Z}} (x \cdot q^i t^j y) q^i t^j.$$

A similar definition gives a pairing

$$\langle \cdot, \cdot \rangle' \colon H_2(\tilde{C}, \tilde{\nu}) \times H_2(\tilde{C}, \partial\tilde{C}) \to \Lambda.$$

To check that these are well defined requires some elementary homology theory. See [**Kaw96**, Appendix E], where the following properties are also proved.

For $x \in H_2(\tilde{C})$, $y \in H_2(\tilde{C}, \partial\tilde{C} \cup \tilde{\nu})$, $\sigma \in B_n$, and $\lambda \in \Lambda$, we have

$$\langle \sigma x, \sigma y \rangle = \langle x, y \rangle,$$

and

$$\langle \lambda x, y \rangle = \lambda \langle x, y \rangle = \langle x, \bar{\lambda} y \rangle,$$

where $\bar{\lambda}$ is the image of λ under the automorphism of Λ taking q to q^{-1} and t to t^{-1}. Similar identities hold for $\langle \cdot, \cdot \rangle'$.

Note that the above definition of $\langle \cdot, \cdot \rangle$ differs from that of [**Big01**] in that the order of the entries is reversed. The above is consistent with [**Kaw96**], and with the usual definition of sesquilinear.

It will frequently be necessary to compute the intersection pairing in specific examples, up to multiplication by a unit in Λ. The rest of this section is devoted to discussion of how to do this.

Let α_1 and α_2 be edges in D with endpoints on puncture points, representing a square $a \in H_2(\tilde{C}, \tilde{\nu})$. Let β_1 and β_2 be edges in D with endpoints on ∂D, representing a square $b \in H_2(\tilde{C}, \partial\tilde{C})$. We discuss how to compute $\langle a, b \rangle'$ up to multiplication by a unit in Λ.

Let A and B be the surfaces in C corresponding to a and b respectively. The intersection $A \cap B$ is the set of pairs of the form $\{x, y\}$ where either

- $x \in \alpha_1 \cap \beta_1$ and $y \in \alpha_2 \cap \beta_2$, or
- $x \in \alpha_1 \cap \beta_2$ and $y \in \alpha_2 \cap \beta_1$.

Assume all edges, intersect transversely, so A intersects B transversely. Let $\tilde{A}$ and $\tilde{B}$ be lifts of A and B to $\tilde{C}$.

If A and B do not intersect then neither will any of their lifts, so $\langle a, b\rangle' = 0$. If A and B intersect at one point $\{x, y\}$ then choose lifts and orientations so that $\tilde{A}$ intersects $\tilde{B}$ with positive sign at a point in the fibre of $\{x, y\}$. Then $\langle a, b\rangle' = 1$.

Now suppose A and B intersect in two points, $\{x, y\}$ and $\{x', y'\}$. Choose lifts and orientations so that $\tilde{A}$ intersects $\tilde{B}$ with positive sign at a point in the fibre of $\{x, y\}$. Let i and j be the integers such that $\tilde{A}$ intersects $q^i t^j \tilde{B}$ at a point in the fibre of $\{x', y'\}$, and let $\epsilon = \pm 1$ be the sign of that intersection. Then $\langle a, b\rangle' = 1 + \epsilon q^i t^j$. It remains to compute i, j and ϵ.

Let γ_A and γ_B be paths from $\{x, y\}$ to $\{x', y'\}$ that lie in A and B respectively. Let $\tilde{\gamma}_A$ and $\tilde{\gamma}_B$ be the lifts that start in $\tilde{A} \cap \tilde{B}$. Then $\tilde{\gamma}_A(1)$ and $\tilde{\gamma}_B(1)$ are the points in the fibre of $\{x', y'\}$ lying in $\tilde{A}$ and $\tilde{B}$ respectively. Thus

$$q^i t^j \tilde{\gamma}_B(1) = \tilde{\gamma}_A(1),$$

so

$$q^i t^j = \Phi(\gamma_A^{-1} \gamma_B).$$

In practice this will be easy to compute. It remains to compute ϵ.

Relabel if necessary so that $x, x' \in \alpha_1$ and $y, y' \in \alpha_2$. Orient the edges α_1, α_2, β_1 and β_2. The sign of the intersection of A with B at $\{x, y\}$ is determined by

- the sign of the intersection of α_1 with $\beta_1 \cup \beta_2$ at x,
- the sign of the intersection of α_2 with $\beta_1 \cup \beta_2$ at y, and
- whether $x \in \beta_2$ or $x \in \beta_1$.

By assumption, the conventions are such that this intersection is positive. Each of the following will reverse the sign of the intersection of A with B at $\{x', y'\}$.

- α_1 intersects $\beta_1 \cup \beta_2$ with different signs at x and x',
- α_2 intersects $\beta_1 \cup \beta_2$ with different signs at y and y',
- either $x \in \beta_1$ and $y \in \beta_2$, or $x \in \beta_2$ and $y \in \beta_1$.

If none or two of these hold then $\epsilon = 1$. If one or three of these hold then $\epsilon = -1$.

We will sometimes need to compute intersection pairings between surfaces that are not squares. First consider the case A is a triangle corresponding to an edge α, B is a square as before, and A and B intersect transversely at two points. Then the above discussion applies almost unchanged. In place of the assumption that $x, x' \in \alpha_1$ and $y, y' \in \alpha_2$, make the assumption that x occurs before y and x' occurs before y' with respect to the orientation of α. We will also compute pairings between surfaces which are neither squares nor triangles, but the same ideas will apply.

Finally, suppose A and B intersect at more than two points. Then each point of intersection contributes a monomial to the pairing. We can assume one of these monomials is 1, and compute the others by the methods discussed above.

5. A basis

In this section, we analyse the Lawrence-Krammer representation using the tools developed in Sections 3 and 4. We give a new proof of the following theorem.

THEOREM 5.1. *$H_2(\tilde{C})$ is a free Λ-module of rank $\binom{n}{2}$.*

This is originally due to Paoluzzi and Paris [**PP01**]. Our proof gives a more explicit description of surfaces representing elements of a free basis for $H_2(\tilde{C})$. The matrices usually given for the Lawrence-Krammer representation use a basis for $\mathbf{Q}(q,t)\otimes H_2(\tilde{C})$, not for $H_2(\tilde{C})$. This subtle distinction will be discussed at the end of this section.

We use the following basic result about $H_2(\tilde{C})$.

LEMMA 5.2. *The vector space $\mathbf{Q}(q,t)\otimes H_2(\tilde{C})$ has dimension $\binom{n}{2}$. The natural map from $H_2(\tilde{C})$ to $\mathbf{Q}(q,t)\otimes H_2(\tilde{C})$ is injective.*

The proof uses a finite 2-complex that is homotopy equivalent to C. Various methods for constructing the required complex can be found in [**Law90**], [**Big01**], [**PP01**], and [**Bud02**]. Since the complex has dimension two, $H_2(\tilde{C})$ is the kernel of a matrix ∂ with entries in Λ. Now $\mathbf{Q}(q,t)\otimes H_2(\tilde{C})$ is the kernel of the matrix ∂, considered as a matrix over $\mathbf{Q}(q,t)$. Thus $H_2(\tilde{C})$ is the submodule of $\mathbf{Q}(q,t)\otimes H_2(\tilde{C})$ consisting of those vectors whose entries lie in Λ. The dimension of $\mathbf{Q}(q,t)\otimes H_2(\tilde{C})$ can be computed from an explicit description of the matrix ∂, which can be found in any of the papers mentioned above.

5.1. Proof of the theorem.

We now prove Theorem 5.1. For $1\le i<j\le n$, we define $v'_{i,j}\in H_2(\tilde{C},\tilde{\nu})$ as follows. If $j-i>2$, let $v'_{i,j}$ be the square corresponding to the edges $[p_i,p_{i+1}]$ and $[p_{j-1},p_j]$. If $j-i=2$ and $i>1$, let $v'_{i,j}$ be the square corresponding to the edge $[p_i,p_{i+1}]$ and an edge from p_{i-1} to p_j whose interior lies in the lower half plane. Let $v'_{1,3}$ be the square corresponding to the edges $[p_1,p_2]$ and $[p_2,p_3]$. If $j-i=1$, let $v'_{i,j}$ be the triangle corresponding to the edge $[p_i,p_j]$.

For $1\le i<j\le n$, let $v_{i,j}$ be an element of $H_2(\tilde{C})$ whose image in $H_2(\tilde{C},\tilde{\nu})$ is

- $(1-q)^2(1+qt)(1-t)v'_{i,j}$ if $j=i+1$,
- $(1-q)^2(1+qt)v'_{i,j}$ if $i=1$ and $j=3$,
- $(1-q)^2v'_{i,j}$ otherwise.

Such an element exists by Section 3. We will show that the $v_{i,j}$ form a free basis for $H_2(\tilde{C})$.

For every $1\le i<j\le n$, let $x_{i,j}\in H_2(\tilde{C},\partial\tilde{C})$ be the square corresponding to a pair of vertical edges, one passing just to the right of p_i, the other passing just to the left of p_j, and both having endpoints on ∂D.

LEMMA 5.3. *The following identities hold up to multiplication by a unit in Λ.*

- $\langle v'_{i,j},x_{i,j}\rangle'=1$ *for* $1\le i<j\le n$,
- $\langle v'_{i,i+2},x_{i-1,i+1}\rangle'=1$ *for* $i=2,\dots,n-2$,
- $\langle v'_{i,i+2},x_{i,i+1}\rangle'=1-t$ *for* $i=2,\dots,n-2$.

All other pairings $\langle v'_{i',j'},x_{i,j}\rangle'$ are zero.

PROOF. Use the methods described in Section 4. □

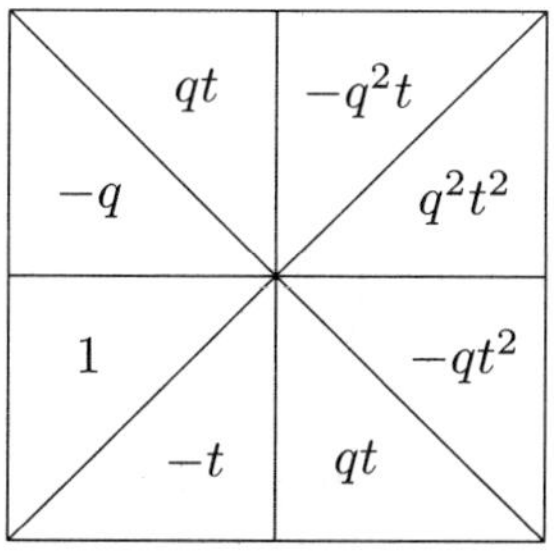

FIGURE 5. The function representing $x_{n-1,n}$ breaks into triangles.

For convenience, choose lifts and orientations so that

$$\langle v'_{i,j}, x_{i,j}\rangle' = 1$$

for $1 \leq i < j \leq n$.

From Lemma 5.3 it follows that the $v'_{i,j}$ are linearly independent. Hence the $v_{i,j}$ are linearly independent. By Lemma 5.2, they span $\mathbf{Q}(q,t) \otimes H_2(\tilde{C})$ as a $\mathbf{Q}(q,t)$-module. It remains to show that they span $H_2(\tilde{C})$ as a Λ-module. In other words, we must prove the following.

LEMMA 5.4. *Let $c_{i,j} \in \mathbf{Q}(q,t)$ for $1 \leq i < j \leq n$ be such that*

$$v = \sum_{1 \leq i < j \leq n} c_{i,j} v_{i,j}$$

lies in $H_2(\tilde{C})$. Then $c_{i,j} \in \Lambda$ for all $1 \leq i < j \leq n$.

We use the following facts about the $x_{i,j}$.

LEMMA 5.5. *The image of $x_{n-1,n}$ in $H_2(\tilde{C}, \partial\tilde{C} \cup \tilde{\nu})$ is $(1-q)(1+qt)(1-t)$ times a triangle.*

PROOF. Let $\alpha\colon I \to D$ be a straight edge from ∂D to p_n. Let α_2 be an edge in an ϵ-neighbourhood of α with endpoints on ∂D, passing anticlockwise around α. Let α_1 be an edge in an ϵ-neighbourhood of α with endpoints on ∂D, passing anticlockwise around α_2. Define $f\colon I \times I \to C$ by $f(x,y) = \{\alpha_1(x), \alpha_2(y)\}$. Then f lifts to a map $\tilde{f}$ which represents $x_{n-1,n}$.

We can assume that for all $s \in I$ the four points $\alpha_1(s)$, $\alpha_2(s)$, $\alpha_1(1-s)$, and $\alpha_2(1-s)$ lie within distance ϵ of each other. Then f maps the lines $\{(x,x)\}$, $\{(x,1-x)\}$, $\{(\frac{1}{2},y)\}$, and $\{(x,\frac{1}{2})\}$ into ν_ϵ. These lines cut $I \times I$ into eight pieces. The restriction of $\tilde{f}$ to each piece represents an element of $H_2(\tilde{C}, \partial\tilde{C} \cup \tilde{\nu}_\epsilon)$. Each of these elements is a multiple of the triangle corresponding to α, as shown in Figure 5. Combining these, we see that $x_{n-1,n}$ is $(1-q)(1+qt)(1-t)$ times the triangle corresponding to α. □

LEMMA 5.6. *For all $1 \leq i < j \leq n$, The image of $x_{i,j}$ in $H_2(\tilde{C}, \partial\tilde{C} \cup \tilde{\nu})$ is $(1-q)^2$ times a linear combination of squares.*

PROOF. Use a similar argument to Lemma 5.5, as suggested by Figure 6. □

PROOF OF LEMMA 5.4. First consider the case $n = 3$. Then

$$\langle v, x_{2,3}\rangle = (1-q)^2(1+qt)(1-t)c_{2,3}.$$

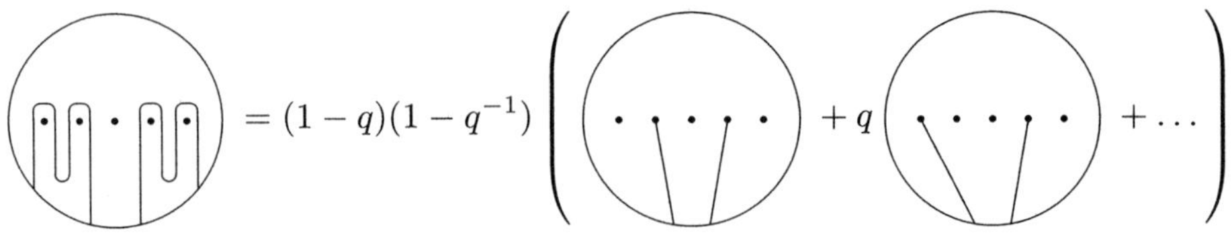

FIGURE 6. $x_{2,4}$ is $(1-q)^2$ times a linear combination of squares.

By Lemma 5.6, $x_{2,3}$ is a multiple of $(1-q)^2$, and hence of $(1-\bar{q})^2$. Since $\langle\cdot,\cdot\rangle$ is sesquilinear, it follows that $(1+qt)(1-t)c_{2,3} \in \Lambda$. Similarly, Lemma 5.5 implies that $(1-q)c_{2,3} \in \Lambda$. Since Λ is a unique factorisation domain, it follows that $c_{2,3} \in \Lambda$. By a symmetrical argument, $c_{1,2} \in \Lambda$.

Now

$$v - (c_{1,2}v_{1,2} + c_{2,3}v_{2,3})$$

still lies in $H_2(\tilde{C})$, so we can reduce to the case $v = c_{1,3}v_{1,3}$. The following holds up to multiplication by a unit in Λ.

$$\langle v'_{1,3}, \sigma_2 x_{2,3}\rangle' = (1-t).$$

Thus

$$\langle v, \sigma_2 x_{2,3}\rangle = (1-t)(1-q)^2(1+qt)c_{1,3}.$$

By Lemma 5.5, $\sigma_2 x_{2,3}$ is a multiple of $(1-t)(1-q)(1+qt)$. Thus $(1-q)c_{1,3} \in \Lambda$. But

$$\langle v, x_{1,3}\rangle = (1-q)^2(1+qt)c_{1,3}.$$

By Lemma 5.6, $(1+qt)c_{1,3} \in \Lambda$. Thus $c_{1,3} \in \Lambda$.

Now suppose $n > 3$. For $i = 1, \dots, n-2$ we have

$$\langle v, x_{i,n}\rangle = (1-q)^2 c_{i,n}.$$

By Lemma 5.6, $c_{i,n} \in \Lambda$. Also

$$\langle v, x_{n-1,n}\rangle = (1-q)^2(1+qt)(1-t)c_{n-1,n}.$$

By Lemma 5.6, $(1+qt)(1-t)c_{n-1,n} \in \Lambda$. By Lemma 5.5, $(1-q)c_{n-1,n} \in \Lambda$. Thus $c_{n-1,n} \in \Lambda$. We can now subtract the terms $c_{i,n}v_{i,n}$, and so reduce to the case $c_{i,n} = 0$ for all $i = 1, \dots, n-1$. Then v represents an element of homology of the preimage in $\tilde{C}$ of the space of unordered pairs of distinct points in an $(n-1)$-times punctured disk. The result now follows by induction on n. □

This completes the proof of Theorem 5.1.

5.2. The Krammer representation. There is some confusion as to the exact definition of the "Lawrence-Krammer representation". In an attempt to clarify the situation, we now compare and contrast a slightly different representation which we will call the "Krammer representation".

Let the *Krammer representation* be the following action of B_n on a free Λ-module V of rank $\binom{n}{2}$ with basis $\{F_{i,j} : 1 \leq i < j \leq n\}$.

$$\sigma_i(F_{j,k}) = \begin{cases} F_{j,k} & i \notin \{j-1, j, k-1, k\}, \\ qF_{i,k} + (q^2 - q)F_{i,j} + (1-q)F_{j,k} & i = j-1, \\ F_{j+1,k} & i = j \neq k-1, \\ qF_{j,i} + (1-q)F_{j,k} + (1-q)qtF_{i,k} & i = k-1 \neq j, \\ F_{j,k+1} & i = k, \\ -tq^2 F_{j,k} & i = j = k-1. \end{cases}$$

This representation is given in [**Kra00**], but with different conventions as explained below. It is also in [**Big01**], but with a sign error. The name "Krammer representation" was chosen because Krammer seems to have initially found this independently of Lawrence and without any use of homology.

The vector spaces $\mathbf{Q}(q,t) \otimes V$ and $\mathbf{Q}(q,t) \otimes H_2(\tilde{C})$ are isomorphic representations of B_n, by [**Big01**, Theorem 4.1]. The isomorphism is given by

$$\phi\colon\ F_{i,j} \mapsto (\sigma_{i-1} \ldots \sigma_2\sigma_1)(\sigma_{j-1} \ldots \sigma_3\sigma_2)v_{1,2}.$$

Thus ϕ maps $F_{i,j}$ to the genus three surface corresponding to an edge from p_i to p_j in the lower half plane.

The basis elements in [**Kra00**] correspond to similar edges in the *upper* half plane. Thus Krammer's $v_{i,j}$ should be identified with my

$$(\sigma_{j-1} \ldots \sigma_2\sigma_1)(\sigma_{k-1} \ldots \sigma_3\sigma_2)F_{1,2}.$$

Also, Krammer's t is my $-t$.

For $n \geq 3$, the map $\phi\colon V \to H_2(\tilde{C})$ defined as above is not an isomorphism. To see this, note that the composition of ϕ with the natural map from $H_2(\tilde{C})$ to $H_2(\tilde{C}, \tilde{\nu})$ sends V into $(1-t)H_2(\tilde{C}, \tilde{\nu})$. However the image of $v_{1,3}$ in $H_2(\tilde{C}, \tilde{\nu})$ does not lie in $(1-t)H_2(\tilde{C}, \tilde{\nu})$. Thus $v_{1,3}$ is not in the image of ϕ.

In fact Paoluzzi and Paris [**PP01**] showed that the modules V and $H_2(\tilde{C})$ are not isomorphic representations of B_n for $n \geq 3$. The distinction becomes important when we specialise q and t to values which are not algebraically independent, as we will in the next section.

It is possible to compute the matrices for the Lawrence-Krammer representation with respect to the basis $\{v_{i,j}\}$. We will not do this since they are quite complicated. They must be conjugate to those of the Krammer representation when considered as matrices over $\mathbf{Q}(q,t)$, but not when considered as matrices over Λ.

The following conjecture would give a nice topological interpretation of the Krammer representation.

CONJECTURE 5.7. *The Krammer representation is isomorphic to the action of B_n on $H_2(\tilde{C}, \tilde{\nu})$.*

The isomorphism should be the map $\psi\colon V \to H_2(\tilde{C}, \tilde{\nu})$ given by

$$\psi(F_{i,j}) = (\sigma_{i-1} \ldots \sigma_2\sigma_1)(\sigma_{j-1} \ldots \sigma_3\sigma_2)v'_{1,2}.$$

This respects the action of B_n by the proof of [**Big01**, Theorem 4.1]. It remains to show that the terms $\psi(F_{i,j})$ form a basis for $H_2(\tilde{C}, \tilde{\nu})$. The proof should follow the same method as that of Theorem 5.1. The only difficulty is showing that $\mathbf{Q}(q,t) \otimes H_2(\tilde{C}, \tilde{\nu})$ has dimension $\binom{n}{2}$.

6. A representation of the Temperley-Lieb algebra

The aim of this section is to prove the following.

THEOREM 6.1. *Let R be a domain containing invertible elements q and t, and let*

$$\iota\colon R \otimes H_2(\tilde{C}) \to R \otimes H_2(\tilde{C}, \tilde{\nu})$$

be induced by the natural map from homology to relative homology. Suppose $1+qt = 0$, q has a square root, $q^2 \neq 1$, and $q^3 \neq 1$. Then the image of ι is the representation of $\mathrm{TL}_n(R)$ corresponding to the partition $(n-2,2)$.

Here, $\mathrm{TL}_n(R)$ is the Temperley-Lieb algebra. We will define this, and the desired representation in Section 6.4.

In [**Law90**, Theorem 5.1], Lawrence constructed the representation of the Hecke algebra corresponding to the partition $(n-k,k)$ for $k = 1, \dots, \lfloor n/2 \rfloor$. These are the representations that factor through the Temperley-Lieb algebra. This was generalised still further in [**Law96**] to give the Hecke algebra representation corresponding to any partition of n. Theorem 6.1 only covers the case $(n-2,2)$, but also has a number of advantages over the work of Lawrence. Firstly, it gives a more elementary description of the required quotient of $R \otimes H_2(\tilde{C})$. Secondly, it works over a fairly general ring, whereas Lawrence worked over $\mathbf{C}$ and used the matrices we gave in Section 5.2 for the "Krammer representation". Finally, our proof gives an explicit isomorphism between the two representations. It is to be hoped that these advantages generalise to arbitrary partitions of n.

I suspect that the requirements on R in Theorem 6.1 can be weakened somewhat. The main example to keep in mind is $R = \mathbf{Z}[q^{\pm\frac{1}{2}}]$ with $t = -q^{-1}$.

6.1. A basis. For $1 \leq i < j \leq n$, let $v_{i,j}$, $v'_{i,j}$ and $x_{i,j}$ be as defined in Section 5. Then

$$\iota(v_{i,j}) = \begin{cases} 0 & \text{if } j = i+1 \text{ or } j = 3, \\ (1-q)^2 v'_{i,j} & \text{otherwise.} \end{cases}$$

The pairing $\langle \cdot, \cdot \rangle'$ can be extended to a map

$$\langle \cdot, \cdot \rangle' \colon (R \otimes H_2(\tilde{C}, \tilde{\nu})) \times H_2(\tilde{C}, \partial \tilde{C}) \to R.$$

Lemma 5.3 still holds, so the $v'_{i,j}$ are linearly independent in $R \otimes H_2(\tilde{C}, \tilde{\nu})$. Thus the image of ι is the free module with basis

$$\{(1-q)^2 v'_{i,j} : j > \max(i+1, 3)\}.$$

Let H be the free module with basis

$$\{v'_{i,j} : j > \max(i+1, 3)\}.$$

To ease the notation, we will work with H instead of the image of ι, since the two are isomorphic.

Let K be the field of fractions of R. Then $K \otimes H$ is a vector space of dimension $n(n-3)/2$, and contains H as an embedded submodule. It is sometimes easier to prove that a relation holds in H by proving that it holds in $K \otimes H$.

6.2. The Hecke algebra. We now prove that H is a representation of the Hecke algebra.

DEFINITION 6.2. The Hecke algebra $\mathcal{H}_n(R)$, or simply $\mathcal{H}_n$, is the R-algebra given by generators $1, \sigma_1, \ldots, \sigma_{n-1}$ and relations

- $\sigma_i \sigma_j = \sigma_j \sigma_i$ if $|i-j| > 1$,
- $\sigma_i \sigma_j \sigma_i = \sigma_j \sigma_i \sigma_j$ if $|i-j| = 1$,
- $(\sigma_i - 1)(\sigma_i + q) = 0$.

Thus $\mathcal{H}_n$ is the group algebra RB_n of the braid group modulo the relations

$$(\sigma_i - 1)(\sigma_i + q) = 0.$$

LEMMA 6.3. *$(\sigma_i - 1)(\sigma_i + q)$ acts as the zero map on H.*

PROOF. Since the σ_i are all conjugate, we need only check the case $i = n-1$. To do this, we find a basis for $K \otimes H$ consisting of eigenvectors of σ_{n-1}, each having eigenvalue 1 or $-q$.

The elements $v'_{i,j}$ for $\max(i+1,3) < j \leq n-2$ are linearly independent eigenvectors of σ_{n-1} with eigenvalue 1. There are $(n-2)(n-5)/2$ of these.

Let α be a circular closed curve based at p_{n-2} and enclosing p_{n-1} and p_n. For $i = 1, \ldots, n-4$, let u_i be the square corresponding to the edges $[p_i, p_{i+1}]$ and α. Let u_{n-3} be the square corresponding to the edge $[p_{n-3}, p_{n-2}]$ and a circular closed curve based at p_{n-4} enclosing $p_{n-3}, \ldots, p_n$. These are eigenvalues of σ_{n-1} with eigenvalue 1. For $i, i' = 1, \ldots, n-3$, the following identities hold up to multiplication by a unit in Λ.

- $\langle u_i, x_{i,n} \rangle' = 1 - q$,
- $\langle u_i, x_{i',n} \rangle' = 0$ for $i' \neq i$,
- $\langle v'_{i,j}, x_{i',n} \rangle' = 0$ for $1 \leq i < j \leq n-2$.

Thus the u_i are linearly independent and not in the span of $\{v'_{i,j} : j \leq n-2\}$. Now let u be the square corresponding to α and a circular closed curve based at p_{n-3} enclosing α. The following identities hold up to multiplication by a unit in Λ.

- $\langle u, x_{n-2,n-1} \rangle' = (1-q^2)(1+q^2 t)(1-t)$,
- $\langle u_i, x_{n-2,n-1} \rangle' = 0$ for $i = 1, \ldots, n-3$,
- $\langle v'_{i,j}, x_{n-2,n-1} \rangle' = 0$ for $j \leq n-2$.

These can be computed using the methods of Section 4. The first is quite difficult, and we will discuss a less direct way to compute it later.

From the above identities, it follows that

$$\mathcal{B}_1 = \{v_{i,j} : \max(i+1,3) < j \leq n-2\} \cup \{u_1, \ldots, u_{n-3}\} \cup \{u\}$$

is a linearly independent set of eigenvectors of σ_{n-1} with eigenvalue 1. Each $v \in \mathcal{B}_1$ lies in H, as can be seen by constructing a surface in $\tilde{C}$ whose image in $H_2(\tilde{C}, \tilde{\nu})$ is $(1-q)^2 v$.

For $i = 1, \ldots, n-3$, note that $v'_{i,n}$ must be an eigenvector of σ_{n-1}, because the function σ_{n-1} can be chosen to fix setwise the square in C whose lift represents $v'_{i,n}$. To find the eigenvalue, note that

$$\sigma_{n-1}^{-1} x_{i,n} = x_{i,n-1} - q^{-1} x_{i,n},$$

as shown in Figure 7, so

$$\langle \sigma_{n-1} v'_{i,n}, x_{i,n} \rangle = \langle v'_{i,n}, \sigma_{n-1}^{-1} x_{i,n} \rangle = -q \langle v'_{i,n}, x_{i,n} \rangle$$

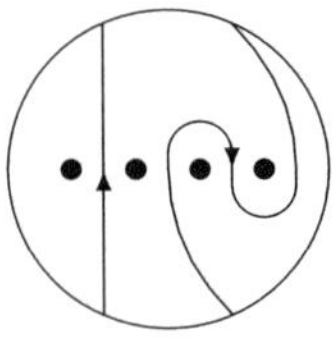

FIGURE 7. Showing that $\sigma_{n-1}^{-1}x_{i,n} = x_{i,n-1} - q^{-1}x_{i,n}$.

Thus

$$\mathcal{B}_{-q} = \{v'_{i,n} : i = 1, \dots, n-3\}.$$

is a set of eigenvectors of σ_{n-1} with eigenvalue $-q$. They are linearly independent, and since $-q \neq 1$, they are not in the span of $\mathcal{B}_1$. By a dimension count, $\mathcal{B}_1 \cup \mathcal{B}_{-q}$ is a basis for $K \otimes H$. Each element of this basis is annihilated by $(\sigma_{n-1}-1)(\sigma_{n-1}+q)$, so we are done. $\square$

In the above proof we used the fact that

$$\langle u, x_{n-2,n-1}\rangle' = (1-q^2)(1+q^2t)(1-t),$$

up to multiplication by a unit in Λ. We now describe an indirect argument to obtain this identity. Let λ denote $(1-q^2)(1+q^2t)(1-t)$.

Let P be an ϵ-neighbourhood of the horizontal edge $[p_{n-1}, p_n]$, where $\epsilon > 0$ is small. Let Z be the set of points $\{x, y\} \in C$ such that either x and y are within distance ϵ of each other, or at least one of them lies in P. Let $\tilde{Z}$ be the preimage of Z in $\tilde{C}$.

Let $x' \in H_2(\tilde{C}, \partial\tilde{C} \cup \tilde{Z})$ be a triangle represented by the horizontal edge $[p_n, 1]$. I claim that the image of $x_{n-2,n-1}$ in $H_2(\tilde{C}, \partial\tilde{C} \cup \tilde{Z})$ is $\lambda x'$. The proof is almost identical to that of Lemma 5.5. This time P is to be treated as one large puncture point. Since P actually contains two puncture points, it is necessary to substitute q^2 for q in the proof of Lemma 5.5.

The surface representing u does not meet $\tilde{Z}$, so it is possible to define an intersection pairing $\langle u, x'\rangle$. The surfaces representing u and x' intersect transversely at one point when projected to C. Thus $\langle u, x'\rangle$ is a unit in Λ. The pairing is sesquilinear, so $\langle u, x\rangle = \bar{\lambda}$. But λ is equal to $\bar{\lambda}$, up to multiplication by a unit in Λ, so we are done.

6.3. The Temperley-Lieb algebra. We now prove that H is a representation of the Temperley-Lieb algebra TL_n.

DEFINITION 6.4. Let R be a domain containing an invertible element q with a square root $q^{\frac{1}{2}}$. The Temperley-Lieb algebra $\mathrm{TL}_n(R)$, or simply TL_n, is the R-algebra given by generators $1, e_1, \dots, e_{n-1}$ and relations

- $e_ie_j = e_je_i$ if $|i-j| > 1$,
- $e_ie_je_i = e_i$ if $|i-j| = 1$,
- $e_i^2 = (-q^{\frac{1}{2}} - q^{-\frac{1}{2}})e_i$.

Consider the map from $\mathcal{H}_n$ to TL_n given by $\sigma_i \mapsto 1 + q^{\frac{1}{2}}e_i$. I claim this is well-defined and has kernel generated by

$$z_{i,j} = \sigma_i\sigma_j\sigma_i - \sigma_i\sigma_j - \sigma_j\sigma_i + \sigma_i + \sigma_j - 1$$

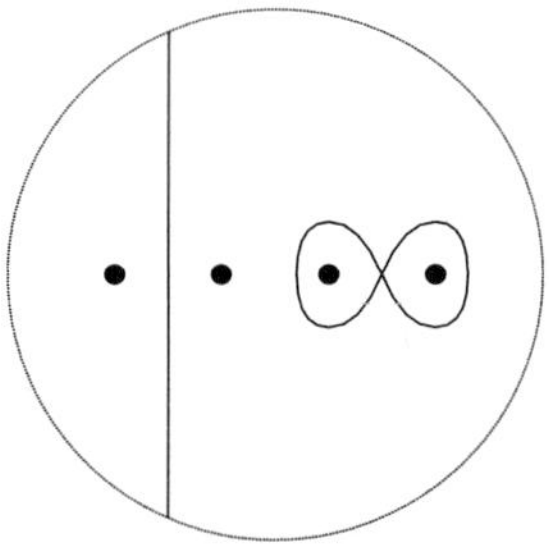

FIGURE 8. The arcs ξ_1 and γ used to define y_1.

for $|i-j|=1$. This is [**Jon87**, Equation 11.6], except using slightly different conventions. It can be verified by adding the relations $z_{i,j}=0$ to the presentation of $\mathcal{H}_n$, substituting $1+q^{\frac{1}{2}}e_i$ for σ_i, and simplifying to obtain the presentation for TL_n.

LEMMA 6.5. *$z_{i,j}$ acts as the zero map on H.*

PROOF. The $z_{i,j}$ are all conjugate to each other, so it suffices to prove that $z_{n-2,n-1}$ acts as the zero map on H. We have

$$\begin{aligned} z_{n-2,n-1} &= (\sigma_{n-2}\sigma_{n-1}-\sigma_{n-1}+1)(\sigma_{n-2}-1) \\ &= (\sigma_{n-1}\sigma_{n-2}-\sigma_{n-2}+1)(\sigma_{n-1}-1). \end{aligned}$$

Thus it suffices to find a basis for $K\otimes H$, each of whose elements is an eigenvector of either σ_{n-1} or σ_{n-2} with eigenvalue 1.

Let $\mathcal{B}_1$ be the linearly independent set of eigenvectors of σ_{n-1} with eigenvalue 1 as defined in the proof of Lemma 6.3.

Let α' be a circular closed curve based at p_n and enclosing p_{n-1} and p_{n-2}. For $i=1,\dots,n-4$, let w_i be the square corresponding to the edges $[p_i,p_{i+1}]$ and α'. Let w_{n-3} be the square corresponding to α' and a circular closed curve based at p_{n-3} and enclosing α'. The w_i all lie in H, and are eigenvectors for σ_{n-2} with eigenvalue 1.

Let $\gamma\colon S^1\to D_n$ be a figure-eight going around p_{n-1} and p_n in a regular neighbourhood of $[p_{n-1},p_n]$. For $i=1,\dots,n-3$, let $\xi_i\colon I\to D_n$ be a vertical edge with endpoints on ∂D, passing just to the right of p_i. See Figure 8. Define

$$f_i\colon I\times S^1\to C$$

by $f_i(x,y)=\{\xi_i(x),\gamma(y)\}$. This lifts to $\tilde{C}$. Let y_i be the corresponding element of $H_2(\tilde{C},\partial\tilde{C})$. For $i,j=1,\dots,n-3$, the following holds up to multiplication by a unit in Λ.

$$\langle w_i,y_j\rangle' = \begin{cases} 1-q^3 & i=j\neq n-3 \\ (1-q^3)(1-q^3t^2) & i=j=n-3 \\ 0 & i\neq j \end{cases}$$

Further, $\langle v,y_i\rangle'=0$ for all $v\in\mathcal{B}_1$. Thus

$$\{w_1,\dots,w_{n-3}\}\cup\mathcal{B}_1$$

is linearly independent. By a dimension count, it forms a basis of $K\otimes H$. Each vector in this basis is annihilated by $z_{n-2,n-1}$, so we are done. □

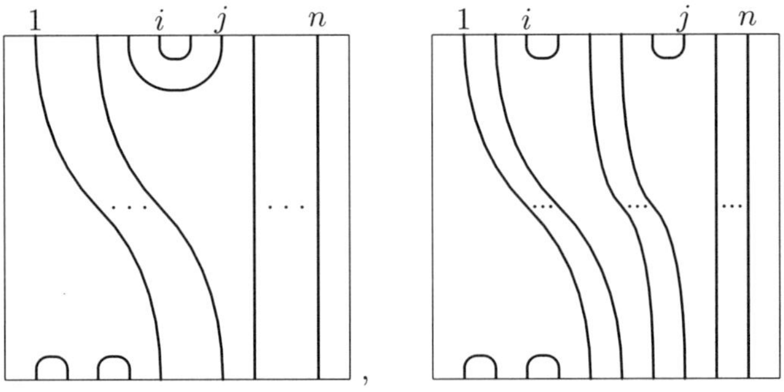

FIGURE 9. $s_{i,j}$ for $j = i+2$ and $j > i+2$.

6.4. The $(n-2,2)$ representation. We have shown that H can be considered as a representation of TL_n. We now define the representation S of TL_n corresponding to the partition $(n-2,2)$, and show that it is isomorphic to H. A good introduction to the representation theory of the Temperley-Lieb algebra is [**Wes95**].

DEFINITION 6.6. Let M be the left-ideal of TL_n generated by e_1e_3, and let N be the left-ideal generated by $\{e_5, \ldots, e_{n-1}\}$. The representation of TL_n corresponding to the partition $(n-2,2)$ is the TL_n-module $S = M/(M \cap N)$.

This is called $S(n,2)$ in [**Wes95**]. It is a free R-module with basis

$$s_{i,j} = (e_i \ldots e_3e_2)(e_{j-1} \ldots e_5e_4)(e_1e_3)$$

for $1 \leq i < j \leq n$ such that $j > \max(i+1,3)$. There is a diagrammatic interpretation of TL_n, in which $s_{i,j}$ corresponds to the diagram shown in Figure 9.

Let $\psi\colon \mathrm{TL}_n \to H$ be the unique map compatible with the action such that $\psi(1) = v'_{1,4}$. Then

$$\psi(e_3) = q^{-\frac{1}{2}}(\sigma_3 - 1)(v'_{1,4}) = (-q^{\frac{1}{2}} - q^{-\frac{1}{2}})v'_{1,4}.$$

Similarly $\psi(e_1) = (-q^{\frac{1}{2}} - q^{-\frac{1}{2}})v'_{1,4}$. Thus

$$\psi(e_1e_3) = (q^{\frac{1}{2}} + q^{-\frac{1}{2}})^2 v'_{1,4}.$$

Let $\psi'\colon M \to H$ be given by

$$\psi' = (q^{\frac{1}{2}} + q^{-\frac{1}{2}})^{-2}\psi|_M.$$

For $i = 5, \ldots, n-1$ we have that

$$\psi(e_i) = q^{-\frac{1}{2}}(\sigma_i - 1)(v'_{1,4}) = 0.$$

Thus $\psi(N) = 0$, so $\psi'(M \cap N) = 0$. We therefore have a map $\phi\colon S \to H$ induced by ψ'.

Recall that S has a basis

$$s_{i,j} = (e_i \ldots e_2)(e_{j-1} \ldots e_4)(e_1e_3)$$

for $1 \leq i < j \leq n$ such that $j > \max(i+1,3)$. We will show that $\phi(s_{i,j}) = v'_{i,j}$ up to multiplication by a unit in Λ.

LEMMA 6.7. *For* $j = 5, \ldots, n$ *we have* $(\sigma_{j-1} - 1)v'_{1,j-1} = v'_{1,j}$ *up to multiplication by a unit in* Λ.

PROOF. Let $\alpha_1 \colon I \to D$ be the edge $[p_1, p_2]$. Let $\alpha_2 \colon I \to D$ be the edge $[p_{j-2}, p_{j-1}]$. Define $f \colon I \times I \to C$ by $f(x, y) = \{\alpha_1(x), \alpha_2(y)\}$. Then f lifts to a map $\tilde{f}$ which represents $v'_{1,j-1}$.

We can assume that $\sigma_{j-1}\alpha_2(\frac{1}{2})$ is within distance ϵ of p_{j-1}. Then f maps the line $\{(x, \frac{1}{2})\}$ into ν_ϵ. The restriction of $\tilde{f}$ to $I \times [0, \frac{1}{2}]$ represents some unit multiple of $v_{1,j-1}$. The restriction of $\tilde{f}$ to $I \times [\frac{1}{2}, 1]$ represents some unit multiple of $v_{1,j}$. Thus

$$\sigma_{j-1} v'_{1,j-1} = \lambda v'_{1,j-1} + \mu v'_{1,j}$$

for some units $\lambda, \mu \in \Lambda$.

It remains to show that $\lambda = 1$. Recall that $\langle v'_{1,j-1}, x_{1,j-1}\rangle' = 1$. Thus $\langle \sigma_{j-1} v'_{1,j-1}, x_{1,j-1}\rangle' = \lambda$. But $\sigma_{j-1} x_{1,j-1} = x_{1,j-1}$, so

$$\begin{aligned} \langle \sigma_{j-1} v'_{1,j-1}, x_{1,j-1}\rangle' &= \langle \sigma_{j-1} v'_{1,j-1}, \sigma_{j-1} x_{1,j-1}\rangle' \\ &= \langle v'_{1,j-1}, x_{1,j-1}\rangle' \\ &= 1. \end{aligned}$$

Thus $\lambda = 1$, so $(\sigma_{j-1} - 1)v'_{1,j-1} = \mu v'_{1,j}$. □

LEMMA 6.8. *For* $i = 2, \ldots, j-1$ *we have* $(\sigma_i - 1)v'_{i-1,j} = v'_{i,j}$ *up to multiplication by a unit in* Λ.

PROOF. In the case $i < j - 1$, the proof is almost identical to that of the previous lemma. Suppose $i = j - 1$. Then

$$\sigma_i v'_{i-1,j} = \lambda v'_{i-1,j} + \mu v'_{i,j}$$

for some $\lambda, \mu \in \Lambda$. To see that $\lambda = 1$, use the same argument as in the proof of the previous lemma. To see that μ is a unit in Λ, note that it is equal to $\langle \sigma_i v'_{i-1,j}, x_{i,j}\rangle'$, and the surfaces representing $\sigma_i v'_{i-1,j}$ and $x_{i,j}$ intersect transversely at one point when projected to C. □

By these two lemmas we have $\phi(s_{i,j}) = v'_{i,j}$ up to multiplication by a unit in Λ. This implies that ϕ is an isomorphism, which completes the proof of Theorem 6.1.

References

[Big01] Stephen J. Bigelow, *Braid groups are linear*, J. Amer. Math. Soc. **14** (2001), no. 2, 471–486 (electronic). MR **2002a:**20043

[Bud02] Ryan Budney, *A note on the image of the Lawrence-Krammer representation*, math.GT/0202246, February 2002.

[Jon87] V. F. R. Jones, *Hecke algebra representations of braid groups and link polynomials*, Ann. of Math. (2) **126** (1987), no. 2, 335–388. MR **89c:**46092

[Kaw96] Akio Kawauchi, *A survey of knot theory*, Birkhäuser Verlag, Basel, 1996, Translated and revised from the 1990 Japanese original by the author. MR **97k:**57011

[Kra00] Daan Krammer, *The braid group B_4 is linear*, Invent. Math. **142** (2000), no. 3, 451–486. MR **2001k:**20078

[Kra02] ———, *Braid groups are linear*, Ann. of Math. (2), **155** (2002), no. 1, 131–156. MR 1 888 796

[Law90] R. J. Lawrence, *Homological representations of the Hecke algebra*, Comm. Math. Phys. **135** (1990), no. 1, 141–191. MR **92d:**16020

[Law96] Ruth J. Lawrence, *Braid group representations associated with sl_m*, J. Knot Theory Ramifications **5** (1996), no. 5, 637–660. MR **98j:**57011

[PP01] Luisa Paoluzzi and Luis Paris, *A note on the Lawrence-Krammer-Bigelow representation*, Algebr. Geom. Topol. **2** (2002), 499–518.

[Son02] Won Taek Song, *The Lawrence–Krammer representation is unitary*, math.GT/0202214, February 2002.

[Wes95] B. W. Westbury, *The representation theory of the Temperley-Lieb algebras*, Math. Z. **219** (1995), no. 4, 539–565. MR **96h:**20029

DEPARTMENT OF MATHEMATICS AND STATISTICS, UNIVERSITY OF MELBOURNE, VICTORIA 3010, AUSTRALIA

Current address: Department of Mathematics, University of California at Santa Barbara, California 93106

E-mail address: `bigelow@math.ucsb.edu`

Proceedings of Symposia in Pure Mathematics
Volume **71**, 2003

Topological Properties of Properly Embedded Minimal Surfaces in $\mathbb{R}^3$

William H. Meeks III

1. Introduction

Our goal here is to present some of the basic theory and conjectures in the classical theory of minimal surfaces in $\mathbb{R}^3$ which have a topological emphasis. Most of the important topological results concerning properly embedded minimal surfaces in $\mathbb{R}^3$ were published during the past twenty-five years. These results break up into three main categories. The first of these categories deals with topological obstructions. For example, a properly embedded surface in $\mathbb{R}^3$ is orientable and there are no compact minimal surfaces without boundary by the maximum principle for harmonic functions. However, there are other interesting deeper topological obstructions than orientability and noncompactness; we will cover all known topological obstructions as well as the main conjectures on topological obstructions. The second category of global results deals with the classification of properly embedded minimal surfaces of fixed topological type. For example, a recent theorem of Meeks and Rosenberg, which depends on earlier results of Collin [**6**] and Lopez-Ros [**20**], states that the plane, the helicoid and the catenoid are the only properly embedded minimal planar domains of finite topology in $\mathbb{R}^3$. Finally, the third category of results deals with the topological uniqueness of the embedding of a minimal surface with a given topological type. An important example of this type of result is the theorem of Meeks and Yau [**32**] that states that if M_1, M_2 are two homeomorphic properly embedded minimal surfaces of finite topology in $\mathbb{R}^3$, then M_1 and M_2 are properly ambiently isotopic. We refer the interested reader to the author's book [**22**] for a long list of unsolved topological problems for minimal surfaces that dates back to 1978; of course, some of the problems in the book have been solved.

2. Basic definitions and theorems.

We first define a minimal surface.

Definition 2.1. An isometric immersion $f\colon M^2 \to \mathbb{R}^3$ of a Riemannian surface $(M^2, <,>)$ is *minimal* if the coordinate functions f_1, f_2, f_2 are harmonic functions on M^2.

This research was supported by NSF grant DMS - 0104044.

We will be interested in noncompact minimal surfaces which are complete in the following sense.

Definition 2.2. An immersed surface $M \subset \mathbb{R}^3$ is *complete* if geodesics on M can be extended indefinitely or, equivalently, if M is a complete metric space with respect to the intrinsic distance function which is defined in terms of the infimum of the lengths of smooth curves on M joining a pair of points in M.

In principle we are interested in understanding the space of all complete embedded minimal surfaces in $\mathbb{R}^3$. All known examples of such surfaces satisfy the stronger hypothesis given in the next definition.

Definition 2.3. An immersion $f\colon M \to \mathbb{R}^3$ is *proper* if for any compact ball B, $f^{-1}(B)$ is compact in M. Let $\mathcal{P}$ denote the space of all properly embedded connected minimal surfaces in $\mathbb{R}^3$ and let $\mathcal{M} \subset \mathcal{P}$ be the subset of examples with more than one end.

Clearly, every smooth properly embedded minimal surface M has *locally bounded* Gaussian curvature in the sense that its Gaussian curvature in balls is bounded. In [**27**], Meeks and Rosenberg prove that if M is a complete embedded minimal surface in $\mathbb{R}^3$ with locally bounded Gaussian curvature, then the closure of M in $\mathbb{R}^3$ has the structure of a $C^{1,\alpha}$-lamination. In the same paper, they also give a structure theorem for minimal laminations in $\mathbb{R}^3$ which, together with curvature estimates from [**4**], can be used to prove the following theorem.

THEOREM 2.1. *(see [**24**] and [**27**]) A complete embedded minimal surface M of finite genus in $\mathbb{R}^3$ which does not have exactly one limit end, is proper if and only if it has locally bounded Gaussian curvature.*

Recent work of Meeks and Rosenberg [**27**], [**29**] shows the following two theorems hold:

THEOREM 2.2. *If $M \in \mathcal{P}$ is simply-connected, then M is a plane or a helicoid. If $M \in \mathcal{P}$ is not a plane and has finite genus and one end, then M is asymptotic to the end of a helicoid.*

THEOREM 2.3. *An $M \in \mathcal{P}$ of finite topology is conformally diffeomorphic to a finitely punctured Riemann surface. In particular, such a minimal surface is recurrent for Brownian motion.*

In order to understand further generalizations of the above theorems, we now discuss the topological notion of "ends" of a surface.

Definition 2.4. Suppose M is a noncompact connected manifold. The space of ends of M, denoted by $\mathcal{E}(M)$, is the set of equivalence classes of proper arcs $\alpha\colon [0,\infty) \to M$ where α_1 is equivalent to α_2 if for every compact subset $C \subset M$, the ends of α_1, α_2 lie in the same component of $M - C$.

A basis for the topology of $\mathcal{E}(M)$ is defined as follows. For each compact set $C \subset M$, define the basis open set $B(C) \subset \mathcal{E}(M)$ to be those equivalence classes of proper arcs in M which have representatives contained in $M - C$. Another way to define $\mathcal{E}(M)$, and its topology, is to consider $\mathcal{E}(M)$ to be the inverse limit of complements of compact subsets of M under the ordering of inclusion.

It can be shown that $\mathcal{E}(M)$ is a totally disconnected compact Hausdorff space that embedds as a compact subspace of the unit interval. Conversely, every compact totally disconnected subset C of the unit interval corresponds to the space of ends of a non-compact surface; namely, consider C to lie on an interval in S^2, then $S^2 - C$ has C as its space of ends.

Definition 2.5. An end $e \in \mathcal{E}(M)$ is a *simple* end of M if it is an isolated point in $\mathcal{E}(M)$. Note that e is simple if and only if there exists a representative proper arc $\alpha \in e$ and a proper subdomain $R \subset M$ containing α such that R is homeomorphic to $S^1 \times [0, \infty)$ or to $S^1 \times [\,0, \infty)$ connect sum of an infinite number of tori where the nth connected sum occurs at the point $(1, n) \in S^1 \times [0, \infty)$. In the first case we refer to e as an *annular* end and in the second case we refer to e as a *simple end of infinite genus.*

Definition 2.6. An end $e \in \mathcal{E}(M)$ is a *limit* end of $\mathcal{E}(M)$ if it is not a simple end. In other words, e is a limit end if it is a limit point of $\mathcal{E}(M)$. As in the previous definition, a limit end has *genus zero* if it can be represented by a proper domain $R \subset M$ with compact boundary and the genus of R is zero. If a limit end e does not have genus zero, we say that e has *infinite genus*; in this case every proper subdomain with compact boundary representing e has infinite genus.

The following theorem by Frohman and Meeks [**11**], called the ordering theorem, is the starting point for describing minimal surfaces in $\mathcal{M}$.

THEOREM 2.4. *Suppose $M \in \mathcal{M}$. After a possible rotation of M, the ends of M can be linearly ordered geometrically by their relative heights over the x_1x_2-plane. Furthermore, this ordering is a topological ordering.*

We make a few comments about the more or less constructive proof of the ordering theorem. By earlier work of Callahan, Hoffman and Meeks [**3**], we know that if $M \in \mathcal{M}$, then there is a simple closed curve α on M that separates $\mathcal{M}$ into two noncompact unstable domains and such that α is not homologous to zero in the closure N of one of the components of $\mathbb{R}^3 - M$. In this case α bounds a least-area properly embedded minimal surface $\Sigma \subset N$ and every such Σ has finite total curvature. Since Σ has finite total curvature and Σ is embedded, then it is well known that each end of Σ is asymptotic to the end of a plane or a catenoid. Thus, each end of Σ has a well-defined limiting normal line and these normal lines coincide since the planar and catenoid ends of Σ do not intersect. One then defines the limit tangent plane of Σ to be the plane in $\mathbb{R}^3$ passing through the origin with normal line corresponding to the limiting normal line of Σ. In fact, as shown in [**3**], this limit tangent plane of Σ does not depend on the choice of Σ but rather it is an invariant of M; this plane is called the *limit tangent plane at infinity* of M. More generally, Frohman and Meeks show that, after rotating $M \in \mathcal{M}$ so that the limit tangent plane at infinity is horizontal, there exists a proper pairwise disjoint family $\Delta = \{E(n) \mid n \in \mathbb{Z}\}$ of ends of horizontal planes and catenoids such that $E(n)$ is a graph over its projection to $\{(x_1, x_2, 0) \mid {x_1}^2 + {x_2}^2 \geq n^2\}$ and such that given any two distinct ends $e_1, e_2 \in \mathcal{E}(M)$, there exists $i, j \in \mathbb{Z}$ and a ball $B(e_1, e_2)$ such that e_1 and e_2 are represented by proper arcs on M in distinct components of $\mathbb{R}^3 - (B(e_1, e_2,) \cup E(i) \cup E(j))$. Since the graphs of Δ are linearly ordered by their relative heights over the x_1x_2-plane and they eventually separate any two distinct ends of M, Δ induces a relative geometric orderings on the ends of $\mathcal{E}(M)$. The property that this ordering is a topological ordering just means that if $f_1, f_2 \colon M \to \mathbb{R}^3$ are two properly minimal embeddings of M into $\mathbb{R}^3$, then the two induced geometric orderings on $\mathcal{E}(M)$ are the same up to possibly reversing the ordering; this property Frohman and Meeks also verify holds.

Definition 2.7. Consider the ordering on the ends $\mathcal{E}(M)$. The end $e_T \in \mathcal{E}(M)$ is the *top* end of M if it is the unique end in $\mathcal{E}(M)$ that is maximal in the ordering. The end $e_B \in \mathcal{E}(M)$ is the *bottom* end of M if it is the minimal in the ordering of the ends. An end of M that is neither the top nor the bottom end of M is called a *middle* end of M.

Definition 2.8. A surface $M \subset \mathbb{R}^3$ has *quadratic area growth* if there exists a positive constant c such that for all positive R, the area of M inside the ball $B(R)$ of radius R centered at the origin is less than cR^2.

By the proof of the ordering theorem [**11**], a middle end of a properly embedded minimal surface M can be represented by a proper subdomain $R \subset M$ with compact boundary such that R "lies between two catenoids." This means that R is contained in a neighborhood S of the x_1x_2-plane, S being topologically a slab, whose width grows logarithmically with the distance from the origin.

One of the fundamental results in the global theory of properly embedded minimal surfaces is that the middle ends of an $M \in \mathcal{M}$ are never limit ends. This is shown by first proving that if R is a properly immersed minimal surface with compact boundary and contained between two catenoids, then R has quadratic area growth [**8**]. By the monotonicity formula for area [**35**], every subend of a representative of a limit end has limiting area growth at least πR^2, and so a limit end never has a representative with quadratic area growth. Thus,

THEOREM 2.5. *(see [**8**]) If $M \in \mathcal{M}$, then a limit end of M must be a top or bottom end of M. In particular, M can have at most two limit ends.*

COROLLARY 2.1. *A properly embedded minimal surface in $\mathbb{R}^3$ has a countable number of ends.*

Now consider a middle end e of $M \in \mathcal{M}$. As we said above, M can be represented by an proper subdomain $R \subset M$ contained in a topological slab S whose width grows at most logarithmically. Furthermore, by Theorem 2.5, we may assume that R has one end. After removing a large round open ball B from S which contains ∂R, $(S - B) - R$ consists of two components, C_1, C_2. The end e is called *odd* if each of the components C_1, C_2 intersects one of the two boundary components ∂S; otherwise, e is called *even.* Note that middle ends which are annular are asymptotic to the end of a plane or a catenoid and so are always odd.

Theorem 2.5 plays a crucial role in the proof of the following topological classification theorem of properly embedded minimal surfaces in $\mathbb{R}^3$ by Frohman and Meeks ([**10**], [**12**]).

THEOREM 2.6. *Suppose $M_1, M_2 \in \mathcal{P}$. There exists an orientation preserving diffeomorphism $f\colon \mathbb{R}^3 \to \mathbb{R}^3$ with $f(M_1) = M_2$ if and only if there exists a homeomorphism $F\colon M_1 \to M_2$ that preserves the ordering on the ends of M_1 and M_2 and the eveness-oddness invariant of each of the middle ends. Thus, if M_1 and M_2 are homeomorphic and have at most two ends, then there exists diffeomorphism $f\colon \mathbb{R}^3 \to \mathbb{R}^3$ with $f(M_1) = M_2$.*

We now recall here some of the arguments needed in the proof the topological classification theorem of Frohman and Meeks. In [**12**] they considered the restricted case of where M_1 and M_2 had one end. The theorem in this case follows from two key results, one geometric and the other topological. The geometric result is that a properly embedded minimal surface with one end in $\mathbb{R}^3$ is a Heegaard surface for $\mathbb{R}^3$; the topological result is that Heegaard surfaces in $\mathbb{R}^3$ depend, up to proper ambient isotopy, only on their genus. A Heegaard surface H in $\mathbb{R}^3$ is a properly embedded surface, where each of the closed complements of H in $\mathbb{R}^3$ is a *one-handlebody* which means it is homeomorphic to the closed regular neighborhood of a properly embedded connected 1-dimensional CW-complex in $\mathbb{R}^3$.

When M has more than one end, then it is never a Heegaard surface for $\mathbb{R}^3$. However, by generalizing the techniques used by Frohman and Meeks, Freedman was able to give a topological characterization of any closed complement of M in $\mathbb{R}^3$ as being a proper union of balls and regions of the type $S^1 \times [0,1) \times \mathbb{R}$ joined by proper collection of one-handlebodies. In the case that $M \in \mathcal{M}$ has a finite number of ends, then it is not difficult to prove that the ends of planes and catenoids used in the proof of ordering theorem can be extended to obtain a finite number of pairwise disjoint topologically parallel planes $P = \{P_1, P_2, \ldots, P_k\}$, $k = |\mathcal{E}(M)| - 1$, that separate the ends of M. The description of the complements given by Freedman allows one to apply Haken moves [**14**] to the collection P thereby producing a bounded ambient isotopy so that at the end of the isotopy each plane in P intersects M in a single simple closed curve. This first step in the proof reduces the general topological uniqueness question to understanding the topology of M in certain topological slabs or halfspaces. Next Frohman and Meeks use Whitehead moves to prove that the embedding of the end of M in such a slab only depends on whether the end is even or odd and in the case a halfspace occurs the genus characterizes the embedding.

One of the most difficult parts of the proof of the general classification theorem occurs in the case where M has an infinite number of ends. In this case, by Theorem 2.5, M must have one or two limit ends and each such limit end is a top or bottom end. From the proof of the ordering theorem one can produce a countable proper collection of topologically parallel planes whose ends are disjoint from M and which separate the ends of M. Again using Freedman's decomposition result and Haken moves, Frohman and Meeks prove that this proper collection of planes can be chosen so that each such plane intersects M in a single simple closed curve. However, this time great care must be taken in the construction of this family of planes to insure that after the resulting isotopy the new planes are a proper collection; the problem is that there are possibly an infinite number of Haken moves to be made.

Motivated by the above result and work of Collin, Hauswirth and Rosenberg [**7**], the author conjectures that properly embedded constant mean curvature $H = 1$ surfaces M in hyperbolic three-space $\mathbb{H}^3$ are unknotted; such surfaces are called Bryant surfaces. When M_1, M_2 are homeomorphic Bryant surfaces with finite topology, the author has shown that M_1 and M_2 are properly ambiently isotopic in $\mathbb{H}^3$, which is similar to the theorem for minimal surfaces in $\mathbb{R}^3$ which was proved by Meeks and Yau [**32**].

Conjecture 2.1. [Topological Uniqueness Theorem for Bryant Surfaces (Meeks)] If M_1, M_2 are two homeomorphic properly embedded constant mean curvature 1 surfaces in $\mathbb{H}^3$, then M_1 and M_2 are properly ambiently isotopic.

The following result in [**8**] gives an interesting topological condition for a properly embedded minimal surface to be recurrent for Brownian motion.

THEOREM 2.7. *If $M \in \mathcal{M}$ has two limit ends, then there exists a proper collection $P = \{P(n) \mid n \in \mathbb{Z}\}$ of parallel planes such that each plane in P intersects M transversely in a compact set and such that every middle end of M is represented uniquely for some $n \in \mathbb{Z}$ by the intersection of M with the slab $S(n)$ between plane $P(n)$ and plane $P(n+1)$. Furthermore, M is recurrent for Brownian motion.*

3. Topological conjectures for minimal surfaces

In this section the author will present a number of topological conjectures in the classical theory of minimal surfaces. For the most part these conjectures are motivated by the author's own research and are not widely known except to classical minimal surface specialists. Hopefully, the presentation of these problems and suggestions for a plan of attack on solving them will speed up their solution and stimulate further interest in this beautiful subject. In the statement of each conjecture the author has included a suggested expected time frame for a solution; only time will tell how accurate this time frame is. The author has listed in the statement of each conjecture the principal researchers to whom the conjecture might be attributed. These conjectures are listed in order according to the author's interest in them or by his personal feeling of their general importance or deepness. For these discussions we again let $\mathcal{P}$ denote the set of all properly embedded connected minimal surfaces in $\mathbb{R}^3$ and let $\mathcal{M} \subset \mathcal{P}$ denote the subset with more than one end.

Conjecture 3.1. [Properness Conjecture (Calabi, Meeks) Time Frame = 100 years] If $f\colon M \to \mathbb{R}^3$ is a complete injective minimal immersion, then $M \in \mathcal{P}$.

Meeks, Perez and Ros [**24**] have conjectured that if a complete embedded minimal $M \subset \mathbb{R}^3$ has finite genus, then M has bounded Gaussian curvature (see also, [**27**]). It follows from work in [**24**] and [**27**] that if such an M has locally bounded Gaussian curvature and finite genus, then M is properly embedded. The suggested time frame for a solution to the finite genus case of the above conjecture is 5 years. The author's suggestion for an attack on the finite genus case is to employ a blow-up argument in the conjunction with the imminent proof of the next conjecture.

While the only planar domains of finite topology in $\mathcal{P}$ are the plane, the helicoid and the catenoid, Riemann defined a one-parameter family of periodic properly embedded minimal planar domains $R(t)$, $t \in \mathbb{R}^+$, with two limit ends. For each t, the surface $R(t)$ is foliated by round circles and straight lines in horizontal planes in $\mathbb{R}^3$. The surface $R(t)$ were defined by Riemann in terms of elliptic functions on the rectangular torus $\frac{\mathbb{C}}{<\sqrt{-1},t>}$ which is the conformal compactifications of $R(t)/\tau(t)$, where $\tau(t)$ is a nontrivial translational symmetry of $R(t)$. In [**26**], Meeks, Perez and Ros proved that the Riemann examples are the only genus zero examples in $\mathcal{P}$ with infinite topology and infinite symmetry group.

Conjecture 3.2. [Genus-zero Conjecture (Meeks-Perez-Ros) Time Frame = 2 years] If $M \in \mathcal{P}$ has genus zero, then M is a plane, a catenoid, a helicoid or a Riemann example.

There is a related conjecture of Meeks-Perez-Ros, which we call the Asymptotic Finite Genus Conjecture, that deals with the expected asymptotic geometry of the ends of an $M \in \mathcal{P}$ with finite genus. From the work of Meeks-Rosenberg [**27**], it is known that each simple end of such an M is asymptotic to the end of a plane, a catenoid or a helicoid. The related conjecture states that if M has an infinite number of ends and finite genus, then M has two limit ends, each of which is asymptotic to a Riemann-example-type limit end. What this means is that, after a rotation of M, for each limit end e of M there is a representative $E \subset M$ with compact boundary, horizontal planar ends and one limit end such that E converges exponentially quickly to a limit end of a fixed Riemann example as a function of the third coordinate function. The solution of this conjecture would complete the desired description of the asymptotic geometry of finite genus surfaces in $\mathcal{M}$. The suggested time frame for solving this related conjecture is also 2 years.

Recently Meeks, Perez and Ros [**24**] have been able to generalize their arguments of their proof of the previous conjecture in the case the surface has infinite symmetry. An important aspect in these generalizations concerns curvature estimates. These curvature estimates imply that if $M \in \mathcal{M}$ is a planar domain with two limit ends, then M has bounded Gaussian curvature. This result in turn implies that such an M is *quasiperiodic* in the sense that there exist vectors $v(i) \in \mathbb{R}^3, |v(i)| \to \infty$, such that the translated surfaces $M(i) = M + v(i)$ converge smoothly on compact subsets of $\mathbb{R}^3$ to another example $M(\infty) \in \mathcal{M}$ which has genus zero and two limit ends. This quasiperiodic property of M hopefully will be sufficient to prove that M is one of Riemann's examples. Furthermore, we hope that the arguments in the proof of these curvature estimates can be adapted to prove that if $M \in \mathcal{M}$ has finite genus, then M does not have one limit end.

Conjecture 3.3. [Finite Topology Conjecture (Hoffman and Meeks) Time Frame = 100 years] An orientable surface M of finite topology with genus g and k ends, $k \neq 2$, occurs in $\mathcal{P}$ if and only if $g \geq k - 2$.

See [**15**], [**17**], [**37**] and [**38**] for many partial existence results which seem to indicate that the Finite Topology Conjecture is true when $k > 2$. There is experimental computer evidence that every orientable surface with finite genus and one end properly minimally embedds in $\mathbb{R}^3$ as a minimal surface of finite type (see [**1**], [**2**]). Recent work of Meeks, Perez and Ros [**24**], [**25**] indicates that for each positive genus g, there exists an upper bound $k(g)$ on the number of ends of an $M \in \mathcal{M}$ with finite topology and genus g. Results of Collin [**6**] and Schoen [**34**] imply that the only examples in $\mathcal{M}$ with finite topology and two ends are catenoids. Results of Collin [**6**] and Lopez-Ros [**20**] imply that if M has finite topology, genus zero and at least two ends, then M is a catenoid.

The Finite Topology Conjecture of Hoffman and Meeks proposes the precise topological conditions under which a noncompact orientable surface of finite topology would properly minimally embedd in $\mathbb{R}^3$. What about the case where the noncompact orientable surface M has infinite topology; i.e., either M has infinite genus or M has an infinite number of ends? By Theorem 2.5, such an M can have at most one or two limit ends. The Asymptotic Finite Genus Conjecture of Meeks, Perez and Ros claims that such an M cannot have one limit end and finite genus. The following conjecture is nothing more than the claim that these restrictions are the only ones.

Conjecture 3.4. [Infinite Topology Conjecture (Meeks) Time Frame = 15 years] A noncompact orientable surface of infinite topology occurs in $\mathcal{P}$ if and only if it has at most one or two limit ends and in the case it has one limit end that it also has infinite genus.

Conjecture 3.5. [One-ended Conjecture (Meeks and Rosenberg) Time Frame = 10 years] For every nonnegative integer g, there exists a unique nonplanar $M \in \mathcal{P}$ with genus g and one end.

There are some partial results on the above conjecture. Meeks and Rosenberg have shown that every example $M \in \mathcal{P}$ of finite genus and one end has a special analytic representation which makes M into a "surface of finite type" (see [**27**]). In the case of genus zero, Meeks and Rosenberg proved that the plane and the helicoid are the only simply-connected genus zero examples. Based on an earlier computational proof of the existence of a helicoid with a handle by Hoffman-Karcher-Wei [**16**], Hoffman, Weber and Wolf [**18**] have given a rigorous mathematical proof of the existence of the genus-one helicoid. Work in progress by Martin and Weber [**21**] indicates that this genus-one helicoid is unique. Other computational work in [**1**], [**2**] and [**36**] indicates the conjecture is true for genus at most five.

The next conjecture appears in [**22**]. Although there are a number of partial results, the conjecture still remains unsolved.

Conjecture 3.6. [Convex Curve Conjecture (Meeks) Time Frame = 30 years] Two convex Jordan curves in parallel planes cannot bound a compact minimal surface of positive genus.

In fact in [**22**], Meeks conjectures that if $D = \{D_1, \dots, D_n\}$ is a finite collection of convex pairwise disjoint disks in the x_1x_2-plane and Γ is a convex Jordan curve in a parallel plane, then $\{\Gamma, \partial D_1, \dots, \partial D_n\}$ does not bound a compact minimal surface of positive genus. The suggested time frame for solving this more general conjecture is 50 years.

There are some partial results on the Convex Curve Conjecture under the assumption of some symmetry on the curves (see [**30**], [**33**], [**34**]). Also, the results in [**30**] and [**31**] indicate that the Convex Curve Conjecture probably holds in the more general case where the two convex planar curves do not necessarily lie in parallel planes but rather lie on the boundary of their convex hull, in this case the planar Jordan curves are called extremal. Recent results by Ekholm, White and Wienholtz [**9**] show that every compact minimal surface that arises as a counterexample is embedded and that for a fixed pair of curves there is a bound on the genus of such a minimal surface.

Conjecture 3.7. [Scherk Uniqueness Conjecture (Meeks) Time Frame = 5 years] If $M \in \mathcal{P}$ and, in every ball $B(R)$ of radius R, $\text{Area}(M \cap B(R)) < 2\pi R^2$, then M is a Scherk singly-periodic minimal surface, a catenoid or a plane.

The above conjecture would imply that if M is a properly embedded minimal surface in $S^1 \times \mathbb{R}^2$ with linear area growth, then M is a quotient of a Scherk surface. M having linear area growth in $S^1 \times \mathbb{R}^2$ is equivalent to the property that M has finite topology and its total curvature is $2\pi\chi(M)$. Theorem 2.3 in Section 2 was motivated by an earlier result of Meeks [**23**] that a properly embedded minimal surface M in $\mathbb{T} \times \mathbb{R}$, $\mathbb{T}$ a flat two-torus, can have only a finite number of ends. Therefore, if M has finite genus, then M has finite topology and the total curvature of M is $2\pi\chi(M)$ (see [**28**]). It turns out that the finite total curvature property of an $M \subset \mathbb{T} \times \mathbb{R}$ can be exploited to obtain a number of geometric restrictions on M and $\mathbb{T}$. For example, if M is a planar domain, then the lattice for $\mathbb{T}$ is commensurable (see [**28**]).

A related conjecture on the uniqueness of Scherk's doubly-periodic minimal surfaces was recently solved by Lazard-Holly and Meeks [**19**]; they proved that if $M \in \mathcal{P}$ is doubly-periodic and the quotient surface has genus zero, then M is one of Scherk's doubly-periodic minimal surfaces. The basic approach used in [**19**] probably could be used to prove the above conjecture under the assumption that the surface is periodic. One approach for solving the general conjecture is first to prove the following conjecture on the uniqueness of the limit tangent cone of M, from which it follows by an unpublished work of Meeks and Ros that M has two Alexandrov-type planes of symmetry. From these planes of symmetry one can describe the Weierstrass representation of M, which hopefully would be useful in completing the proof of the conjecture. Recently Meeks and Wolf have obtained a proof of Conjecture 3.7 under the additional hypothesis that the surface have infinite symmetry.

In the case of singly-periodic minimal surfaces, there is one known topological obstruction which we now describe. A screw motion symmetry S of $\mathbb{R}^3$ is the composition of a rotation around an axis followed by a nontrivial translation along the axis. Meeks has shown that if the rotational part of S has order greater than two, then every properly embedded minimal surface in $\mathbb{R}^3/S$ has a finite number of ends; this result always fails if the order of the rotational part of S is not greater than two.

Let $\mathbb{T}^3$ be a flat three-torus. In this case, the Gauss-Bonnet theorem implies that if M is a minimal surface of genus g in $\mathbb{T}^3$, then the total curvature of $M = 4\pi(1-g)$. In particular the degree of the Gauss map, which can be considered to be a meromorphic function on M, has degree $g-1$; hence, the genus of such an M cannot be two. The next conjecture represents the belief that this is the only topological obstruction for orientable surfaces.

Conjecture 3.8. [Triply-Periodic Conjecture (Meeks) Time Frame = 2 years] Let $\mathbb{T}^3$ be a flat three-torus and M a closed orientable surface of positive genus different from two. Then there exists an infinite sequence of minimal embeddings $f(n)\colon M \to \mathbb{T}^3$ such that Area $(f(n)) \geq n$.

Conjecture 3.9. [Standard Middle End Conjecture (Meeks) Time Frame = 8 years] If $M \in \mathcal{M}$ and $E \subset M$ is a simple end representative for a middle end of M, then E is C^0-asymptotic to the end of a plane or catenoid. In particular, if M has two limit ends, then each middle end is C^0-asymptotic to a plane.

By the work in [**8**], $\lim_{t\to\infty}(\frac{1}{t})E$ is a plane P passing through the origin with positive integer multiplicity. Also, if M has two limit ends, then, for such a middle end representative E, a sequence of translates of E has a positive finite number of parallel planes for a limit; in this case, it remains to prove there is only one plane in this limit.

Gulliver and Lawson [**13**] proved that if Σ is a stable orientable minimal surface with compact boundary that is properly embedded in the punctured unit ball in $\mathbb{R}^3$, then its closure is an embedded surface. If Σ is not stable, then the corresponding result is not known. Recent results in [**5**] and [**25**] motivate the following conjecture.

Conjecture 3.10. [Isolated Singularities Conjecture (Gulliver, Lawson, and Meeks) Time Frame = 8 years] A two-dimensional minimal lamination with simply-connected leaves except possibly one leaf of finite topology cannot have an isolated singularity. Furthermore, there does not exist a properly embedded minimal surface in the punctured ball $B - \{(0,0,0)\}$ whose closure is not a surface at $(0,0,0)$.

One of the outstanding problems in three-manifold theory is to prove that if $G \subset \mathrm{Diff}(S^3)$ is a finite group acting freely on the three-sphere S^3, then G is conjugate to a subgroup of the orthogonal group $O(4) \subset \mathrm{Diff}(S^3)$. If $G \subset \mathrm{Diff}(S^3)$ is a finite group which acts nonfreely on S^3, then it is known that G is conjugate to a subgroup of the orthogonal group $O(4) \subset \mathrm{Diff}(S^3)$. A positive solution of the following variant of a conjecture of Pitts-Rubenstein would solve the just described "Spherical Space Form Problem". The Pitts-Rubenstein conjecture motivates the study of properly embedded minimal surfaces in $\mathbb{R}^3$ with finite genus for the following reason: understanding the geometry of examples in $\mathcal{P}$ gives, by blow up arguments, a good understanding of the local geometry of compact embedded minimal surfaces in Riemannian three-manifolds near points of large Gaussian curvature.

Conjecture 3.11. [Pitts-Rubenstein Conjecture Time Frame = 20 years] For every compact three-manifold M^3 with finite fundamental group, there exists a Riemannian metric $<,>$ which is generic in the sense that on the universal cover $(\tilde{M}^3, \widetilde{<,>})$ every compact embedded minimal surface in $\tilde{M}^3$ has no Jacobi vector fields and such that for any fixed genus g there is a bound on the stability index of embedded compact minimal surfaces of genus g in $(\tilde{M}^3, \widetilde{<,>})$.

References

[1] A. I. Bobenko. Helicoids with handles and Baker-Akhiezer spinors. *Math. Z.*, (1):9–29, 1998.

[2] A. I. Bobenko and M. Schmies. Computer graphics experiments for helicoids with handles. Personal communication.

[3] M. Callahan, D. Hoffman, and W. H. Meeks III. The structure of singly-periodic minimal surfaces. *Invent. Math.*, 99:455–481, 1990.

[4] T. H. Colding and W. P. Minicozzi II. The space of embedded minimal surfaces of fixed genus in a 3-manifold III; locally simply-connected. Preprint.

[5] T. H. Colding and W. P. Minicozzi II. The space of embedded minimal surfaces of fixed genus in a 3-manifold IV; planar domains. Preprint.

[6] P. Collin. Topologie et courbure des surfaces minimales de $\mathbb{R}^3$. *Annals of Math. 2nd Series*, 145–1:1–31, 1997.

[7] P. Collin, L. Hauswirth, and H. Rosenberg. The geometry of finite topology surfaces properly embedded in hyperbolic space with constant mean curvature one. *Annals of Math.*, 2001.

[8] P. Collin, R. Kusner, W. H. Meeks III, and H. Rosenberg. The geometry, conformal structure and topology of minimal surfaces with infinite topology. Preprint.

[9] T. Ekholm, B. White, and D. Wienholtz. Embeddedness of minimal surfaces with total curvature at most 4π. SFB 256, Preprint 688 Bonn (2000).

[10] C. Frohman and W. H. Meeks III. The topological classification of minimal surfaces in $\mathbb{R}^3$. Preprint.

[11] C. Frohman and W. H. Meeks III. The ordering theorem for the ends of properly embedded minimal surfaces. *Topology*, 36(3):605–617, 1997.

[12] C. Frohman and W. H. Meeks III. The topological uniqueness of complete one-ended minimal surfaces and Heegaard surfaces in $\mathbb{R}^3$. *J. of the Amer. Math. Soc.*, 10(3):495–512, 1997.

[13] R. Gulliver and H. B. Lawson. The structure of minimal hypersurfaces near a singularity. *Proc. Symp. Pure Math.*, 44:213–237, 1986.

[14] W. Haken. Some results on surfaces in 3-manifolds. In *Studies in Modern Topology*, volume 5, pages 39–98, Englewood Cliffs, NJ, 1968.

[15] D. Hoffman and H. Karcher. Complete embedded minimal surfaces of finite total curvature. In R. Osserman, editor, *Encyclopedia of Mathematics, Vol. 90, Geometry V*, pages 5–93. Springer Verlag, 1997.

[16] D. Hoffman, H. Karcher, and F. Wei. The genus one helicoid and the minimal surfaces that led to its discovery. In *Global Analysis and Modern Mathematics*. Publish or Perish Press, 1993. K. Uhlenbeck, editor, p. 119–170.

[17] D. Hoffman and W. H. Meeks III. Embedded minimal surfaces of finite topology. *Annals of Math.*, 131:1–34, 1990.

[18] D. Hoffman, M. Weber, and M. Wolf. The existence of the genus-one helicoid. Preprint.

[19] H. Lazard-Holly and W. H. Meeks III. The classification of embedded doubly–periodic minimal surfaces of genus zero. *Invent. Math.*, 143:1–27, 2001.

[20] F. J. Lopez and A. Ros. On embedded complete minimal surfaces of genus zero. *J. of Differential Geometry*, 33(1):293–300, 1991.

[21] F. Martin and M. Weber. The uniqueness of the genus-one helicoid. Work in progress.

[22] W. H. Meeks III. *Lectures on Plateau's Problem*. Instituto de Matematica Pura e Aplicada (IMPA), Rio de Janeiro, Brazil, 1978.

[23] W. H. Meeks III. The theory of triply-periodic minimal surfaces. *Indiana Univ. Math. J.*, 39(3):877–936, 1990.

[24] W. H. Meeks III, J. Pérez, and A. Ros. The geometry of minimal surfaces of finite genus I; curvature estimates and quasiperiodicity. Preprint.

[25] W. H. Meeks III, J. Pérez, and A. Ros. The geometry of minimal surfaces of finite genus II; nonexistence of one limit end examples. Preprint.

[26] W. H. Meeks III, J. Pérez, and A. Ros. Uniqueness of the Riemann minimal examples. *Invent. Math.*, 131:107–132, 1998.

[27] W. H. Meeks III and H. Rosenberg. The uniqueness of the helicoid and the asymptotic geometry of properly embedded minimal surfaces with finite topology. Preprint.

[28] W. H. Meeks III and H. Rosenberg. The global theory of doubly periodic minimal surfaces. *Invent. Math.*, 97:351–379, 1989.

[29] W. H. Meeks III and H. Rosenberg. The geometry and conformal structure of properly embedded minimal surfaces of finite topology in $\mathbb{R}^3$. *Invent. Math.*, 114:625–639, 1993.

[30] W. H. Meeks III and B. White. Minimal surfaces bounded by convex curves in parallel planes. *Comment. Math. Helvetici*, 66:263–278, 1991.

[31] W. H. Meeks III and B. White. The space of minimal annuli bounded by an extremal pair of planar curves. *Communications in Analysis and Geometry*, 1(3):415–437, 1993.

[32] W. H. Meeks III and S. T. Yau. The topological uniqueness of complete minimal surfaces of finite topological type. *Topology*, 31(2):305–316, 1992.

[33] A. Ros. Embedded minimal surfaces: forces, topology and symmetries. *Calc. Var.*, 4:469–496, 1996.

[34] R. Schoen. Uniqueness, symmetry, and embeddedness of minimal surfaces. *J. of Differential Geometry*, 18:791–809, 1983.

[35] L. Simon. Lectures on geometric measure theory. In *Proceedings of the Center for Mathematical Analysis*, volume 3, Canberra, Australia, 1983. Australian National University.

[36] M. Traizet. The genus 2 helicoid. Personal Communication.

[37] M. Traizet. An embedded minimal surface with no symmetries. *J. Differential Geometry*, 60(1):103–153, 2002.

[38] M. Weber and M. Wolf. Teichmuller theory and handle addition for minimal surfaces. *Annals of Math.* To appear.

Department of Mathematics and Statistics, University of Massachusetts, Amherst, MA 01003

E-mail address: `bill@gang.umass.edu`

Proceedings of Symposia in Pure Mathematics
Volume **71**, 2003

Introductory Lectures on Contact Geometry

John B. Etnyre

1. Introduction

Though contact topology was born over two centuries ago, in the work of Huygens, Hamilton and Jacobi on geometric optics, and been studied by many great mathematicians, such as Sophus Lie, Elie Cartan and Darboux, it has only recently moved into the foreground of mathematics. The last decade has witnessed many remarkable breakthroughs in contact topology, resulting in a beautiful theory with many potential applications. More specifically, as a coherent – though sketchy – picture of contact topology has been developed, a surprisingly subtle relationship arose between contact structures and 3- (and 4-) dimensional topology. In addition, the applications of contact topology have extended far beyond geometric optics to include non-holonomic dynamics, thermodynamics and more recently Hamiltonian dynamics [**25, 40**] and hydrodynamics [**12**].

Despite its long history and all the recent work in contact geometry, it is not overly accessible to those trying to get into the field for the first time. There are a few books giving a brief introduction to the more geometric aspects of the theory. Most notably the last chapter in [**1**], part of Chapter 3 in [**34**] and an appendix to the book [**2**]. There have not, however, been many books or survey articles (with the notable exception of [**20**]) giving an introduction to the more topological aspects of contact geometry. It is this topological approach that has lead to many of the recent breakthroughs in contact geometry and to which this paper is devoted. I planned these lectures when asked to give an introduction to contact geometry at the Georgia International Topology Conference in the summer of 2001. My idea was to give an introduction to the "classical" theory of contact topology, in which the *characteristic foliation* plays a central roll, followed by a hint at the more modern trends, where specific foliations take a back seat to *dividing curves.* This was much too ambitious for the approximately one and a half hours I had for these lectures, but I nonetheless decided to follow this outline in preparing these lecture notes. These notes begin with an introduction to contact structures in Section 2, here all the basic definitions are given and many examples are discussed. In the following section we consider contact structures near a point and near a surface. It is in

2000 *Mathematics Subject Classification.* Primary 53C15; Secondary 57M50.
Key words and phrases. tight, contact structure, Legendrian, convex surface.
Supported in part by NSF Grant # DMS-9705949.

this section that the fundamental notion of characteristic foliation on a surface first appears. In an appendix to Section 3, I briefly describe Moser's method, which is a technique for understanding families of contact structures. Section 4 is devoted to the all pervasive dichotomy in contact geometry: tight vs. overtwisted. Here we see that overtwisted contact structures are not so interesting from a topological point of view and that tight contact structures have and intimate and subtle relationship with topology. Then, in Section 5, we consider special knots in contact structures. The study of these knots sheds light on the tight vs. overtwisted dichotomy and allows us to prove a general existence theorem for contact structures. We end with a brief introduction to convex surfaces. Though this section is short we will be able to indicate the power of convex surfaces in contact geometry and point the interested reader to recent literature on the subject.

These lectures are written in an informal style with many exercises, which are usually not too difficult and copious hints are provided. The proofs of most results are left to the exercises, however for more complicated proofs the outline is given with the details left as exercises. I am assuming the reader is familiar with basic differential topology (manifolds, vector fields, Lie derivatives, forms, ..., see [**39**]) and has a passing knowledge of 3–manifold topology (as can be gleaned from a glance or two at [**37**] or [**24**]).

As these notes have a bias for topological techniques in contact geometry, many exciting and important recent developments have been left out, specifically in regards to the use of holomorphic curves in contact geometry. Here we refer the reader to [**7, 11, 23**]. For connections with Seiberg-Witten Theory see [**29, 30**]. Finally for an interesting historical overview the reader should consult [**18**].

2. Definitions and Examples

A *plane field* ξ on M is a subbundle of the tangent bundle TM such that $\xi_p = T_pM \cap \xi$ is a 2-dimensional subspace of T_pM for each $p \in M$.

EXAMPLE 2.1. Consider the 3-manifold $M = \Sigma \times S^1$ where Σ is a surface. Then for each $p = (x, \theta) \in \Sigma \times S^1$ let $\xi_p = T_x\Sigma \subset T_pM$. Clearly ξ is a plane field on M.

EXAMPLE 2.2. Let α be a 1-form on M. So at each point $p \in M$ we have a linear map

$$\alpha_p\colon T_pM \to \mathbb{R}. \tag{1}$$

Thus $\ker \alpha_p$ is either a plane or all of T_pM. If we assume the 1-form never has all of T_pM as its kernel, then $\xi = \ker \alpha$ is a plane field. Note in the previous example the 1-form $\alpha = d\theta$ defines ξ.

It turns out that, locally, you can always represent a plane field as the kernel of a 1-form.

EXERCISE 2.3. Prove this. In other words, given a plane field ξ on M and a point $p \in M$ show you can find a neighborhood U of p and a 1-form α_U defined on the neighborhood such that $\xi|_U = \ker \alpha_U$.

EXERCISE 2.4. If M and ξ are both oriented show that you can find a 1-form α defined on all of M such that $\xi = \ker \alpha$.

A plane field ξ is called a *contact structure* if for any 1-form α with $\xi = \ker \alpha$ (α can be locally or globally defined) we have

$$\alpha \wedge d\alpha \neq 0. \tag{2}$$

EXERCISE 2.5. Show that $\alpha \wedge d\alpha \neq 0$ if and only if $d\alpha|_\xi \neq 0$.

Before we look at some examples of contact structures note that our first example of a plane field is not a contact structure. Indeed the plane field is defined by the 1-form $\alpha = d\theta$ so $d\alpha = d(d\theta) = 0$.

EXAMPLE 2.6. Consider the manifold $\mathbb{R}^3$ with standard Cartesian coordinates (x, y, z) and the 1-form

$$\alpha_1 = dz + x dy. \tag{3}$$

Note that $d\alpha_1 = dx \wedge dy$ so $\alpha_1 \wedge d\alpha_1 = dz \wedge dx \wedge dy \neq 0$. Thus α_1 is a contact form and $\xi_1 = \ker \alpha_1$ is a contact structure. At a point (x, y, z) the contact plane ξ_1 is spanned by $\{\frac{\partial}{\partial x}, x\frac{\partial}{\partial z} - \frac{\partial}{\partial y}\}$. So at any point in the yz-plane (*i.e.* where $x = 0$) ξ_1 is horizontal. If we move to the point $(1, 0, 0)$ then ξ_1 is spanned by $\{\frac{\partial}{\partial x}, \frac{\partial}{\partial z} - \frac{\partial}{\partial y}\}$. So the plane is tangent to the x-axis but has been tilted clockwise by 45%. In general, if we start at $(0, 0, 0)$ we have a horizontal plane and as we move out along the x-axis the plane will twist in a left handed manner (*i.e.* clockwise). The twist will be by 90% when x "gets to" ∞. There is similar behavior on all rays perpendicular to the yz-plane. See Figure 1.

REMARK 2.7. Many authors prefer to use the form $dz - y dx$ to define the "standard" contact structure on $\mathbb{R}^3$. There is really no difference between these structures. (Rotating about the z-axis will take one of these structures to the other.)

EXAMPLE 2.8. Consider $\mathbb{R}^3$ with cylindrical coordinates (r, θ, z) and the 1-form

$$\alpha_2 = dz + r^2 d\theta. \tag{4}$$

Since $\alpha_2 \wedge d\alpha_2 = 2r dr \wedge d\theta \wedge dz \neq 0$, $\xi_2 = \ker \alpha_2$ is a contact structure. At the point (r, θ, z) the contact plane ξ_2 is spanned by $\{\frac{\partial}{\partial r}, r^2\frac{\partial}{\partial z} - \frac{\partial}{\partial \theta}\}$. So when $r = 0$ (*i.e.* in the z-axis) ξ_2 is horizontal. As you move out on any ray perpendicular to

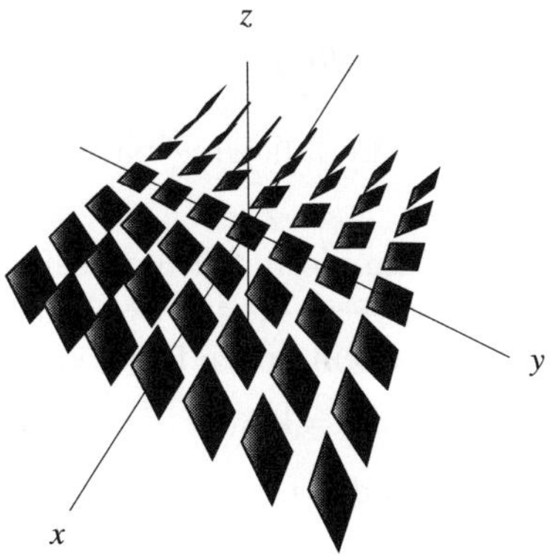

FIGURE 1. The contact structure $\ker(dz + x dy)$.

the z-axis the planes ξ_2 will twist in a clockwise manner. So this example is just like the previous one except that everything is symmetric about the z-axis.

Two contact structures ξ_0 and ξ_1 on a manifold M are called *contactomorphic* if there is a diffeomorphism $f\colon M \to M$ such that f send ξ_0 to ξ_1 :

$$f_*(\xi_0) = \xi_1.$$

EXERCISE 2.9. Show that a diffeomorphism $f\colon M \to M$ is a contactomorphism if and only if there are contact forms α_0 and α_1 for ξ_0 and ξ_1, respectively, and a non-zero function $g\colon M \to \mathbb{R}$ such that $f^*\alpha_1 = g\alpha_0$.

EXERCISE 2.10. Check that Examples 2.6 and 2.8 are contactomorphic. If you are having trouble coming up with the contactomorphism then first try to write down the contactomorphism implied in Remark 2.7

EXAMPLE 2.11. Once again consider $\mathbb{R}^3$ with cylindrical coordinates, but this time take the 1-form $\alpha_3 = \cos r dz + r \sin r d\theta$. One may compute that

$$\alpha_3 \wedge d\alpha_3 = (1 + \frac{\sin r \cos r}{r}) d\text{vol}. \tag{5}$$

Thus to see that α_3 is a contact form you only have to check that

$$1 + \frac{\sin r \cos r}{r} > 0. \tag{6}$$

Note that $\xi_3 = \ker \alpha_3$ is horizontal along the z-axis and as you move out on any ray perpendicular to the z-axis the planes will twist in a clockwise manner. This time, however, the planes will twist 90% by the time you get to $r = \pi/2$. In fact, as you move out on any ray ξ_3 will make infinitely many full twists as r goes to ∞!

This example certainly looks different from our previous two examples, but it is not exactly obvious how one would actually show it is different. In the early 1980's Bennequin [**3**] did distinguish this example from the previous ones and in the process ushered in a new era in contact geometry. We will indicate Bennequin's proof in Section 5.

So far all our examples are on $\mathbb{R}^3$. We now give an example on a closed manifold.

EXAMPLE 2.12. Consider the unit 3-sphere, S^3, in $\mathbb{R}^4$. Let

$$\alpha = (x_1 dy_1 - y_1 dx_1 + x_2 dy_2 - y_2 dx_2)|_{S^3}, \tag{7}$$

where (x_1, y_1, x_2, y_2) are standard Cartesian coordinates on $\mathbb{R}^4$ and set $\xi = \ker \alpha$.

EXERCISE 2.13. Check that $\alpha \wedge d\alpha \neq 0$ and thus ξ is a contact structure on S^3.
Hint: It might be helpful to read the following paragraph before attempting this exercise.

In anticipation of the next example it will be useful to describe ξ in another way. If we let $f(x_1, y_1, x_2, y_2) = x_1^2 + y_1^2 + x_2^2 + y_2^2$ then $S^3 = f^{-1}(1)$. Moreover at a point (x_1, y_1, x_2, y_2) in S^3 the tangent space is given by

$$T_{(x_1,y_1,x_2,y_2)} S^3 = \ker df_{(x_1,y_1,x_2,y_2)} = \ker(2x_1 dx_1 + 2y_1 dy_1 + 2x_2 dx_2 + 2y_2 dy_2). \tag{8}$$

Now we can think of $\mathbb{R}^4$ as $\mathbb{C}^2$. Under this identification we denote the complex structure (*i.e.* multiplication by i) by J. In other words, $Jx_i = y_i$, $Jy_i = -x_i$ for $i = 1, 2$. The complex structure J induces a complex structure on each tangent space: $J\frac{\partial}{\partial x_i} = \frac{\partial}{\partial y_i}$ and $J\frac{\partial}{\partial y_i} = -\frac{\partial}{\partial x_i}$ for $i = 1, 2$.

CLAIM. *The plane field ξ is the set of complex tangencies to S^3. By this we mean*

$$\xi = T_{(x_1,y_1,x_2,y_2)}S^3 \cap J(T_{(x_1,y_1,x_2,y_2)}S^3). \tag{9}$$

Indeed one may easily check that

$$J(T_{(x_1,y_1,x_2,y_2)}S^3) = \ker(df_{(x_1,y_1,x_2,y_2)} \circ J) \tag{10}$$

and

$$df_{(x_1,y_1,x_2,y_2)} \circ J = 2x_1dy_1 - 2y_1dx_1 + 2x_2dy_2 - 2y_2dx_2. \tag{11}$$

Thus we have $\alpha = (df \circ J)|_{S^3}$ and the claim is proved.

EXERCISE 2.14. Show that $(S^3 \setminus \{p\}, \xi|_{S^3\setminus\{p\}})$ is contactomorphic to $(\mathbb{R}^3, \xi_2)$. Hint: Pick the point p carefully and use stereographic coordinates.

It turns out that many contact structures can be described as the set of complex tangencies to a real hypersurface in a complex manifold.

EXAMPLE 2.15. Let X be a complex manifold with boundary and denote the induced complex structure on TX by J. We can find a function ϕ defined in a neighborhood of the boundary such that $\phi^{-1}(0) = \partial X$. Now as in the previous example we can see that the complex tangencies to $M = \partial X$ are given by $\ker(d\phi \circ J)$. Thus the complex tangencies ξ to M form a contact structure if and only if $d(d\phi \circ J)$ is a non-degenerate 2-form on ξ.

A fruitful way to construct such manifolds has been through the use of Stein surfaces. To define Stein surfaces we need some preliminary notions. Let X be a complex manifold of complex dimension 2 (real dimension 4). Again let J denote the induced complex structure on TX. From a function $\phi\colon X \to \mathbb{R}$ we can define a 2-form $\omega = d(d\phi \circ J)$ and a symmetric form $g(v, w) = \omega(v, Jw)$. If this symmetric form is positive definite (*i.e.* defines a metric on X) the function ϕ is called *strictly plurisubharmonic*. The manifold X is a *Stein surface* if X admits a proper strictly plurisubharmonic function $\phi\colon X \to \mathbb{R}$. It is easy to see that in this situation the complex tangencies to $M_c = \phi^{-1}(c)$ form a contact structure whenever c is not a critical value. We will call such a contact structure *Stein fillable*. Later we will see that this implies ξ is a special type of contact structure. See [**22**] to learn how to construct many Stein surfaces and hence many contact structures.

3. Local Structure

In this section we discuss the nature of contact structures near a point (Darboux's Theorem) and near a surface. You can find further discussion of all these local theorems in [**1, 34**].

3.1. Darboux's Theorem.

Darboux's Theorem essentially says that all contact structures look the same near a point. So contact structures do not have interesting local structure (this should be compared with Riemannian geometry, where the curvature is an obstruction to metrics being locally the same). This is an indication that any interesting phenomena in contact geometry should be of a global nature (*i.e.* be related to the global topology of the manifold supporting the contact structure).

THEOREM 3.1. *Let (M, ξ) be any contact 3-manifold and p any point in M. Then there exist neighborhoods N of p in M, and U of $(0,0,0)$ in $\mathbb{R}^3$ and a contactomorphism*

$$f\colon (N, \xi|_N) \to (U, \xi_1|_U).$$

The current modern proof of Darboux's Theorem uses "Moser's Method." We will discuss this more in the appendix to this section. The classical proofs of this theorem are more elementary in nature and we encourage the reader to try and come up with an elementary proof.

EXERCISE 3.2. Find an elementary proof of Darboux's theorem.

3.2. The Characteristic Foliations. Let Σ be an embedded oriented surface in a contact manifold (M, ξ). At each point x of Σ consider

$$l_x = \xi_x \cap T_x\Sigma.$$

For most x, the subspace l_x will be a line in $T_x\Sigma$, but at some points, which we call singular points, $l_x = T_x\Sigma$.

EXERCISE 3.3. Show that l_x cannot equal $T_x\Sigma$ for all x in some open subset of Σ.
Hint: If this is true, then you can show that the contact condition is violated. Consider two vectors fields v and w tangent to Σ defined along this open subset. Using the formula $d\alpha(v,w) = v\alpha(w) - w\alpha(v) + \alpha([v,w])$ compute $\alpha \wedge d\alpha$.

It is not hard to show (see the next exercise) that we may find a singular foliation $\mathcal{F}$ of Σ tangent to l_x at each x. By this we mean the complement of the singularities is the disjoint union of 1-manifolds, called leaves of $\mathcal{F}$, and the leaf through x is tangent to l_x. This singular foliation is called the *characteristic foliation* of Σ and is denoted Σ_ξ (some authors prefer $\xi\Sigma$).

EXERCISE 3.4. Show there is a singular foliation tangent to l_x.
Hint: Locally on the surface find a vector field tangent to l_x at the nonsingular points and 0 at the singular points. Then use basic existence results from ordinary differential equations to construct the singular foliation locally. Finally make sure your local foliations fit together to give a global foliation.

EXAMPLE 3.5. Let Σ be the unit sphere in $(\mathbb{R}^3, \xi_2)$. The only singularities in Σ_{ξ_2} are at the north and south poles. See Figure 2.

EXAMPLE 3.6. Let Σ be the disk of radius π in the $r\theta$-plane in $(\mathbb{R}^3, \xi_3)$. As shown on the left hand side of in Figure 3, the center of Σ is a singular point and

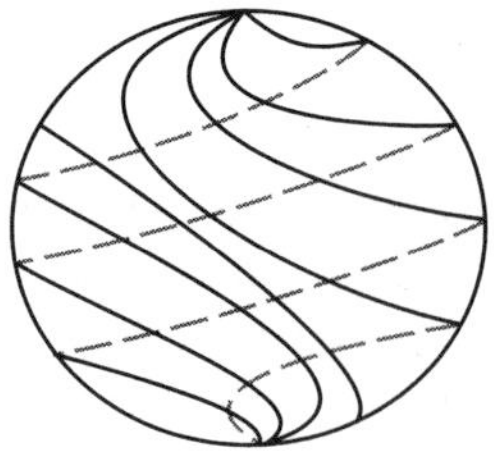

FIGURE 2. The characteristic foliation on S^2.

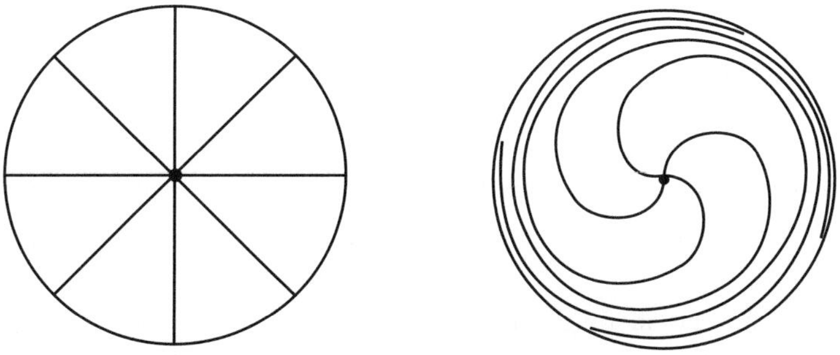

FIGURE 3. The degenerate and non-degenerate characteristic foliation on D^2.

each point on the boundary of Σ is also a singular point. Let Σ' be Σ with its interior pushed up slightly. Now the only singularity in the characteristic foliation is at the center point. The boundary of Σ' is now a closed leaf in the foliation. See Figure 3.

This last example illustrates an important point: any surface Σ may be perturbed by a C^∞-small isotopy so that its characteristic foliation has only "generic" isolated singularities. A singularity is "generic" if it looks like one in Figure 4. On the left hand side is an elliptic point and on the right hand side is a hyperbolic point.

Recall that Σ is oriented and we chose an orientation on ξ. We can orient $l_x = \xi_x \cap T_x\Sigma$ as follows: the vector $v \in l_x$ orients l_x if when we choose vectors $v_\xi \in \xi_x$ and $v_\Sigma \in T_x\Sigma$ such that (v, v_ξ) orients ξ_x and (v, v_Σ) orients $T_x\Sigma$ then (v, v_ξ, v_Σ) orients M. In this manner the leaves of the characteristic foliation inherit an orientation, so we can draw arrows on the leaves of the foliation and think of the foliation as a "flow." Moreover, at each singular point x we can assign a sign: the singularity is $+$ (respectively, $-$) if the orientation on $T_x\Sigma$ agrees (respectively, disagrees) with the one on ξ_x. With these conventions, a positive elliptic point is a source, a negative elliptic point is a sink. Note the sign if a hyperbolic point is not obvious at first glance as it is related to the ratio of the eigenvalues associated to the linearized flow at the singularity.

EXERCISE 3.7. Determine the sign of a hyperbolic point.

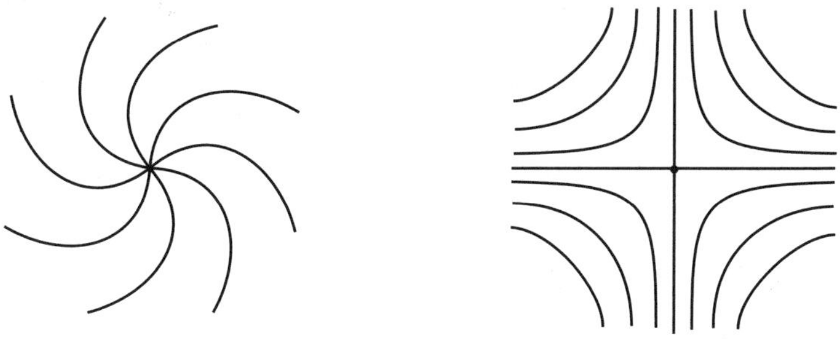

FIGURE 4. Generic singularities in the characteristic foliation.

THEOREM 3.8. *Let (M_i, ξ_i) be a contact manifold and Σ_i an embedded surface for $i = 0, 1$. If there is a diffeomorphism $f\colon \Sigma_0 \to \Sigma_1$ that preserves the characteristic foliation:*

$$f((\Sigma_0)_{\xi_0}) = (\Sigma_1)_{\xi_1},$$

then f may be extended to a contactomorphism in some neighborhood of Σ_0. Moreover, if f was already defined on a neighborhood of Σ_0 then we can isotop f so as to be a contactomorphism in some (possibly) smaller neighborhood.

So the characteristic foliation of on a surface determines (the germ of) the contact structure near the surface. Once again this theorem my be proved using "Moser's method".

APPENDIX TO SECTION 3: MOSER'S METHOD

There are several good references for Moser's method and its many corollaries [**1, 34**]. One of the most general theorems one can prove using these techniques is

THEOREM 3.9. *Let M be an oriented three manifold and $N \subset M$ a compact subset. Suppose ξ_0 and ξ_1 are contact structures on M for which $\xi_0|_N = \xi_1|_N$. Then there is a neighborhood U of N such that the identity map on a neighborhood of N is isotopic, rel. N, to a contactomorphism when restricted to U.*

EXERCISE 3.10. Show Theorems 3.1, 3.8 and 5.20 follow from this theorem. Hint: Consider Darboux's theorem. Write down a diffeomorphism from a neighborhood N' of the point p in M to a neighborhood U' of $(0, 0, 0)$ in $\mathbb{R}^3$ so that the contact plane at p is sent to the contact plane at $(0, 0, 0)$. Push the contact structure ξ forward to U. Now you have two contact structures on U that agree on $(0, 0, 0)$, so use the above theorem to finish the proof. Theorem 5.20 is similar to this and Theorem 3.8 is similar but it is not so obvious you can write down the correct initial diffeomorphism.

The proof of this theorem follows essentially from the above mentioned references, but we will outline the proof in the following exercises.

EXERCISE 3.11. Let α_i be a contact form for $\xi_i, i = 0, 1$, that determines the orientation on $\xi_i|_N$. Let $\alpha_t = (1 - t)\alpha_0 + t\alpha_1$. Show that on some neighborhood U' of N all the $\xi_t = \ker \alpha_t$'s are contact structures.

EXERCISE 3.12. We now wish to find a family of diffeomorphisms $\phi_t\colon U \to U''$ (U and U'' are possibly smaller neighborhood of N) such that $\phi_t^*\xi_t = \xi_0$. (Here $\phi_t^* = (\phi_t^{-1})_*$.) This will of course finish the proof of the theorem. We will find the ϕ_t's as the flow of a vector field. Suppose v_t is a time dependent vector field whose flow generates the ϕ_t's. Show that if $v_t \in \xi_t$ then the ϕ_t's satisfy $\phi_t^*\xi_t = \xi_0$ if and only if $\iota_{v_t} d\alpha_t|_{\xi_t} = (h_t\alpha_t - \frac{d\alpha_t}{dt})|_{\xi_t}$, where $h_t = \frac{d\alpha_t}{dt}(X_t)$ and X_t is the unique vector field satisfying $\alpha_t(X_t) = 1$ and $\iota_{X_t} d\alpha_t = 0$. (Here ι_{v_t} means contraction with v_t).

EXERCISE 3.13. Given α_t above, prove there is a v_t as described in the previous exercise.

4. Tight and Overtwisted Contact Structures

There is a fundamental dichotomy in 3-dimensional contact geometry. A contact structure ξ on M is called *overtwisted* if there is an embedded disk D whose characteristic foliation is homeomorphic to either one shown in Figure 3. Such a disk is called an *overtwisted disk*. A contact structure is called *tight* if it does not contain an overtwisted disk. Though tight vs. overtwisted is obviously a dichotomy, it is not clear that it is a useful one. Throughout the rest of these lectures we will indicate that overtwisted contact structures are somewhat "easy" to deal with, whereas tight contact structures are quite a bit more difficult to understand. Moreover, a tight contact structure is capable of detecting subtle properties of the manifold supporting it.

Later we will see directly that overtwisted contact structures are fairly simple to construct and work with. This is all reflected in the following theorem.

THEOREM 4.1 (Eliashberg [**6**]). *Given a closed compact 3-manifold M, let $\mathcal{H}$ be the set of homotopy classes of (oriented) plane fields on M and $\mathcal{C}_o$ be the set of isotopy classes of (oriented) overtwisted contact structures on M. The natural inclusion map $\mathcal{C}_o$ into $\mathcal{H}$ induces a homotopy equivalence.*

This theorem basically reduces the classification of overtwisted contact structures on a 3-manifold to the classification of homotopy classes of plane fields. This latter problem is algebraic in nature and can be understood through the Thom-Pontryagin construction, see [**35**]. In addition, see [**22**] for a discussion with contact geometry in mind.

One thing, among many, that this theorem implies is that any 3-manifold has an overtwisted contact structure on it! Moreover, any $c \in H^2(M, \mathbb{Z})$ that is the Euler class of an oriented plane field is also the Euler class of an overtwisted contact structure.

EXERCISE 4.2. Show that $c \in H^2(M, \mathbb{Z})$ is the Euler class of an oriented plane field if and only if its mod 2 reduction is 0. (You might need to review a few facts about characteristic classes to do this.)

Tight contact structures are not understood nearly as well and they do not always exist.

THEOREM 4.3 (Etnyre-Honda [**13**]). *There exists a closed compact 3-manifold that does not support any tight contact structure.*

Despite this theorem, it seems that in some sense "most" 3-manifolds do admit tight contact structures and when they do they reveal interesting things about the manifold, see Section 4.2 below. The easiest, and most common, way to construct tight contact structures is via symplectic geometry. Recall a closed two form ω on a 4–manifold X is a *symplectic form* if $\omega \wedge \omega \neq 0$. A compact symplectic 4–manifold (X, ω) is said to fill a contact 3–manifold (M, ξ) if $\partial X = M$ (as oriented manifolds!) and $\omega|_\xi$ is an area form on ξ. Note that all Stein fillable contact structures (Example 2.15) are filled by a symplectic 4–manifold (since $\omega = d(d\phi \circ J)$ is a symplectic form).

THEOREM 4.4 (Eliashberg, Gromov [**23, 7**]). *If a contact structure can be filled by a compact symplectic manifold then it is tight.*

We will not go into what is known about the classification of tight contact structures, see [**21, 26, 27**], but we do mention the method most commonly used to understand them. The key ingredient in all classification results is the following:

THEOREM 4.5 (Eliashberg [**8**]). *If $\mathcal{F}$ is a singular foliation on S^2 that is induced by some tight contact structure, then there is a unique (up to isotopy fixing the boundary) tight contact structure ξ on B^3 such that $(\partial B^3)_\xi = \mathcal{F}$.*

Now to understand tight contact structures on a manifold M one "merely" removes pieces from M on which you understand the contact structure (*e.g.* neighborhoods of surfaces on which the characteristic foliation is known) until all that is left of M is a collection of 3-balls. Then apply the previous theorem to conclude you understand the contact structure. This is, of course, quite vague but to understand the strategy better try the following exercise.

EXERCISE 4.6. By Theorem 4.4 the contact structure on S^3 described in Example 2.12 is tight. Use the above strategy to show there is only one tight contact structure on S^3. Specifically, fill in the details and understand the following argument: If you have two tight contact structures on S^3 use Darboux's Theorem to say they agree in a neighborhood of a point. Then use Theorem 4.5 to conclude that they agree in the complement of the neighborhood.

4.1. Manipulations of the Characteristic Foliations. Since any 3–manifold can be cut up along surfaces into a collection of 3–balls (in many ways, *e.g.* Heegaard decompositions, Haken decompositions, ...) it is clear, from the strategy discussed above, that to understand tight contact structures on a 3–manifold we should understand tight contact structures in the neighborhood of surfaces better. A first step in this direction is to develop techniques to manipulate characteristic foliations. One of the most important theorems along these lines is:

LEMMA 4.7 (Elimination Lemma: Giroux, Fuchs [**9**]). *Suppose γ is a leaf in a characteristic foliation Σ_ξ connecting an elliptic and hyperbolic point of the same sign. Then given any neighborhood N of γ, we may find an isotopy, supported in N, of Σ to Σ' so that $\Sigma'_\xi \cap N$ contains no singularities and, of course, Σ_ξ and Σ'_ξ agree outside of N. See Figure 5.*

Thus this theorem says we may eliminate singularities of the same sign that are connected by an arc!

EXERCISE 4.8. Visualize Figure 5 in $(\mathbb{R}^3, \xi_1)$ (recall $\xi_1 = \ker dz + x dy$) as follows. Start with a embedded rectangle containing the y-axis and tilted slightly out of the xy-plane (*e.g.* a piece of the graph of $f(x, y) = \epsilon x$). The characteristic foliation on this is nonsingular. Now create two singularities by rotating the middle part of the rectangle past the xy-plane (*e.g.* rotate a bit of the rectangle to agree with the graph of $-f(x, y)$). If you did this correctly then the characteristic

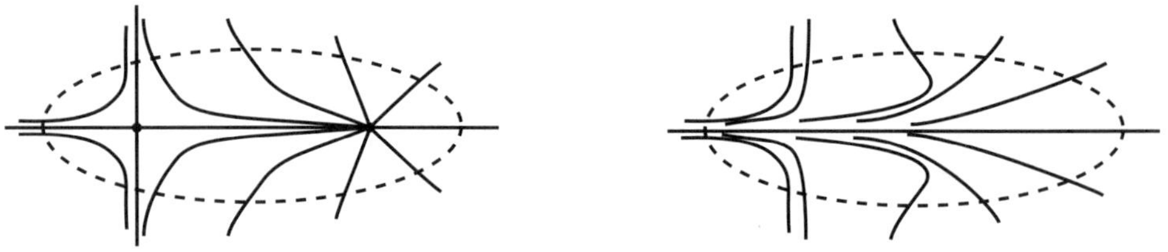

FIGURE 5. The cancellation of singularities with the same sign.

foliation should look like the left hand side of Figure 5. From the construction we know how to remove the singularities in this example. Use Theorem 3.8 to prove Lemma 4.7. This argument is explicitly worked out in [**1**].

There is an important strengthening of the Elimination Lemma. Note that in the Elimination Lemma the arc γ is part of some leaf of the new characteristic foliation on Σ'. The strengthened lemma give some control over this new leaf.

LEMMA 4.9. *Suppose γ is as in the Elimination Lemma. Let γ' be any leaf (distinct from γ) that limits to the same elliptic point as γ. Then we may assume that after the cancellation of the singularities γ and γ' are on the same leaf of the new characteristic foliation.*

Note that there is no flexibility over which two leaves limiting to a hyperbolic point will end up on the same leaf after the cancellation.

As Exercise 4.8 indicates, it is much easier to create singularities that eliminate them. In particular we have

LEMMA 4.10. *Let γ be a segment of a leaf in Σ_ξ and N be a neighborhood of γ such that $\Sigma_\xi \cap N$ contains no singularities. Then we may find an isotopy, supported in N, of Σ to Σ' so that $\Sigma'_\xi \cap N$ contains an elliptic and hyperbolic singularity of the same sign and Σ_ξ and Σ'_ξ agree outside of N.*

Up to this point the careful reader might have been concerned that we discuss "elliptic" singularities as if there were only one type of elliptic singularity. (A similar discussion applies to hyperbolic singularities.) Topologically this is true (*i.e.* up to homeomorphism) but up to diffeomorphism this is not true and Theorem 3.8 needs a diffeomorphism!

EXERCISE 4.11. Show that any two elliptic sources singularities are topologically equivalent (and similarly for sinks).
Hint: This is a small extension of the Hartman-Grobman Theorem which you can find most books on dynamical systems [**36**].

EXERCISE 4.12. Show that (generically) up to (C^1) diffeomorphism an elliptic singularity is determined by the eigenvalues of its linearization (this is not so easy, you might want to consult [**36**]).

So how is it that we can ignore this subtlety? It turns out that we may perturb a surface near an elliptic singularity so that the singularity will be diffeomorphic to a preassigned elliptic singularity.

EXERCISE 4.13. Verify this statement.
Hint: Use Darboux's theorem to reduce the problem to considering disks in $(\mathbb{R}^3, \xi_1)$ which are tangent to the xy-plane at the origin. Such disk can be represented as graphs of functions $\{(x, y, f(x, y))\}$. Now use the previous exercise and perturbations of f to prove the statement.

So as long as we are willing to perturb our surfaces (by a C^∞-small isotopy) we may ignore this problem of smooth equivalence of elliptic singularities. More precisely, we actually have

LEMMA 4.14. *Suppose there is a homeomorphism from Σ_ξ and $\Sigma'_{\xi'}$ (both characteristic foliations should be generic), then there is a C^∞-small isotopy of Σ' to*

Σ'' such that Σ_ξ and $\Sigma''_{\xi'}$ are diffeomorphic by a diffeomorphism that is isotopic to the original homeomorphism.

Thus we can just "look at" the characteristic foliation and do not need to worry about the subtleties of the singularities.

4.2. Tightness and Genus Bounds. We use the above manipulations of the characteristic foliation to show

THEOREM 4.15 (Eliashberg [**8**]). *Let (M,ξ) be a tight contact 3-manifold and Σ an embedded surface in M. If $e(\xi) \in H^2(M,\mathbb{Z})$ denotes the Euler class of ξ, then*

$$|e(\xi)([\Sigma])| \leq \begin{cases} -\chi(\Sigma) & \text{if } \Sigma \neq S^2, \\ 0 & \text{if } \Sigma = S^2, \end{cases} \tag{12}$$

where $[\Sigma]$ denotes the homology class of Σ.

Though it may not be apparent at first, this theorem begins to indicate the delicacy of tight contact structures. For example, we have

COROLLARY 4.16. *There are only finitely many elements in $H^2(M,\mathbb{Z})$ that can be realized as the Euler class of a tight contact structure.*

PROOF. There is no torsion in $H_2(M,\mathbb{Z})$.

EXERCISE 4.17. Show this.
Hint: Use Poincaré Duality and the Universal Coefficients Theorem.

Now let $g_1,\ldots,g_n$ be generators for $H_2(M,\mathbb{Z})$

EXERCISE 4.18. Show that any element in $H_2(M,\mathbb{Z})$ can be represented by a surface.

Let $S_1,\ldots,S_n$ be embedded surfaces such that the homology class of S_i is g_i, for $i=1,\ldots,n$. We can assume that none of the S_i are 2-spheres (Why?) and then for each of the S_i, Inequality (12) gives a region between two parallel hyperplanes in $H^2(M,\mathbb{Z})$ in which an Euler class for a tight contact structure can live.

EXERCISE 4.19. Show that all the hyperplanes coming from the S_i define a compact convex polytope in $H^2(M,\mathbb{Z})$.

There can clearly be only finitely many Euler classes of tight contact structures since they have to live in this polytope. □

Note that this corollary clearly shows the difference between tight and overtwisted contact structures, since any element in $H^2(M,\mathbb{Z})$ whose mod 2 reduction equals 0 is the Euler class of an overtwisted contact structure by Theorem 4.1.

Inequalities like (12) have shown up in other places too. For example, Thurston [**38**] proved that the inequality in Theorem 4.15 is true for the Euler class of a taut foliation. Due in part to this inequality, and many interesting constructions, foliation theory has found a central place in 3–manifold topology.

PROOF OF THEOREM 4.15. Note that it suffices to prove the theorem when Σ is connected. We begin by trying to understand how to calculate $e(\xi)([\Sigma])$ and $\chi(\Sigma)$ in terms of Σ_ξ. First perturb Σ so that the characteristic foliation is generic. (By generic, we mean that the singularities are isolated elliptic or hyperbolic points and no two hyperbolic points are connected by a leaf in the foliation.) Then let $e_\pm$ be the number of $\pm$ elliptic points in Σ_ξ and $h_\pm$ be the number of $\pm$ hyperbolic points in Σ_ξ. We first have the following simple observation:

$$\chi(\Sigma) = (e_+ + e_-) - (h_+ + h_-). \tag{13}$$

This should be clear since we may take a vector field v that directs the characteristic foliation (*i.e.* is tangent to Σ_ξ at non-singular points, is zero at the singularities and induces the orientation on Σ_ξ). The Poincaré-Hopf Theorem [**35**] now says that $\chi(\Sigma)$ can be computed in terms of the zeros of v.

EXERCISE 4.20. Check that the Poincaré-Hopf Theorem implies Equation (13).

We now claim that

$$e(\xi)([\Sigma]) = (e_+ - h_+) - (e_- - h_-). \tag{14}$$

To see this recall the the Euler class of a bundle is the obstruction to finding a non-zero section of the bundle. Moreover, $e(\xi)([\Sigma])$ is just the Euler class of the restriction of ξ to Σ (since everything behaves well with respect to pull back). Thus to compute the Euler class of $\xi|_\Sigma$ we just need to take a generic section of $\xi|_\Sigma$ and calculate the intersection of its graph (in $\xi|_\Sigma$) with the zero section. More specifically, take v from above as our section then the graph of v is

$$\Gamma = \{(x,p) \in \xi|_\Sigma : p = v(x)\},$$

where x is a point in Σ and $p \in \xi_x$. So Γ is a surface in the 4-manifold $\xi|_\Sigma$, this is a 4-manifold since it is the total space of a 2-dimensional vector bundle over a surface. The zero section, $\Gamma_0 = \{(x,0) \in \xi|_\Sigma\}$, is another surface. Now the Euler class of $\xi|_\Sigma$ is just the (oriented) intersection number of these two surfaces.

EXERCISE 4.21. Show that the contribution to the intersection number of each zero of v is a $+1$ for a positive elliptic or negative hyperbolic point and a -1 for a negative elliptic or positive hyperbolic point.
Hint: It might be helpful to think about $\chi(\Sigma)$ in these terms and remember $\xi_x = \pm T_x\Sigma$ at the singularities.

Now to prove Equation (12) when $\Sigma \neq S^2$ we need to see that $\pm e(\xi)([\Sigma]) \leq -\chi(\Sigma)$. Adding Equations (13) and (14) we see that

$$\chi(\Sigma) + e(\xi)([\Sigma]) = 2(e_+ - h_+). \tag{15}$$

So if we can show that, after isotoping Σ, $e_+ = 0$, then we will know $e(\xi)([\Sigma]) \leq -\chi(\Sigma)$. To this end, we first arrange that there are no closed leaves in Σ_ξ by using Lemma 4.9 to creating negative elliptic-hyperbolic pairs along any closed leaf. (Note we will of course have to isotop Σ to do this, but we still call the resulting surface Σ.) Now if there are any positive elliptic points x then let U_x be the set of all leaves in Σ_ξ that limit to x and B_x be the closure of U_x. Denote $B_x \setminus U_x$ by ∂B_x. Ultimately we will show that ∂B_x contains a positive hyperbolic point y (which is clearly connected to x by an arc) and use the Elimination Lemma to cancel x and y. To do this we need to understand the structure of B_x better.

Refer to Figure 6 as we discuss B_x. First note that U_x does not contain any singularities (other than x), so all the singularities in B_x, except x, are in ∂B_x. Now if p is an elliptic point in ∂B_x then it must be negative (Why?). Also note that if p is a hyperbolic point in ∂B_x then its unstable manifolds (the two curves in Σ_ξ that limit to p in backwards time when we think if Σ_ξ as a flow) are also in ∂B_x and they limit (in forward time) to negative elliptic points in ∂B_x.

We claim that ∂B_x contains a positive hyperbolic point. To see this, we assume that there are no positive hyperbolic points in ∂B_x and derive a contradiction. Note that U_x is embedded in Σ and is diffeomorphic to an open disk.

EXERCISE 4.22. Show that if B_x is embedded then it is diffeomorphic to a closed disk, with piecewise smooth boundary, and the boundary of the disk contains only negative elliptic and hyperbolic points connected by arcs.

Thus if B_x is embedded then we may use the (strengthened) Elimination Lemma to cancel all the singularities in ∂B_x resulting in an overtwisted disk. Thus B_x cannot be embedded.

EXERCISE 4.23. If there are hyperbolic points in Σ_ξ, then convince yourself that we can think of B_x as the image of an immersed polygon $f\colon P \to M$ such that f is an embedding on the interior of P. Moreover, it can be arranged that each edge maps to the union of a hyperbolic point and its unstable manifolds and each vertex maps to an elliptic point. See Figure 6. Though $B_x \subset \Sigma$ we will sometimes talk as if $B_x = P$, this will simplify notation and (hopefully) clarify what is going on. Just remember that B_x can refer to the image of an immersed polygon or the polygon itself depending on context. If there are no hyperbolic singularities in Σ_ξ then convince yourself that $\Sigma = S^2$, x is the unique positive elliptic point in Σ_ξ and ∂B_x is the unique negative elliptic point in Σ_ξ.

If B_x is not embedded then f identifies vertices, or vertices and edges, of P. Suppose f identifies only vertices. In this case one may refine Lemma 4.10 to create a negative elliptic–hyperbolic pair near each non-embedded vertex, as shown in Figure 7, so as to make B_x embedded for this new characteristic foliation. Thus we are back in the embedded case and can construct an overtwisted disk.

We are left to consider the case when f identifies edges of P. We consider the simplest case first. Suppose the image of f is as shown on the left hand side of Figure 8. If we cancel the hyperbolic point with the upper elliptic point then the new B_x will be related to the old B_x as shown in Figure 8. Thus the new B_x has only vertices identified, but we know from here we can get to an embedded B_x and thus an overtwisted disk!

Note the right hand side of Figure 8 is not correct if the top and bottom vertices on the left hand side are identified. In this case a periodic obit will be formed. If this

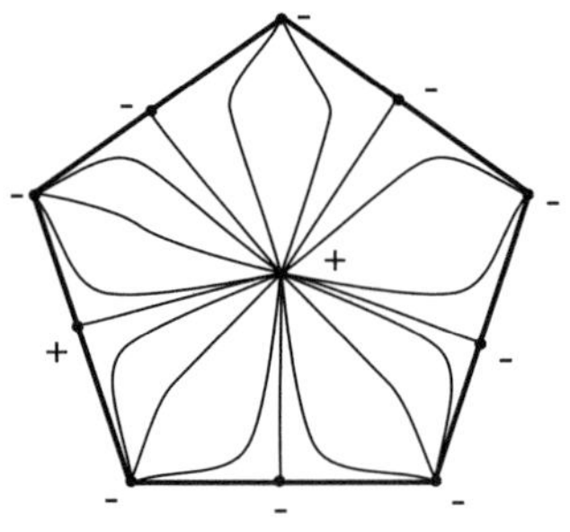

FIGURE 6. A typical B_x.

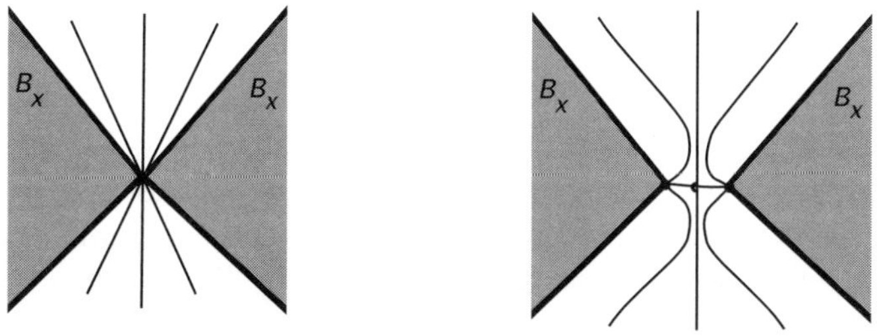

FIGURE 7. Making vertices disjoint.

happens then we loose the structure of B_x. To prevent this from happening always cancel edges with distinct vertices first (making vertices disjoint whenever possible). Since ∂B_x is connected we will eventually get to the situation where there is only one vertex in ∂B_x. If there are no edges left then $\Sigma = S^2$ as we discussed above. If there is only one edge then B_x is embedded and after canceling the boundary singularities we have an overtwisted disk. If ∂B_x has two or more edges that are not identified then we may use the move depicted in Figure 7 to create two distinct vertices on $\partial B_x \subset \Sigma$ and cancel more edges. So if we have simplified B_x as far as possible and have not found an overtwisted disk (or $\Sigma = S^2$) then $\partial B_x \subset \Sigma$ has only one vertex and all but possibly one edge is identified with some other edge. Thus the image of B_x in Σ is a closed surface or a subsurface with one boundary component. We show how to construct an overtwisted disk when $B_x = \Sigma = T^2$. In this case $\partial B_x \subset \Sigma$ has two edges E and F. Let N be a neighborhood of E in Σ and N' (respectively N'') a copy of N pushed slightly up off of Σ (respectively slightly down off of Σ). We can now cut Σ along E and push one side up to agree with N' and the other edge down to agree with N''. Note since the foliation on N is generic it is also stable so $N'_\xi = N''_\xi = N_\xi$. If we consider B_x sitting the new surface then we now have two copies of E and two vertices.

EXERCISE 4.24. Find an overtwisted disk associated to B_x on this new surface.

Even though this new B_x is not on Σ this is not a problem since we are arguing by contradiction — using our assumption that ∂B_x has no positive hyperbolic singularities to construct an overtwisted disk in M. It is irrelevant that the overtwisted disk is not actually on Σ.

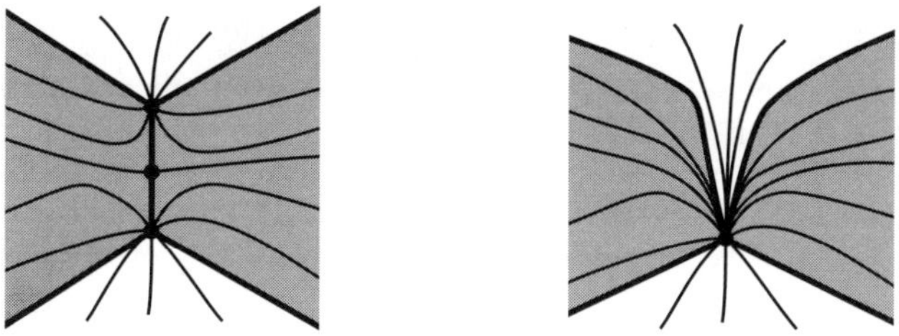

FIGURE 8. A possible B_x when edges are identified (right) and the resulting B_x when two singularities are canceled (left).

EXERCISE 4.25. Expanding the above argument find an overtwisted disk when Σ is a surface of genus greater than one and when the image of $B_x \subset \Sigma$ has a boundary component.

So we can find an overtwisted disk unless there is some positive hyperbolic singularity on ∂B_x (or $\Sigma = S^2$ and $e(\xi)([\Sigma]) = 0$). Thus we can cancel x against a hyperbolic point. Continuing in this way we eventually show that $e(\xi)([\Sigma]) \leq -\chi(\Sigma)$. (Note you should be careful since when canceling x a new closed leaf may be born. If this happens do not cancel them until all the positive elliptic points have been paired with positive hyperbolic points. Note this will not affect the above arguments.)

EXERCISE 4.26. Finish the proof by showing that $-e(\xi)([\Sigma]) \leq -\chi(\Sigma)$. Note you can do this by showing that Σ may be perturbed so that $e_- = 0$. (Why is this sufficient?)

□

EXERCISE 4.27. Using the ideas in the proof of Theorem 4.15 show: If there is an embedded disk D in (M, ξ) such that D_ξ contains a closed leaf, then ξ is overtwisted. The original definition of *tight* was the absence of embedded disks D whose characteristic foliation contains closed leaves. So it was not clear that a contact structure must be tight or overtwisted. But this exercise shows that the original definition of tight is equivalent to not being overtwisted.

5. Legendrian and Transverse Knots

Just as studying surfaces in a contact 3-manifolds can illuminate the contact structure so can studying curves. Two particularly interesting types of curves to study are Legendrian curves and transverse curves. If (M, ξ) is a contact manifold then a curve $\gamma\colon S^1 \to M$ is called *Legendrian* (respectively *transverse*) if $\gamma(S^1)$ is always tangent (respectively transverse) to ξ, that is for every $x \in S^1$, $d\gamma(T_xS^1)$ is contained in (respectively, is transverse to) $\xi_{\gamma(x)}$. As is the custom in knot theory, we will frequently confuse γ with its image. When we try to classify Legendrian or transverse knots we will always be trying to classify them up to isotopies through knots of the same type.

Let's begin by considering Legendrian and transverse knots in the standard contact structure on $\mathbb{R}^3$. Recall, the contact structure is $\xi = \ker(dz + x\,dy)$. Now suppose γ is a Legendrian curve in $(\mathbb{R}^3, \xi)$. To picture γ we will project it to the yz-plane. This is called the *front projection* of γ. The projection of γ will "look like" Figure 9. What we mean by "look like" is two things:

(1) at all the crossings the strand of γ with the smaller slope lies in front of the strand with the larger slope, and
(2) there are no vertical tangencies; instead there are cusps.

EXERCISE 5.1. If γ is Legendrian then show that

$$x = -\frac{dz}{dy}. \tag{16}$$

That is the x-coordinate of γ is determined by the slope of its front projection. Thus the Legendrian knot γ can be recovered from its front projection.

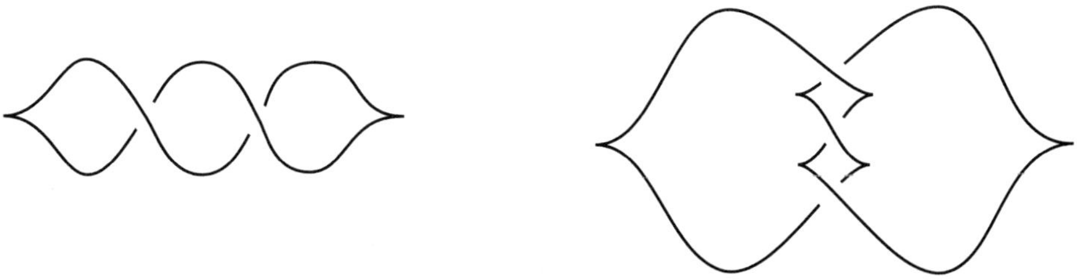

FIGURE 9. Examples of Legendrian knots.

EXERCISE 5.2. Convince your self that the restrictions above on the front projection are the only restrictions on a Legendrian knot and that any projection satisfying these restrictions is the projection of a Legendrian knot.

From this exercise we see that the study of Legendrian knots in $\mathbb{R}^3$ reduces to the study of their front projections. In particular if two Legendrian knots are isotopic (through Legendrian knots) then you can get from the front projection of one to the front projection of the other by a sequence of Legendrian Reidemister moves shown in Figure 10 (and the moves obtained from these by rotating the pictures 180° around the y or z-axes).

LEMMA 5.3. *Any knot in* $\mathbb{R}^3$ *can be* C^0 *approximated by a Legendrian knot.*

EXERCISE 5.4. Prove this lemma.
Hint: Consider the projection of the knot into the yz-plane. You will have to consider how the projection fails to satisfy the two condition discussed above for a

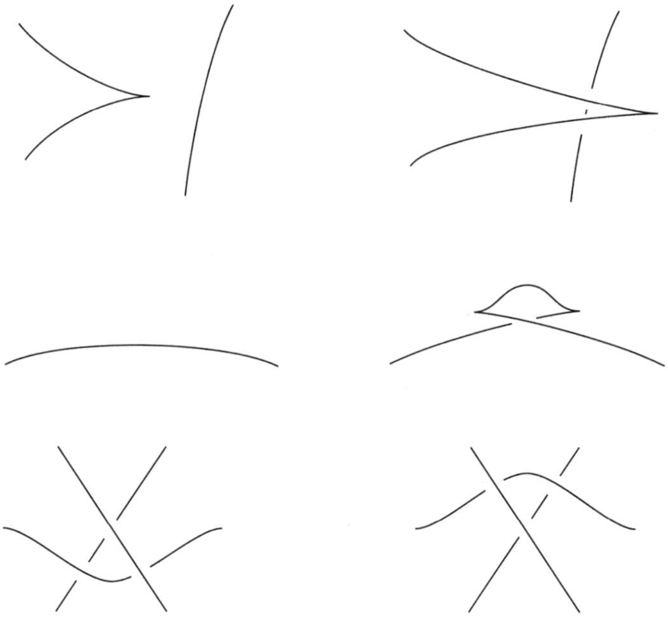

FIGURE 10. Legendrian Reidemister moves.

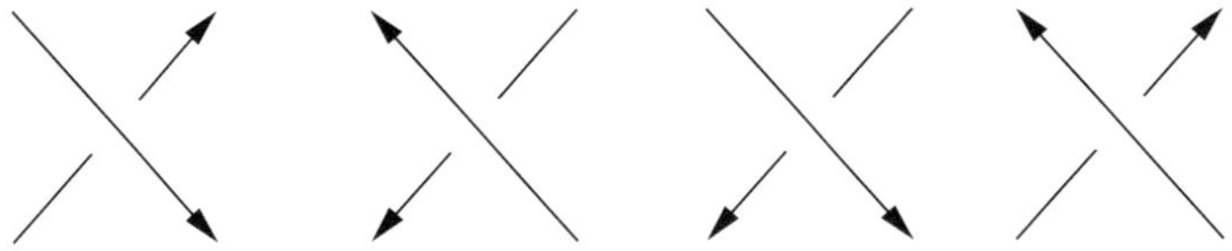

FIGURE 11. Right handed crossings (left) contribute +1 to the writhe while left handed crossings (right) contribute −1.

front projection and how it fails to satisfy Equation (16). To fix this you can use "zig-zags."

Even though we have been only discussing knots in $(\mathbb{R}^3, \xi)$ we can actually use this and Darboux's Theorem to show

LEMMA 5.5. *Any curve in a contact manifold may be* C^0 *approximated by a Legendrian curve.*

EXERCISE 5.6. Prove this lemma.

EXERCISE 5.7. Try to carry out a discussion, similar to the one above, for transverse knots. In other words understand their "front projections" and prove the transverse versions of Lemmas 5.3 and 5.5. If you get stuck take a look at [**14**].

5.1. The Classical Invariants of Legendrian and Transverse Knots. The first step in trying to classify something is to find invariants that can help you distinguish the objects under consideration (*e.g.* the Euler characteristic for surfaces). For Legendrian knots there are two easily defined invariants. Let γ be a Legendrian knot and Σ a surface bounded by it. (If no such surface exists the situation is a bit more complicated, see [**13**].) Take a vector field v along γ that is transverse to ξ, then form γ' by pushing γ in the direction of v. Now the *Thurston-Bennequin invariant* of γ, $\text{tb}(\gamma)$, is the signed intersection number of γ' with Σ (*i.e.* the linking number of γ and γ'). If we orient γ then we can take a vector field u along γ that induces the chosen orientation on γ. Note that u is in ξ (since γ is Legendrian). The *rotation number* of γ, $r(\gamma)$, is Euler number $\xi|_\Sigma$ relative to u. By this we mean $r(\gamma)$ is the obstruction to extending u to a non-zero vector field in $\xi|_\Sigma$.

EXERCISE 5.8. Choose any trivialization of ξ over Σ. (Why can you always find such a trivialization?) Using this trivialization u rotates some number of times as we traverse γ positively (*i.e* in the direction of the orientation). Prove that this number of rotations is the rotation number of γ.

It is easy to compute these invariants in $\mathbb{R}^3$ using the front projection. Let γ be an oriented Legendrian knot in $\mathbb{R}^3$. Recall the *writhe* of a knot diagram is the sum (over the crossings in a diagram) of a ± 1 at each crossing, where the sign of the crossing is determined by its handedness. See Figure 11. Denote by $w(\gamma)$ the writhe of the front projection of γ. Let $c(\gamma), c_u(\gamma)$ and $c_d(\gamma)$ be the number of cusps, upward oriented cusps and downward oriented cusp (respectively) in the front projection.

LEMMA 5.9. *With the notation above*

$$\text{tb}(\gamma) = w(\gamma) - \frac{1}{2}c(\gamma) \tag{17}$$

and

$$r(\gamma) = \frac{1}{2}(c_d(\gamma) - c_u(\gamma)). \tag{18}$$

EXERCISE 5.10. Prove this Lemma.
Hint: A global trivialization of ξ is given by $\{\frac{\partial}{\partial x}, x\frac{\partial}{\partial z} - \frac{\partial}{\partial y}\}$. Use this trivialization to compute the rotation number. Moreover, the writhe of a diagram is the difference between the "blackboard" framing of a knot (*i.e.* the obvious one coming from the diagram) and the framing coming from a Seifert surface. Now use the vector $\frac{\partial}{\partial z}$ to compute the Thurston-Bennequin invariant.

EXERCISE 5.11. Show tb and r are invariants of Legendrian knots in $\mathbb{R}^3$ by using the Legendrian Reidemister moves.

Now suppose γ is a transverse knot with Seifert surface Σ. We choose a nonzero vector field v in $\xi|_\Sigma$ and form a copy of γ' of γ by pushing γ in the direction of v. The *self-linking number* of γ is the signed intersection number of γ' with Σ (once again it is just the linking number of γ and γ'). The self-linking number of a knot in $\mathbb{R}^3$ may also be computed via its projection onto the yz-plane. Specifically, one can show

$$l(\gamma) = w(\gamma). \tag{19}$$

5.2. The Bennequin Inequality. We may now state the fundamental Bennequin Inequality.

THEOREM 5.12. *If γ is a transverse knot in a tight contact structure then*

$$l(\gamma) \leq -\chi(\Sigma), \tag{20}$$

where Σ is any Seifert surface for γ.

EXERCISE 5.13. Prove this theorem.
Hint: Since γ is oriented it induces an orientation on its Seifert surface Σ. With these orientations the characteristic foliation is oriented so that, thought of as a flow, it flows transversely out of $\partial\Sigma = \gamma$. Thus $\partial\Sigma$ "acts like a negative elliptic point." With this observation the proof of this theorem is very similar to the proof of Theorem 4.15. It will be helpful to interpret $l(\gamma)$ as a relative Euler class and then show (using notation from the proof of Theorem 4.15) that $l(\gamma) = -[(e_+ - h_+) - (e_- - h_-)]$.

This inequality provides a lower bound on the genus of a Seifert surface for γ. In general, it is difficult to determine the smallest possible genus of a Seifert surface for a given knot. (You should convince yourself that you can always find Seifert surfaces of arbitrarily large genus for a given knot.) The Bennequin inequality can sometimes help in determining this smallest genus.

EXERCISE 5.14. Look at the table of knots in [**37**] and see which of them have transverse realizations realizing the upper bound in Inequality (20).
Hint: It might be easier to consider Legendrian knots (see below). You will need to be able to construct Seifert surfaces for the knots. The most common algorithm for this can be found in [**37**].

It is interesting to note that Bennequin proved Inequality (20) for any transverse knot in the standard contact structure on $\mathbb{R}^3$. But he did it *without knowing that the contact structure was tight!* This, in fact, was the first hint that there was more than one type of contact structure, but it still took several years for the notions of "tight" and "overtwisted" to be developed. So, in modern language, Bennequin proved the standard contact structure on $\mathbb{R}^3$ was tight by proving Inequality (20). Indeed, being able to prove this inequality for a contact structure is equivalent to showing it is tight.

EXERCISE 5.15. Prove a contact structure is tight if and only if Inequality (20) is true. It might be better to read below about the Legendrian version of Bennequin's Inequality and then think in terms of Legendrian knots.

He did this by examining relations between transverse knots and braid theory. See [**4**] for more on this relationship. Nowadays, our understanding of the inequality is somewhat different. In stead of using it to prove a contact structure is tight, we usually prove the contact structure is tight using other techniques and then use the inequality to study transverse knots in the contact structure. The current preferred method to show contact structures are tight is to use Theorem 4.4

We now consider the Legendrian version of Bennequin's Inequality.

THEOREM 5.16. *Let γ be a Legendrian knot in a tight contact structure. Then*

$$\text{tb}(\gamma) + |r(\gamma)| \leq -\chi(\Sigma), \tag{21}$$

where Σ is a Seifert surface for γ.

To prove this we just need to notice a simple relation between Legendrian and transverse knots. Let γ be a Legendrian knot and $A = [-\epsilon, \epsilon] \times S^1$ an embedded annulus with $\gamma = \{0\} \times S^1$ and twisting so as never to be tangent to ξ along γ. Note that γ is a closed leaf in A_ξ and for a generic choice of A there will be no singularities and no other closed orbits. In fact, we can assume (Why?) that away from $\{0\} \times S^1$ the curves $\gamma_\pm = \{\pm x\} \times S^1$, where $0 < x < \epsilon$, are transverse to ξ. Furthermore, one can show that

$$l(\gamma_\pm) = \text{tb}(\gamma) \mp r(\gamma). \tag{22}$$

EXERCISE 5.17. Understand the relation between γ and $\gamma_\pm$ in the front projection. Use this to prove Equation (22) for knots in the standard contact structure on $\mathbb{R}^3$. You might want to try to prove the equation in general, or see [**14**].

EXERCISE 5.18. Show how Theorem 5.16 follows from Equations (20) and (22).

We have seen that the study of Legendrian and transverse knots can illuminate the nature of contact structures (such as the tight vs. overtwisted dichotomy), but their study is also quite interesting in its own right. Legendrian and transverse unknots [**10**], torus knots and figure eight knots [**14**] have been classified and are essentially determined by their knot types and the invariants described above. However, there are Legendrian knots that are topologically isotopic, have the same Thurston-Bennequin invariants and rotation numbers but are not Legendrian isotopic. Such examples were first found in a tight contact structure on $S^2 \times S^1 \# S^2 \times S^1$ (see [**16**]). Here a geometric argument very specific to the situation was used to distinguish the knots. Shortly after these examples were found an exciting new invariant was discovered [**5, 11, 15**] that allowed one to find many such "non-simple" Legendrian knots in the standard tight contact structure on S^3. The situation for transverse knots is not so well understood: it is unknown whether transverse knots are determined by their topological knot type and their self-linking number.

5.3. Transverse Knots and the Existence of Contact structures. Dehn surgery is an important tool in understanding topological 3-manifolds. We wish to show that Dehn surgery can be used in the world of contact 3-manifolds too. First let us recall the relevant definition. If γ is a knot in a 3-manifold M then it has a neighborhood, N, diffeomorphic to $S^1 \times D^2$. Fix an embedded curve α on $\partial N \subset \partial \overline{M \setminus N}$. Now choose any diffeomorphism f of $T^2 = \partial(S^1 \times D^2)$ that sends the meridian, $\{p\} \times \partial D^2$, to α and define the α *Dehn surgery along* γ to be the manifold obtained from $\overline{M \setminus N}$ by gluing in a solid torus via f :

$$M(\gamma, \alpha) = (\overline{M \setminus N}) \cup_f (S^1 \times D^2). \tag{23}$$

EXERCISE 5.19. Show that any choice of f sending the meridian to α will produce the same 3-manifold (up to diffeomorphism).

We would now like to consider doing Dehn surgery on a transverse knot. To this end we observe that another application of Moser's method yields

LEMMA 5.20. *Let γ_i be a transverse knot in (M_i, ξ_i) for $i = 1, 2$. Then any smooth map from γ_1 to γ_2 may be extended to a contactomorphism from a neighborhood of γ_1 to a neighborhood of γ_2.*

Let's construct a standard model for the neighborhood of a transverse curve. For this consider the contact structure $\xi = \ker(\cos r d\phi + r \sin r d\theta)$ on $S^1 \times \mathbb{R}^2$, where ϕ is the coordinate on S^1 and (r, θ) are polar coordinates on $\mathbb{R}^2$. (Note this contact structure is just the one in Example 2.11 with the z-axis wrapped around the S^1. Said another way, $\mathbb{R}^3$ is the universal cover of $S^1 \times \mathbb{R}^2$ and the contact structure in Example 2.11 is just the pull back of this one under the covering map.) Note that $T_a = \{(\phi, r, \theta) | r = a\}$ is a torus, and $(T_a)_\xi$ is a non-singular foliation by lines of slope $-a \tan a$. Lemma 5.20 implies that any transverse knot γ has a neighborhood N contactomorphic to $S_a = \{(\phi, r, \theta) | r \leq a\}$ for some a.

Now if γ is some transverse knot in S^3, with the standard contact structure, then it has a neighborhood N contactomorphic to S_a for some a. If we remove N from S^3 and then glue in a solid torus $S^1 \times D^2$ via a map f, the resulting manifold M has a contact structure defined on all but the $S^1 \times D^2$ part. Note that on $\partial(S^1 \times D^2) \subset M$ we have a characteristic foliation. This foliation is a linear foliation with some slope s (when measured with respect to the $S^1 \times D^2$ product structure).

EXERCISE 5.21. Determine what s is in terms of a and the slope of the curve α. Is s uniquely determined? If not what are the possible s's.

Now we can find a model contact structure on $S^1 \times D^2$ whose characteristic foliation is also linear with slope s.

EXERCISE 5.22. Check that this model contact structure on $S^1 \times D^2$ and the contact structure on $\overline{S^3 \setminus N}$ induced from S^3 define a contact structure on M.

We can clearly perform this construction on a link in S^3. Thus since any 3-manifold can be obtained from S^3 by Dehn surgery on a link we have proved:

THEOREM 5.23 (Martinet [**33**]). *All closed compact 3-manifolds support a contact structure.*

Note that there are many choices for a so that S_a has the appropriate slope to be used in the above construction. However, it is clear that if we choose any a except the smallest possible a then we automatically get an overtwisted structure. (Find the overtwisted disk!) Even choosing the smallest possible a we will frequently get an overtwisted structure on the surgered manifold. If you are sufficiently careful with this construction you can show

THEOREM 5.24 (Lutz [**31**]). *In every homotopy class of oriented plane fields on a closed compact 3-manifold there is an overtwisted contact structure.*

EXERCISE 5.25. Try to prove this theorem on S^3.
Hint: There are $\mathbb{Z}$ homotopy classes of oriented plane fields. (To see this trivialize the tangent bundle and choose a metric. Now given a plane field you can use the unit vector orthogonal to the plane field to get a map to S^2, well defined up to homotopy. Thus homotopy classes of plane fields are in one-to-one correspondence with homotopy classes of maps $S^3 \to S^2$. That is $\pi_3(S^2) = \mathbb{Z}$.) The standard contact structure on S^3 is orthogonal to the Hopf fibration of S^3. So if γ is a fiber in the Hopf fibration then it has a neighborhood contactomorphic to S_a for some a. Now replace S_a with S_{a+b}, where b is chosen so that S_a and S_{a+b} have characteristic foliations with the same slope. Note that when we do this we are still on S^3 but the contact structure is (possibly) different.

Unfortunately it is much harder to construct tight contact structures.

EXERCISE 5.26. Try to understand how the constructions above relate to "Legendrian surgery" in which tight contact structures are produced. See [**22**] for a discussion of Legendrian surgery and [**17**] for part of its relation to the surgery described above.

6. Introduction to Convex Surfaces

In the previous sections we have been discussing a classical approach to contact geometry. By classical, I mean concentrating on specific characteristic foliations. In [**19**], Giroux initiated the use of convex surfaces in contact geometry. Using the theory of convex surfaces one can ignore specific characteristic foliations when studying surfaces in a contact structure and concentrate on a few curves on the surface (the so called "dividing curves"). In this section we will indicate how to use convex surfaces in the study of contact geometry. For applications of this to the classification of contact structures see [**26**] and [**21**], to the classification of Legendrian knots see [**14**] and to the nature of tightness see [**13**].

Given a contact manifold (M, ξ) a vector field is called *contact* if its flow preserves the contact structure. A surface Σ is called *convex* if there is a contact vector field transverse to it.

EXERCISE 6.1. Show a surface Σ is convex if and only if there is a neighborhood $N = \Sigma \times I$ such that $\xi|_N$ is invariant in the I direction. (Note this exercise implies that convex is a not such a great term for such a surface but we are stuck with it.)

The first question one should ask is: Are there any convex surfaces? In [**19**] it was shown that any closed surface is C^∞-close to a convex surface. Moreover, in [**28**] it was shown that this is also true for a surface with boundary so long as the surface has Legendrian boundary and the twisting of the contact planes relative to the surface is not positive.

Now let Γ be the set of points on a convex surface Σ where the contact vector field is tangent to ξ. In [**19**] it was shown that (generically) Γ is a multi-curve, that is collection of curves, on Σ. The multi-curve Γ satisfies:

(1) $\Sigma \setminus \Gamma = \Sigma_+ \coprod \Sigma_-$,
(2) Σ_ξ is transverse to Γ, and
(3) there is a vector field w and volume form ω on Σ such that
 (a) w is directs Σ_ξ (*i.e.* w is tangent to Σ_ξ where it is nonsingular and is zero where it is singular),
 (b) the flow of w expands ω on Σ_+ and contracts ω on Σ_-,
 (c) w points transversely out of Σ_+.

EXERCISE 6.2. Verify these properties for Γ.
Hint: Use Exercise 6.1 to show that in a neighborhood of Σ, ξ is the kernel of $\beta + u dt$ where β is a 1–form on Σ and u is a function on Σ. Now try to understand Γ and Σ_ξ in terms of this 1–form.

If $\mathcal{F}$ is any singular foliation of Σ then a multi-curve Γ on Σ is said to *divide* $\mathcal{F}$ if they satisfy the above conditions where Σ_ξ is replaced by $\mathcal{F}$. Moreover, the curves Γ are called the *dividing curves* for $\mathcal{F}$. We now have the first major theorem about convex surfaces.

THEOREM 6.3 (Giroux [**19**]). *Suppose $\mathcal{F}$and Σ_ξ are both divided by the same multi-curve Γ. Then inside any neighborhood N of Σ there is an isotopy $\Phi_t\colon \Sigma \to N, t \in [0,1]$ of Σ such that*

(1) *$\Phi_0 =$ inclusion of Σ into N,*
(2) *$\Phi_t(\Sigma)$ is a convex surface for all t,*
(3) *Φ_t does not move Γ,*
(4) *$(\Phi_1(\Sigma))_\xi = \Phi_1(\mathcal{F})$.*

This theorem basically says that given a convex surface we can assume the characteristic foliation is anything we wish it to be as long as it is divided by the appropriate curves. Or said another way, it is really the dividing curves that carry the essential information about the contact structure in a neighborhood of a convex surface and not the specific characteristic foliation. One needs to be very careful with this heuristic statement but it is a useful way to think about convex surfaces.

EXERCISE 6.4. Show that if $\Sigma \neq S^2$ is any convex surface in (M,ξ) and it has a closed contractible dividing curve then ξ is overtwisted.
Hint: Try to write down some foliation respecting the dividing curves in which it is easy to see an overtwisted disk. Why is it important that $\Sigma \neq S^2$?

EXAMPLE 6.5. Consider $\mathbb{R}^3$ with the contact structure $\xi = \ker(dz + x dy)$. If we quotient $\mathbb{R}^3$ by $z \mapsto z+1$ and $y \mapsto y+1$ we will get $M = \mathbb{R} \times T^2$ and since the contact structure is preserved by this action ξ will induce a contact structure on M. The characteristic foliation on $\{0\} \times T^2$ is by horizontal lines (*i.e.* by the lines $z =$ constant). You can check that this is *not* a convex surface, but it is easy to perturb into a convex surface. Let $f\colon [0,1] \to \mathbb{R}$ be the function whose graph is given in Figure 12, then set $\Sigma = \{(f(z), y, z)\}$. Clearly Σ is a small perturbation of $T^2 \times \{0\}$, moreover, the characteristic foliation on Σ is as shown on the left hand side of Figure 13. It is easy to check that the dotted lines in the figure give a set of dividing curves for Σ_ξ, and thus Σ is convex. Now using Theorem 6.3 we can perturb

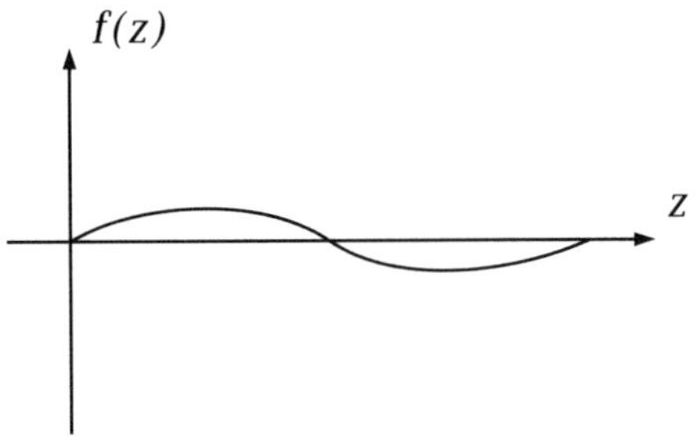

FIGURE 12. The graph of f.

Σ so as to realize any characteristic foliation that respects these dividing curves. In particular, we can arrange for the foliation to look like the one the right hand side of Figure 13. This foliation has two lines of singularities along $\{z = 0\} \cup \{z = \frac{1}{2}\}$ and all of the nonsingular leaves have slope $s \neq 0$. The nonsingular leaves are called *ruling curves* and the singular curves are called *Legendrian divides*. Note the Legendrian divides must be parallel to the dividing curves, but we may choose the ruling curves to have any slope except 0, the slope of the Legendrian divides. Any torus with a foliation like this will be said to be in *standard form*.

The power of convex surfaces is contained largely in Theorem 6.3 in conjunction with

LEMMA 6.6 ([**28, 26**]). *Suppose that Σ and Σ' are convex surfaces, with dividing curves Γ and Γ', and $\partial\Sigma' \subset \Sigma$ is Legendrian. Let $S = \Gamma \cap \partial\Sigma'$ and $S' = \Gamma' \cap \partial\Sigma'$. Then between each two adjacent points in S there is one point in S' and vice verse. See Figure 14. (Note the sets S and S' are cyclically ordered since they sit on $\partial\Sigma'$)*

In this lemma clearly Σ' is not a closed surface. All of our previous discussion goes through for surface with boundary as long as the boundary is Legendrian and the twisting of the contact planes relative to the surface in not positive. See [**28**].

We can now give a simple proof of the following result which is essentially due to Makar-Limanov [**32**], but for the form presented here see Kanda [**28**]. Though this theorem seems easy, it has vast generalizations which we indicate below.

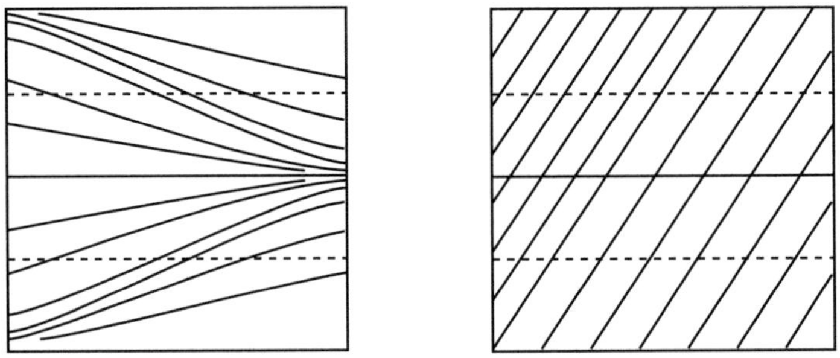

FIGURE 13. The characteristic foliation on Σ (left), with dividing curves (dotted lines). Another foliation on Σ with the same dividing set (right).

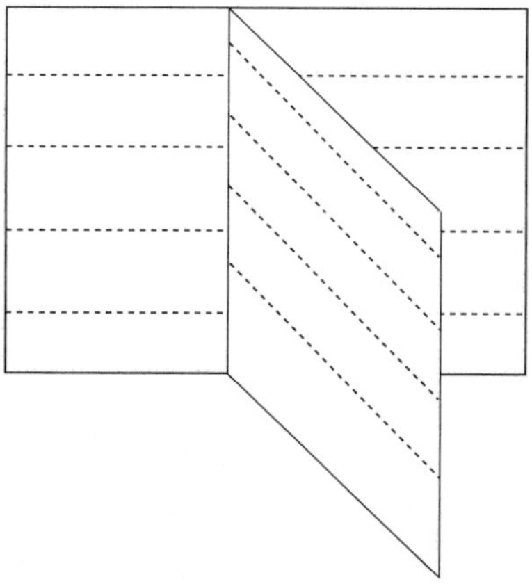

FIGURE 14. Transferring information about dividing curves from one surface to another. The top and bottom of the picture are identified.

THEOREM 6.7. *Suppose $M = D^2 \times S^1$ and $\mathcal{F}$ is a singular foliation on ∂M that is divided by two parallel curves with slope $\frac{1}{n}$ (here slope $\frac{1}{n}$ means that the curves are homotopic to $n[\partial D^2 \times \{p\}] + [\{q\} \times S^1]$ where $p \in S^1$ and $q \in \partial D^2$). Then there is a unique tight contact structure on M whose characteristic foliation on ∂M is $\mathcal{F}$.*

PROOF. Suppose we have two tight contact structures ξ_0 and ξ_1 on M inducing $\mathcal{F}$ as the characteristic foliation on ∂M. We will find a contactomorphism from ξ_0 to ξ_1 (in fact this contactomorphism will be isotopic to the identity). Let $f\colon M \to M$ be the identity map. By Theorem 3.8 we can isotop f rel. ∂M to be a contactomorphism in a neighborhood N of ∂M. Now let T be a convex torus in N isotopic to ∂M. Moreover we can assume that the characteristic foliation on T is in standard form. We know the slope of the Legendrian divides is $\frac{1}{n}$ and we choose the slope of the ruling curves to be 0. Let D be a meridianal disk whose boundary is a ruling curve. We can perturb D so that it is convex and using Lemma 6.6 we know that the dividing curves for D intersect the boundary of D in two points. Moreover, since there are no closed dividing curves on D (since the contact structure is tight, see Exercise 6.4) we know that Γ_D consists of one arc. We may isotop $f(D)$ (rel. boundary) to D' so that all of this is true for D' with respect to ξ_1. Now using Theorem 6.3 we can arrange that the characteristic foliations on D and D' agree; and further, we can isotop f (rel. N) so that f takes D to D' and preserves the characteristic foliation on D. Thus another application of Theorem 3.8 says we can isotop f so as to be a contactomorphism on $N' = N \cup U$, where U is a neighborhood of D. Note that $B = \overline{M \setminus N'}$ is a 3–ball, so Theorem 4.5 tells us that we can isotop f on B so that it is a contactomorphism on B too. Thus f is a contactomorphism on all of M and we are done with the proof. □

EXERCISE 6.8. Suppose that $\mathcal{F}$ is a convex foliation on $\partial(S^1 \times D^2)$ with $2n$ dividing curves of slope $\frac{p}{q}$. Find an upper bound on the number of tight contact structures on $S^1 \times D^2$ which induce this foliation. For $\frac{p}{q} = \frac{2}{q}$ or $\frac{3}{q}$ (and $n = 1$) classify the corresponding tight contact structures. If you are feeling bold you might want to try and prove the upper bound you found is not in general sharp and then actually find the sharp upper bound. This second part is not particularly easy; if you would like to see the answer consult [**26**].

EXERCISE 6.9. Try to generalize Exercise 6.8 to a genus g-handle body. Which configurations of dividing curves correspond to a unique contact structure? Can their ever be infinitely many tight structures with a fixed foliation?

EXERCISE 6.10. Prove the well known folk theorem of Eliashberg: There is a unique positive tight contact structure on $S^1 \times S^2$.
Hint: The argument is similar to the one in the proof of Theorem 6.7. First see that you can normalize the contact structure in a neighborhood of $\{p\} \times S^2$. The complement of this neighborhood is $I \times S^2$, where I is an interval. Now take a Legendrian arc isotopic to $I \times \{p\}$ and remove a standard neighborhood of this arc. Warning: Be careful here, you need to find a way to deal with the fact that the dividing curves on the boundary of the standard neighborhood can spin around the S^1 factor many times. The complement of the normalized regions is a 3–ball which has a unique tight contact structure (with given boundary data).

These exercises only give a hint at the power of convex surfaces. For further developments see [**13, 14, 21, 26, 27**].

References

[1] B. Aebisher, et. al., *Symplectic Geometry*, Progress in Math. **124**, Birkhäuser, Basel, Boston and Berlin, 1994.

[2] V.I. Arnold, *Mathematical Methods of Classical Mechanics*, GTM **60**, Springer-Verlag New York, Berlin, Heidelberg, 1978.

[3] D. Bennequin, *Entrelacements et équations de Pfaff*, Astérisque **107–108** (1983), 87–161.

[4] J. Birman and E. Finkelstein, *Studying surfaces via closed braids*, Jour. of Knot Theory and its Ramific., **7** (1998), 267–334.

[5] Yu. Chekanov, *Differential Algebras of Legendrian Links*, Preprint, 1997.

[6] Y. Eliashberg, *Classification of overtwisted contact structures on 3-manifolds*, Invent. Math. **98** (1989), 623–637.

[7] Y. Eliashberg, *Filling by holomorphic discs and its applications*, Geometry of Low-Dimensional Manifolds, Vol. II (Donaldson and Thomas eds.), Cambridge, 1990.

[8] Y. Eliashberg, *Contact 3-manifolds twenty years since J. Martinet's work*, Ann. Inst. Fourier **42** (1992), 165–192.

[9] Y. Eliashberg, *Legendrian and transversal knots in tight contact 3-manifolds*, Topological Methods in Modern Mathematics (1993), 171–193.

[10] Y. Eliashberg and M. Fraser, *Classification of topologically trivial Legendrian knots*, in *Geometry, topology, and dynamics* (Montreal, PQ, 1995), 17–51, CRM Proc. Lecture Notes, 15, Amer. Math. Soc., Providence, RI, 1998.

[11] Y. Eliashberg, A. Givental, and H. Hofer, *Introduction to symplectic field theory*, GAFA 2000 (Tel Aviv, 1999). Geom. Funct. Anal. 2000, Special Volume, Part II, 560–673.

[12] J. Etnyre and R. Ghrist, *Contact topology and hydrodynamics I: Beltrami fields and the Seifert conjecture*, Nonlinearity **13** (2000) 441–458.

[13] J. Etnyre and K. Honda, *On the nonexistence of tight contact structures*, Annals of Math. **153** (2001), 749–766.

[14] J. Etnyre and K. Honda, *Knots and Contact Geometry*, to appear in Symplectic Geometry.

[15] J. Etnyre, L. Ng and J. Sabloff, *Invariants of Legendrian Knots and Coherent Orientations*, preprint 2001.

[16] M. Fraser, *Example of nonisotopic Legendrian curves not distinguished by the invariants tb and r*, Internat. Math. Res. Notices **19** (1996), 923–928
[17] D. Gay, *Symplectic 2-Handles and Transverse Links*, preprint 2000.
[18] H. Geiges, *A brief history of contact geometry and topology*, Expo. Math. **19** (2001), no. 1, 25–53.
[19] E. Giroux, *Convexité en topologie de contact*, Comment. Math. Helv. **66** (1991), no. 4, 637–677.
[20] E. Giroux, *Topologie de contact en dimension 3 (autour des travaux de Yakov Eliashberg)*, Séminaire Bourbaki, Astérisque **216** (1993), Exp. No. 760, 3, 7–33.
[21] E. Giroux, *Structures de contact en dimension trois et bifurcations des feuilletages de surfaces*, Invent. Math. **141** (2000), 615–689.
[22] R. Gompf, *Handlebody construction of Stein surfaces*, Annals of Math. **148** (1998), 619–693.
[23] M. Gromov, *Pseudoholomorphic curves in symplectic manifolds*, Invent. Math. **82** (1985), 307–347.
[24] J. Hempel, *3-Manifolds*, Annals of Mathematics Studies **86** (1976), Princeton University Press.
[25] H. Hofer, *Pseudoholomorphic curves in symplectizations with applications to the Weinstein conjecture in dimension three*, Invent. Math. **114** (1993) 515–563.
[26] K. Honda, *On the classification of tight contact structures I*, Geom. Topol. **4** (2000), 309–368.
[27] K. Honda, *On the classification of tight contact structures II*, to appear in J. Diff. Geom.
[28] Y. Kanda, *The classification of tight contact structures on the 3-torus*, Comm. in Anal. and Geom. **5** (1997) 413–438.
[29] P. Kronheimer and T. Mrowka, *Monopoles and contact structures*, Invent. Math. **130** (1997), 209–255.
[30] P. Lisca, *Symplectic fillings and positive scalar curvature*, Geom. Topol. **2** (1998), 103–116.
[31] R. Lutz, *Structures de contact sur les fibrés principaux en cercles de dimension 3*, Ann. Inst. Fourier **3** (1977), 1–15.
[32] S. Makar-Limanov, *Tight contact structures on solid tori*, Trans. Amer. Math. Soc. **350** (1998), pp. 1013-104.
[33] J. Martinet, *Formes de contact sur les variétés de dimension 3*, pages 142–163, Number 209 in Springer Lecture Notes in Math. Springer-Verlag, 1971.
[34] D. McDuff and D. Salamon, *Introduction to Symplectic Topology*, Oxford University Press, 1995.
[35] J. Milnor, *Topology from the differentiable viewpoint*, The University of Virginia Press, Charlottesville 1965.
[36] C. Robinson, *Dynamical Systems*, CRC Press, Inc., Boca Raton, 1995.
[37] D. Rolfson, *Knots and Links*, Mathematics Lecture Series **7** (1976), Publish or Perish Inc.
[38] W.P. Thurston, *Norm on the homology of 3-manifolds*, Memoirs of the AMS **339** (1986), 99–130.
[39] F. Warner, *Foundations of Differentiable Manifolds and Lie Groups*, Springer-Verlag, New York, Berlin, Heidelberg, 1971.
[40] A. Weinstein, *On the hypothesis of Rabinowitz's periodic orbit theorem*, J. Diff. Equ., **33** (1979), 353–358.

UNIVERSITY OF PENNSYLVANIA, PHILADELPHIA, PA 19104
E-mail address: `etnyre@math.upenn.edu`
URL: `http://www.math.upenn.edu/~etnyre`

Proceedings of Symposia in Pure Mathematics
Volume **71**, 2003

On the coarse classification of tight contact structures

Vincent Colin, Emmanuel Giroux, and Ko Honda

ABSTRACT. We present a sketch of the proof of the following theorems: (1) Every 3-manifold has only finitely many homotopy classes of 2-plane fields which carry tight contact structures. (2) Every closed atoroidal 3-manifold carries finitely many isotopy classes of tight contact structures.

In this article we explain how to normalize tight contact structures with respect to a fixed triangulation. Using this technique, we obtain the following results:

THEOREM 0.1. *Let M be a closed, oriented 3-manifold. There are finitely many homotopy classes of 2-plane fields which carry tight contact structures.*

THEOREM 0.2. *Every closed, oriented, atoroidal 3-manifold carries a finite number of tight contact structures up to isotopy.*

P. Kronheimer and T. Mrowka [**KM**] had previously shown Theorem 0.1 for (weakly) symplectically (semi-)fillable contact structures. Our theorem is a genuine improvement of the Kronheimer-Mrowka theorem because there exist tight structures which are not fillable [**EH**].

Now, since every Reebless foliation is a limit of tight contact structures [**Co4, ET**], we obtain a new proof of a recent result of D. Gabai [**Ga**].

COROLLARY 0.3 (Gabai). *There are finitely many homotopy classes of plane fields which carry Reebless foliations.*

Next, shifting our attention to isotopy classes of contact structures, we see that Theorem 0.2 complements the following theorem [**Co2, Co3, HKM**]:

THEOREM 0.4 (Colin, Honda-Kazez-Matić). *Every closed, oriented, irreducible, toroidal 3-manifold carries infinitely many tight contact structures up to isomorphism.*

Summarizing, we have:

THEOREM 0.5. *A closed, oriented, irreducible 3-manifold carries infinitely many tight contact structures (up to isotopy or up to isomorphism) if and only if it is toroidal.*

2000 *Mathematics Subject Classification.* Primary 53D35; Secondary 53C15.
Key words and phrases. tight, contact structure, branched surface.

A more complete account of the proofs will appear in [**CGH**].

Acknowledgements. This project was started in the fall of 2000 when the authors visited the American Institute of Mathematics and Stanford University during the Workshop on Contact Geometry. We wholeheartedly thank the organizers Yasha Eliashberg and John Etnyre. Parts of this work were also realized at the Université de Nantes and the IHÉS. Special thanks go to François Laudenbach for his careful reading of the paper.

1. Contact geometry in dimension 3

In dimension 3, there are exactly two types of locally homogeneous 2-plane fields $\xi = \ker \alpha$. The plane field ξ is a *contact structure* when $\alpha \wedge d\alpha$ is nowhere vanishing, and is a *foliation* when $\alpha \wedge d\alpha \equiv 0$. Almost by definition, the classification of contact structures and foliations must reflect global properties of the ambient manifold M. Contact structures, unlike foliations, are very stable objects. For example, two C^0-close contact structures are isotopic [**Co1**]. Thus classifying contact structures up to isomorphism or isotopy appears to be a reasonable project and has been the subject of numerous studies in the past twenty years.

In what follows, we only consider oriented manifolds and oriented contact structures which are *positive*, i.e., satisfy $\alpha \wedge d\alpha > 0$. Also, denote the metric completion of each component of $A \setminus B$ by $\overline{A \setminus B}$.

1.1. Tight vs. overtwisted. The study of contact 3-manifolds reveals a dichotomy in the world of contact structures: *tight* vs. *overtwisted* [**Be, El1, El2**]. A contact structure ξ is *overtwisted* if there exists an embedded 2-disk D which is tangent to ξ at all its boundary points; otherwise ξ is *tight*. The boundary of D plays a role analogous to vanishing cycles in foliation theory, and there are many analogies between tight contact structures and Reebless (or taut) foliations [**ET**].

The classification of overtwisted contact structures up to isotopy coincides with the classification of plane fields up to homotopy by the work of Y. Eliashberg [**El1**]; contact structures therefore need tightness to become geometrically significant. A similar observation holds in foliation theory: Reeb components should be avoided.

1.2. Convex surfaces. The main tool for analyzing contact manifolds is the theory of convex surfaces, first introduced in [**Gi1**].

Let S be an oriented, properly embedded surface in a contact manifold (M, ξ). We assume that S is either closed or compact with *Legendrian* boundary (i.e., tangent to ξ at every point). Then the *characteristic foliation* ξS on S is the singular foliation obtained by integrating the singular line field $TS \cap \xi$. It is directed by the orientations of S and ξ, and is singular precisely when $\xi = TS$. The singularities are generically isolated and of two types: elliptic of index 1 or hyperbolic of index -1. They are positive (or negative) if the orientation of ξ and TS coincide (or not). The germ of a contact structure near a surface S is completely determined by the characteristic foliation ξS.

A surface S is said to be *convex* if it is transversal to a vector field which preserves ξ, i.e., a *contact vector field.* Surprisingly, this notion of convexity is generic [**Gi1**]: every closed surface admits a C^∞-small perturbation into a convex surface. Moreover, the convexity can be read off from the characteristic foliation ξS; for example, if ξS is singular Morse-Smale, then S is convex [**Gi1**].

Let γ be a closed embedded Legendrian curve on an embedded surface S. We define the *relative Thurston-Bennequin invariant* $tb(\gamma, S)$ (also called the *twisting number* relative to S) to be half the number of algebraic intersections of ξ and TS along γ.

If S is a surface with Legendrian boundary and if each component of ∂S has nonpositive relative Thurston-Bennequin invariant, then S can be made convex by an isotopy which fixes ∂S, is C^0-small in a neighborhood of ∂S, and is C^∞-small outside.

Let S be a surface transverse to a contact vector field X. Define the *dividing set* of S to be

$$\Gamma_S = \{x \in S | X(x) \in \xi(x)\}.$$

It is a smooth embedded multicurve transverse to ξS and does not depend on X up to isotopy through multicurves transverse to ξS. Γ_S inherits a natural orientation from that of ξS. We say that the characteristic foliation ξS is *adapted* to Γ_S.

The multicurve Γ_S captures all of the essential information on ξ in a neighborhood of S, according to the following Flexibility Theorem [**Gi1**]:

THEOREM 1.1 (Giroux). *Let S be a closed surface, $f_0 : S \hookrightarrow (M, \xi)$, $g : S \hookrightarrow (M', \xi')$ be two embeddings with $f_0(S)$ and $g(S)$ convex, and X a contact vector field transverse to $f_0(S)$. If the oriented multicurves $f_0^{-1}(\Gamma_{f_0(S)})$ and $g^{-1}(\Gamma_{g(S)})$ coincide, then there exists an isotopy $(f_t)_{t\in[0,1]}$ transverse to X such that $f_1^*(\xi f_1(S)) = g^*(\xi g(S))$. The same holds if $\partial S \neq \emptyset$ and the characteristic foliations coincide near ∂S; in this case we get an isotopy rel ∂S.*

If S is a surface with Legendrian boundary, an arc δ of Γ_S is said *boundary-parallel* if one component of $S \setminus \delta$ is a half disk which has no other intersections with Γ_S.

1.3. Bypasses. We now introduce the notion of a *bypass*, as in [**Ho1**] and [**Et**], which allows us to investigate the contact structure outside the neighborhood of a convex surface.

Let D be a half-disk embedded in a contact manifold and $\alpha \cup \beta$ be the decomposition of ∂D into two smooth arcs which meet only at their endpoints. The disk D is called a *bypass* if it satisfies the following:

- ∂D is Legendrian.
- D is convex, without singularities in its interior.
- Along ∂D, we have two positive elliptic singularities at $\partial\alpha = \partial\beta$, a positive hyperbolic singularity in the interior of β and a negative elliptic singularity in the interior of α.

The Thurston-Bennequin invariant of ∂D is -1, with a contribution of 0 from β and -1 from α. In other words, β is a more "efficient" Legendrian arc than α.

LEMMA 1.2. *Let S be a convex surface with Legendrian boundary. Assume there exists a boundary-parallel component δ of Γ_S which cuts off a half-disk $D_0 \subset S$. (If S is a disk, we assume in addition that Γ_S is not connected.) Let D_1 be a tubular neighborhood of D_0 inside S, with $D_1 \cap \Gamma_S = \delta$. Denote $\alpha_1 = D_1 \cap \partial S$. If X is a contact vector field transverse to S, then there exists an isotopy of S rel ∂S through surfaces transversal to X, which leads to a surface S' such that the image $D' \subset S'$ of D_1 is a bypass.*

The α_1 in the lemma plays the role of α in the definition of a bypass. If S has been nicely normalized along α_1, the isotopy can be chosen to be C^∞-small.

PROOF. One can draw a foliation on S which is adapted to Γ_S, and where D_1 is a bypass. The desired isotopy is then given by the Flexibility Theorem. □

Here is a fundamental result of Y. Eliashberg [**El2**] which classifies tight contact structures on the 3-ball B^3.

THEOREM 1.3. *Let Γ be a connected dividing set on ∂B^3, and $\mathcal{F}$ a foliation adapted to Γ. Then there exists a unique tight contact structure on $\mathcal{B}^3$, up to isotopy rel ∂B^3, whose characteristic foliation on ∂B^3 is $\mathcal{F}$.*

1.4. The generalized Lutz modification. Let $T \subset (M, \xi)$ be an embedded 2-torus which meets ξ transversely. Then there exists a closed tubular neighborhood $U = (\mathbb{R}^2/\mathbb{Z}^2) \times [0, 2\pi]$ of $T = (\mathbb{R}^2/\mathbb{Z}^2) \times \{\pi\}$ fibered by Legendrian intervals $\{pt\} \times [0, 2\pi]$. Then, with respect to the coordinate system (x, y, t) for $(\mathbb{R}^2/\mathbb{Z}^2) \times [0, 2\pi]$, ξ is given by:

$$\cos f(x, y, t)dx - \sin f(x, y, t)dy = 0,$$

where f is a circle-valued function $(\mathbb{R}^2/\mathbb{Z}^2) \times [0, 2\pi] \to \mathbb{R}/2\pi\mathbb{Z}$. For every $n \in \mathbb{N}$, choose a smooth increasing function $g_n : [0, 2\pi] \to \mathbb{R}$ with $g_n(0) = 0$ and $g_n(2\pi) = 2n\pi$. The plane field defined on U as the kernel of $\alpha_n = \cos(f+g_n)dx - \sin(f+g_n)dy$ is a contact structure. It coincides with ξ on ∂U, and thus can be extended by ξ to give a contact structure ξ_n on M. It is called a *generalized Lutz modification of ξ of index n along T*. (While the contact structure ξ_n depends on the choice of tubular neighborhood U, its isotopy class only depends on T.)

Consider the function $\chi : U \to [0, 1]$ which is 0 on ∂U and is strictly positive on the interior of U. The path of 1-forms $(1 - t)\alpha_0 + t\alpha_n + t(1 - t)\chi dt$ defines a path of plane fields rel ∂U between $\xi|_U$ and $\ker \alpha_n$. Thus ξ and ξ_n are homotopic through plane fields.

Note that if (M, ξ) is overtwisted, then many generalized Lutz modifications do not alter the isotopy class of ξ. For example, if T bounds a solid torus and the characteristic foliation ξT is a suspension, then the new contact structure ξ_n is also overtwisted, and is therefore isotopic to ξ by Eliashberg's overtwisted classification theorem. In general, it is not clear whether a generalized Lutz modification on an overtwisted contact manifold remains overtwisted.

Exercise: Let (M, ξ) be an overtwisted contact manifold and $T \subset M$ a pre-Lagrangian incompressible torus. If ξ_n is obtained from ξ by a generalized Lutz modification along T, then (M, ξ_n) has a finite cover which is overtwisted.

Question: Is (M, ξ_n) itself overtwisted?

1.5. Branched surfaces. A *branched surface* $\mathcal{B}$ is a topological space such that every $p \in \mathcal{B}$ has a neighborhood which is given by one of the three possibilities in Figure 1. A *branched surface with boundary* $\mathcal{B}$ is a topological space locally modeled on the three diagrams in Figure 1, plus "halves" of the diagrams to the left and in the middle, so that $\partial\mathcal{B}$ is a *train track*.

The branch locus L of a branched surface $\mathcal{B}$ is the set of points $p \in \mathcal{B}$ for which no neighborhood is modeled on a plane or a half-plane. It is a collection of embedded curves which intersect transversally at double points. If $\mathcal{B}$ is embedded inside a 3-manifold, the neighborhood $N(\mathcal{B})$ of $\mathcal{B}$ admits a fibration by intervals and

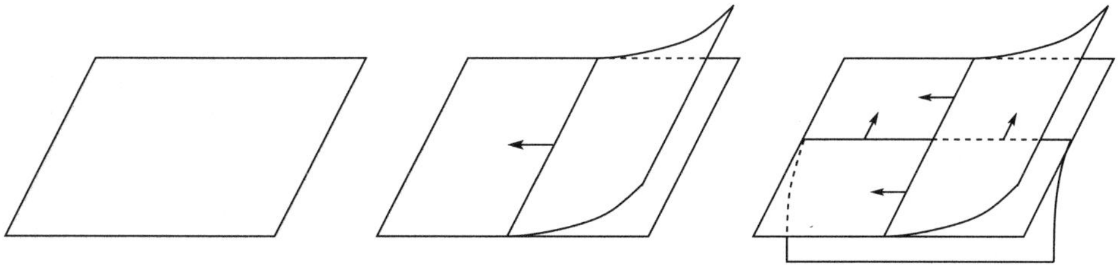

FIGURE 1. The three local models for a branched surface. The arrows indicate the branching direction.

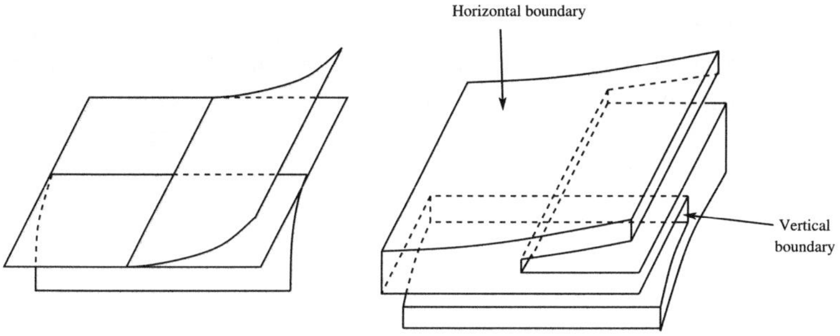

FIGURE 2. The fibered neighborhood of a branched surface.

a projection $\pi : N(\mathcal{B}) \to \mathcal{B}$ (see Figure 2). The boundary of $N(\mathcal{B})$ is divided into two pieces, the horizontal boundary $\partial_h N(\mathcal{B})$ which is transversal to the fibration and the vertical boundary $\partial_v N(\mathcal{B})$ which is tangent to the fibration.

We can now strengthen Theorem 0.1:

THEOREM 1.4. *Given a closed 3-manifold M, there exists a finite collection of contact structures $\{\xi_1, \ldots, \xi_k\}$ and, for each $1 \leq i \leq k$, a finite set of tori $\mathcal{T}_i = \{T_i^1, \ldots, T_i^{k_i}\}$ transverse to ξ_i, whose union is a branched surface, such that every tight contact structure ξ on M is obtained from one of the ξ_i by performing a generalized Lutz modification on a subset of $\mathcal{T}_i$.*

Theorem 0.1 follows immediately from Theorem 1.4, since we have already shown that a generalized Lutz modification does not alter the homotopy class of the contact structure.

In [**Co2, Co3, Gi2, HKM**], generalized Lutz modifications were performed along pre-Lagrangian incompressible tori to produce infinitely many nonisomorphic tight contact structures. More precisely, if ξ_n is the tight contact structure given by the equation $\cos(nt)dx - \sin(nt)dy = 0$ on the thickened torus $(\mathbb{R}/\mathbb{Z})^2 \times [0, 2\pi]$ with coordinates (x, y, t), then one can define the *torsion* of a contact manifold (M, ξ) as the supremum over $n \in \mathbb{N}$ for which there exists a contact embedding $\phi_n : (T^2 \times [0, 2\pi], \xi_n) \hookrightarrow (M, \xi)$. This invariant of (M, ξ), introduced in [**Gi2**], is intended to measure "how large" the contact manifold is. Notice that if the torsion is finite, then it can be increased by a Lutz modification.

In [**Co2, Co3, HKM**], we essentially prove that every closed, oriented, irreducible, and toroidal 3-manifold carries tight contact structures with arbitrarily large and finite torsion.

2. Normalizing a triangulation in a tight contact manifold

We now describe the main tool (Proposition 2.4) for proving Theorems 1.4 and 0.2, which echoes the theory of normal surfaces of Haken and Kneser [**Ha, Kn**]. This technique is also very similar to Gabai's in [**Ga**].

Let τ be a fixed triangulation of a 3-manifold M. Denote the i-skeleton of τ by τ^i.

2.1. Maximal triangulations. Given a tight contact structure ξ on M, we can isotop τ so that the 1-skeleton is a Legendrian graph and each face is convex. Such a triangulation will be called a *contact triangulation*. This is easy to accomplish because:

- every embedded graph can be C^0-approximated by a Legendrian graph;
- the Thurston-Bennequin number $tb(\gamma)$ along each edge γ, computed with the trivialization given by any adjacent face, can be made strictly negative by a C^0-small isotopy of τ^1 rel τ^0;
- by genericity, after a C^0-small isotopy in a neighborhood of τ^1 rel τ^1, each face can then be made convex by a C^∞-small isotopy rel τ^1.

Note that the triangulation is now singular: at each vertex x, the tangent lines to the adjacent edges are all contained in $\xi(x)$.

Denote $TB(\xi,\tau) = \sum_{F\in\tau^2} tb(\partial F)$. If $[\tau]$ is the set of all contact triangulations isotopic to τ, then define

$$TB(\xi,[\tau]) = \max_{\tau'\in[\tau]} TB(\xi,\tau').$$

We now assume, after a possible change of notation, that τ realizes the maximum value $TB(\xi,[\tau])$. Such a triangulation is said to be a *maximal triangulation* for ξ in the class $[\tau]$. In that case, we have the following:

LEMMA 2.1.

(a) *For every face F of τ, each component of Γ_F is an arc whose endpoints are on different edges, with the possible exception of six arcs which are boundary-parallel components and whose endpoints are close to a vertex (see Figure 3).*

(b) *The "holonomy" of the dividing curves around the boundary of a 3-simplex B is as in Figure 5.*

PROOF. We only sketch the proof of (a). Closed components of Γ_F are not allowed because the contact structure ξ is tight. The existence of a boundary-parallel component "far" from the vertices would imply the existence of a bypass which does not meet the vertices. Let $\alpha\cup\beta$ be the boundary of the bypass, where $\alpha\subset\partial F$, and isotop the edge l containing α to $(l\setminus\alpha)\cup\beta$. It strictly increases $TB(\xi,\tau)$. See Figure 4.

The proof of (b) also relies on the existence of bypasses inside B in case the holonomy is not as in Figure 5 (see [**Ho3**]). □

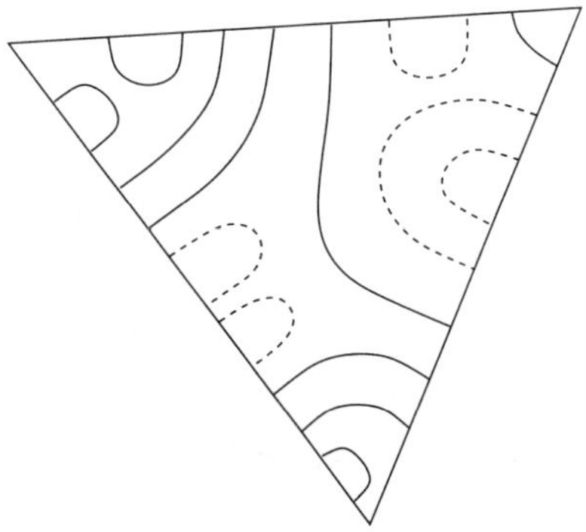

FIGURE 3. The dotted dividing curves can be isotoped across, leaving the solid dividing curves.

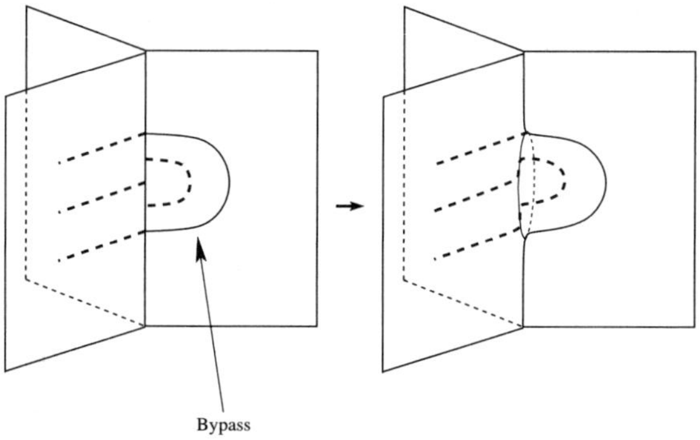

FIGURE 4

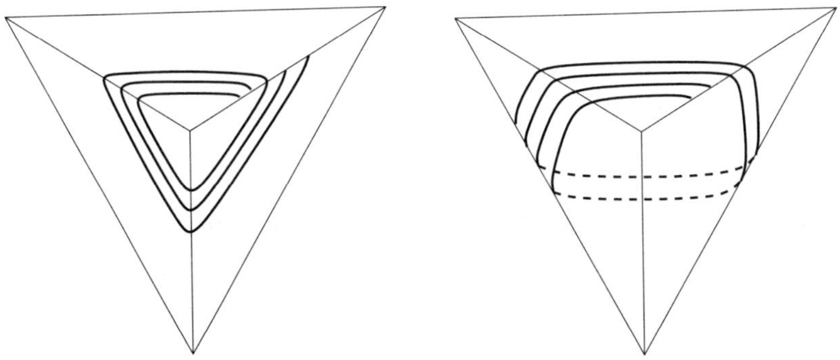

FIGURE 5. The "holonomy", after the edges are rounded. In the left-hand diagram, every time the dividing curve wraps around the vertex in a counterclockwise manner, it moves two units away from the vertex. The right-hand diagram is similar.

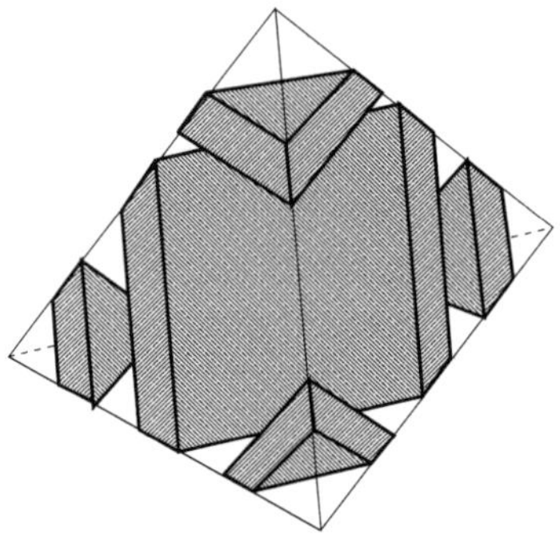

FIGURE 6. Fibered prisms.

2.2. Fibered prisms. A *fibered prism* is a polyhedron P diffeomorphic to $R \times [-1,1]$, where R is a triangle or a rectangle, and $\{pt\} \times [-1,1]$ are the fibers. The set $R \times \{-1,1\}$ is called the *horizontal boundary*, and the set $(\partial R) \times [-1,1]$ the *vertical boundary.*

Let ξ be a tight contact structure on M, which has been isotoped so that τ is a maximal triangulation for ξ. Then we partition families of parallel arcs of Γ_F (F is a face of τ) inside fibered prisms:

LEMMA 2.2. *For each 3-simplex B of τ, there are at most 5 embeddings $\phi_i : P_i \hookrightarrow B$, $i = 1, \dots, 5$, of fibered prisms P_i, such that the following are satisfied.*

- *The vertical edges (resp. faces) of P_i are sent into edges (resp. faces) of B.*
- *$\phi_i(P_i \setminus (\partial P \times [-1,1])) \subset int(B)$.*
- *$\phi_i(P_i \times \{-1,1\}) \cap \Gamma_F = \emptyset$, for every face F of B.*
- *The fibered prisms coming from two different 3-simplices intersect along rectangles.*
- *For every face F of B, at most C components of Γ_F are not contained in the image of the ϕ_i's. Here C is a universal constant which does not depend on ξ or on τ.*

Moreover, by a combination of the Flexibility Theorem and Theorem 1.3, we can normalize $\xi|_{\phi_i(P_i)}$ as in the following lemma, after isotoping ξ. (For notational convenience, we will not distinguish P_i from its image from now on.)

LEMMA 2.3. *Fix a nonsingular foliation $\mathcal{F}$ on the horizontal boundary of each prism (the same for each face) which coincides with the characteristic foliation near its* 1*-skeleton. Then ξ can be isotoped rel τ^1 so that the characteristic foliation of ξ equals $\mathcal{F}$ on the horizontal boundary and each vertical fiber $\{pt\} \times [-1,1]$ is Legendrian.*

2.3. The branched surfaces.

PROPOSITION 2.4. *There exists a finite number of pairs $(\mathcal{B}_1, \zeta_1), \dots, (\mathcal{B}_n, \zeta_n)$, where $\mathcal{B}_i$ is a branched surface and ζ_i a contact structure on $M \setminus N(\mathcal{B}_i)$, such that every tight contact structure ξ on M, up to isotopy, is generated by one of the $(\mathcal{B}_i, \zeta_i)$, i.e.,*

(1) $\xi|_{M \setminus N(\mathcal{B}_i)} = \zeta_i$;

(2) *$\xi|_{N(\mathcal{B}_i)}$ is tangent to the fibers of $N(\mathcal{B}_i)$.*

Moreover, one may assume that the branched surfaces have empty boundary.

PROOF. Let ξ be a tight contact structure on M, isotoped so that τ is a maximal triangulation for ξ. Apply Lemmas 2.2 and 2.3 to obtain the collection $\{P_1, \dots, P_k\}$ of fibered prisms $P_i = R_i \times [-1, 1]$ ranging over all the 3-simplices B of τ. For every edge $a \in \tau^1$, denote a' the arc obtained from a by deleting a small neighborhood of its endpoints, and $a' \times D^2$ a small tubular neighborhood of a' such that the fibration $a' \times \{pt\}$ is Legendrian and coincides with the one given on the P_i's.

LEMMA 2.5. *There exists a branched surface $\mathcal{B}$ with boundary, together with a fibered neighborhood $N(\mathcal{B})$ fibered by Legendrian intervals, which is obtained by smoothing*

$$K = \left(\bigcup_{1 \le i \le k} P_i \right) \cup \left(\bigcup_{a \in \tau^1} (a' \times D^2) \right).$$

Moreover, every Legendrian fiber of $N(\mathcal{B})$ lies inside a Legendrian fiber of K.

PROOF OF LEMMA 2.5. Up to reparametrization, one can always assume that $Q_i \cap Q_j = \emptyset$ for $i \neq j$, where we denote $Q_i = R_i \times \{0\}$. Now let

$$Q_i' = \overline{Q_i \setminus \cup_{a \in \tau^1} (a' \times D^2)},$$

i.e., we remove from Q_i a neighborhood of its vertices. If P_i intersects P_j, we join the corresponding edges of Q_i' and Q_j' by a small band transverse to the fibration. Denote by $\mathcal{B}'$ the branched surface with boundary consisting of the union of the Q_i' and the bands. Note that $\mathcal{B}'$ does not have any triple branch points.

Take a disk $\{pt\} \times D^2 \subset a' \times D^2$ for each edge a of τ^1, and squash each point of $\mathcal{B}' \cap (a' \times D^2)$ to the point of $\{pt\} \times D^2$ situated on the same fiber. We obtain a "singular surface" that can easily be perturbed into a branched surface $\mathcal{B}$; this is where we need to introduce triple branch points. Once the edges of K are smoothed, we naturally obtain a fibered neighborhood $N(\mathcal{B})$ of $\mathcal{B}$ with the desired properties. □

The construction of Lemma 2.5 gives rise to a finite number of branched surfaces, since each simplex contains at most 5 prisms, and for each simplex there are 7 total prism positions to choose from (see Figure 6). Here we are considering two prisms P_i, P_j in a 3-simplex B to be equivalent if there is an isotopy from P_i to P_j which restricts to each edge a to be an isotopy of $P_i \cap a$ to $P_j \cap a$ and on each face F to be an isotopy of $P_i \cap F$ to $P_j \cap F$.

We may assume that all the tight contact structures ξ with the same set of fibered prisms (and hence the same branched surface $\mathcal{B}$) agree on $\partial N(\mathcal{B})$, because we can impose the same nonsingular characteristic foliation on the horizontal boundaries of the prisms by Lemma 2.3. On the complementary regions of $N(\mathcal{B})$, we have a natural decomposition into polyhedra which is inherited from τ. On the faces of the polyhedra, the number of components of the dividing set is universally bounded by Lemma 2.2. This leads to a finite number of possible dividing curve configurations, and hence to finitely many characteristic foliations by the Flexibility Theorem. But now, according to Theorem 1.3, a tight contact structure on each polyhedron is determined by its restriction to the boundary.

Now the branched surface $\mathcal{B}$ may have nonempty boundary. We then use an *amputation principle* to excise the boundary: if a vertical face of a prism is entirely contained in $\partial_v N(\mathcal{B})$, then all tight contact structures which coincide outside of $N(\mathcal{B})$ with ζ and are tangent to the fibers of $N(\mathcal{B})$ coincide on this face. In particular, they all have the same number of components of dividing curves on this face, and the number of dividing curves on the other vertical faces of the prism is bounded. Thus, all the tight contact structures which coincide with ζ on $M \setminus N(\mathcal{B})$ coincide, up to isotopy, with a finite set of models on this prism, by taking into account the finite repartition of the parallel dividing curves inside adjacent prisms. Therefore, we can amputate this prism from $N(\mathcal{B})$, at the expense of possibly increasing the number of pairs $(\mathcal{B}_i, \zeta_i)$. This is a finite process which must eventually yield a branched surface without boundary (or an empty branched surface). □

3. Tight contact structures carried by the fibered neighborhood of a branched surface

We now analyze contact structures which are *carried* by the fibered neighborhood of a branched surface $N(\mathcal{B})$, i.e., are tangent to the fibers and are fixed on the boundary.

Let $\mathcal{S}$ be the set of tight contact structures tangent to the fibers of $N(\mathcal{B})$ (with a given germ along $\partial N(\mathcal{B})$), and let $B_1, \dots, B_d$ be the components of $\overline{\mathcal{B} \setminus L}$ (recall L is the singular locus of $\mathcal{B}$). If we fix $\xi_0 \in \mathcal{S}$, then for any given $\xi \in \mathcal{S}$ one can define a weight system w_ξ on $\mathcal{B}$: $w_\xi(B_i)$, $i = 1, \dots, d$, is the difference of rotation between ξ and ξ_0 along a fiber $\pi^{-1}(p)$, where $p \in B_i$. It is a set of integers which does not depend on the choice of $p \in B_i$. Note that, unlike the relative Thurston-Bennequin invariant, a left twist contributes positively to the weight. It is easy to verify that $\xi \in \mathcal{S}$ is determined up to isotopy by its weight $w_\xi = (w_\xi(B_1), \dots, w_\xi(B_d)) \in \mathbb{Z}^d$. Also, each $w_\xi(B_i)$ is bounded below by the positive contact structure condition. By amputating the B_i for which $w_\xi(B_i)$ is negative, we may assume that $w_\xi(B_i)$ is nonnegative for each $\xi \in \mathcal{S}$.

To each smooth edge A of L, we associate a linear equation

$$\phi_A(x_1, \dots, x_d) = x_i - (x_j + x_k) = 0,$$

with $(x_1, \dots, x_d) \in \mathbb{Z}^d$, where B_i, B_j and B_k are the components of $\mathcal{B} \setminus L$ which meet along A, and B_j and B_k branch out of B_i. Let V be the subspace of $\mathbb{R}^d$ of solutions to the system $\{\phi_A = 0 \mid \forall \text{ smooth edges } A \text{ of } L\}$. Each weight w is an element of $V \cap \mathbb{Z}^d_{\geq 0}$, where $\mathbb{Z}^d_{\geq 0} = \{(x_1, \dots, x_d) \in \mathbb{Z}^d | x_i \geq 0 \ \forall i\}$.

Define a partial order $\leq$ on $\mathbb{Z}^d$ by:

$$(x_1, \dots, x_d) \leq (x'_1, \dots x'_d) \text{ if } x_i \leq x'_i \text{ for } i = 1, \dots, d.$$

There is only a finite number of minimal elements of $V \cap \mathbb{Z}^d_{\geq 0}$ with respect to $\leq$, and we denote them $u_1, \dots, u_k$. They generate the monoid $V \cap \mathbb{Z}^d_{\geq 0}$. On the other hand, we have the following standard fact:

Lemma 3.1. *The isotopy classes of embedded surfaces carried by $N(\mathcal{B})$ (i.e., transverse to the fibers) are in bijection with elements of $V \cap \mathbb{Z}^d_{\geq 0}$.*

Remark. Every surface carried by $N(\mathcal{B})$ is transverse to a contact structure and is thus either a torus or a Klein bottle.

Let $\{T_1, \ldots, T_k\}$ be the collection of surfaces carried by $N(\mathcal{B})$ and determined by the weights $u_1, \ldots, u_k$. The following lemma, together with Proposition 2.4 and the remark below, yields Theorem 1.4.

Assume all the T_i's are tori. Then we have:

LEMMA 3.2. *Every contact structure $\xi \in \mathcal{S}$ is obtained from ξ_0 by a collection of generalized Lutz modifications along $T_1, \ldots, T_k$.*

PROOF. Given $\xi \in \mathcal{S}$, we can write $w_\xi = \sum_{1\leq i\leq k} n_i u_i$, $n_i \in \mathbb{Z}_{\geq 0}$. The contact structure ξ' obtained from ξ_0 by generalized Lutz modifications of index n_i along T_i is tangent to the fibers of $N(\mathcal{B})$ and has the same weight as ξ. Therefore ξ and ξ' are isotopic rel $\partial N(\mathcal{B})$. □

Remark. If T_i is a Klein bottle, then we can replace it by the torus $T_i' = \partial N(T_i)$, where $N(T_i)$ is the tubular neighborhood of T_i. Denote $T_i' = T_i$ if T_i is a torus. Then the set $\mathcal{S}$ is still generated by Lutz modifications along the T_i', but starting from a finite set of structures (not only ξ_0), obtained from ξ_0 by performing π-twists along the Klein bottles.

4. Isotopy and beyond

The proof of Theorem 0.2 follows from a more in-depth analysis of tight contact structures carried by the fibered neighborhood of these branched surfaces. We use the classification of tight contact structures on the thickened torus (see [**Gi4, Ho1**]) to build a process which allows us to decrease the weights in case the manifold is atoroidal. In the general case, one should be able to prove that there exists a finite number of tight contact structures of bounded torsion up to isomorphism. Moreover, these techniques apply to manifolds M with boundary, provided ∂M is not a torus and we prescribe a fixed characteristic foliation along ∂M. We then obtain relative versions of Theorems 0.1 and 0.2. Finally, we have the following:

THEOREM 4.1. *Given an oriented topological knot type $\mathcal{K}$ in the standard tight contact 3-sphere (S^3, ξ_{std}) and an integer n, there exists, up to contact isotopy, a finite number of Legendrian knots in $\mathcal{K}$ with Thurston-Bennequin invariant equal to n.*

References

[Be] D. Bennequin, *Entrelacements et équations de Pfaff*, Astérisque **107-108** (1983), 87–161.

[Co1] V. Colin, *Stabilité topologique des structures de contact en dimension* 3, Duke Math. J. **99** (1999), 329–351.

[Co2] V. Colin, *Sur la torsion des structures de contact tendues*, Ann. Sci. École Norm. Sup. (4), **34** (2001), 267–286.

[Co3] V. Colin, *Une infinité de structures de contact tendues sur les variétés toroïdales*, Comment. Math. Helv. **76** (2001), 353–372.

[Co4] V. Colin, *Structures de contact tendues sur les variétés toroïdales et approximation de feuilletages sans composante de Reeb*, to appear in Topology.

[CGH] V. Colin, E. Giroux, and K. Honda, *Finitude homotopique et isotopique des structures de contact tendues*, in preparation.

[El1] Y. Eliashberg, *Classification of overtwisted contact structures on 3-manifolds*, Invent. Math. **98** (1989), 623–637.

[El2] Y. Eliashberg, *Contact 3-manifolds twenty years since J. Martinet's work*, Ann. Inst. Fourier (Grenoble) **42** (1992), 165–192.

[ET] Y. Eliashberg and W. Thurston, *Confoliations*, University Lecture Series **13**, Amer. Math. Soc., Providence (1998).

[Et] J. Etnyre, *Tight contact structures on lens spaces*, Commun. Contemp. Math. **2** (2000), 559–577.

[EH] J. Etnyre and K. Honda, *Tight contact structures with no symplectic fillings*, Invent. Math. **148** (2002), 609–626.

[Ga] D. Gabai, *Essential laminations and Kneser normal form*, J. Differential Geom. **53** (1999), 517–574.

[Gi1] E. Giroux, *Convexité en topologie de contact*, Comment. Math. Helv. **66** (1991), 637–677.

[Gi2] E. Giroux, *Une structure de contact, même tendue, est plus ou moins tordue*, Ann. Sci. École Norm. Sup. (4) **27** (1994), 697–705.

[Gi3] E. Giroux, *Une infinité de structures de contact tendues sur une infinité de variétés*, Invent. Math. **135** (1999), 789–802.

[Gi4] E. Giroux, *Structures de contact en dimension trois et bifurcations des feuilletages de surfaces*, Invent. Math. **141** (2000), 615–689.

[Ha] W. Haken, *Theorie der Normalflächen. Ein Isotopiekriterium für den Kreisknoten*, Acta Math. **105** (1961), 245–375.

[Ho1] K. Honda, *On the classification of tight contact structures I*, Geom. Topol. **4** (2000), 309–368.

[Ho2] K. Honda, *On the classification of tight contact structures II*, J. Differential Geom. **55** (2000), 83–143.

[Ho3] K. Honda, *Gluing tight contact structures*, to appear in Duke Math. J.

[HKM] K. Honda, W. Kazez and G. Matić, *Convex decomposition theory*, Internat. Math. Res. Notices **2002**, 55–88.

[Kn] H. Kneser, *Geschlossene Flächen in dreidimensionalen Mannigfaltigkeiten*, Jahres. der Deut. Math.-Verein. **38** (1929), 248–260.

[KM] P. Kronheimer and T. Mrowka, *Monopoles and contact structures*, Invent. Math. **130** (1997), 209–255.

Université de Nantes
E-mail address: `Vincent.Colin@math.univ-nantes.fr`

École Normale Supérieure de Lyon
E-mail address: `Emmanuel.GIROUX@umpa.ens-lyon.fr`

University of Southern California
E-mail address: `khonda@math.usc.edu`
URL: `http://math.usc.edu/~khonda`

Proceedings of Symposia in Pure Mathematics
Volume **71**, 2003

On the classification of tight contact structures

Paolo Ghiggini and Stephan Schönenberger

ABSTRACT. Recently, there have been several breakthroughs in the classification of tight contact structures. We give an outline on how to exploit methods developed by Ko Honda and John Etnyre to obtain classification results for specific examples of small Seifert manifolds.

1. Introduction

After Eliashberg proved a classification for so-called overtwisted contact structures [**4**], work concentrated on the classification of tight contact structures, which turned out to be much more subtle and provide interesting relations to the topology of the underlying manifold. See [**11**] for an introduction to contact geometry and further references.

Until recently, the main tool to show that a contact structure on a manifold is tight is to show it is fillable. A contact structure is holomorphically fillable if it is the oriented boundary of a compact Stein 4-manifold. Gromov and Eliashberg showed that a fillable contact structure is tight [**21, 5**]. Moreover, fillability is preserved by Legendrian surgery [**35, 6**], thus providing a rich source of tight contact structures. Gompf's extensive study on Legendrian surgery [**19**] enables one in particular to construct holomorphically fillable contact structures on many Seifert manifolds. Using Legendrian surgery and techniques from Seiberg-Witten-theory, Lisca and Matić [**31**] proved that for every integer $n > 1$ there exist at least $[\frac{n}{2}]$ tight contact structures on the Brieskorn homology spheres with reversed orientation, $-\Sigma(2,3,6n-1) = M(-\frac{1}{2}, \frac{1}{3}, \frac{n}{6n-1})$. Later, they improved this lower bound to $n-1$ in [**32**].

Contact structures induce a singular foliation on embedded surfaces and these are often easier to study than the contact structure itself. Motivated by work of Eliashberg and Gromov [**9**], Giroux introduced the notion of convex surfaces, i.e. surfaces whose characteristic foliation is cut transversely by a certain multicurve, called the dividing set [**15**]. This dividing set essentially determines the contact structure in a neighbourhood of the surface and is a convenient tool to study contact structures.

Exploiting this idea, Kanda gave a complete classification of tight contact structures on the 3-torus [**27**]; see also [**16**]. This led Honda to study so-called bypasses

2000 *Mathematics Subject Classification.* Primary 54C40, 14E20; Secondary 46E25, 20C20.

attached along convex surfaces, which provide a systematic tool for altering the dividing set of a convex surface. By splitting a contact 3-manifold along convex surfaces into simpler pieces and studying the possibilities of tight contact structures, Honda gave a complete classification of tight contact structures on solid tori, toric annuli, Lens spaces in [**24**], as well as torus bundles over the circle and circle bundles over closed Riemannian surfaces; see [**25**]. Many of these were independently obtained by Giroux [**16, 17, 18**], and on some Lens spaces by Etnyre [**10**].

Furthermore, Lisca proved in [**29**] that the Poincaré homology sphere with reverse orientation $-\Sigma(2,3,5)$ (this corresponds to the Seifert manifold $M(-\frac{1}{2},\frac{1}{3},\frac{1}{5})$ in the notation below) has no symplectically (weakly) semi-fillable contact structure, thus proving a conjecture of Gompf in [**19**]. Using the bypass technique in contact topology, Etnyre and Honda finally proved [**14**] the nonexistence of a tight contact structure on $M(-\frac{1}{2},\frac{1}{3},\frac{1}{5})$, thereby providing the first example of a closed 3-manifold which admits no tight contact structure.

Lisca [**30**] went further and proved (among other things) that the Seifert manifolds $M(-\frac{1}{2},\frac{1}{3},\frac{1}{4})$, $M(-\frac{1}{2},\frac{1}{3},\frac{1}{3})$ admit no (weakly) symplectically semi-fillable contact structure. From Lisca's examples, Etnyre and Honda proved that on the Seifert manifolds $M(-\frac{1}{2},\frac{1}{4},\frac{1}{4})$ and $M(-\frac{2}{3},\frac{1}{3},\frac{1}{3})$ there exist tight contact structures without symplectic fillings. These examples belong to the handful of Seifert manifolds which can be defined as torus bundles over the circle; see [**3**] for fillability results on these manifolds.

Furthermore, the examples above are Seifert manifolds over the sphere S^2 with three singular fibres. On 'larger' Seifert manifolds it is recently proven by Colin [**1**] that every orientable Seifert manifold over a surface of genus $g \geq 1$ has infinitely many non-isomorphic tight contact structures. Moreover, Colin [**2**], see also Honda, Kazez, Matić [**26**], proved that every closed irreducible orientable toroidal 3-manifold carries infinitely many contact structures.

Therefore the classification of tight contact structures on Seifert manifolds may provide interesting new insight to the topology of tight contact structures on 3-manifolds. In this work, we will demonstrate on two examples how to apply bypass techniques to obtain upper bounds on the number of tight contact structures on Seifert manifolds over the sphere with three singular fibres. In the examples below, tight contact structures are constructed using Legendrian surgery.

2. Basic contact geometry

A positive contact structure on an oriented 3-manifold M is a 2-plane field $\xi = \ker\alpha \subset TM$, defined by a 1-form α satisfying $\alpha \wedge d\alpha > 0$. According to this definition, ξ is co-oriented by α and oriented by $d\alpha$ such that the orientation on M coincides with the orientation defined by $\alpha \wedge d\alpha$.

2.1. Legendrian curves and twisting. A curve γ in a contact manifold (M,ξ) everywhere tangent to ξ is called *Legendrian*. Throughout this paper, we assume curves to be closed, and we will refer to 'arcs' otherwise. Recall that every diffeomorphism between Legendrian curves extends to a contactomorphism of their neighbourhoods. A Legendrian curve γ in a contact manifold (M,ξ) is endowed with a natural framing defined by a vector field along γ transverse to ξ, called the *contact framing*. The *twisting number* $\mathtt{t}(\gamma,\mathcal{F})$ is defined as the number of right 2π twists of the contact framing with respect to a preassigned framing $\mathcal{F}$ of γ. In case γ is a Legendrian boundary component of an oriented surface S, let $\mathcal{F}_S$ denote the framing of γ defined by S. In this case we will write $\mathtt{t}(\gamma) := \mathtt{t}(\gamma, \mathcal{F}_S)$. When γ is null-homologous and S is a Seifert surface for γ, the twisting number is called *Thurston-Bennequin invariant* and denoted by $\mathtt{tb}(\gamma)$.

We have another classical invariant for Legendrian knots, the *rotation number* $\mathtt{r}$: If γ is the Legendrian boundary of a Seifert surface S, we define $\mathtt{r}(\gamma)$ as the number of revolution of its tangent $\dot{\gamma}$ with respect to a trivialization of $\xi|_S$. Note that, for any relative homology class $\beta \in H_2(M, \gamma)$ and $S \in \beta$ a representing surface, the rotation number of γ is independent of the choice of a trivialization but depends on β, and reversing the orientation of γ reverses the sign of $\mathtt{r}$. We refer the reader to [**11**] for a detailed discussion of the following

PROPOSITION 2.1 (Bennequin's Inequality). *If γ is a Legendrian knot in a tight contact manifold (M, ξ) and S a Seifert surface for γ with Euler characteristic $\chi(S)$, then*

$$\mathtt{tb}(\gamma) + |\mathtt{r}(\gamma)| \leq -\chi(S).$$

2.2. Convex surface theory. Assume S is a compact oriented surface embedded in a contact manifold (M, ξ). The line field $l_x = \xi_x \cap T_x S$, $x \in S$, integrates to a singular foliation S_ξ of S called *characteristic foliation.* Recall that the singularities of S_ξ are exactly the points in S where the contact plane is tangent to S. The characteristic foliation determines the contact structure in a tubular neighbourhood and one has a certain freedom to alter the characteristic foliation by perturbing the surface; see [**11**]. Generically, the amount of information needed to locally determine the contact structure can be reduced to a collection of curves on the surface S.

A properly embedded orientable surface S in a contact manifold (M, ξ) is called *convex*, if there exists a collection of curves Γ on S satisfying the following conditions:

(1) $S \setminus \Gamma = S^+ \sqcup S^-$
(2) Γ is transverse to the characteristic foliation S_ξ of S
(3) There exists a vector field v and a volume form θ on S such that the characteristic foliation is directed by v, the flow of v expands θ on S^+, contracts θ on S^- and v points transversely out of S^+.

Recall that the existence of *dividing curves* Γ is equivalent to the existence of a contact vector field v transverse to the surface S, determining the contact structure in a neighbourhood of the surface up to admissible isotopy, i.e. an isotopy ϕ : $S \times [0, 1] \to M$ such that $\phi(S \times \{t\})$ is transverse to v for all $t \in [0, 1]$.

In [**15**], Giroux proved that every closed surface can be perturbed by a C^∞-small isotopy to be convex. More generally, a compact surface with Legendrian boundary can be perturbed to be convex provided the twisting number of each boundary component is not positive. Moreover, the twisting number of a boundary component ∂S of a convex surface S determines the dividing set in a tubular neighbourhood of ∂S. This follows from a relative version of Gray's Theorem in dimension three; see Theorem 3.7 in [**11**]. We describe a standard tubular neighbourhood of a Legendrian boundary component γ as follows: After perturbing S we find a neighbourhood N of a boundary component $\gamma \subset \partial S$ so that a collar neighbourhood $A = N \cap S$ of γ in S has the form $A = S^1 \times [0, 1] = (\mathbb{R}/\mathbb{Z}) \times [0, 1]$ with coordinates (x, y) where $\gamma = S^1 \times \{0\}$. In a neighbourhood $A \times [-1, 1]$ of A with coordinates (x, y, z) the contact 1-form is defined by $\alpha = \sin(2\pi n x)dy + \cos(2\pi n x)dz$

for $n = |\mathtt{t}(\gamma)| \in \mathbb{Z}^+$. Note that on this annulus the characteristic foliation consists of circles parallel to γ, called *Legendrian rulings* and the dividing set consists of arcs transverse to the boundary, leading from one boundary component to another. Between two dividing arcs lies an arc of singularities, which we call *Legendrian divides*; see Fig. 1. If $\mathtt{t}(\gamma) = 0$ then the contact structure is defined by by $\alpha = dz - ydx$. In particular the twisting number of γ is related to the number of intersections of the dividing set Γ with γ.

PROPOSITION 2.2. *Suppose S is a convex surface with Legendrian boundary in a contact manifold (M, ξ) and $\gamma \in \partial S$ is a boundary component of S. Then*

$$\mathtt{t}(\gamma) = -\frac{1}{2}\#(\gamma \cap \Gamma), \tag{2.1}$$

where $\#(\gamma \cap \Gamma)$ denotes the cardinality of the intersection $\gamma \cap \Gamma$. Moreover, if γ is null-homologous and S a Seifert surface, then

$$\mathtt{r}(\gamma) = \chi(S^+) - \chi(S^-), \tag{2.2}$$

where $S^\pm$ is as in the definition of dividing set and $\chi(S^\pm)$ denotes the Euler characteristic.

If γ is a Legendrian curve contained in a convex surface S, i.e. not necessarily a boundary component, γ can be made to have a standard collar neighbourhood as depicted in Fig. 1 (where γ is a ruling curve in the interior); see [**27**]. Formula (2.1) is also valid in this case.

Giroux pointed out that for convex surfaces, the dividing set, not the particular characteristic foliation, essentially determines the contact structure in a neighbourhood. Namely:

THEOREM 2.3 (Giroux's Flexibility Theorem, [**15**]). *Consider a surface S, closed or compact with Legendrian boundary, in a contact manifold (M, ξ). Assume Γ is a dividing set for the characteristic foliation S_ξ and $\mathcal{F}$ is another singular foliation on S divided by Γ. Then there is an isotopy $\phi : S \times [0, 1] \to M$ of S such that $\phi_0 = id$, $(\phi_1(S))_\xi = \phi_1(\mathcal{F})$, $(\phi_t(S))_\xi$ is divided by Γ for all $t \in [0, 1]$ and Γ is fixed.*

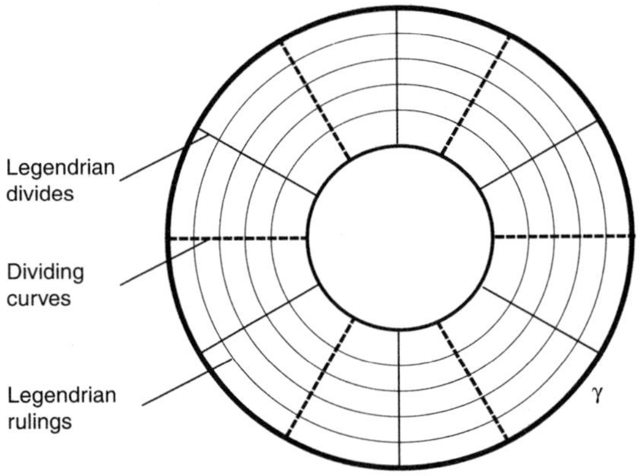

FIGURE 1. Convex collar neighbourhood of a Legendrian boundary component with negative twisting.

On the other hand, on a convex surface in a tight contact manifold, two dividing sets of a characteristic foliation are isotopic. We will then, by slightly abusing language, refer to Γ as 'the' dividing set.

As a consequence of Giroux's Flexibility Theorem, one can realize curves or arcs in a convex surface to be Legendrian:

THEOREM 2.4 (Legendrian Realization, [**24**]). *Consider a collection of disjoint properly embedded closed curves and arcs C on a convex surface S, which satisfies the following properties:*

(i) *C is transverse to the dividing set Γ of S and every arc in C begins and ends on Γ,*

(ii) *every component of $S \setminus (\Gamma \cup C)$ has a boundary component which intersects Γ,*

then there exists an isotopy $\phi : S \times [0,1] \to M$ such that $\phi_0 = id$, $\phi_t(S)$ are all convex, $\phi_1(\Gamma) = \Gamma$ and $\phi_1(C)$ is Legendrian.

In a tight contact structure, the possibilities of the dividing sets are rather restricted. Namely:

THEOREM 2.5 (Giroux's criterion, [**24**]). *A convex surface (closed or compact with Legendrian boundary) S other than the sphere S^2 has a tight neighbourhood if and only if no component of Γ_S bounds a disc. A convex sphere S^2 has a tight neighbourhood if and only if $\#\Gamma_{S^2} = 1$, i.e. if there exists exactly one dividing curve.*

2.2.1. *Edge-rounding.* Next, we describe how to smooth out two convex surfaces intersecting transversally along a common Legendrian curve with a negative twisting number and moreover, to relate the dividing set of the two surfaces to the dividing set on the smoothed surface.

PROPOSITION 2.6 (Edge-Rounding Lemma, [**24**]). *Assume S_1 and S_2 are convex surfaces, with convex collar boundary, intersecting transversely inside a contact manifold (M, ξ) along a common Legendrian boundary curve γ with negative twisting number. Suppose that S_1 and S_2 are oriented such that smoothing the edge yields an oriented surface S'. Then the edge may be smoothed so that the dividing set of S' is obtained from the dividing sets on S_1 and S_2, the dividing curves connect such that positive (negative) regions $S_1^{\pm}$ of S_1 connect to positive (negative) regions $S_2^{\pm}$ of S_2 as indicated in figure 2.*

PROOF. After possibly a small perturbation of a neighbourhood N of γ in M, we can consider the following situation: Consider $\mathbb{R}^2 \times (\mathbb{R}/\mathbb{Z})$ with coordinates (x, y, z), and contact 1-form $\alpha = \sin(2\pi nz)dx + \cos(2\pi nz)dy$ for some $n \in \mathbb{Z}^{>0}$. Locally, a neighbourhood of γ is contactomorphic to $N_\varepsilon = \{x^2 + y^2 \leq \varepsilon\}$ and γ is given by $x = y = 0$. The convex surfaces become $S_1 \cap N_\varepsilon = \{x = 0, 0 \leq y < \varepsilon\}$ and $S_2 \cap N_\varepsilon = \{y = 0, 0 \leq x < \varepsilon\}$ oriented by ∂_x and ∂_y respectively. The transverse vector field for $\{(x-\delta)^2 + (y-\delta)^2 = \delta^2\} \cap N_\delta$, $0 < \delta < \varepsilon$, is the inward-pointing radial vector $-\partial_r$ for the circle $\{(x-\delta)^2 + (y-\delta)^2 = \delta^2\}$. Take $S' = (S_1 \cup S_2 \setminus N_\delta) \cup \{(x-\delta)^2 + (y-\delta)^2 = \delta^2\} \cap N_\delta$, the dividing curve $z = \frac{k}{2n}$ on S_1 then connects to the dividing curve $z = \frac{k}{2n} - \frac{1}{4n}$ on S_2, for $k = 0, \ldots, 2n-1$; see Fig. 2. □

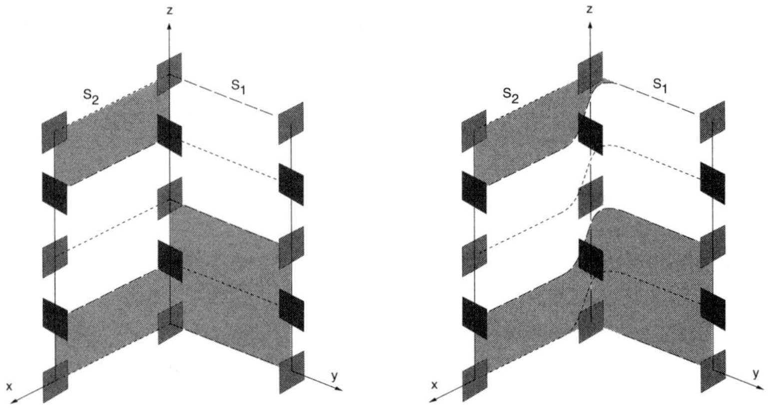

FIGURE 2. Rounding edges. (Shaded regions are positive, dashed lines are dividing curves and Legendrian divides are dotted.)

2.2.2. *Bypasses and alteration of the dividing set.* By perturbing a surface, one can alter the characteristic foliation to assume certain normal forms. On convex surfaces, we want to alter directly the dividing set. A crucial tool for this is the use of *bypasses*, first exploited by Honda (see e.g. [**24, 25**]) with precursors in [**27**].

DEFINITION 2.7. Assume $S \subset M$ is a convex surface. A *bypass* for S is an oriented embedded disc D whose Legendrian boundary satisfies the following:

(1) ∂D is the union of two arcs γ_1, γ_2 which intersect at their endpoints.
(2) D intersects S transversely along γ_1.
(3) along γ_1, there are three elliptic tangencies in the characteristic foliation D_ξ, two of the same sign at the endpoints and one of different sign in the interior of γ_1
(4) along γ_2 there are at least three tangencies, all have the same sign but alternating indices.
(5) there are no interior singular points of D_ξ.

See Fig. 3 for an illustration.

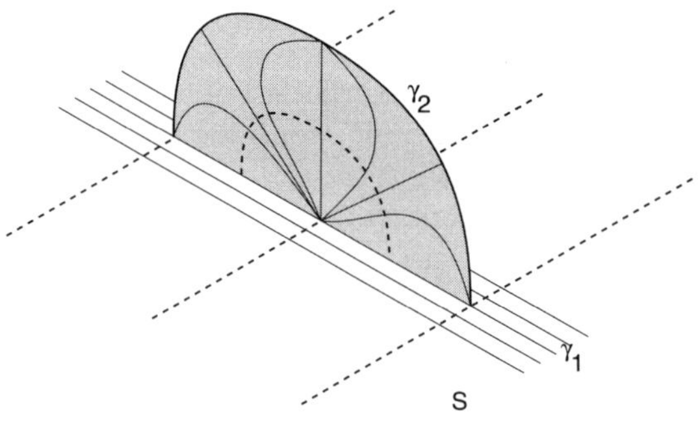

FIGURE 3. A bypass. (Dashed lines are dividing curves on S.)

Observe that all singular points on ∂D have the same sign except the one elliptic point in the interior of γ_1. We call this the *sign* of the bypass. The endpoints of γ_1 may be the same elliptic point, in this case we call D a *degenerate* bypass.

We first explain how to find bypasses and then give a discussion regarding how bypasses are used to alter the dividing set of a convex surfaces. We discuss the cases used later in Section 4, for a more complete discussion including applications, the reader may refer to the literature; see [**24, 25, 23, 14, 13, 12**].

Assume S is a convex surface with Legendrian boundary. After possibly perturbing S we can further assume that all boundary tangencies are half-elliptic (see Lemma 3.2 in [**24**]). If $\mathtt{t}(\gamma) = -n \leq 0$ for $\gamma \subset \partial S$, then the dividing curves intersect γ exactly $2n$ times. Suppose one of these dividing arcs is boundary-parallel, i.e. the arc cuts off a half-disc which has no further intersections with Γ_S. A nearby arc in the complement, parallel to this dividing arc, can be made Legendrian using the Realization Principle. After this, the arc bounds a bypass. Thus, we have as a general principle:

PROPOSITION 2.8. *Let S be a compact surface having one Legendrian boundary with non-positive twisting number, other than D^2 with $\mathtt{t}(\partial D^2) = -1$. After possibly a small perturbation we can assume that S is convex and all singular points of S_ξ on ∂S are half-elliptic. Suppose further γ is a boundary-parallel dividing curve. Then there exists a bypass which contains the half-disc cut off by γ.*

In the sequel, we need bypass existence for two special surfaces: discs and annuli. We therefore consider these special cases as discussed in [**24**]; see Fig. 4 for examples.

PROPOSITION 2.9. *Assume D^2 is a convex disc with Legendrian boundary lying inside a tight contact manifold and $\mathtt{t}(\partial D^2) = -n < 0$. After possibly a small perturbation we can assume that all tangencies at the boundary are half-elliptic. Then Γ_{D^2} consists of arcs which begin and end on ∂D^2. If $\mathtt{t}(\partial D^2) < -1$, then there exists a bypass along ∂D^2.*

PROPOSITION 2.10 (Imbalance Principle). *Assume $A = S^1 \times [0,1]$ is a convex annulus with Legendrian boundary in a tight contact manifold. After possibly a small perturbation we can assume that all tangencies at the boundary are half-elliptic. If $\mathtt{t}(S^1 \times \{0\}) < \mathtt{t}(S^1 \times \{1\}) \leq 0$, then there exists a bypass along $S^1 \times \{0\}$.*

Once we have a bypass for a convex surface, it can be used to manipulate the dividing set. The basic attachment process is described as follows:

PROPOSITION 2.11 (Bypass attachment). *Assume $Q = [0,1] \times [0,1]$ is a convex square in a convex surface with three horizontal dividing arcs as in Fig. 5 (a). If there exists a bypass D for Q, along a vertical Legendrian arc δ, we can isotope Q (fixing the boundary) by pushing Q across D such that the characteristic foliation has a dividing set as shown in Fig. 5 (b).*

PROOF. Because Q is convex, we can consider an I-invariant one-sided neighbourhood $Q \times [0,\varepsilon]$, for some $\varepsilon > 0$, such that $Q = Q \times \{\varepsilon\}$. Then, $A' = \delta \times [0,\varepsilon]$ is a rectangle in standard form, (i.e. with horizontal linear characteristic foliation and parallel dividing arcs in $[0,\varepsilon]$-direction), transverse to Q. Then, $A = A' \cup D$ is convex with piecewise smooth boundary. The endpoints of the arc $\delta = A' \cap D$ are half-elliptic corners. In order to smooth the corners of A, we convert half-elliptic

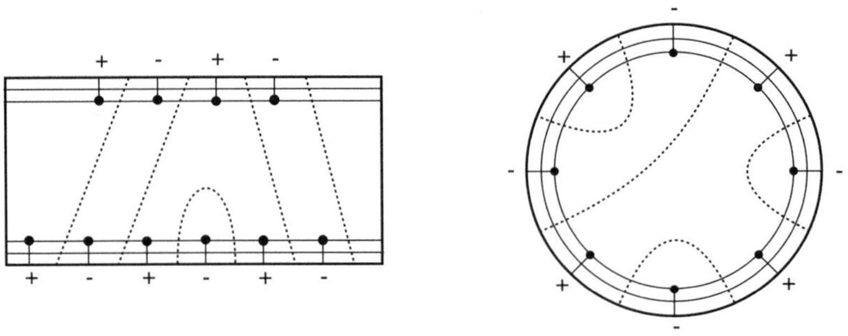

FIGURE 4. A possible constellation of dividing curves on an annulus (left and right edge of the rectangle are identified) and a disc. (Dividing curves are dashed lines, on the right, we have $\mathtt{tb}(\partial D) = -4$)

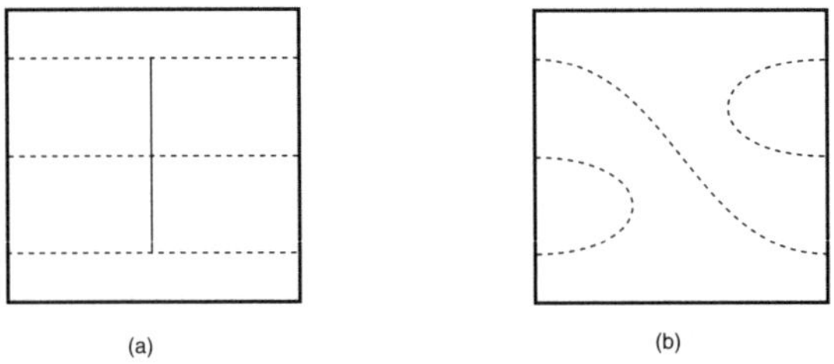

FIGURE 5. Bypass attachment. Dividing curves on Q before (a) and after (b) isotopy. The Legendrian line where the bypass is attached is drawn as thin vertical arc in (a).

points to full elliptic points. Finally, we apply the Pivot Lemma[1] to smooth the corners; see [**24**]. Because A is convex, we can take an I-invariant neighbourhood $N(A) = A \times [0,1]$. The boundary components $A_i = A \times \{i\}$, $i = 0, 1$ are copies of A, i.e. have the same dividing set Γ. Both A_0 and A_1 are oriented as boundary of $N(A)$ and therefore corresponding regions of $A_0 \setminus \Gamma$ and $A_1 \setminus \Gamma$ have different signs. Now, using the Edge-Rounding Lemma 2.6, we smooth out the four edges of $N(A) \cup Q \times [0, \varepsilon]$ to obtain a surface Q' with dividing set as in Fig. 5 (b), which completes the proof. See Fig. 6 for an illustration. □

In the sequel of this paper, we frequently encounter the situation where a bypass is attached along a torus. We first describe a standard normal form for a convex torus in a contact structure and explain then the consequences of the bypass attachment in this situation. On a convex torus T^2 in a tight contact manifold, we know by Giroux's criterion (Theorem 2.5) that no dividing curve bounds a disc. Therefore, the dividing set Γ_{T^2} consists of $2n$ homotopically essential parallel dividing curves and the number $n = \frac{1}{2}\#\Gamma_{T^2}$ is called the *torus division number*. Using some identification of T^2 with $\mathbb{R}^2/\mathbb{Z}^2$, the dividing curves have slope s, called the *boundary slope* of the torus. Due to Giroux's Flexibility Theorem 2.3, we can

[1]The Pivot Lemma allows to perturb a surface near an elliptic point such that any two transverse trajectories become smooth; see [**8**].

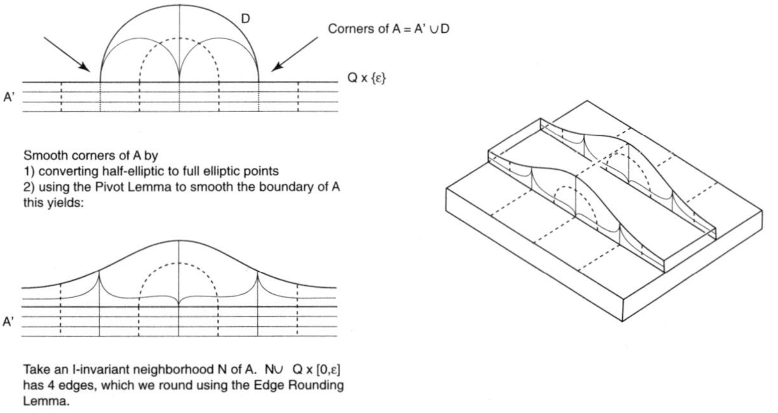

FIGURE 6. Bypass attachment. An illustration of the proof. (Dashed lines are dividing curves.)

deform the torus T^2 inside a neighbourhood of $T^2 \subset M$, fixing the dividing set Γ_{T^2} so that the characteristic foliation T^2_ξ consists of a 1-parameter family of closed curves, called *Legendrian rulings*, of the same slope r, called *ruling slope*. Each component of $T^2 \setminus \Gamma$ contains a line of singular points of slope s, called *Legendrian divide*. A convex torus in this (non-generic) form is said to be in *standard form* (see Fig. 7 for an illustration). An immediate consequence of Giroux's Flexibility Theorem is the following:

PROPOSITION 2.12 (flexibility of Legendrian rulings, [**24**]). *Assume T^2 is a convex torus in standard form, and, using coordinates in $\mathbb{R}^2/\mathbb{Z}^2$, has boundary slope s and ruling slope r. Then by a C^0-small perturbation near the Legendrian divides, we can modify the ruling slope from $r \neq s$ to any other $r' \neq s$ (∞ included).*

If a bypass is attached along some Legendrian ruling on T^2, we can push the torus across the bypass, using the bypass attachment (Proposition 2.11), which yields a new torus with different boundary conditions. If the torus division number n of T^2 is greater than one, this will yield a torus with division number $n-1$. In the

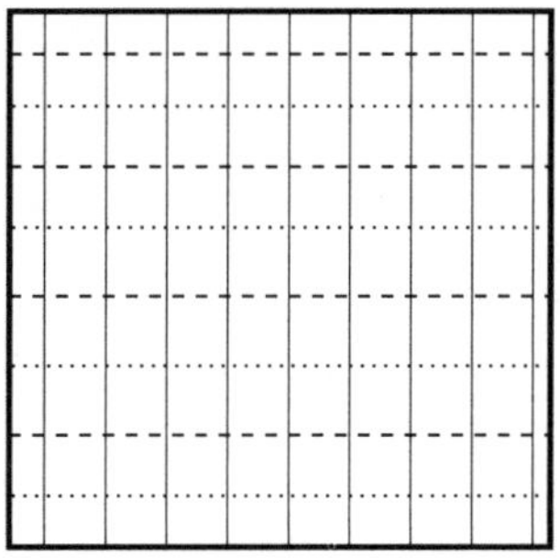

FIGURE 7. A torus T^2 in standard form, in some identification with $\mathbb{R}^2/\mathbb{Z}^2$, (Dividing curves are dashed, Legendrian divides are dotted horizontal and Legendrian rulings are vertical solid lines) i.e. the sides are identified and the bottom and top are identified.

case $n = 1$ attaching a bypass does not change the torus division number but the boundary slope of the torus; see [**24**]. In order to describe how the new boundary conditions are obtained from the old, we first recall the Farey tessellation of the hyperbolic disc: Consider the hyperbolic unit disc $\mathbb{H} = \{(x,y) : x^2 + y^2 \leq 1\}$. We label the point $(1,0)$ as $0 = \frac{0}{1}$, the point $(-1,0)$ as $\infty = \frac{1}{0}$ and join the two points by an arc. Now label inductively points on $S^1 = \partial\mathbb{H}$ as follows (for $y > 0$): assume we have labelled two points $\infty \geq \frac{p}{q}, \frac{p'}{q'} \geq 0$ (where numerators and denominators are relatively prime). Label the point half way between $\frac{p}{q}$ and $\frac{p'}{q'}$ along the shorter arc on S^1 by $\frac{(p+p')}{(q+q')}$. Connect two points $\frac{p}{q}$ and $\frac{p'}{q'}$ by an arc if the corresponding shortest integral vectors form an integral basis of $\mathbb{Z}^2$; see Fig. 8.

THEOREM 2.13 (Honda, [**24**]). *Assume T is a convex torus in standard form with $\#\Gamma = 2$ and boundary slope $s = s(T)$. If a bypass D is attached to T along a Legendrian ruling curve of slope $r \neq s$, then the resulting convex torus T' will have $\#\Gamma_{T'} = 2$ and boundary slope s' which is obtained as follows: take the arc $[r,s] \subset \partial\mathbb{H}$ obtained by starting from r and moving counterclockwise until we hit s. On this arc, s' is the point which is closest to r and has an edge from s' to s.*

2.3. Tight contact structures on basic blocks. A key principle in the classification of tight contact structures on 3-manifolds is to cut along convex surfaces to obtain simpler pieces on which the classification is known. In this subsection, we review the basic properties of tight contact structures on various simple pieces referred to as *basic blocks*.

2.3.1. *The 3-ball.* The following key Theorem was proven by Eliashberg in [**7**]:

THEOREM 2.14. *Assume there exists a contact structure ξ on a neighbourhood of ∂B^3 such that ∂B^3 is convex and $\#\Gamma_{\partial B^3} = 1$. Then there exists a unique extension of ξ to a tight contact structure on B^3, up to an isotopy relative to ∂B^3.*

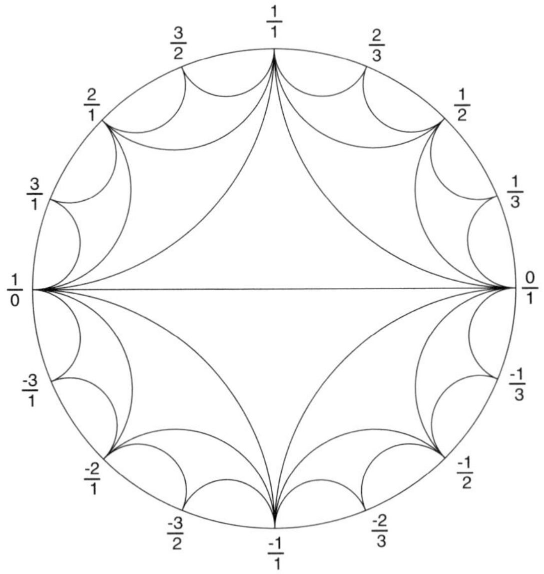

FIGURE 8. The Farey tessellation of the hyperbolic disc.

2.3.2. *The solid torus* $S^1 \times D^2$. Assume $\gamma \subset M$ is a Legendrian curve with a negative twisting number $\mathtt{t}(\gamma) = n$ with respect to some fixed framing. The *standard tubular neighbourhood* $N(\gamma)$ of γ is defined as solid torus $S^1 \times D^2$ with coordinates $(z,(x,y))$ and contact 1-form $\alpha = \sin(2\pi nz)dx + \cos(2\pi nz)dy$ and $\gamma = \{(z,(x,y)) : x = y = 0\}$. With respect to the fixed framing of γ, we may identify $\partial N(\gamma) = \mathbb{R}^2/\mathbb{Z}^2$ such that the meridian is $(1,0)^T$ and the longitude (fixed by the framing) is $(0,1)^T$. Then the boundary slope is $s(\partial N(\gamma)) = \frac{1}{n}$.

In standard neighbourhoods of Legendrian curves the model standard tubular neighbourhood provides a unique tight contact structure. This fact was used extensively by Kanda [**27**], and proved (in a slightly different form) by Makar-Limanov in [**33**]; we refer to Theorem 6.7 in [**11**].

Proposition 2.15. *There exists a unique tight contact structure on* $S^1 \times D^2$ *with a fixed convex boundary with* $\#\Gamma_{\partial(S^1\times D^2)} = 2$ *and slope* $s(\partial(S^1 \times D^2)) = \frac{1}{n}$, *where* n *is a negative integer. With the possibility of modifying the characteristic foliation on the boundary using the Flexibility Theorem (Proposition 2.12), the tight contact structure is isotopic to the standard neighbourhood of a Legendrian curve with twisting number* n.

Decreasing the twisting number of a Legendrian curve is feasible, as commonly understood, by adding a 'zigzag' in the front projection; see [**11**]. Increasing the twisting number is not an easy task, but possible in the presence of bypasses.

Proposition 2.16 (Twist Number Lemma, [**24**]). *Consider a Legendrian curve* γ *in a contact manifold* (M,ξ) *with twisting number* n *relative to a fixed framing and* N *a standard tubular neighbourhood of* γ. *If there exists a bypass attached to a Legendrian ruling curve of* ∂N *of slope* r *and* $\frac{1}{r} \geq n+1$, *then there exists a Legendrian curve with twisting number* $n+1$ *isotopic to* γ. *Notice that this isotopy cannot be a Legendrian isotopy because the twisting number changes.*

Suppose we have given a solid torus $S^1 \times D^2$ and an oriented identification of the boundary torus $\partial(S^1 \times D^2)$ with $\mathbb{R}^2/\mathbb{Z}^2$ so that $(1,0)^T$ corresponds to the meridian and $(0,1)^T$ corresponds to a longitude. Assuming the boundary of $S^1 \times D^2$ is a torus in standard form with torus division number one, the number of tight contact structures are determined by the boundary slope, i.e. the slope of the dividing curves. More precisely:

Theorem 2.17 (Theorem 2.3, [**24**]). *Let* $S^1 \times D^2$ *be a solid torus with convex boundary* T^2 *in standard form. If* $\#\Gamma_{T^2} = 2$ *and the boundary slope* $s(T^2) = -\frac{p}{q}$, $p \geq q > 1$, $(p,q) = 1$ *and continued fraction expansion*

$$-\frac{p}{q} = r_0 - \cfrac{1}{r_1 - \cfrac{1}{r_2 \cdots - \frac{1}{r_k}}},$$

with all $r_i < -1$, *then there are exactly* $|(r_0+1)(r_1+1)\dots(r_{k-1}+1)r_k|$ *tight contact structures on* $S^1 \times D^2$ *up to isotopy fixing the boundary.*

Proposition 2.18 (Lemma 3.16, [**12**]). *Assume* $S^1 \times D^2$ *has convex boundary with boundary slope* $s < 0$. *Then we can find a convex torus parallel to the boundary* $\partial(S^1 \times D^2)$ *with any boundary slope in* $[s,0)$.

2.3.3. *Toric annuli* $T^2 \times [0,1]$. Assume a toric annulus $T^2 \times I$ is given in coordinates $(x, y, z) \in \mathbb{R}^2/\mathbb{Z}^2 \times [0,1]$. A standard tight contact structure is given by $\alpha = \sin(\frac{\pi}{2}z)dx + \cos(\frac{\pi}{2}z)dy$. Note that the boundary $\partial(T^2 \times I) = T_0 - T_1$ consists (after a perturbation) of convex tori with boundary slope 0 and ∞ respectively. It is not hard to see that the tori $T^2 \times \{z\}$ are linearly foliated and the boundary slopes decrease as z increases. More generally, one obtains models for tight contact structures on toric annuli with different slopes on the boundary T_0, T_1 by changing the chosen interval I on the z-axis.

PROPOSITION 2.19 (Proposition 4.16, [**24**]). *Assume a toric annulus $T^2 \times I$ has convex boundary in standard form and the boundary slope on $T_i = T^2 \times \{i\}$ is s_i, $i = 0, 1$ respectively. Then we can find convex tori parallel to T_0 with any boundary slope s in $[s_1, s_0]$ (if $s_0 < s_1$ this means $[s_1, \infty] \cup [-\infty, s_0]$).*

On the other hand, consider a tight contact structure ξ on a toric annulus $T^2 \times I$ with convex boundary and boundary slope $s_i = s(T_i)$, $i = 0, 1$. We say ξ is *minimally twisting* (in the I-direction) if every convex torus parallel to the boundary has slope $s \in [s_1, s_0]$.

We will outline the classification of tight contact structures in thickened tori $T^2 \times I$. For a detailed description, we refer to Honda [**24**]. To state the Theorems, we first recall the notion of the relative Euler class. Consider a complex line bundle ξ on a 3-manifold M with boundary ∂M. Assume $\xi|_{\partial M}$ has a nowhere vanishing section s. We may define the *relative Euler class* $e(\xi, s) \in H^2(M, \partial M)$ as the obstruction to extending s to the whole manifold. It is related to the Euler class by the following exact sequence:

$$\begin{array}{ccccccc} H^1(\partial M) & \xrightarrow{\Delta} & H^2(M, \partial M) & \to & H^2(M) & \to & H^2(\partial M) \\ & & e(\xi, s) & \mapsto & e(\xi) & \mapsto & 0 \end{array}$$

The following two Lemmas are useful for the calculation of the relative Euler class of contact structures. The proofs are found in Section 4.2 of [**24**].

LEMMA 2.20. *Let (M, ξ) be a contact manifold with convex boundary, and s a fixed section of $\xi|_{\partial M}$.*

(1) *If $\Sigma \subset M$ is a closed convex surface with positive (resp. negative) region R_+ (resp. R_-) divided by Γ_Σ, then $\langle e(\xi), \Sigma \rangle = \chi(R_+) - \chi(R_-)$.*
(2) *If $\Sigma \subset M$ is a compact convex surface with Legendrian boundary on ∂M and regions R_+ and R_-, and s is homotopic to s' which coincides with $\dot{\gamma}$ for every oriented connected component γ of $\partial\Sigma$, then $\langle e(\xi, s), \Sigma \rangle = \chi(R_+) - \chi(R_-)$.*

LEMMA 2.21. *Let (M, ξ) be a tight contact manifold with convex boundary consisting of tori. Then the relative Euler class $\langle e(\xi, s), \Sigma \rangle$ is independent of the slope of the Legendrian rulings, where s is a nonzero section of ξ tangent to the ruling curves.*

In the following we use the Euler class only in its relative form. Thus, when ∂M is a union of convex tori in standard form, we will write $e(\xi)$ instead of $e(\xi, s)$ with an abuse of notations if the section s comes from a vector field tangent to the Legendrian rulings.

We say that two rational slopes in $\mathbb{Q} \cup \{\infty\}$ are *consecutive* if they are joined by an edge in the Farey tessellation. The very basic building blocks for contact structures are the minimally twisting tight contact structures on $T^2 \times I$ whose boundary slopes s_0 and s_1 are consecutive. Such contact structures are called *basic slices*. We have the following classification result for basic slices.

THEOREM 2.22 ([**24**], Section 4.3). *Given consecutive s_0 and s_1, there are, up to isotopy fixed on the boundary, two minimally twisting tight contact structures on $T^2 \times I$ with boundary in standard form, $\#\Gamma_{T_i} = 2$ for $i = 0, 1$, and boundary slopes s_0 and s_1. The two contact structures are distinguished by their relative Euler class, and both can be contact-embedded in a tight contact structure on T^3.*

Let v_0 and v_1 be shortest integer vectors representing the slopes s_0 and s_1 of a basic slice, such that (v_1, v_0) is a positively oriented basis. The possible relative Euler classes of the basic slices are the Poincaré duals of the homology classes represented by $\pm(v_1 - v_0)$. For this reason, in what follows we will refer to the isotopy class of a basic slice as its *sign*.

The basic slices are basic in the sense that any minimally twisting tight contact structure on $T^2 \times I$ can be decomposed into basic slices, as explained in the following Theorem:

THEOREM 2.23 (Lemma 4.12, [**24**]). *Given a minimally twisting, tight contact structure on $T^2 \times I$ with boundary slopes s_0 and s_1, we can find a partition $0 = t_0 < \ldots < t_k = 1$ of $[0,1]$ such that $T^2 \times \{t_i\}$ is a convex torus in standard form with slope s_{t_i} for any $i = 0, \ldots, k$. These slopes form a counterclockwise sequence in the arc $[s_1, s_0]$ with the property that s_{t_i} and $s_{t_{i+1}}$ are consecutive and there is no edge in the Farey tessellation joining s_{t_i} and $s_{t_{i+2}}$.*

Moreover such a basic slices decomposition is minimal in the sense that any other basic slices decomposition of the given tight contact structure on $T^2 \times I$ is a further decomposition of this one.

Note that the intermediate slopes of the basic slices decomposition depend only on s_0 and s_1 and are independent of the isotopy class of the contact structure.

Conversely, given a decomposition into basic slices, we have the following gluing Theorem, which is a particular case of the more general Theorem 4.24 in [**24**].

THEOREM 2.24. *Let $s_1 < s_0$ be rational slopes. Then every choice of signs for the basic slices in the basic slices decomposition associated to s_0 and s_1 realizes a minimally twisting, tight contact structure on $T^2 \times I$ with slopes s_0 and s_1.*

We observe that, unlike basic slices, in general these tight contact structures cannot be contact-embedded into a tight contact structure on T^3. This is possible if and only if the relative Euler class is $\pm(v_1 - v_0)$.

We may ask when two different choices of signs for the basic slices give the same contact structure. We say the basic slices in $T^2 \times [t_j, t_l]$ form a *continued fraction block* if there is a slope r such that there is an edge in the Farey tessellation joining r and t_i for all $j \leq i \leq l$. To understand the origin of the name, see [**24**], where this concept appeared for the first time in this context. The importance of this notion comes from the fact that the sign of basic slices belonging to the same continued fraction block can be shuffled without affecting the isotopy type of the contact structure on $T \times I$. This is a nontrivial result whose proof can be found in [**24**].

In this paper we will use the property of continued fraction blocks only in the following case. Let $T^2 \times I$ carry a minimally twisting tight contact structure with boundary slopes $s_0 = -\frac{1}{n}$, for $n > 0$, and $s_1 = \infty$. Then all the basic slices of its decomposition belong to the same continued fraction block, and therefore their signs can be shuffled.

As a result of the classification of basic slices, the basic slices decomposition, and this last fact about continued fraction blocks, we can derive the following classification Theorem for tight contact structures on $T^2 \times I$:

THEOREM 2.25 (Proposition 4.22, [**24**]). *The minimally twisting tight contact structures on $T^2 \times I$ with standard boundary, $\#\Gamma_{T_i} = 2$ and boundary slopes s_0 and s_1 are distinguished up to isotopy fixed on the boundary by their relative Euler class.*

Actually, in this Theorem Honda proves more than what he states: in fact he shows that, after normalising the boundary slopes to $s_0 = -1$ and $s_1 < -1$, the tight contact structures are distinguished by the value their relative Euler class takes at a horizontal annulus with Legendrian boundary. He proves this fact by showing that each basic slice in the decomposition gives a contribution to the value of the relative Euler class which is bigger than the sum of the contributions of the basic slices belonging to all the preceding continued fraction blocks. The same arguments prove the following Corollary.

COROLLARY 2.26. *The minimally twisting tight contact structures on $T^2 \times I$ with standard boundary, $\#\Gamma_{T_i} = 2$ and boundary slopes $s_0 = 0$ and $s_1 < -1$ are distinguished up to isotopy fixed on the boundary by the value their relative Euler class takes at a horizontal annulus with Legendrian boundary.*

The number of minimally twisting tight contact structures on $T^2 \times I$ with fixed dividing set on the boundary is finite, and is expressed as a function of the continued fraction representation of s_1, after normalising s_0 to -1 by a change of coordinates, as in the case of solid tori in Theorem 2.17.

We mention also the case where the boundary slopes are equal. Then, either any convex intermediate torus $T \subset T^2 \times I$ has the same slope as the boundary tori, or there is a convex torus of slope s for any $s \in \mathbb{Q}$. This is a consequence of Proposition 2.19.

The tight contact structures on $T^2 \times I$ with boundary slopes $s_0 = s_1$ such that all the intermediate convex tori have the same slope are called *non-rotative*.

3. Legendrian (-1) surgery

A very useful method to construct contact structures on a manifold is given by Legendrian surgery. Together with the fact that, in particular, Legendrian (-1) surgery preserves fillability one can construct tight contact structures. Suppose γ is a Legendrian curve in a contact manifold (M, ξ) and we fix a framing $\mathcal{F}$ of γ such that the twisting number is zero $\mathtt{t}(\gamma, \mathcal{F}) = 0$; for example take $\mathcal{F}$ to be the contact framing of γ. We then find a standard neighbourhood $N(\gamma) = S^1 \times D^2$ with convex boundary so that the dividing set $\Gamma_{\partial N(\gamma)}$ consists of two parallel curves. Take an oriented identification $-\partial(M \setminus N(\gamma)) = \partial N(\gamma) \simeq \mathbb{R}^2/\mathbb{Z}^2$ so that $(1,0)^T$ corresponds to the meridian and $(0,1)^T$ to the longitude given by a dividing curve. Thus the boundary slope $s(\partial N(\gamma))$ is infinite and the meridian has slope zero. Let $M' = (M \setminus N(\gamma)) \cup_f N(\gamma)$, where $f : \partial N(\gamma) \to -\partial(M \setminus N(\gamma))$ is a diffeomorphism corresponding to

$$\begin{bmatrix} 1 & 0 \\ -1 & 1 \end{bmatrix} \in SL_2(\mathbb{Z}).$$

Topologically this corresponds to a (-1) Dehn surgery along γ with respect to the chosen framing, and the contact structure can be glued together, after possibly adjusting the characteristic foliation, since the dividing sets on $-\partial(M \setminus N(\gamma))$ and $f(\partial N(\gamma))$ are isotopic. More generally one can define Legendrian (r) surgery, for $r \in \mathbb{Q}$, as described in [**3**].

Legendrian (-1) surgery corresponds to a handlebody construction in the sense of [**19, 35**] and thus preserves fillability. Recall that a contact manifold (M, ξ) is called *holomorphically fillable* if it is the oriented boundary of a compact Stein surface. For example the standard tight contact structure on the three-sphere $S^3 \subset \mathbb{C}^2$, given as the oriented plane field of tangent complex lines, is holomorphically fillable. It is a remarkable result of Gromov [**21**] and Eliashberg [**5**] that fillable contact structures are tight. Furthermore:

THEOREM 3.1 (Eliashberg, [**6**]). *If (M', ξ') is obtained from a holomorphically fillable contact manifold (M, ξ) by Legendrian (-1) surgery as described above, then (M', ξ') is holomorphically fillable.*

In fact both Theorems remain true for any notion of fillability (see [**13**] for a survey), but Legendrian (-1) surgery does not preserve tightness in case of contact manifolds with boundary, as pointed out in [**23**].

4. Applications

In this section, we will show how to obtain the classification of tight contact structures in two specific cases: the Brieskorn homology spheres $\pm\Sigma(2, 3, 11)$. Both manifolds are Seifert fibred spaces over the sphere with three singular fibres.

Consider a Seifert manifold M with three singular fibres over S^2. M is described by Seifert invariants $(\frac{\beta_1}{\alpha_1}, \frac{\beta_2}{\alpha_2}, \frac{\beta_3}{\alpha_3})$, we refer to [**22**] for an introduction.

Assume V_i are solid tori $S^1 \times D^2$ with core curves F_i, $i = 1, 2, 3$. We identify ∂V_i with $\mathbb{R}^2/\mathbb{Z}^2$ by choosing $(1, 0)^T$ as the meridional direction and $(0, 1)^T$ as a longitudinal direction. Furthermore consider $S^1 \times \Sigma$, where Σ is a three-punctured sphere, i.e. a pair of pants. We identify each boundary component of $-\partial(S^1 \times \Sigma)$ with $\mathbb{R}^2/\mathbb{Z}^2$ by setting $(0, 1)^T$ as the direction of the S^1–fibre and $(1, 0)^T$ as the direction given by $-\partial(\{pt\} \times \Sigma)$. Then we obtain the Seifert manifold $M(\frac{\beta_1}{\alpha_1}, \frac{\beta_2}{\alpha_2}, \frac{\beta_3}{\alpha_3})$ by attaching the solid tori V_i to $S^1 \times \Sigma$, where the attaching maps $A_i : \partial V_i \to -\partial((S^1 \times \Sigma)_i)$ are given by

$$A_i = \begin{bmatrix} \alpha_i & \gamma_i \\ -\beta_i & \delta_i \end{bmatrix} \in SL(2, \mathbb{Z})$$

REMARK 4.1. Note that we often refer to the same surface by different names. For example ∂V_i and $\partial(M \setminus V_i)$ denote the same torus, but the identification with $\mathbb{R}^2/\mathbb{Z}^2$ is different.

REMARK 4.2. Unless stated otherwise, properly embedded surfaces in a contact manifold are understood to be convex, if possible.

Note that the three singular fibres F_i may be isotoped to be Legendrian so that their twisting numbers n_i are particularly negative. Recall that a standard neighbourhood V_i of F_i with convex boundary has boundary slope $\frac{1}{n_i}$. Furthermore we may assume that the ruling slope on $-\partial(M \setminus V_i)$ is infinite, thereby using the flexibility of Legendrian rulings (Proposition 2.12). Starting with this initial configuration, we do the following:

(1) In the first step, we try to maximise the twisting numbers of the singular fibres. For this, consider a vertical annulus $A = S^1 \times I$ with Legendrian boundary along ruling curves of two different tori V_i, V_j. If the Imbalance Principle forces a bypass on A we may apply the Twist Number Lemma (Proposition 2.16) to increase one of the twisting numbers n_i or n_j. Repeating this process, two different situations might occur.
  (a) Either there exists a bypass on A however we cannot apply the Twist Number Lemma. In this case, we can thicken the tori by attaching the bypass. In the cases below this yields an infinite boundary slope on a $-\partial(M \setminus V_i)$. Consider a vertical annulus from a Legendrian divide to the other two tori, we can thicken all three tori so that $s(-\partial(M \setminus V_i)) = \infty$, $i = 1, 2, 3$.
  (b) There exists no bypass on A. In this case a tubular neighbourhood of $V_i \cup V_j \cup A$ is a piecewise smooth torus with exactly four edges. Rounding the edges using the Edge-Rounding Lemma (Proposition 2.6), we obtain a torus with boundary slope s, which can be thought of as the boundary of a neighbourhood of the third singular fibre F_k. In case $s < s(\partial(M \setminus V_k))$ we can eventually increase the twisting number of F_k.

  In either case, this process ends in a configuration with fixed boundary conditions on the basic blocks V_i, $i = 1, 2, 3$ and $M \setminus \{V_1 \cup V_2 \cup V_3\}$. Combinations of tight contact structures on the basic blocks give a possible tight contact structure on M. Since there are finitely many tight contact structures on each basic piece, we obtain an upper bound on the number of tight contact structures on M.
(2) In the second step we try to further analyse combinations of tight contact structures on the basic blocks. Observe that if we were able to find further bypasses to thicken one V_i such that V_i contains a neighbourhood V_i' of F_i so that the boundary slope $s(\partial V_i')$ is zero, we would find an overtwisted disc as meridional disc with boundary a Legendrian divide, and thus reduce the number of potentially tight contact structures on M.

  In the examples below, we are able to find further bypasses and eventually find an overtwisted disc in case there exists a thickening so that $-\partial(M \setminus V_i)$ has infinite boundary slope.
(3) We finally construct tight contact structures by Legendrian surgery to show that the upper bound is sharp.

REMARK 4.3. Note that the strategy of constructing tight contact structures by Legendrian surgery yields fillable contact structures. Hence this strategy is not successful in every case; see [**13**] and section 5.

4.1. The case $\Sigma(2,3,11)$. In this subsection, M denotes the Brieskorn homology sphere $\Sigma(2,3,11)$. This corresponds to the manifold $M(\frac{1}{2}, -\frac{1}{3}, -\frac{2}{11})$ in the notation above.

THEOREM 4.4. *On the Seifert manifold $\Sigma(2,3,11)$ there exist, up to isotopy, exactly two tight contact structures, which are both holomorphically fillable.*

The attaching maps are given by

$$A_1 = \begin{bmatrix} 2 & 1 \\ -1 & 0 \end{bmatrix}, \qquad A_2 = \begin{bmatrix} 3 & -1 \\ 1 & 0 \end{bmatrix}, \qquad A_3 = \begin{bmatrix} 11 & -6 \\ 2 & -1 \end{bmatrix}.$$

Assume the singular fibres F_i are (simultaneously) isotoped to Legendrian and further isotoped such that their twisting numbers n_i are particularly negative. The standard neighbourhood of F_i is denoted by V_i and the slope of the dividing curves on ∂V_i is $\frac{1}{n_i}$. Because $A_1 \cdot (n_1, 1)^T = (2n_1 + 1, -n_1)^T$, $A_2 \cdot (n_2, 1)^T = (3n_2 - 1, n_2)^T$ and $A_3 \cdot (n_3, 1)^T = (11n_3 - 6, 2n_3 - 1)^T$, we calculate the boundary slopes on $-\partial(M \setminus V_i)$ $(i = 1, 2, 3)$ to be $\frac{-n_1}{2n_1+1}$, $\frac{n_2}{3n_2-1}$, and $\frac{2n_3-1}{11n_3-6}$, respectively.

4.1.1. *Increasing twisting numbers of singular fibres.* We try to increase the twisting numbers of the singular fibres as far as possible. As described above, we start by assuming the twisting numbers $n_i < 0$ are particularly negative.

LEMMA 4.5. *We can increase the twisting numbers n_i of the singular fibres F_i, $i = 1, 2, 3$, up to $n_1 = -1$, $n_2 = n_3 = 0$.*

PROOF. Using the flexibility of Legendrian rulings, we modify the Legendrian rulings on each $\partial(M \setminus V_i)$ to have infinite slope. Consider a vertical annulus $S^1 \times I$ from $\partial(M \setminus V_1)$ to $\partial(M \setminus V_2)$ such that the boundary consists of Legendrian ruling curves on the tori. Observe that the boundary of this annulus intersects the dividing curves on $\partial(M \setminus V_i)$ exactly $2(2n_1 + 1)$ and $2(3n_2 - 1)$ times respectively.

If $2n_1 + 1 \neq 3n_2 - 1$, then, due to the Imbalance Principle (Proposition 2.10), there exists a bypass along a Legendrian ruling curve either on $\partial(M \setminus V_1)$ or $\partial(M \setminus V_2)$. The Legendrian rulings on ∂V_1 have slope -2 and we can apply the Twist Number Lemma (Proposition 2.16) to increase the twisting number of a singular fibre by one as long as $n_1 < -1$. A similar argument shows that we can use the Twist Number Lemma to increase n_2 as long as $n_2 < 0$.

Assume $2n_1 + 1 = 3n_2 - 1$ and there exists no bypass on a vertical annulus $A = S^1 \times I$ between $\partial(M \setminus V_1)$ and $\partial(M \setminus V_2)$. We cut along the tori connected by A and round the corners using the Edge-Rounding Lemma: For this, observe that a neighbourhood of $M \setminus (V_1 \cup V_2 \cup S^1 \times I)$ is a piecewise smooth solid torus with four edges. Using the Edge-Rounding Lemma (Proposition 2.6), each rounding changes the slope by an amount $-\frac{1}{4}\frac{1}{2n_1+1}$. Because there are four edges to round, we get on the boundary $\partial(M \setminus V_1 \cup V_2 \cup S^1 \times I)$ the slope

$$\frac{-n_1}{2n_1+1} + \frac{\frac{2}{3}(n_1+1)}{2n_1+1} + \frac{-1}{2n_1+1} = -\frac{1}{3}\frac{n_1+1}{2n_1+1}.$$

Note that we identified this torus with $\mathbb{R}^2/\mathbb{Z}^2$ in the same way as $\partial(M \setminus V_3)$. Since $A_3^{-1} \cdot (6n_1+3, n_1+1)^T = (3, -n_1+5)^T$, this corresponds to slope $s = -\frac{1}{3}n_1 + \frac{5}{3}$ when measured using ∂V_3. Now $s > 1$ for $n_1 < 0$ and we find a standard neighbourhood V_3 of F_3 with infinite boundary slope, corresponding to $n_3 = 0$. We remark that the boundary slope becomes $\frac{1}{6}$, when measured with respect to $-\partial(M \setminus V_3)$.

Next, to increase the twisting number n_2, take a vertical annulus $S^1 \times I$ from a Legendrian ruling on $\partial(M \setminus V_2)$ to a Legendrian ruling on $\partial(M \setminus V_3)$. Observe that if $n_2 < -2$, we have $|3n_2 - 1| > 6$ and thus there exists a bypass on the V_2 side along a vertical Legendrian ruling, which allows us to increase n_2 up to -1 by the Twist Number Lemma. A similar argument shows that we can increase n_1 up to -2.

Now the slopes on $-\partial(M \setminus V_i)$ are $-\frac{2}{3}$, $\frac{1}{4}$ and $\frac{1}{6}$ respectively. Taking, once more, a vertical annulus between V_1 and V_2, we find a bypass due to the Imbalance Principle and are finally able to increase n_2 to 0 and n_1 to -1. □

We have now arrived at $n_1 = -1$, $n_2 = n_3 = 0$. Note that the boundary slopes on $-\partial(M \setminus V_i)$ are -1, 0 and $\frac{1}{6}$ respectively. Take again a vertical annulus between V_1 and V_2. There are two possibilities: Either there exists a bypass along both boundary components or not. If there is a bypass, the cutting and rounding construction yields a torus of infinite slope. We use vertical annuli from a Legendrian divide of this torus to thicken each V_i to V_i' s.t. $-\partial(M \setminus V_i')$ has infinite boundary slope. In case there is no bypass, we perform a cutting and rounding construction on V_1 and V_2 as in the proof of Lemma 4.5 to obtain a further thickening of V_3 to V_3' such that $-\partial(M \setminus V_3')$ has boundary slope 0. We have shown that there are two possibilities, distinguished by whether or not there exists a thickening of all V_i such that the boundary slope with respect to $-\partial(M \setminus V_i)$ is infinite for $i = 1, 2, 3$. We will primarily be concerned with

4.1.2. *The case when a thickening to infinite slope exists.* We will now show that all possible tight contact structures arising in this case are overtwisted. We do this by patching together meridional discs of two solid tori thus obtaining a surface with boundary on the third torus and relate its dividing set to the dividing curves given on the discs. This may produce a bypass which allows a further thickening, i.e. increasing the twisting and eventually becoming overtwisted. In order to do this patching we have to examine the possible tight contact structures on the complement of the singular fibres $S^1 \times \Sigma \cong M \setminus (\cup_i V_i')$. Consider $\Sigma = \{1\} \times \Sigma$. Each boundary component of Σ intersects the dividing set of the corresponding tori twice and therefore contains exactly two half-elliptic points. The following two Lemmata are proven by Etnyre and Honda in [**14**]. We enclose the proofs for the reader's convenience.

LEMMA 4.6. *The dividing set on Σ consists of arcs, each connecting two different boundary components.*

PROOF. Assume there is a boundary-parallel dividing arc as shown, for example, in cases (A) and (B) of Fig. 9. This implies the existence of a bypass along some $\partial(M \setminus V_i')$. Attaching this bypass yields a thickening V_i'' of V_i' with slope 0. Take a vertical annulus from a Legendrian dividing curve on a $\partial(M \setminus V_j')$ $(j \neq i)$ to $\partial(M \setminus V_i'')$. We find a bypass on this annulus producing a further thickening to V_i''' such that $\partial(M \setminus V_i''')$ has infinite boundary slope. Therefore, by Proposition 2.18 we find a neighbourhood V_i of F_i so that the boundary slope of ∂V is zero. A meridional disc in V whose boundary is a Legendrian divide on ∂V is an overtwisted disc. Possible configurations of the dividing set on Σ without boundary-parallel arcs are as shown in Fig. 9 (C), up to twisting as shown in Fig. 9 (D). □

LEMMA 4.7. *There exists a unique tight contact structure on $S^1 \times \Sigma$, up to isotopy moving the boundary, where the configuration of the dividing set on Σ is given as in Lemma 4.6.*

PROOF. We cut along Σ and round the edges using the Edge-Rounding Lemma thus obtaining a solid two-handlebody. We can arrange the dividing set on the boundary so that two meridional discs intersect the dividing set exactly twice; see Fig. 10. Cutting along these two discs we obtain a three-ball. Since there is a unique tight contact structure on the three-ball (Theorem 2.14) and the dividing curves on the surface we cut along are determined by the initial data, we must have a unique tight contact structure on $S^1 \times \Sigma$. □

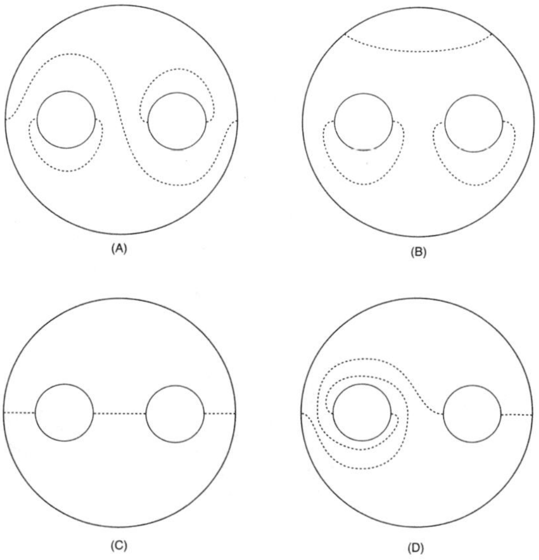

FIGURE 9. Possible constellations of dividing curves on the pair of pants Σ. (Dividing curves are dashed lines.)

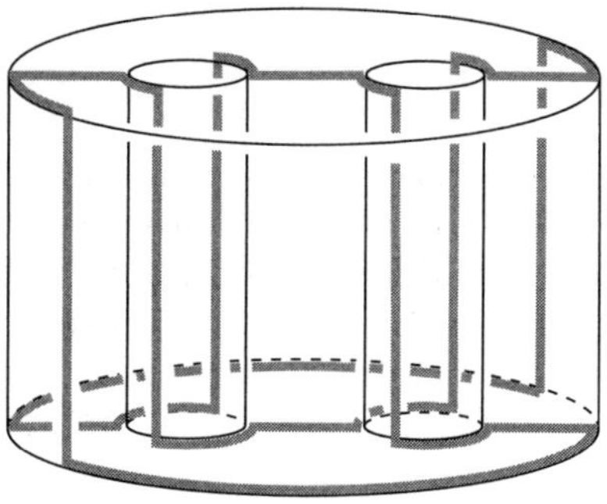

FIGURE 10. The dividing set on the boundary of the two-handle. (The dividing curve, after rounding edges, is drawn as a thick curve.)

Using the flexibility of Legendrian rulings, we can choose the slopes on the V_i' to be 0, this is possible since the boundary slopes, measured on $\partial V_i'$ are -2, 3, and $\frac{11}{6}$, respectively. Take a meridional disc D_i in V_i' such that ∂D_i consists of a Legendrian ruling curve. On $-\partial(M \setminus V_i')$, these discs have slopes $-\frac{1}{2}$, $\frac{1}{3}$ and $\frac{2}{11}$ respectively.

Since $\#(\partial D_i \cap \Gamma_{\partial(M\setminus V_i')}) = 4$, 6 and 22, we obtain $\mathtt{tb}(\partial D_i) = -2$, -3 and -11 and the possible constellations of dividing curves on D_i, are distinguished by their relative Euler number, i.e. the rotation number of ∂D_i according to $\mathtt{r}(\partial D_i) = \chi(D_i^+) - \chi(D_i^-)$.

The two Lemmata above imply that the tight contact structure on $S^1 \times \Sigma$ is contactomorphic to an I-invariant contact structure on a $T^2 \times I$ with a neighbourhood of a vertical Legendrian curve of zero twisting removed. View the $T^2 \times I$

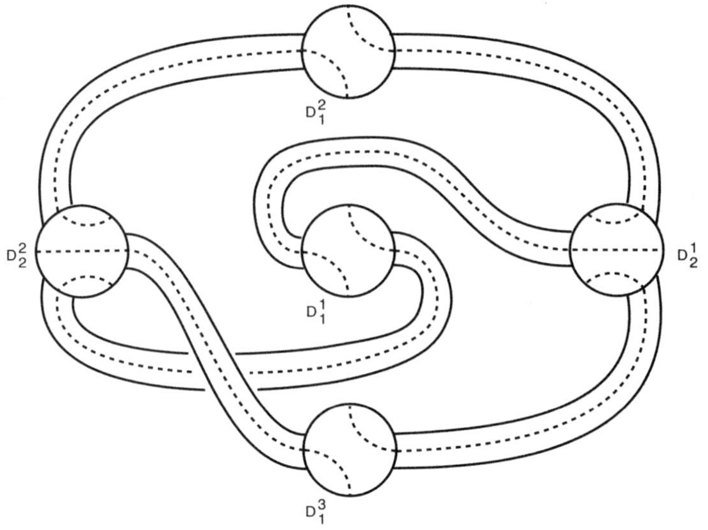

FIGURE 11. Possible constellations of dividing curves on the punctured torus P, determined by the 'signs' of the bypasses on the meridional discs. (Dividing curves are dashed lines.)

(minus $S^1 \times D^2$) from the above Lemma as the region between $\partial V_1'$ and $\partial V_2'$ (minus $\partial V_3'$) i.e. assume $T_0 = \partial V_2'$ and $T_1 = -\partial V_1'$. We write $T_t = T^2 \times \{t\}$, $t \in [0,1]$.

Now pick three copies of meridional discs D_1 in V_1 and two copies of meridional discs D_2 in V_2. Due to the I-invariance of ξ, we have a 1-parameter family of positive regions $(T_t)^+ = (T^2)^+ \times \{t\}$. Consider D_1 and D_2 such that $(D_2 \cap T_0)^+ = \delta \times \{0\}$ and $(D_1 \cap T_1)^+ = \delta \times \{1\}$, where δ is a union of Legendrian arcs on $(T^2)^+$. Now $P = D_1 \cup D_2 \cup \delta \times [0,1]$ is a punctured torus. After smoothing the corners using the Pivot-Lemma, P has smooth boundary $\partial P \subset \partial V_3'$ with slope $-\frac{1}{2} + \frac{1}{3} = -\frac{1}{6}$, measured using $\partial(M \setminus V_3')$.

Case (1) D_1 and D_2 have bypasses of the same sign. Then the dividing set on P contains a boundary-parallel curve, i.e. there exists a bypass on P. We can think of this bypass as attached on $-\partial(M \setminus V_3')$, along a ruling curve of slope $\frac{1}{6}$. Attaching this bypass (Theorem 2.13) yields a thickening of V_3' to V_3'' such that $-\partial(M \setminus V_3'')$ has boundary slope 1. Repeating the argument of Lemma 4.5 shows that we can increase the twisting numbers n_1 and n_2 to -1 and 0 respectively. Thus we can thicken V_3'' further to V_3''' such that the boundary slope of $-\partial(M \setminus V_3''')$ is 0. Recall that we started in a stage where the boundary slope of $-\partial(M \setminus V_3)$ is zero. We assumed to find a thickening that this slope becomes infinite and showed that we are then able to thicken further to obtain slope zero again. Thus we find a neighbourhood V of F_3 so that $s(\partial V) = 0$, and a meridional disc in V bounding a Legendrian divide is an overtwisted disc.

Case (2) D_1 and D_2 have different sign. Then there is a bypass on D_3 of the same sign as a bypass on D_1 or D_2. Assume D_1 and D_3 contain bypasses of the same sign. A similar argument as in Case (1) shows that patching eleven copies of D_1 and two copies of D_3 yields a surface whose boundary is contained in $\partial(M \setminus V_2')$ with slope $-\frac{1}{2} + \frac{2}{11} = -\frac{7}{22}$. A bypass on each D_1 and D_3 joins to a bypass on the patched surface, if both have the same sign. Thus we find a bypass along V_2' and its attaching yields a thickening to V_2'' such that the boundary slope of $-\partial(M \setminus V_2'')$ is 1. Repeating the argument of Lemma 4.5 shows that we can again arrange $n_1 = -1$ and $n_3 = 0$. A cutting and rounding construction gives a further thickening of V_3' to V_3'' such that $-\partial(M \setminus V_3'')$ has boundary slope -1 and hence contains an overtwisted disc.

Similarly, if D_2 and D_3 contain a bypass of the same sign, we patch together eleven copies of D_2 and three copies of D_3 to obtain a surface whose boundary is contained in $\partial(M \setminus V_1')$ with slope $\frac{1}{3} + \frac{2}{11} = \frac{17}{33}$. We find a bypass and its attaching yields a thickening of V_1 to V_1'' such that $-\partial(M \setminus V_1'')$ has boundary slope 0. Make n_2 and n_3 again particularly negative and the same argument as in Lemma 4.5 shows that we can increase both n_2 and n_3 to 0. Then, cutting and rounding along a vertical annulus between V_1 and V_2 gives neighbourhood V_3'' of F_3 with boundary slope 1 when measured using $-\partial(M \setminus V_3'')$. Since the boundary slope on $-\partial(M \setminus V_1'')$ is zero, we find by Proposition 2.18 a neighbourhood V of F_1 so that $-\partial(M \setminus V)$ has infinite boundary slope. Take a vertical annulus $A = S^1 \times I$ from a Legendrian divide on $\partial(M \setminus V)$ to $\partial(M \setminus V_3'')$. There exists a bypass on A whose attachment yields a further thickening of V_3'' to V_3''' where $\partial(M \setminus V_3''')$ has slope zero. Therefore, V_3''' contains, by Proposition 2.18 a neighbourhood V of F_3 so that the boundary slope of $-\partial(M \setminus V)$ is $\frac{2}{11}$. A meridional disc in V with boundary a Legendrian divide on ∂V is overtwisted. Hence we have eliminated all possibilities in case there exists a thickening of the V_i such that $s(-\partial(M \setminus V_i))$ is infinite.

4.1.3. *The case when no thickening exists.* We are left now with the case when there exists no thickening of the standard neighbourhoods so that the boundary slopes of the complements is infinite. We have the following conditions: for the first singular fibre F_1 we obtained twisting number $n_1 = -1$, hence a standard neighbourhood V_1 has boundary slope -1. Measured using $\partial(M \setminus V_1)$, the boundary slope is 1, because $A_1 \cdot (-1,1)^T = (-1,1)^T$. For the second singular fibre F_2 we obtained twisting number $n_2 = 0$ and hence a standard neighbourhood of F_2 has infinite boundary slope, which corresponds to slope 0, when measured using $\partial(M \setminus V_i)$. Lastly, for the third singular fibre, the twisting number is $n_3 = 0$ and the slope on $-\partial(M \setminus V_3)$ is $\frac{1}{6}$. A cutting and rounding construction along a vertical annulus between V_1 and V_2 yields a further thickening of V_3 such that $-\partial(M \setminus V_3)$ has boundary slope 0.

In the first and second solid torus V_1 and V_2 there exists exactly one tight contact structure as standard neighbourhood of Legendrian fibres. Because $A_3^{-1} \cdot (1,0)^T = (-1,-2)^T$ we find two tight contact structures on V_3.

The remaining block is $\Sigma \times S^1 = M \setminus (V_1 \cup V_2 \cup V_3)$, where we can arrange the boundary components of the pair of pants Σ to be Legendrian along the boundary components of $\Sigma \times S^1$. With this boundary conditions there exists exactly one tight contact structure on this block. This is due to the following Lemma, as part of Lemma 5.1. in **[25]**

LEMMA 4.8. *If for $S^1 \times \Sigma$, where Σ is a pair of pants, we have on the boundary tori $\partial(S^1 \times \Sigma) = T_1 + T_2 + T_3$ slopes* 1, 0, 0 *respectively, then there exists exactly one tight contact structure on $S^1 \times \Sigma$ with no vertical Legendrian curve.*

Thus there are at most two tight contact structures on M. In the next section, we use Legendrian surgery to see that there are two Stein fillable contact structures on M.

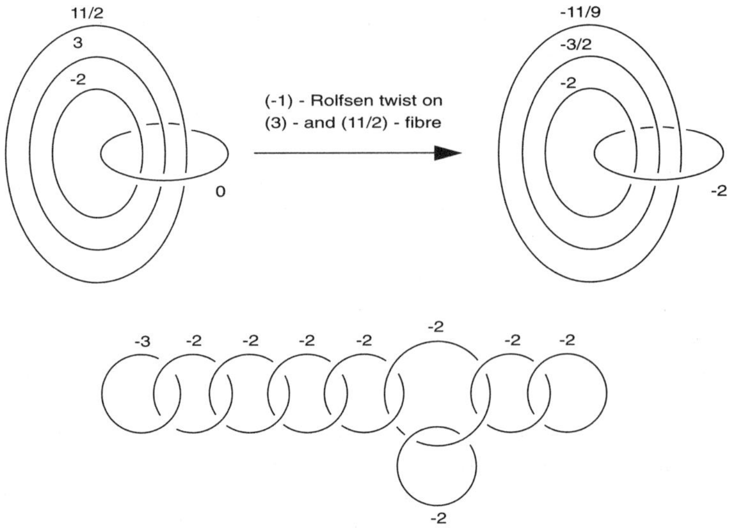

FIGURE 12. Surgery description of the Seifert manifold. By Kirby calculus we obtain the description where all components of the link have coefficients < -1.

4.1.4. *Construction of a tight contact structure.* We will describe now how to establish a tight contact structure on M by Legendrian surgery. The Seifert manifold $M(\frac{1}{2}, -\frac{1}{3}, -\frac{2}{11})$ has a surgery description as the left hand side of Fig. 12. By performing one (-1)-Rolfsen twist on the (3) and ($\frac{11}{2}$) fibre, we obtain the surgery description as shown on the right hand side; see [**20**]. Observe that we have the continued fraction expansions $-\frac{3}{2} = [-2, -2]$ and $-\frac{11}{9} = [-2, -2, -2, -2, -3]$, thus the surgery description as at the bottom in Fig. 12. Because the surgery coefficients are -2 or -3 we may conclude that there are exactly two Legendrian realizations of this link, where each component of the link will have $\mathtt{tb} = -1$ and hence $r = 0$, except one with $\mathtt{tb} = -2$ and $\mathtt{r} = \pm 1$. Note that the difference in the rotation number in a component of the two Legendrian realizations implies that the Chern classes of the corresponding Stein surfaces are different, thus that the contact structures on the boundary are non isotopic; see Theorem 4.20 in [**24**] and [**31, 32**] for details.

4.2. The case $-\Sigma(2, 3, 11)$. In this subsection M will denote the Seifert manifold over S^2 with three singular fibres with invariants $(-\frac{1}{2}, \frac{1}{3}, \frac{2}{11})$, corresponding to the Brieskorn homology sphere $-\Sigma(2, 3, 11)$.

THEOREM 4.9. *On the Seifert manifold $-\Sigma(2, 3, 11)$ there exists, up to isotopy, exactly one tight contact structure, which is holomorphically fillable.*

Assume V_i are tubular neighbourhoods of the singular fibres F_i, $i = 1, 2, 3$, and identify $M \setminus \cup V_i$ with $\Sigma \times S^1$, where Σ is a pair of pants. We identify ∂V_i and $-\partial(M \setminus V_i)$ with $\mathbb{R}^2/\mathbb{Z}^2$ as in the previous example. The gluing maps $A_i : \partial V_i \to -\partial(M \setminus V_i)$ are given by

$$A_1 = \begin{bmatrix} 2 & -1 \\ 1 & 0 \end{bmatrix}, \qquad A_2 = \begin{bmatrix} 3 & 1 \\ -1 & 0 \end{bmatrix}, \qquad A_3 = \begin{bmatrix} 11 & 6 \\ -2 & -1 \end{bmatrix}.$$

4.2.1. *Increasing the twisting number of the singular fibres.* We begin by increasing the twisting number of the singular fibres as far as possible in a similar way as in the previous example. We start by assuming the singular fibres F_i are simultaneously isotoped to Legendrian curves with very negative twisting numbers n_i. The slopes of ∂V_i are $\frac{1}{n_i}$, while the slopes of $-\partial(M \setminus V_i)$ are $\frac{n_1}{2n_1-1}$, $-\frac{n_2}{3n_2+1}$, and $-\frac{2n_3+1}{11n_3+6}$ respectively.

LEMMA 4.10. *We can increase the twisting numbers n_1 and n_2 up to -2, and the twisting number n_3 up to -1.*

PROOF. Using Proposition 2.12, we modify the Legendrian rulings on each $-\partial(M \setminus V_i)$ to have infinite slope. Consider a convex annulus A whose boundary consists of Legendrian rulings of $-\partial(M \setminus V_1)$ and $-\partial(M \setminus V_2)$. If $2n_1-1 \neq 3n_2+1$ the Imbalance Principle provides a bypass along a Legendrian ruling either in $\partial(M \setminus V_1)$ or in $\partial(M \setminus V_2)$. Using such a bypass we can apply the Twist Number Lemma to increase the twisting number n_i of a singular fibre by one as long as $n_1 < 0$ and $n_2 < -1$.

If $2n_1 - 1 = 3n_2 + 1$, and there exist no bypasses on A, we get stuck in this operation. Suppose we are in this case: we cut along A and round the edges, obtaining a torus with slope $\frac{n_2}{6n_2+2}$ isotopic to $\partial(M \setminus V_3)$. This slope corresponds to $-\frac{1}{2}n_2 - 2$ in ∂V_3, and is non-negative when $n_2 \leq -4$, therefore we can find a standard neighbourhood V_3 of F_3 with infinite boundary slope. This boundary slope becomes $-\frac{1}{6}$ if measured with respect to $-\partial(M \setminus V_3)$. To further increase n_2 take an annulus between $\partial(M \setminus V_2)$ and $\partial(M \setminus V_3)$. If $n_2 < -2$, we have $|3n_2+1| > 6$ and thus there exists a bypass attached to $\partial(M \setminus V_2)$ which allows us to increase n_2 by one so that we can start again. In this way we can increase n_1 and n_2 up to -2. When $n_1 = n_2 = -2$ the boundary slopes are $\frac{2}{5}$ on $-\partial(M \setminus V_1)$ and $-\frac{2}{5}$ on $-\partial(M \setminus V_2)$. By the Imbalance Principle, a convex annulus between $\partial(M \setminus V_2)$ and $\partial(M \setminus V_3)$ produces a bypass attached to $\partial(M \setminus V_3)$ as long as $5 < |11n_3 + 6|$, thus we can use the Twist Number Lemma to increase n_3 up to -1. □

LEMMA 4.11. *Let us suppose $n_1 = n_2 = -2$, $n_3 = -1$ and A is a convex vertical annulus whose boundary consists of Legendrian rulings of $\partial(M \setminus V_1)$ and $\partial(M \setminus V_2)$. If A has a boundary-parallel dividing curve, then the twisting numbers can be increased up to $n_1 = 0$, $n_2 = n_3 = -1$ and moreover there is a regular fibre with twisting number zero.*

PROOF. If there is a boundary-parallel dividing curve, then A carries a bypass on each side after perturbing its characteristic foliation. Using these bypasses, we can further increase n_1 and n_2 up to -1. By the Imbalance Principle, we can find one more bypass in an annulus between $\partial(M \setminus V_1)$ and $\partial(M \setminus V_2)$ on the side of $\partial(M \setminus V_1)$. This bypass increases the twisting number n_1 up to 0. The slope of $-\partial(M \setminus V_1)$ is 0, and the slope of $-\partial(M \setminus V_2)$ is $-\frac{1}{2}$, therefore two possibilities for an annulus A between $\partial(M \setminus V_1)$ and $\partial(M \setminus V_2)$ are given: either A carries a bypass for $\partial(M \setminus V_1)$, or not. If such a bypass exists, then all the boundary slopes can be made infinite, and we can decrease the twisting n_3 to -1. If there is no such bypass, cutting along A and rounding edges yields a torus with slope 0, which is -2 when measured in ∂V_3. In V_3 we find a convex torus with slope $-\frac{11}{6}$, which corresponds to infinite slope in $-\partial(M \setminus V_3)$. □

4.2.2. *The case when a thickening to infinite slope exists.* In this subsection we will show that there are no tight contact structures on $-\Sigma(2,3,11)$ with a regular fibre with twisting number zero. We suppose such a fibre exists and argue by contradiction.

Let V_i be a standard neighbourhood of F_i. Then $M \setminus \cup V_i$ is diffeomorphic to $\Sigma_0 \times S^1$ for a pair of pants Σ_0, and has boundary slopes 0, $-\frac{1}{2}$, and $-\frac{1}{5}$. We use vertical annuli from a regular fibre with twisting number zero to thicken each V_i to V_i' such that $-\partial(M \setminus V_i')$ has infinite slope. We can find another pair of pants $\Sigma_1 \subset \Sigma_0$ such that $\Sigma_1 \times S^1$ is diffeomorphic to $M \setminus \cup V_1'$. The arguments in Lemma 4.6 apply to show that the dividing set of Σ_1 looks like in Figure 9 (C). Note that $(\Sigma_0 \setminus \Sigma_1) \times S^1$ is the disjoint union of three thickened tori $T_i \times I$ such that $T_i \times \{0\} = -\partial(M \setminus V_i)$ and $T_i \times \{1\} = -\partial(M \setminus V_i')$ for $i = 1, 2, 3$; see Fig. 13. $T_1 \times I$ is a basic slice, $T_2 \times I$ is a union of two basic slices $T_2 \times [0, \frac{1}{2}]$ with slopes $-\frac{1}{2}$ and -1, and $T_2 \times [\frac{1}{2}, 1]$ with slopes -1 and ∞, and $T_3 \times I$ is a union of five basic slices $T_3 \times [\frac{i}{5}, \frac{i+1}{5}]$ for $i = 0, \dots, 4$ with slopes $-\frac{1}{5-i}$ and $-\frac{1}{4-i}$. We observe that the basic slices which compose the tight contact structures on the $T_i \times I$ belong to the same continued fraction block.

The contact structures of the previous Lemma are described by the three numbers p_i of positive basic slices in $T_i \times I$, for $i = 1, 2, 3$.

Our strategy to find overtwisted discs is cutting (a suitable sub-manifold of) $M \setminus V_i \cup V_j$, for some i and j, along a vertical annulus A and rounding the edges of the cut open manifold to obtain a neighbourhood of the third singular fibre with boundary slope zero. Slope zero on V_i corresponds to $\frac{1}{2}$ on $-\partial(M \setminus V_1)$, to $-\frac{1}{3}$ on $-\partial(M \setminus V_2)$, and to $-\frac{2}{11}$ on $-\partial(M \setminus V_3)$. We will call these slopes *critical slopes.*

REMARK 4.12. This technique differs from the 'disc patching' used in section 4.1.2 of the previous example. We remark that both techniques work for each example and encourage the reader to perform both proofs with the technique not used.

The following Lemma allows us to control the dividing set of such an annulus A.

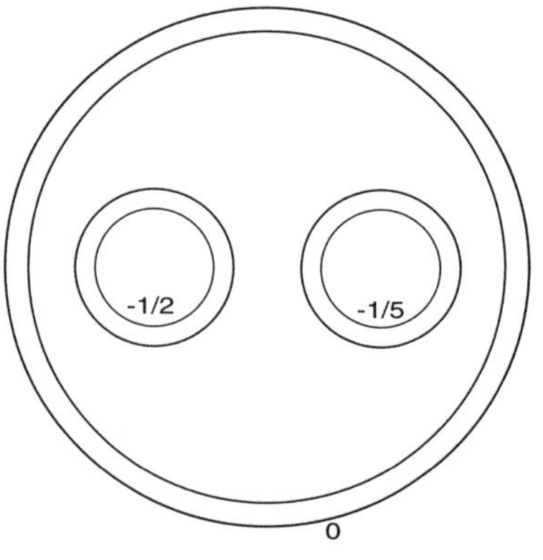

FIGURE 13. Σ_0 with Σ_1 inside. The numbers refer to the slopes of the corresponding tori. The tori without indication of the slope are meant to have infinite slope.

LEMMA 4.13. *Let Σ be a pair of pants and ξ a tight contact structure on $\Sigma\times S^1$. Suppose that the boundary $-\partial(\Sigma\times S^1)=T_1\cup T_2\cup T_3$ consists of tori in standard form with $\#\Gamma_{T_i}=2$ for $i=0,1,2$, and slopes $s(T_0)=\frac{p_0}{q}$, $s(T_1)=\infty$, $s(T_2)=\frac{p_2}{q}$. Suppose also that there exists a pair of pants $\Sigma'\subset\Sigma$ such that $\Sigma\times S^1$ decomposes as $\Sigma\times S^1=\Sigma'\times S^1\cup(T_0\times I)\cup(T_2\times I)$, where $\xi|_{\Sigma'\times S^1}$ is the tight contact structure with infinite boundary slopes described in Lemma 4.7, and $\xi|_{T_i\times I}$ is minimally twisting for $i=0,2$.*

Suppose that one of the following holds:

(1) *$s(T_0)=s(T_1)=-\frac{1}{q}$ and $\xi|_{T_0\times I}$ is isotopic to $\xi|_{T_2\times I}$*

(2) *$s(T_2)<0$ and $\xi|_{T_i\times I}$, for $i=0,2$, decomposes into basic slices of the same sign (i.e. with relative Euler class $\pm(q,p_i-1)$)*

Then there exists a convex annulus with Legendrian boundary consisting of vertical Legendrian rulings of T_0 and T_2 without boundary parallel dividing curves.

PROOF. Let ξ' be obtained in the following way from a tight contact structure on $T^2\times I$ isotopic to $\xi|_{T_0\times I}$. Remove a standard neighbourhood U'' of a vertical Legendrian ruling of a standard torus parallel to $T^2\times\{0\}$ and contained in its invariant neighbourhood, then thicken U'' to U' with infinite boundary slope attaching the bypasses coming from the annulus between a vertical Legendrian ruling of U'' and a Legendrian divide of $T^2\times\{1\}$. By Proposition 2.19, there is a solid torus U between U'' and U' with boundary slope $\frac{p_2}{q}$ because $\frac{p_2}{q}\in[-\frac{1}{q},-\infty)$. In case (1) this operation is not necessary and we can simply take $U=U''$. In a similar way we can find a collar C of $T^2\times\{0\}$ in $T^2\times I\setminus U$ with boundary slopes $\frac{p_0}{q}$ and ∞.

We identify $T^2\times I\setminus U$ to $\Sigma\times S^1$ so that $T^2\times\{0\}$ corresponds to T_0, $T^2\times\{1\}$ corresponds to $-T_1$, ∂U corresponds to T_2, C corresponds to $T_0\times I$, and $U'\setminus U$ corresponds to $T_2\times I$. A convex annulus with Legendrian boundary $A\subset(\Sigma\times S^1,\xi')$ between $T_0=T^2\times\{0\}$ and $T_2=\partial U$ contained in the invariant neighbourhood of $T^2\times\{0\}$ has no boundary parallel dividing arcs. To prove the Lemma we only need to show that ξ and ξ' are isotopic.

The dividing set of $\Sigma'\subset(\Sigma\times S^1,\xi)$ cannot contain a boundary parallel dividing arc, otherwise there would be a convex torus with slope 0 in $\Sigma'\times S^1$ (see the proof of Lemma 4.6). By construction, $(\Sigma\times S^1,\xi)$ contact embeds in $T^2\times S^1$ with a minimally twisting tight contact structure. A convex torus with slope 0 around T_2 would give an overtwisted disc in $T^2\times I$, while between $T_0\times\{1\}$ and T_1, both with infinite boundary slope, would contradict minimally twisting. This proves that $\xi|_{\Sigma'\times S^1}$ and $\xi'|_{\Sigma'\times S^1}$ induce the same dividing set on Σ, therefore they are isotopic by Lemma 4.7.

Take two vertical annuli with Legendrian boundary A_i, for $i=0,2$, between T_i and T_1. We find

$$\langle e(\xi'),A_0\rangle=\langle e(\xi'),A_2\rangle$$

because $[A_0]=[A_2]+[A]$ in $H_2(\Sigma\times S^1,\partial(\Sigma\times S^1))$ and $\langle e(\xi'),A\rangle=0$ by Lemma 2.20. Decompose $A_i=B_i\cup B_i'$, such that $B_i=A_i\cap T_i\times I$, and $B_i'=A_i\cap\Sigma'\times S^1$ for $i=0,2$. We can suppose that all these annuli have Legendrian boundary. Since

$$\langle e(\xi'|_{\Sigma'\times S^1}),B_0'\rangle=\langle e(\xi'|_{\Sigma'\times S^1}),B_2'\rangle=0$$

we get

$$\langle e(\xi'|_{T_0\times I}),B_0\rangle=\langle e(\xi'|_{T_2\times I}),B_2\rangle$$

and hence

$$\langle e(\xi'|_{T_i \times I}), B_i \rangle = \langle e(\xi|_{T_i \times I}), B_i \rangle$$

for $i = 0, 2$ because $\langle e(\xi|_{T_2 \times I}), B_2 \rangle = \langle e(\xi|_{T_0 \times I}), B_0 \rangle = \langle e(\xi'|_{T_0 \times I}), B_0 \rangle$. Applying Corollary 2.26 after a change of coordinates, we conclude $\xi|_{T_i \times I}$ is isotopic to $\xi'|_{T_i \times I}$ for $i = 0, 2$, and this ends the proof of the Lemma. □

THEOREM 4.14. *$M(-\frac{1}{2}, \frac{1}{3}, \frac{2}{11})$ carries no tight contact structure with a Legendrian regular fibre with twisting number zero.*

PROOF. For any choice of the number p_i of positive basic slices in $T_i \times I$, for $i = 0, 1, 2$, we will find a convex torus in $M \setminus \bigcup V_i$ with critical slope. Most of the possible tight contact structures fall into one of the following cases.

Case (1) We work between V_2 and V_3. If in $T_3 \times I$ there are two basic slices with the same signs as the two basic slices in $T_2 \times I$, we can arrange them so that $T_3 \times [\frac{3}{5}, 1]$ is isotopic to $T_2 \times I$. The manifold $M \setminus (V_1' \cup V_2 \cup V_3 \cup (T_3 \times [0, \frac{3}{5}]))$ has boundary slopes ∞, $-\frac{1}{2}$, $-\frac{1}{2}$, and by Lemma 4.13, we can find a convex annulus A between $T_2 \times \{0\}$ and $T_3 \times \{\frac{3}{5}\}$ whose dividing curves go from a component of the boundary to the other one. See Fig. 14. After cutting along A and rounding the edges, we obtain a torus with slope $-\frac{1}{2}$ isotopic to $\partial(M \setminus V_1)$. Because the critical slope for $-\partial(M \setminus V_1)$ is $\frac{1}{2}$, M is overtwisted. This case excludes all the candidate tight contact structures except for the ones with $p_2 = 0$, $p_3 = 4, 5$, or $p_2 = 1$, $p_3 = 0, 5$, or $p_2 = 2$, $p_3 = 0, 1$.

Case (2) We work between V_1 and V_3. Suppose that all the three basic slices in $T_3 \times [\frac{2}{5}, 1]$ have the same sign as $T_1 \times I$. We decrease the twisting number n_1 to -1 and take a standard neighbourhood V_1'' so that $-\partial(M \setminus V_1'') = T_1 \times \{-1\}$ has slope $\frac{1}{3}$. We can choose the sign of the basic slice $T_1 \times [-1, 0]$ so that it is the same as $T_1 \times [0, 1]$. To show this fact, embed V_1 in the standard S^3, and perform stabilisation there, see [**12**]. We add $+1$ or -1 to the rotation number, according to the sign of the stabilisation, and it turns out that the relative Euler class of the contact structure on $V_1 \setminus V_2$ calculated on a vertical annulus is the difference between the rotations.

The manifold $M \setminus (V_1'' \cup V_2' \cup V_3 \cup (T_3 \times [0, \frac{2}{5}]))$ has boundary slopes $\frac{1}{3}$, $-\frac{1}{3}$, ∞, and by Lemma 4.13 we can find a convex annulus A between $T_1 \times \{-1\}$ and

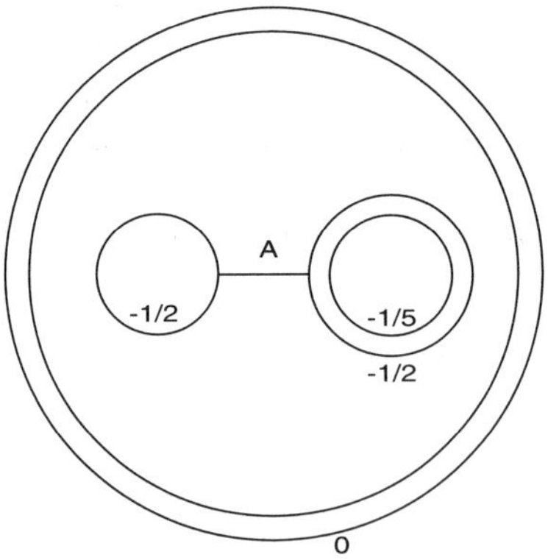

FIGURE 14. Case 1. The tori without indication of the slope are meant to have infinite slope.

$T_3 \times \{\frac{2}{5}\}$ as in figure 15 without boundary-parallel arcs, and by cutting along A and rounding the edges we obtain a torus with slope $\frac{1}{3}$ isotopic to $\partial(M \setminus V_2)$. This gives an overtwisted disk because $-\frac{1}{3}$ is the critical slope for $-\partial(M \setminus V_2)$. This case excludes the contact structures with $p_1 = 0$ and $p_3 \leq 2$, or the contact structures with $p_1 = 1$ and $p_3 \geq 3$.

Case (3) Now we work between V_1 and V_2. Suppose that the basic slices which compose $T_2 \times I$ have the same sign as $T_1 \times I$. We decrease the twisting number of the singular fibres F_1 and F_2 to -2 and take standard neighbourhoods V_i'', for $i = 1, 2$, so that $-\partial(M \setminus V_1'') = T_1 \times \{-1\}$ has slope $\frac{2}{5}$, $-\partial(M \setminus V_2'') = T_2 \times \{-1\}$ has slope $-\frac{2}{5}$, and all the basic slices in $T_1 \times [-1, 1]$ and $T_2 \times [-1, 1]$ have the same sign. The manifold $M \setminus (V_1'' \cup V_2'' \cup V_3')$ has boundary slopes $\frac{2}{5}$, $-\frac{2}{5}$, ∞, therefore by Lemma 4.13 we can find a convex annulus $A \subset M \setminus V_3 \cup (T_3 \times I)$ between $T_1 \times \{-1\}$ and $T_2 \times \{-1\}$ without boundary-parallel dividing curves. See figure 16. Then, after cutting along A and rounding the edges, we find a torus V_3'' isotopic to $\partial(M \setminus V_3)$ with slope $\frac{1}{5}$. The thickened torus between $-\partial(M \setminus V_3)$ and $-\partial(M \setminus V_3'')$ has both boundary slopes $-\frac{1}{5}$, and contains a torus with infinite slope, hence, by the classification Theorem for thickened tori, it contains an intermediate convex torus for each slope. In particular, it contains a convex torus with slope $-\frac{2}{11}$, which is the critical slope for $-\partial(M \setminus V_3)$, therefore it gives an overtwisted disk around F_3. This case excludes the contact structures with $p_1 = 0$, $p_2 = 0$ and $p_1 = 1$, $p_2 = 2$.

We can exclude the remaining candidate tight contact structures with $p_1 = 1$, $p_2 = 1$, $p_3 = 0$, or $p_1 = 0$, $p_2 = 1$, $p_3 = 5$ in the following way. We go back to V_2 and V_3 and arrange the basic slices in $T_2 \times I$ such that $T_2 \times [\frac{1}{2}, 1]$ has the same sign as the basic slices in $T_3 \times I$ and consider an annulus A between $T_2 \times \{\frac{1}{2}\}$ and $T_3 \times \{\frac{4}{5}\}$ without boundary parallel dividing curves. Cutting along A and rounding the edges we obtain a torus with slope -1 isotopic to $\partial(M \setminus V_1)$. This torus has slope 1 in the basis of ∂V_1, which corresponds to increasing the twisting number of the singular fibre F_1 up to $n_1 = 1$. Now we can decrease n_1 again choosing the sign of $T_1 \times I$, showing that the contact structures with $p_1 = 1$, $p_2 = 1$, $p_3 = 0$, or $p_1 = 0$, $p_2 = 1$, $p_3 = 5$ are isotopic to the contact structures with $p_1 = 0$, $p_2 = 1$, $p_3 = 0$, or $p_1 = 1$, $p_2 = 1$, $p_3 = 5$ respectively, which have already been shown to be overtwisted in case (2). □

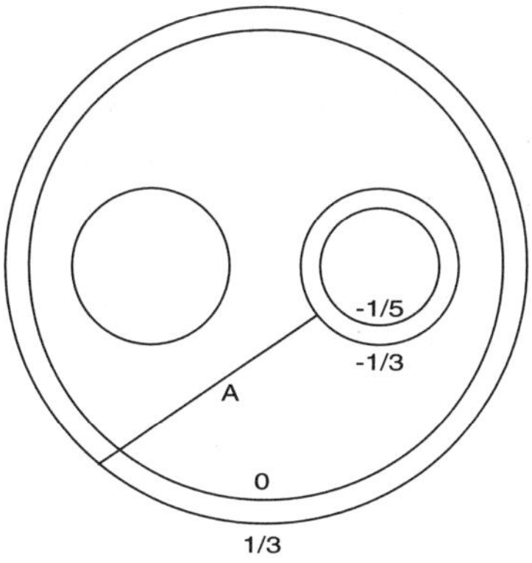

FIGURE 15. Case 2. The tori without indication of the slope are meant to have infinite slope.

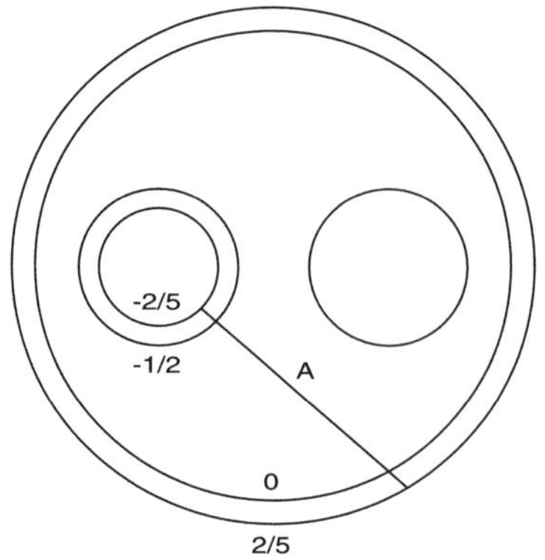

FIGURE 16. Case 3. The tori without indication of the slope are meant to have infinite slope.

4.2.3. *The case when no thickening to infinite slope exists.* It remains to analyse only the case when the convex annulus A between $\partial(M \setminus V_1)$ and $\partial(M \setminus V_2)$ of Lemma 4.11 carries no bypasses. Since $-\partial(M \setminus V_1)$ has slope $\frac{2}{5}$ and $-\partial(M \setminus V_2)$ has slope $-\frac{2}{5}$, the dividing set of A consists of 10 dividing arcs going from one side of the annulus to the other. We say that an arc in A is horizontal if its algebraic intersection with $\Sigma \subset M \setminus \cup V_i \cong \Sigma \times S^1$ is zero.

PROPOSITION 4.15. *The manifold $M(-\frac{1}{2}, \frac{1}{3}, \frac{2}{11})$ carries at most one tight contact structure.*

PROOF. Let ϕ_t be an isotopy of A such that $\phi_0 = id$ and $\phi_1(\Gamma_A)$ is a collection of horizontal arcs, and extend it to an isotopy on the whole M. We can consider $\phi_{1*}\xi$ instead of ξ and suppose without loss of generality that the dividing arcs of A are horizontal.

Cutting $M \setminus V_1 \cup V_2$ along A and rounding the edges yields a solid torus with boundary slope $-\frac{1}{5}$, calculated with respect to the basis of $-\partial(M \setminus V_3)$, which corresponds to slope -1 in the basis of ∂V_3 . By Proposition 2.15, there exists only one tight contact structure up to isotopy on V_3 with this boundary condition, thus the candidate tight contact structure on $M(-\frac{1}{2}, \frac{1}{3}, \frac{2}{11})$ is unique up to isotopy. □

4.2.4. *Construction of the tight contact structure.* We will use Kirby calculus to show that this manifold can be represented as Legendrian surgery on a link in S^3. This will prove that M has at least one holomorphically fillable, and therefore tight, contact structure.

The structure of Seifert fibration of the manifold $M(-\frac{1}{2}, \frac{1}{3}, \frac{2}{11})$ gives the surgery presentation shown by (a) in Fig. 17. By a slam-dunk on the (2) component of the link, we obtain the link (b). Next, we perform a Rolfsen twist on the $(-\frac{1}{2})$ component to obtain the link (c). After another Rolfsen twist around the (-1) component, we obtain diagram (d) in Fig. 17 and finally, after one inverse slam-dunk, we obtain diagram (e). This link can be made Legendrian with the Thurston-Bennequin invariant of each component one more than the surgery coefficient, see (f) for a Legendrian realization. This proves that $M(-\frac{1}{2}, \frac{1}{3}, \frac{2}{11})$ carries at least one tight contact structure, and therefore ends the classification.

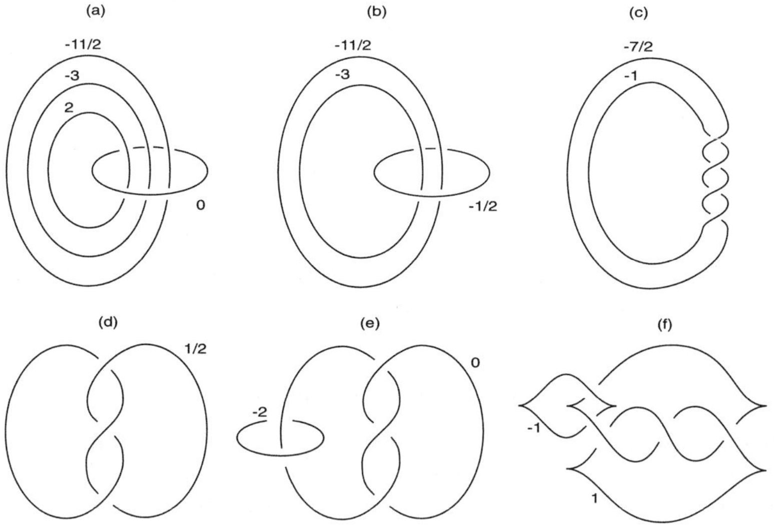

FIGURE 17. Kirby calculus on M. The link as shown in (f) gives a Legendrian realization of the one in (e), the coefficients meaning the tb invariant of the respective component.

5. What we can do, and what we can't do

In this section we give a quick overview of similar results on the classification of tight contact structures on Seifert manifolds we are able to do with the techniques exposed in this paper, and point out some problems we have met. In addition to the Poincaré homology sphere with reversed orientation studied by Etnyre and Honda in [**14**] and the two examples treated in this paper, we are able to give a classification in several other cases. Among them there are the manifold $M(\frac{1}{2}, -\frac{1}{2}, -\frac{1}{2})$, where the existence of a unique tight contact structure has been proven by S. Schönenberger; see [**34**], the Poincaré homology sphere $M(\frac{1}{2}, -\frac{1}{3}, -\frac{1}{5})$, and the manifold $M(-\frac{1}{2}, \frac{1}{3}, \frac{3}{17})$. Moreover P. Ghiggini in his Ph.D. thesis is working towards the classification of tight contact structures on Seifert manifolds over the torus.

We would like to point out that in the examples computed in this paper, like in most of the other results mentioned above, the tight contact structures one obtains happen to be holomorphically fillable. In fact, the presence of possibly tight contact structures for which no Stein filling is known is a major source of difficulties in achieving a complete classification. The problems one has to face when dealing with non-holomorphically fillable contact structures are twofold.

The first problem is the proof of tightness, in fact we have very few techniques to do that, the main ones being holomorphic or symplectic fillability and the gluing techniques developed by Colin [**2**] and Honda [**23**]. In particular, the gluing Theorems require the presence of incompressible surfaces, which do not exist in manifolds whose fundamental group is finite, such as small Seifert manifolds. An example is $M(-\frac{1}{2}, \frac{1}{3}, \frac{1}{4})$, which has been proved by Lisca to be not fillable in any sense, and where only an upper bound for the number of tight contact structures is known.

The second problem is distinguishing the tight contact structures. For Stein fillable contact structures this is made easy by a result from Seiberg-Witten theory due to Lisca and Matić [**31**] and Kronheimer and Mrowka [**28**], which gives necessary conditions for holomorphically fillable contact structures to be isotopic. Another way to distinguish contact structures is through their homotopy classification as 2-planes fields given by Gompf [**19**] in terms of algebraic topological invariants. An example of what happens outside the range of applicability of both methods is given by the family of manifolds $M(-\frac{1}{2}, \frac{1}{3}, \frac{2}{6n-1})$ for $n > 3$, where we are able to give both an upper and a lower bound on the number of tight contact structures, but the problem of determining whether some of them are isotopic or not remains open.

6. Acknowledgement

Both authors are very grateful to John Etnyre and Ko Honda for their encouragement, steady support and interest. This work started during the Contact Geometry Quarter held at AIM and Stanford University during Fall 2000. We would like to thank both institutions for providing support.

References

1. V. Colin, *Sur la torsion des structures de contact tendues*, Ann. Sci. École Norm. Sup. (4) **34** (2001), no. 2, 267–286. MR 1 841 879
2. ———, *Une infinité de structures de contact tendues sur les variétés toroï dales*, Comment. Math. Helv. **76** (2001), no. 2, 353–372. MR 1 839 351
3. F. Ding and H. Geiges, *Symplectic fillability of tight contact structures on torus bundles*, Algebr. Geom. Topol. **1** (2001), 153–172 (electronic). MR 1 823 497
4. Y. Eliashberg, *Classification of overtwisted contact structures on 3-manifolds*, Invent. Math. **98** (1989), no. 3, 623–637. MR **90k**:53064
5. ———, *Filling by holomorphic discs and its applications*, Geometry of low-dimensional manifolds, 2 (Durham, 1989), Cambridge Univ. Press, Cambridge, 1990, pp. 45–67. MR **93g**:53060
6. ———, *Topological characterization of Stein manifolds of dimension* > 2, Internat. J. Math. **1** (1990), no. 1, 29–46. MR **91k**:32012
7. ———, *Contact 3-manifolds twenty years since J. Martinet's work*, Ann. Inst. Fourier (Grenoble) **42** (1992), no. 1-2, 165–192. MR **93k**:57029
8. Y. Eliashberg and M. Fraser, *Classification of topologically trivial Legendrian knots*, Geometry, topology, and dynamics (Montreal, PQ, 1995), Amer. Math. Soc., Providence, RI, 1998, pp. 17–51. MR **99f**:57007
9. Y. Eliashberg and M. Gromov, *Convex symplectic manifolds*, Several complex variables and complex geometry, Part 2 (Santa Cruz, CA, 1989), Amer. Math. Soc., Providence, RI, 1991, pp. 135–162. MR **93f**:58073
10. J. Etnyre, *Tight contact structures on lens spaces*, Commun. Contemp. Math. **2** (2000), no. 4, 559–577. MR **2001k**:57017
11. ———, *Introductory lectures on contact geometry*, AMS/IP Studies in Advanced Mathematics, volume 35, 2003.
12. J. Etnyre and K. Honda, *Knots and Contact Geometry*, arXiv:math.GT/0006112.
13. ———, *Tight contact structures with no symplectic fillings*, AIM 2000-10, arXiv:math.GT/0010044.
14. ———, *On the nonexistence of tight contact structures*, Ann. of Math. (2) **153** (2001), no. 3, 749–766. MR 1 836 287
15. E. Giroux, *Convexité en topologie de contact*, Comment. Math. Helv. **66** (1991), no. 4, 637–677. MR **93b**:57029
16. ———, *Une infinité de structures de contact tendues sur une infinité de variétés*, Invent. Math. **135** (1999), no. 3, 789–802. MR **2000d**:53133
17. ———, *Structures de contact en dimension trois et bifurcations des feuilletages de surfaces*, Invent. Math. **141** (2000), no. 3, 615–689. MR **2001i**:53147

18. ———, *Structures de contact sur les variétés fibrées en cercles audessus d'une surface*, Comment. Math. Helv. **76** (2001), no. 2, 218–262. MR 1 839 346
19. R. Gompf, *Handlebody construction of Stein surfaces*, Ann. of Math. (2) **148** (1998), no. 2, 619–693. MR **2000a**:57070
20. R. Gompf and A. Stipsicz, *4-manifolds and Kirby calculus*, American Mathematical Society, Providence, RI, 1999. MR **2000h**:57038
21. M. Gromov, *Pseudoholomorphic curves in symplectic manifolds*, Invent. Math. **82** (1985), no. 2, 307–347. MR **87j**:53053
22. A. Hatcher, *Notes on Basic 3-Manifold Topology*, `http://www.math.cornell.edu/ hatcher/`.
23. K. Honda, *Gluing tight contact structures*, arXiv:math.GT/0102029.
24. ———, *On the classification of tight contact structures. I*, Geom. Topol. **4** (2000), 309–368 (electronic). MR **2001i**:53148
25. ———, *On the classification of tight contact structures. II*, J. Differential Geom. **55** (2000), no. 1, 83–143. MR 1 849 027
26. K. Honda, W. Kazez, and G. Matić, *Tight contact structures and taut foliations*, Geom. Topol. **4** (2000), 219–242 (electronic). MR **2001i**:53149
27. Y. Kanda, *The classification of tight contact structures on the 3-torus*, Comm. Anal. Geom. **5** (1997), no. 3, 413–438. MR **99c**:57054
28. P. B. Kronheimer and T. S. Mrowka, *Monopoles and contact structures*, Invent. Math. **130** (1997), no. 2, 209–255. MR **98h**:57058
29. P. Lisca, *Symplectic fillings and positive scalar curvature*, Geom. Topol. **2** (1998), 103–116 (electronic).
30. ———, *On symplectic fillings of 3-manifolds*, Proceedings of 6th Gökova Geometry-Topology Conference, vol. 23, 1999, pp. 151–159. MR **2001f**:53178
31. P. Lisca and G. Matić, *Tight contact structures and Seiberg-Witten invariants*, Invent. Math. **129** (1997), no. 3, 509–525. MR **98f**:57055
32. ———, *Stein 4-manifolds with boundary and contact structures*, Topology Appl. **88** (1998), no. 1-2, 55–66. MR **99f**:57037
33. S. Makar-Limanov, *Tight contact structures on solid tori*, Trans. Amer. Math. Soc. **350** (1998), no. 3, 1013–1044. MR **98e**:53046
34. S. Schönenberger, *A manifold with unique tight contact structure*, 2001, Diplomarbeit ETH Zürich.
35. A. Weinstein, *Contact surgery and symplectic handlebodies*, Hokkaido Math. J. **20** (1991), no. 2, 241–251. MR **92g**:53028

SCUOLA NORMALE SUPERIORE, PISA
E-mail address: `ghiggini@mail.dm.unipi.it`

UNIVERSITY OF PENNSYLVANIA, PHILADELPHIA
E-mail address: `stephans@math.upenn.edu`

Proceedings of Symposia in Pure Mathematics
Volume **71**, 2003

Four Dimensions from Two in Symplectic Topology

Margaret Symington

ABSTRACT. This paper introduces two-dimensional diagrams that are slight generalizations of moment map images for toric four-manifolds and catalogs techniques for reading topological and symplectic properties of a symplectic four-manifold from these diagrams. The paper offers a purely topological approach to toric manifolds as well as methods for visualizing other manifolds, including the K3 surface, and for visualizing certain surgeries. Most of the techniques extend to higher dimensions.

CONTENTS

1. Introduction

The goal of this paper is to establish techniques for "seeing" certain symplectic four-manifolds: a method for drawing two dimensional diagrams (base diagrams) that determine symplectic four-manifolds and rules for reading topological and symplectic features of the manifold from the diagrams. The inspiration comes

2000 *Mathematics Subject Classification.* Primary 57R15, 55R55; Secondary 53D05, 57M99.
Key words and phrases. Lagrangian fibrations, almost toric manifolds, moment map.
The author was supported in part by NSF grant DMS0204368.

from closed toric manifolds, i.e. closed symplectic $2n$-manifolds equipped with an effective Hamiltonian T^n action. Delzant's Theorem (Theorem 3.3) says that the moment map image, a polytope in $\mathbb{R}^n$, determines the symplectic manifold and the torus action. We are concerned here with the symplectic manifold and not the action. The philosophy therefore is to forget the torus action and then try to extend Delzant's theorem to a broader class of manifolds using diagrams that generalize moment map images. Without the torus action, what remains is a Lagrangian fibration by tori with singular fibers. We define the class of almost toric fibrations as Lagrangian fibrations with certain very simple singularities allowed in the fibers (Definition 4.5); accordingly, we call symplectic manifolds equipped with such fibrations almost toric manifolds. This emphasis on the fibration rather than the torus action gives a new perspective on toric manifolds in arbitrary dimension (including a simple topological construction from the moment map image) and, generalizing to the almost toric case, gives ways to visualize four-manifolds such as the K3 surface and rational balls.

The definition of almost toric fibrations is guided by a classification, due to Zung [**39**], of Lagrangian fibrations that have the local structure of an integrable system with non-degenerate topologically stable singularities. This classification, up to fiber-preserving symplectomorphism, is in terms of the base, its geometry, the structure of fibered neighborhoods of the singular fibers and a Lagrangian Chern class. The almost toric fibrations are those with no hyperbolic singularities, and hence in dimension four the topology of the singular fibers is determined by the geometry of the base. Therefore, whenever the Lagrangian Chern class must be trivial (when the base has the homotopy type of a one-dimensional complex), the base determines the manifold. The two features that distinguish an almost toric fibration from a toric one are a base that does not immerse, preserving its geometry, in $\mathbb{R}^2$ and the presence of nodal singularities (the positive transverse self-intersection points that also arise in Lefschetz fibrations).

While many of the results and techniques of this paper can be generalized to higher dimensions, unless otherwise specified we restrict attention to dimension four so that we can actually draw the base diagrams. The set of closed toric four-manifolds consists of $\mathbb{C}P^2$, $S^2 \times S^2$ and blow-ups of these manifolds. The set of closed almost toric manifolds includes the toric ones as well as sphere bundles over a torus and blow-ups of them, certain torus bundles over a torus or Klein bottle, the K3 surface and the Enriques surface (a quotient of the K3 surface by a $\mathbb{Z}_2$ action). See Theorem 5.19. However, some of the most interesting almost toric manifolds are manifolds with boundary. For instance rational balls that are useful in surgery constructions (Section 9.3) admit almost toric fibrations that show these surgeries can be done symplectically. Furthermore, an exotic symplectic four-ball (Example 3.27) – diffeomorphic to the standard four-ball but not symplectically embeddable in $(\mathbb{R}^4, dx \wedge dy)$ – admits a simple almost toric (even toric) fibration.

The topological and symplectic features of an almost toric manifold that we explain how to see include

- certain surfaces (called visible surfaces), their symplectic area and self-intersections, and how to detect if they are symplectic or Lagrangian (Section 7);
- the Euler characteristic and first Chern class of the manifold (Section 8); and

- embedded lens spaces, including when they admit induced contact structures and when said contact structures are tight or overtwisted (Section 9.2).

We also explain how to blow up and down in the almost toric category (which is more general than the toric blowing up and down) and how toric and almost toric fibrations assist in understanding symplectic surgeries such as generalized rational blowdowns and symplectic sums. These are all discussed in Section 10. As an application of visible surfaces we explain in Section 11 how to "see" surfaces in a K3 surface that reveal the intersection form $E8 \oplus E8 \oplus 3\left(\begin{smallmatrix} 0 & 1 \\ 1 & -2 \end{smallmatrix}\right)$.

In addition to an introduction to almost toric manifolds, the material presented here includes some well known facts from toric geometry, but explained from a non-standard perspective with no reference to torus actions and moment maps. For instance, Section 3.2 includes a purely topological construction of toric manifolds that is more direct and more general than the usual quotient construction (cf. [**6**]). In Section A we provide a simple proof of the well-known classification of closed toric (symplectic) four-manifolds. Most symplectic proofs appeal to the classification of complete toric (complex) surfaces (cf. [**16**]); ours is purely symplectic, relying only on the notions in this paper. In Section 5.5 we outline the classification of almost toric four-manifolds [**30**].

1.1. Conventions. Throughout this paper we assume manifolds are smooth, except possibly on their boundaries where they may have corners. We assume that the fibers of a fibration are connected, compact and without boundary.

We use $\mathcal{A}$ to denote an integral affine structure (Section 2.2) and $\mathcal{S}$ to denote a stratification. We write $\phi : (B_1, \mathcal{A}_1, \mathcal{S}_1) \to (B_2, \mathcal{A}_2, \mathcal{S}_2)$ to indicate that the map ϕ preserves the affine structures and stratifications, while we write $\phi : (B_1, \mathcal{A}_1) \to (B_2, \mathcal{A}_2)$ if only the affine structure is necessarily preserved by ϕ.

We use two sets of *standard coordinates* and *standard symplectic forms* on two different spaces: (x, y) and $\omega_0 = dx \wedge dy$ on $\mathbb{R}^{2n}$ and (p, q) and $\omega_0 = dp \wedge dq$ on $\mathbb{R}^n \times T^n$ where we are employing vector and summation notation so $(p, q) = (p_1, q_1 \dots p_n, q_n)$, $dp \wedge dq = \sum_i dp_i \wedge dq_i$ and $v\frac{\partial}{\partial p} = \sum_i v_i \frac{\partial}{\partial p_i}$, $v \in \mathbb{R}^n$. We assume the q-coordinates are cyclic with period 2π. By $a[\frac{\partial}{\partial q}]$ we mean the homology class in $H_1(\mathbb{R}^n \times T^n)$ represented by integral curves of the vector field $\sum_i a_i \frac{\partial}{\partial q_i}$.

The meanings of light vs. heavy lines and light lines with heavy dashes are explained in Sections 3.2 and 4.4.

The following common notations are employed: A^{-T} denotes the inverse transpose of the matrix A; I is the unit interval $[0, 1]$; and given a fibration $\pi : M \to B$, the vertical bundle of TM means the subbundle of TM that is the kernel of π_*. By a lattice in a vector bundle of rank n we mean a smoothly varying lattice (isomorphic to $\mathbb{Z}^n$) in each fiber of the bundle.

Acknowledgments. I am grateful to John Etnyre for encouraging me to write this paper and to Eugene Lerman for bringing the work of Nguyen Tien Zung to my attention. I am also thankful for the helpful comments of John Etnyre, Andy Wand and the referee on early versions of the paper.

2. Background

2.1. Lagrangian fibrations. A smooth $2n$ dimensional manifold M is symplectic if it is equipped with a closed non-degenerate two-form ω (i.e. $d\omega = 0$ and

ω^n is nowhere-vanishing). A submanifold L of a symplectic manifold (M, ω) is *Lagrangian* if $\dim L = \frac{1}{2}\dim M$ and $\omega|_L = 0$.

DEFINITION 2.1. A locally trivial fibration of a symplectic manifold is a *regular Lagrangian fibration* if the fibers are Lagrangian. More generally, a projection $\pi : (M^{2n}, \omega) \to B^n$, is a *Lagrangian fibration* if it restricts to a regular Lagrangian fibration over an open dense set $B_0 \subset B$, the set of regular values. (We assume throughout this paper that the fibers of a Lagrangian fibration are connected, compact and without boundary.)

A basic feature of symplectic manifolds is expressed in Darboux's theorem which states that in a neighborhood of any point of a symplectic $2n$-manifold one can find *Darboux coordinates* (x, y) such that the symplectic form is standard: $\omega_o = dx \wedge dy$. Given a regular Lagrangian fibration we can instead demand that the coordinates be adapted to the fibration, as in the following basic example.

EXAMPLE 2.2. Consider standard coordinates (p, q) on $\mathbb{R}^n \times T^n$ where the q coordinates have period 2π. Then projection on the first factor $\pi : (\mathbb{R}^n \times T^n, \omega_0) \to \mathbb{R}^n$ is a Lagrangian fibration with respect to the standard symplectic structure $\omega_0 = dp \wedge dq$.

Example 2.2 captures the local behavior near any fiber of a regular Lagrangian fibration. This is a consequence of the Arnold-Liouville theorem [**4**] about integrable systems (where the coordinates (p, q) are called *action-angle coordinates*). Rephrased in terms of Lagrangian fibrations we have:

THEOREM 2.3. *(Arnold-Liouville) Let F be a fiber of a regular Lagrangian fibration $\pi : (M^{2n}, \omega) \to B^n$. Then there is a open neighborhood of F that is fiber-preserving symplectomorphic to $(V \times T^n, \omega_0) \to V$ where $V \subset \mathbb{R}^n$.*

Notice in particular that Theorem 2.3 implies that the regular fibers of a Lagrangian fibration are tori. For a proof of Theorem 2.3 the reader can consult [**9**] or [**4**]. The key steps of the proof are the construction of a natural torus action in the neighborhood of the regular fiber.

The same argument can be carried out globally (see Duistermaat [**9**]) in which case one constructs a natural fiberwise action of T^*B on M, namely $\alpha \cdot x = \phi_\alpha(x)$ where ϕ_α is the time-one flow of the vector field X_α defined by

$$\omega(\cdot, X_\alpha) = \pi^*\alpha. \tag{2.1}$$

Following this line of reasoning the isotropy subgroup of this action is a lattice Λ^* in T^*B such that $(T^*B/\Lambda^*, d\lambda)$ and (M, ω) are locally fiberwise symplectomorphic. (Here, λ is the canonical 1-form on T^*B, the Liouville form.) This construction yields another lattice: Λ^{vert} belonging to the vertical bundle in TM defined by the fibration, namely $\{X_\alpha | \alpha \in \Lambda^*\}$ where X_α is defined by Equation 2.1. The lattice $\Lambda^* \subset T^*B$ also defines a dual lattice $\Lambda \subset TB$ according to $u \in \Lambda_b$ if $v^*u \in \mathbb{Z}$ for every $v^* \in \Lambda_b^*$.

Here we take a different approach to the globalization in which the lattice $\Lambda \subset TB$ appears directly. Our emphasis on this lattice rather than its dual follows from a naïve preference for visualizing tangent vectors to the base rather than covectors. To find the lattice Λ we appeal to the Arnold-Liouville theorem, identify Λ in local standard coordinate charts and then show that this lattice is well defined globally.

The following example covers the local case and serves in part to set notation.

EXAMPLE 2.4. Given the fibration $\pi : (\mathbb{R}^n \times T^n, \omega_0) \to \mathbb{R}^n$ defined by projection onto the first factor, the vectors $\{\frac{\partial}{\partial p_i}\}$ and $\{\frac{\partial}{\partial q_i}\}$ provide bases for the *standard lattice* Λ_0 on $T\mathbb{R}^n$ and the *standard vertical lattice* Λ_0^{vert} in the vertical bundle of $T(\mathbb{R}^n \times T^n)$ induced by the projection π. Indeed, the dual lattice Λ_0^* on $T^*\mathbb{R}^n$ is spanned by $\{dp_i\}$ and each lattice covector $vdp \in T_p\mathbb{R}^n$ defines (by Equation 2.1) the vertical vector field $v\frac{\partial}{\partial q} \subset \{p\} \times T(T^n) \subset T(\mathbb{R}^n \times T^n)$ since $dp \wedge dq(u\frac{\partial}{\partial p} + u'\frac{\partial}{\partial q}, v\frac{\partial}{\partial q}) = u \cdot v = vdp(u\frac{\partial}{\partial p} + u'\frac{\partial}{\partial q})$ for all u, u'. (Here u, u' are vectors in $\mathbb{R}^n$ and $v \in \mathbb{Z}^n$.)

To globalize we rely on the following lemma about the local structure of fiber-preserving symplectomorphisms.

LEMMA 2.5. *Let φ be a fiber-preserving symplectomorphism of $\pi : (\mathbb{R}^n \times T^n, \omega_0) \to \mathbb{R}^n$ where π is projection on the first factor. Then $\varphi(p, q) = (A^{-T}p + c, Aq + f(p))$ for some $A \in GL(n, \mathbb{Z})$, $c \in \mathbb{R}^n$ and smooth map $f : \mathbb{R}^n \to \mathbb{R}^n$ such that $A^{-1}\frac{\partial f}{\partial p}$ is symmetric.*

PROOF. Since φ is fiber-preserving we can write $\varphi(p, q) = (\varphi_1(p), \varphi_2(p, q))$. Requiring φ to be a symplectomorphism amounts to imposing

$$\varphi_*^T J \varphi_* = J \qquad \text{where } J = \begin{pmatrix} 0 & -I \\ I & 0 \end{pmatrix}. \tag{2.2}$$

In terms of φ_1, φ_2 this is equivalent to

$$\varphi_{1,p} = \varphi_{2,q}^{-T} \quad \text{and} \quad \varphi_{1,p}^T \varphi_{2,p} \text{ is symmetric.} \tag{2.3}$$

Since $\varphi_{1,p}$ does not depend on q, neither does $\varphi_{2,q}$. Therefore, $\varphi_2 = A(p)q + f(p)$ where $A(p)$, for each p, is an element of $GL(2, \mathbb{Z})$ and $f(p)$ gives rise to a translation in each fiber. Finally, $GL(2, \mathbb{Z})$ being a discrete group implies that $\varphi_1, p = A^{-T}(p)$ is constant. Thus we can write $\varphi(p, q) = (A^{-T}p + c, Aq + f(p))$ for $c \in R^2$ where the condition on f follows from (2.3). □

The above simple lemma captures the rigidity of a Lagrangian fibration and hints at why, at least in the toric case, one might expect to see a linear object such as a polytope appear as the base of a Lagrangian fibration. Specifically, it leads to:

THEOREM 2.6. *If $\pi : (M, \omega) \to B$ is a regular Lagrangian fibration then there are lattices $\Lambda \subset TB$, $\Lambda^* \subset T^*B$ and Λ^{vert} in the vertical bundle of TM (induced by π) that, with respect to standard local coordinates, are the standard lattice, its dual, and the standard vertical lattice (of the previous example).*

PROOF. The Arnold-Liouville theorem implies that we can find a cover $\{U_i\}$ of B and fiber-preserving symplectomorphisms $\Phi_i : \pi^{-1}(U_i) \to V_i \times T^n \subset (\mathbb{R}^n \times T^n, \omega_0)$. The Φ_i cover smooth maps $\phi_i : U_i \to V_i$ which, by Lemma 2.5, satisfy $(\phi_j \circ \phi_i^{-1})_* \in GL(n, \mathbb{Z})$ on any $U_j \cap U_i$. Therefore the pullbacks via the $(\phi_i)_*$ of the standard lattices in $T\mathbb{R}^n$ and $T^*\mathbb{R}^n$ give well-defined lattices $\Lambda \subset TB$ and $\Lambda^* \subset T^*B$. Since the pairing that relates the lattices Λ^* and Λ^{vert} is defined locally by the symplectic form ω it is automatically defined globally. Therefore the vertical lattice Λ^{vert} is well-defined globally. □

The rigidity of a regular Lagrangian fibration is in stark contrast with a symplectic fibration where, for instance, the fiber can be any symplectic manifold and there is no interdependence of diffeomorphisms of the base and fiber components of a local fiber-preserving symplectomorphism.

2.2. Integral affine structures.

DEFINITION 2.7. An *integral affine structure* $\mathcal{A}$ on an n-manifold B is a lattice Λ in its tangent bundle, in which case $(B, \mathcal{A})$ is called an *integral affine manifold.*

EXAMPLE 2.8. The standard lattice Λ_0 generated by the unit vectors tangent to the coordinate axes in $\mathbb{R}^n$ defines the *standard integral affine structure* $\mathcal{A}_0$ on $\mathbb{R}^n$.

EXAMPLE 2.9. The induced lattice Λ on the tangent bundle of the base B of a regular Lagrangian fibration provides B with an integral affine structure $\mathcal{A}$.

PROPOSITION 2.10. *An n-manifold B admits an integral affine structure if and only if it can be covered by coordinate charts $\{U_i, h_i\}$, $h_i : U_i \to \mathbb{R}^n$ such that the map $h_j \circ h_i^{-1}$, wherever defined, is an element of $AGL(n, \mathbb{Z})$, i.e. a map of the form $\Phi(x) = Ax + b$ where $A \in GL(n, \mathbb{Z})$ and $b \in \mathbb{R}^n$.*

PROOF. Given an affine structure $\mathcal{A}$ on B, affine coordinate charts $h_i : (U_i, \mathcal{A}) \to (\mathbb{R}^n, \mathcal{A}_0)$ exist since all affine structures (in a given dimension) are isomorphic. Then $(h_j \circ h_i^{-1})_* \in GL(n, \mathbb{Z})$ so $h_j \circ h_i^{-1} \in AGL(n, \mathbb{Z})$. Given such a cover, the integral affine structure $\mathcal{A}$ is defined as the pull-back (via the h_i) of the standard affine structure $\mathcal{A}_0$ on R^n. □

REMARK 2.11. Note that the tangent bundle of an integral affine manifold has a flat connection, namely the pull-back via the coordinate charts of the flat connection on $T\mathbb{R}^n$ (since it is preserved under the transition maps).

DEFINITION 2.12. A k-dimensional submanifold $P \subset (B, \mathcal{A})$ is *affine planar* if at each point $b \in P$ the subspace T_bP is spanned by k vectors in Λ, or equivalently, the subspace in T_b^*B consisting of covectors that vanish on T_bP is spanned by $n-k$ covectors in Λ^*. Accordingly, a one dimensional submanifold is *affine linear* if at every point it has a tangent vector in Λ.

In general there is no metric compatible with an integral affine structure. However, we can still define the following $AGL(n, \mathbb{Z})$-invariant quantity:

DEFINITION 2.13. Consider a parameterized curve $\gamma : [0, 1] \to (B, \mathcal{A})$ and define $s(t), v(t)$ so that $s(t) > 0$ is a scaler, $v(t) \in \Lambda_{\gamma(t)}$ is a primitive tangent vector and $\gamma'(t) = s(t)v(t)$. Then the *affine length* of γ is

$$\int_0^1 s(t)dt. \tag{2.4}$$

2.3. Monodromy – topological and affine. An important measure of the non-triviality of a torus bundle is its monodromy. Interestingly, any topological monodromy of a regular Lagrangian fibration is reflected in the global geometry of the base. In this section we review the definition of (topological) monodromy, define the affine monodromy of an integral affine structure and then give the relation between the topological monodromy of a regular Lagrangian fibration and the affine monodromy of its base.

Let $\mathcal{M}(T^n)$ denote the *mapping class group* of the n-torus, namely the set of isotopy classes of self-diffeomorphisms of the torus. Let ζ_M be an n-torus bundle given by $\pi : M \to B$. Then a *monodromy representation* of ζ_M is a homomorphism $\Psi : \pi_1(B, b) \to \mathcal{M}(T^n)$ such that for any $[\gamma] \in \pi_1(B, b)$ represented by a based loop $\gamma : S^1 \to B$, the total space of $\gamma^*\zeta_M$ is diffeomorphic to $I \times T^n/((0,q) \sim (1, \Psi[\gamma](q))$. We can identify each isotopy class $\Psi[\gamma]$ with the induced automorphism on $H_1(F_b, \mathbb{Z})$ where F_b is the fiber over the base point b. Choosing a basis for $H_1(F_b, \mathbb{Z})$ we then get an induced map from $\pi_1(B, b)$ into $GL(n, \mathbb{Z})$. In general, by the *monodromy* of a regular torus fibration we mean this induced map for some choice of monodromy representation and some choice of basis for $H_1(F_b, \mathbb{Z})$.

REMARK 2.14. Note that the base of a regular Lagrangian fibration need not be orientable and hence the fiber need not be oriented.

Similar to a torus fibration, an integral affine structure $\mathcal{A}$ on a manifold B has an *affine monodromy representation* $\Psi_B : \pi_1(B, b) \to \mathrm{Aut}(\Lambda_b)$ where Λ_b is the restriction to T_bB of the lattice Λ that defines $\mathcal{A}$. Specifically, if we identify (T_bB, Λ_b) with $(\mathbb{R}^n, \mathbb{Z}^n)$ and let γ be a based loop, $\Psi_B[\gamma]$ is the automorphism of $(\mathbb{R}^n, \mathbb{Z}^n)$ such that $\gamma^*(TB, \Lambda)$ is isomorphic to $I \times (\mathbb{R}^n, \mathbb{Z}^n)/((0,p) \sim (1, \Psi_B[\gamma](p))$.

Now recall that a regular Lagrangian fibration $\pi : (M, \omega) \to B$ induces lattices $\Lambda \subset TB$, $\Lambda^* \subset T^*B$ and $\Lambda^{\mathrm{vert}} \subset TM$ (see Section 2.1). This allows us to relate the topological monodromy of the underlying torus fibration to the affine monodromy of the induced affine structure on the base.

PROPOSITION 2.15. *Suppose that, with respect to local standard coordinates near a regular fiber, the topological monodromy around a loop in the base of a regular Lagrangian fibration is given by $A \in GL(n, \mathbb{Z})$. Then the affine monodromy is given by its inverse transpose A^{-T}.*

EXERCISE 2.16. Prove this. Note that the monodromy of the lattice Λ^* would be A.

PROPOSITION 2.17. *Suppose $\Phi[\gamma], \Phi_B[\gamma]$ are the topological and affine monodromies along γ with respect to standard local coordinates near the regular fiber over the base point b. Then $\Phi[\gamma]$ has an invariant sublattice spanned by $v\frac{\partial}{\partial q}$ for some $v \in \mathbb{Z}^n$ if and only if $\Phi_B[\gamma]$ has an invariant sublattice spanned by $u\frac{\partial}{\partial p}$, $u \in \mathbb{Z}^n$ such that $v \cdot u = 0$.*

EXERCISE 2.18. Prove this.

3. Toric manifolds

3.1. More background. If a group G acts on a symplectic manifold (M, ω) then each element ξ of the Lie algebra $\mathcal{G}$ induces a vector field X_ξ, the infinitesimal action. A smooth function $f : M \to \mathbb{R}$ also defines a (Hamiltonian) vector field X_f by $\omega(\cdot, X_f) = df$.

DEFINITION 3.1. A group action on a symplectic manifold (M, ω) is *Hamiltonian* if there is a Lie algebra homomorphism from $\mathcal{G}$ to the smooth functions on M (with the Poisson bracket) taking ξ to f_ξ so that $X_\xi = X_{f_\xi}$.

If a group action is Hamiltonian then there is an associated *moment map* $\mu : M \to \mathcal{G}^*$ defined implicitly by $< \xi, \mu(x) >= f_\xi(x)$ for $\xi \in \mathcal{G}$ and $x \in M$. Thus for k-torus actions the image of the moment map is in $\mathbb{R}^k$.

A fundamental theorem about Hamiltonian torus actions is:

THEOREM 3.2 (Atiyah [**5**], Guillemin and Sternberg [**24**]). *The moment map image $\Delta := \mu(M)$ for a Hamiltonian k-torus action on a closed symplectic manifold (M, ω) is a convex polytope.*

When the dimension of the torus acting on (M^{2n}, ω) is maximal, i.e. $k = n$, then M is called *toric*. In the toric case the regular fibers of μ are tori of dimension n and singular fibers are tori of lower dimension. In this case the moment map has a simple interpretation as the orbit map. However, the geometry of Δ reveals much more than just the orbit structure:

THEOREM 3.3 (Delzant [**8**]). *If a closed symplectic manifold (M^{2n}, ω) is equipped with an effective Hamiltonian n-torus action, then the image of the moment map Δ determines the manifold M, its symplectic structure ω and the torus action.*

REMARK 3.4. Both the Atiyah, Guillemin and Sternberg convexity theorem and Delzant's theorem remain true if the hypothesis that the manifold be closed is replaced by the moment map being proper and the fibers of the moment map being connected (cf. [**29**]).

Not all polytopes can be the moment map image for a closed manifold; the following is standard:

PROPOSITION 3.5. *Suppose Δ is the moment map image of a closed symplectic manifold. Then each $(n-1)$-dimensional face of Δ has a primitive integral normal vector and each $(n-k)$-dimensional face, $k \geq 2$, lies at the intersection of k faces of dimension $n-1$. Furthermore, smoothness of the manifold implies that the k primitive integral vectors normal to the k faces intersecting at an $(n-k)$-dimensional face span a k-dimensional sublattice of the standard integral lattice Λ_0. (If not, M has orbifold singularities.)*

For further information on Hamiltonian torus actions consult [**6**].

Now forget the torus action but remember the projection to $\mathbb{R}^n$ defined by the moment map μ. All fibers of the moment map are tori, but their dimensions vary: the preimage of a point on the interior of Δ is an n-torus while the preimage of a point in the interior of a k-dimensional face of Δ is a k-torus. A standard fact from toric geometry is that the symplectic form vanishes on fibers of the moment map. Therefore regular fibers are Lagrangian submanifolds and we can think of the moment map μ as providing a Lagrangian fibration.

If (M, ω) is toric and either is non-compact or has non-empty boundary then preimages under the moment map may be disconnected, a union of tori of possibly varying dimensions (as in Example 3.27). In this case Δ is no longer the orbit space; rather we have $\mu = \Phi \circ \pi$ where $\pi : (M, \omega) \to B$ is the induced Lagrangian fibration and $\Phi : B \to \mathbb{R}^n$ is an immersion. The immersion Φ respects the geometry of the base in the following sense: the induced affine structure on the regular values of π extends in a natural way an affine structure $\mathcal{A}$ on B and we have $\Phi : (B, \mathcal{A}) \to (\mathbb{R}^2, \mathcal{A}_0)$. (This will become clear in the next section.) Furthermore, the base B has a natural stratification $\mathcal{S}$: the l-stratum of $(B, \mathcal{S})$ is the set of points in $b \in B$ such that $\pi^{-1}(b)$ is a torus of dimension l. (That this defines a stratification follows from the normal forms for the fibered neighborhoods of l-dimensional fibers.) Notice that when M is a closed manifold the l-stratum of Δ is the union of the interiors of the l-dimensional faces, $l > 0$, and the 0-stratum is the union of the vertices.

DEFINITION 3.6. A Lagrangian fibration $\pi : (M^{2n}, \omega) \to (B, \mathcal{A}, \mathcal{S})$ is a *toric fibration* if there is a Hamiltonian n-torus action and an immersion $\Phi : (B, \mathcal{A}) \to$

$(\mathbb{R}^n, \mathcal{A}_0)$ such that $\Phi \circ \pi$ is the corresponding moment map and $\mathcal{S}$ is the induced stratification. In this case we call $(B, \mathcal{A}, \mathcal{S})$ a *toric base.*

Our goal to generalize part of Delzant's theorem can then be summarized by:

QUESTION 3.7. What assumptions about a triple $(B, \mathcal{A}, \mathcal{S})$ ensure that it determines a unique symplectic manifold?

An answer is provided, for two-dimensional bases, by Theorem 5.2 and Corollary 5.4. As a warm-up, in the next section we give criteria for recognizing whether or not $(B, \mathcal{A}, \mathcal{S})$ is the base of a toric fibration (Theorem 3.19).

3.2. Construction via boundary reduction. In this section we explain how to construct a toric fibration from a toric base $(B, \mathcal{A}, \mathcal{S})$. The idea is to start with a regular Lagrangian fibration over the base $(B, \mathcal{A})$ and then collapse certain fibers to lower dimensional tori so as to obtain the stratification $\mathcal{S}$.

The collapsing of fibers is achieved by boundary reduction, an operation originally defined in [**35**]:

PROPOSITION 3.8. *Let (M, ω) be a symplectic manifold with boundary such that a smooth component Y of ∂M is a circle bundle over a manifold Σ. Suppose also that the tangent vectors to the circle fibers lie in the kernel of $\omega|_Y$. Then there is a projection $\rho : (M, \omega) \to (M', \omega')$ and an embedding $\phi : \Sigma \to M'$ such that $\rho(Y) = \phi(\Sigma)$, $\rho|_{M-Y}$ is a symplectomorphism onto $M' - \phi(\Sigma)$ and $\phi(\Sigma)$ is a symplectic submanifold.*

PROOF. The circle bundle structure of Y gives the existence of the manifold M' and the embedding ϕ. Using a local model for the symplectic neighborhood of $Y \subset M$ it is not hard to see the induced symplectic structure on a neighborhood of $\rho(Y) \subset M'$. Essentially, it arises from a fiberwise application of Example 3.10. That $\rho(Y)$ is a symplectic submanifold can be seen in the local models or else by noting that, for points $x \in Y$, $\rho_*(T_xY)$ is the quotient of T_xY by the kernel of $\omega|_{T_xY}$, thereby showing that ω' is non-degenerate on $\rho(Y)$. □

DEFINITION 3.9. With the notation of Proposition 3.8, $(M', \omega') = \rho(M, \omega)$ is the *(symplectic) boundary reduction* of (M, ω) along Y.

Boundary reduction can be viewed as "half" of a symplectic cut (introduced by Lerman [**27**]). Its essence is captured in the following two-dimensional example:

EXAMPLE 3.10. Consider a cylinder with linear coordinate $0 \leq p \leq 1$, cyclic coordinate q of period 2π and symplectic form $dp \wedge dq$. Let D be a disk of radius $\sqrt{2}$ in $(\mathbb{R}^2, dx \wedge dy)$. Then the change of coordinates

$$x = \sqrt{2p}\cos(q), \qquad y = \sqrt{2p}\sin(q) \tag{3.1}$$

defines a symplectomorphism from the cylinder minus the boundary component defined by $p = 0$ to the punctured disk. Boundary reduction of the cylinder along the $p = 0$ boundary component yields the disk $(D, dx \wedge dy)$.

Notice that a hypersurface in $(\mathbb{R}^n \times T^n, \omega_0)$ that projects to a hypersurface in $\mathbb{R}^n$ is fibered by circles in the kernel of the symplectic form only if the hypersurface in $\mathbb{R}^n$ is affine planar. In higher dimensions one can perform boundary reduction along different smooth components of a boundary simultaneously. The following example reveals the local convexity of (toric) moment map images as well as the properties listed in Proposition 3.5.

EXAMPLE 3.11. For $0 \leq k \leq n$, let $U_k = \{(p_1, \ldots, p_n) | p_j \geq 0, j \leq k\} \subset (\mathbb{R}^n, \mathcal{A}_0)$ and let P_j be the hyperplane through the origin defined by $dp_j = 0$. Let $(M, \omega) = (U_k \times T^n, dp \wedge dq)$ so we have a Lagrangian fibration $\pi : (M, \omega) \to (U_k, \mathcal{A}_0, \mathcal{S})$ in which the stratification is trivial – all fibers are top dimensional. Then the boundary reduction of (M, ω) along $\pi^{-1}(\cup_{j=1}^k P_j \cap U_k)$ is (M', ω') where $M' = \mathbb{R}^{2k} \times (\mathbb{R} \times S^1)^{n-k}$ with symplectic form $\omega' = \sum_{i=1}^k dx_i \wedge dy_i + \sum_{i=k+1}^n dp_i \wedge dq_i$. The torus action on (M, ω) given by $t \cdot (p, q) = (p, q + t)$ descends to the action $t \cdot (x, y, p, q) = (x \cos(t') - y \sin(t'), x \sin(t') + y \cos(t'), p, q + t'')$ on (M', ω'), $t = (t', t'')$. This yields the toric fibration $\pi' : (M', \omega') \to (U_k, \mathcal{A}_0, \mathcal{S}')$ where $\mathcal{S}'$ is the stratification such that p belongs to the k-stratum if it is in the intersection of precisely $n - k$ of the P_j, $1 \leq j \leq k$.

OBSERVATION 3.12. A neighborhood of $\pi^{-1}(0)$ in Example 3.11 provides a normal form for a neighborhood of a k-dimensional fiber of a toric fibration. (Compare [**6**]).

In working with manifolds that arise via boundary reduction it is helpful to have the following two definitions:

DEFINITION 3.13. The *reduced boundary* of a toric base $(B, \mathcal{A}, \mathcal{S})$, denoted by $\partial_R B$, is the set of points in B that belong to lower-dimensional strata, i.e. k-strata for $k < n$. (Notice that a point can belong to a lower-dimensional stratum only if it belongs to ∂B.)

DEFINITION 3.14. Given a toric fibration $\pi : (M, \omega) \to (B, \mathcal{A}, \mathcal{S})$, the *boundary recovery* is the unique regular Lagrangian fibered manifold $(B \times T^n, \omega_0)$ that yields (M, ω) via boundary reduction.

REMARK 3.15. A local model for the neighborhood of a point in the boundary of a toric manifold can be extracted from Example 3.11. Simply let W be a top dimensional submanifold with boundary in $\mathbb{R}^n$ such that the origin belongs to ∂W and ∂W is transverse to the $\{P_j\}_{j=1,\ldots k}$; then take $B = W \cap U_k$ and perform the boundary reduction along $\pi^{-1}(\cup_{j=1}^k P_j \cap B)$. Notice that the image of the boundary of the resulting manifold is the closure of $\partial B - \partial_R B$ and $B_0 = B - \partial_R B$.

The geometric structure of B and the stratification $\mathcal{S}$ together determine what circles get collapsed during boundary reduction of $(B \times T^n, \omega_0)$.

DEFINITION 3.16. Let $\pi : (M, \omega) \to (B, \mathcal{A}, \mathcal{S})$ be a toric fibration and γ a compact embedded curve with one endpoint b_1 in the $(n-1)$-dimensional stratum of $\partial_R B$ and such that $\gamma - \{b_1\} \subset B_0$. Let b_0 be the other endpoint of γ. The *collapsing class*, with respect to γ, for the smooth component of $\partial_R B$ containing b_1 is the primitive class $\mathbf{a} \in H_1(F_{b_0}, \mathbb{Z})$ that spans the kernel of $\iota_* : H_1(F_{b_0}, \mathbb{Z}) \to H_1(\pi^{-1}(\gamma), \mathbb{Z})$ where ι is the inclusion map. Accordingly, the *collapsing covector* with respect to γ is the primitive covector $v^* \in T^*_{b_0} B$ that determines vectors $v(x) \in T_x^{\text{vert}} M$ for each $x \in \pi^{-1} b_0$ such that the integral curves of this vector field represent $\mathbf{a}$. (Notice that $\mathbf{a}$ and v^* are well defined up to sign.)

REMARK 3.17. The collapsing class and covector are independent of γ in a toric fibration, but will not be when this definition is extended to almost toric fibrations.

REMARK 3.18. We can also define the collapsing covector in $T^*_{b_1} B$ as the parallel transport of v^* along γ.

Now we can give a characterization of toric bases that yields a simple topological construction of all toric manifolds.

THEOREM 3.19. *A triple* $(B, \mathcal{A}, \mathcal{S})$ *is a toric base if and only if there is an immersion* $\Phi : (B, \mathcal{A}) \to (B, \mathcal{A}_0)$ *and the stratification* $\mathcal{S}$ *is such that*

- *each smooth component of* $\partial_R B$ *is affine planar,*
- $(B, \mathcal{A})$ *is locally convex in a neighborhood of any point of* $\partial_R B$*,*
- *each component* K *of the* $(n-k)$*-stratum of* $\partial_R B$*,* $k \geq 2$*, is the intersection of the closures of* k *smooth components of the* $(n-1)$*-stratum,*
- *the* k *primitive integral covectors defining those planes span a* k*-dimensional sublattice of* $\Lambda|_K$ *and*
- *any point in the boundary of a* k*-stratum of* $\partial_R B$ *has a neighborhood isomorphic to a model neighborhood as described in Remark 3.15.*

Furthermore, the boundary reduction of $(B \times T^n, \omega_0)$ *according to* $\mathcal{S}$ *is the unique toric manifold whose base is* $(B, \mathcal{A}, \mathcal{S})$

REMARK 3.20. Bear in mind that in general B is not the image of the moment map.

PROOF. Given any toric manifold (M, ω), the moment map μ factors through the orbit space yielding the map $\Phi : B \to \mathbb{R}^n$ and an induced affine structure $\mathcal{A} = \Phi^* \mathcal{A}_0$. We have already noted that Example 3.11 and Remark 3.15 provide normal forms for the fibration; therefore they provide normal forms for the base. The listed conditions are all properties implied by these normal forms.

Conversely, the immersion Φ defines a toric fibration of $(B \times T^n, \omega_0)$ given by projection to the first factor. The first hypothesis about the base $(B, \mathcal{A}, \mathcal{S})$ guarantees that we can perform boundary reduction according to the stratification $\mathcal{S}$ and the remaining hypotheses guarantee that the resulting manifold is smooth. The uniqueness of the toric manifold fibering over $(B, \mathcal{A}, \mathcal{S})$ follows from the Boucetta and Molino's classification of locally toric fibrations [**7**]. □

Since we will be concerned for the most part with bases that are two-dimensional, to simplify exposition we use the following natural language: an *edge* is the closure of a component of the 1-stratum and a *vertex* is a component of the 0-stratum.

While a toric base $(B, \mathcal{A}, \mathcal{S})$ gives a well-defined toric fibration, to specify a torus action one needs to choose an affine immersion Φ, at least up to translation. The moment map then is $\Delta = \Phi(B)$. Meanwhile, an immersion Φ is useful for visualizing $(B, \mathcal{A})$ as a subset of $(\mathbb{R}^n, \mathcal{A}_0)$.

We will use a bit more information than just the image of Φ: A *base diagram* for a Lagrangian fibered four-manifold $\pi : (M, \omega) \to (B, \mathcal{A}, \mathcal{S})$ is a subset $\Delta = \Phi(B, \mathcal{A}) \subset (\mathbb{R}^2, \mathcal{A}_0)$, for some choice of Φ, together with coding that enables reconstruction of the base $(B, \mathcal{A}, \mathcal{S})$ from Δ. For instance, for a toric base we use heavy and light lines to represent reduced and non-reduced parts of the boundary respectively. Then, unless the base multiply covers of enough of the image in $(\mathbb{R}^2, \mathcal{A}_0)$ so as to obscure the structure of the preimage, we can construct the fibered manifold by boundary reduction. (If the immersion does obscure the structure of the base one would merely have to add extra information to the base diagram.) To do this, we need to know the collapsing classes. They can be determined from the base diagram as follows, assuming for simplicity of notation that $(B, \mathcal{A}, \mathcal{S}) \subset (\mathbb{R}^2, \mathcal{A}_0)$ so $\mathcal{A} = \mathcal{A}_0$:

FIGURE 1. (a) B^4 (b) $\mathbb{C}P^2$

OBSERVATION 3.21. If v is a primitive integral vector that is normal to a smooth component of the reduced boundary of $(B, \mathcal{A}_0, \mathcal{S}) \subset (\mathbb{R}^2, \mathcal{A}_0)$ then $v^* =< v, \cdot >$ is a collapsing covector for that component. (Here we are using the standard inner product on $\mathbb{R}^2$ to provide an isomorphism between its tangent and cotangent bundles.)

Of course, we can build up new toric bases from existing ones by identifying open subsets in each that are isomorphic. Furthermore, it is easy to determine when bases can be joined along smooth components of their boundaries in such a way that the affine structures agree. We are most interested in when we can "see" the isomorphism, therefore with base diagrams in mind we note:

PROPOSITION 3.22. *Consider two toric bases $(B_i, \mathcal{A}_0, \mathcal{S}_i) \subset (\mathbb{R}^n, \mathcal{A}_0)$. Suppose there is an element $(A, u) \subset AGL(n, \mathbb{Z})$ such that on the intersection $B_1 \cap (AB_2 + u)$ the n-stratum is non-empty, the stratifications $\mathcal{S}_1, \mathcal{S}_2$ agree on the intersection thereby defining $\mathcal{S}_3$ on $B_1 \cup (AB_2 + u)$, and each connected component of the strata of $\mathcal{S}_3$ is smooth. Then $(B_1 \cup (AB_2 + u), \mathcal{A}_0, \mathcal{S}_3)$ is a toric base.*

PROOF. The hypotheses of Theorem 3.19 remain satisfied because the union of two affine planar submanifolds of the same dimension (that have non-empty intersection) can be smooth only if the union itself is affine planar. The condition that the n-stratum be non-empty then ensures that the convexity condition is preserved. □

3.3. Examples. In this section, we present examples of toric fibered four-manifolds $(M, \omega) \to (B, \mathcal{A}, \mathcal{S})$ via base diagrams corresponding to immersions $\Phi : (B, \mathcal{A}) \to (\mathbb{R}^2, \mathcal{A}_0)$. Accordingly, the preimage (under $\Phi \circ \pi$) in M of a point in the interior of a light edge is a 2-torus while the preimage of a point in the interior of a heavy edge is a circle. Similarly, a vertex can have a preimage in M that is a torus, circle or point depending upon whether it is the intersection of two light lines, one light and one heavy, or two heavy lines.

EXAMPLE 3.23. A toric fibration of a four-ball B^4 of radius $\sqrt{2}$ in $(\mathbb{R}^4, dx \wedge dy)$ is given by $\pi(x, y) = (\frac{1}{2}(x_1^2 + y_1^2), \frac{1}{2}(x_2^2 + y_2^2))$ whose image is the triangle with vertices $(0,0)$, $(1,0)$ and $(0,1)$. See Figure 1(a). Notice that if B^4 is identified with a ball in $\mathbb{C}^2$ by $z_j = x_j + iy_j$, the preimage of the horizontal and vertical edges are the disks in complex coordinate axes $z_2 = 0$ and $z_1 = 0$ respectively.

EXAMPLE 3.24. The triangle with vertices $(0,0)$, $(1,0)$ and $(0,1)$ shown in Figure 1(b) represents a Lagrangian fibered symplectic manifold diffeomorphic to

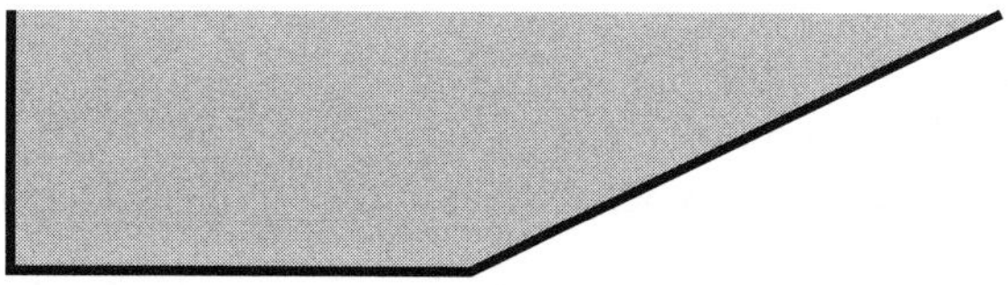

FIGURE 2. Neighborhood of a sphere: $U_{2,1}$

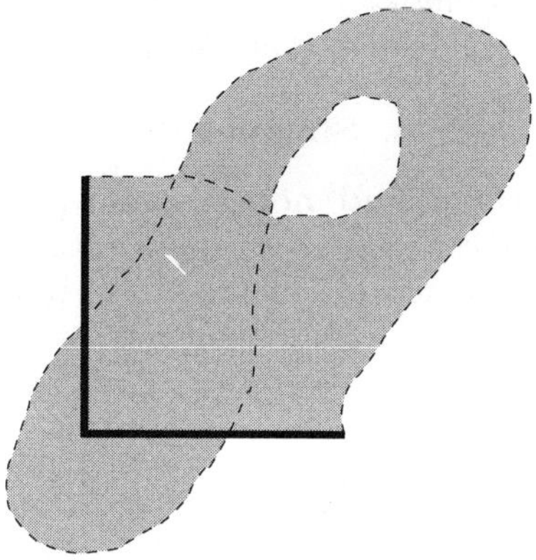

FIGURE 3. Exotic $\mathbb{R}^4$

$\mathbb{C}P^2$. This reflects the fact that $\mathbb{C}P^2$ is a boundary reduction of (B^4, ω_0) in which the circles of the Hopf fibration are collapsed to points.

Figure 1(b) also illustrates the standard construction of $\mathbb{C}P^2$ from three affine coordinate charts:

EXERCISE 3.25. Use Example 3.23 and Proposition 3.22 to find the fiber-preserving symplectomorphisms that yield the Lagrangian fibered $\mathbb{C}P^2$ defined by Figure 1(b) as the union of three open Lagrangian fibered four-balls.

EXAMPLE 3.26. For each $n \in \mathbb{Z}$ consider the domain

$$U_{n,1} := \{(p_1, p_2) | p_1 \geq 0, 0 \leq p_2 < t, np_2 \geq p_1 - s\}. \tag{3.2}$$

for some $s, t > 0$. Figure 2 shows $U_{2,1}$. Then $U_{n,1}$ defines a toric fibration of a neighborhood of a sphere – the sphere that is the preimage of the heavy horizontal line segment. The sphere has coordinates (p_1, q_1) on the complement of two points (the preimages of the vertices), so with respect to the symplectic form $dp \wedge dq$ it clearly has (symplectic) area $2\pi s$. Furthermore, the sphere has self-intersection $-n$. There are several ways to see this. For instance, one can check that the fiber-preserving symplectomorphism needed to present it as the union of two Lagrangian fibered four-balls implies it is diffeomorphic to a disk bundle over the sphere with Chern class $-n$. In Section 7.3 we explain an alternative way to see this that generalizes beyond spheres that are invariant under a torus action.

EXAMPLE 3.27. Figure 3 is a base diagram for a toric fibration of an exotic symplectic $\mathbb{R}^4$, i.e. $(\mathbb{R}^4, \omega)$ where ω is exotic in the sense that there is no symplectic embedding of $(\mathbb{R}^4, \omega)$ into $(\mathbb{R}^4, \omega_0)$ where ω_0 is the standard symplectic structure. This example, due to Zung [**39**], generalizes in an obvious way to higher dimensions yielding an exotic $\mathbb{R}^{2n}$ for any $n \geq 2$.

To see that ω is exotic observe that the immersion of the base into $(\mathbb{R}^2, \mathcal{A}_0)$ gives local coordinates (p, q) on the complement of the reduced boundary with respect to which the symplectic structure is $dq \wedge dq$ with primitive $\alpha = pdq$. Placing the vertex of the base diagram at the origin, the regular torus that projects to the origin is an exact Lagrangian torus (i.e. α restricted to this torus is cohomologous to 0, in this case actually equal to 0). Gromov [**21**] proved that there are no exact Lagrangian tori in $(\mathbb{R}^{2n}, \omega_0)$, thereby establishing that this $\mathbb{R}^4$ is symplectically exotic.

4. Singularities

4.1. Integrable systems and non-degenerate singularities. When we pass from regular Lagrangian fibrations to those with singularities we need to restrict the class of allowable singularities in order to keep control of a relation between the geometry of the base and the topology of the total space. A natural place to turn for a classification of singularities of a Lagrangian fibration is the theory of integrable systems.

DEFINITION 4.1. An *integrable system* on a symplectic manifold (M^{2n}, ω) is a collection of n functionally independent Poisson commuting functions $F_i : M \to \mathbb{R}^n$.

An equivalent definition is a Lagrangian fibration whose base can be immersed in $\mathbb{R}^n$ (ignoring affine structures).

Abusing slightly the integrable systems definition of non-degenerate singularities, we have

DEFINITION 4.2. A critical point of a Lagrangian fibration $\pi : (M^{2n}, \omega) \to B^n$ is called a *non-degenerate singular point of rank k* if it has a Darboux neighborhood (with symplectic form $dx \wedge dy$) in which the projection is $\pi = (\pi_1, \dots \pi_k, \pi_{k+1}, \dots \pi_n)$ where $\pi_j(x, y) = x_j$ for $j \leq k$ and the other components have some combination of the following forms:

$$\pi_j(x, y) = (x_j^2 + y_j^2) \quad \textit{elliptic} \tag{4.1}$$

$$\pi_j(x, y) = (x_j^2 - y_j^2) \quad \textit{hyperbolic} \tag{4.2}$$

$$(\pi_i, \pi_j)(x, y) = (x_i y_i + x_j y_j, x_i y_j - x_j y_i) \quad \textit{nodal,} \text{ or } \textit{focus-focus.} \tag{4.3}$$

In other words, a non-degenerate singularity splits as a product of elliptic, hyperbolic and nodal singularities. Notice that the elliptic singularities are precisely the singularities that occur in a toric fibration.

The nodal singularity is a complex hyperbolic singularity: in terms of complex coordinates

$$\mathrm{x} = x_1 + ix_2, \ \mathrm{y} = y_1 + iy_2, \quad \omega = \operatorname{Re} d\bar{\mathrm{x}} \wedge d\mathrm{y} \tag{4.4}$$

the projection from a four-manifold to $\mathbb{C}$ is

$$\pi(\mathrm{x}, \mathrm{y}) = \bar{\mathrm{x}}\mathrm{y}. \tag{4.5}$$

This is the same singularity that appears in *Lefschetz fibrations* of symplectic manifolds (cf. [**19**]).

REMARK 4.3. The precise definition of a non-degenerate singular point is as follows. Given an integrable system $F = (F_1, \ldots, F_n) : (M^{2n}, \omega) \to \mathbb{R}^n$, a point $x \in M$ such that $DF(x) = 0$ is called a *non-degenerate fixed point* if there are symplectic coordinates near x with respect to which the quadratic parts of the functions F_i form a Cartan subalgebra of the Lie algebra of quadratic functions on $\mathbb{R}^{2n} = T_xM$ with respect to the Poisson bracket. To generalize to singular points x of rank k (where the image of DF has dimension k) replace T_xM with the symplectic vector space $\mathrm{Ker}(DF(x))/\mathcal{I}$ where $\mathcal{I}$ is the subspace of T_xM spanned by the Hamiltonian vector fields X_i of the functions F_i (so $\iota_{X_i}\omega = -dF_i$). Eliasson [**10**] proved that such singularities have neighborhoods that are fiberwise diffeomorphic to the singularities of linearized integrable systems – which in turn were classified by Williamson [**38**]. Since we are interested in questions of global symplectic topology rather than the intricacies of a given integrable system, our abuse is to assume the singularity has a neighborhood fiberwise symplectomorphic to a linearized integrable system.

The structure imposed on a fibration by the presence of a hyperbolic singularity is fundamentally different from the structure imposed by elliptic or nodal singularities. The key features of the fibration near an elliptic or nodal singularity are that the base of the fibration is a manifold and the structure of the singular fiber is very simple. For instance, an elliptic singular point in a two-manifold has a neighborhood that is, on the complement of the singular point, fibered by circles that project to a half-open interval in the base. In dimension four a nodal singular point has a neighborhood that is fibered by cylinders and two transversely intersecting disks; since this singular point is isolated the cylinders belong to tori, thereby forcing the singular fiber to be a pinched torus (if it has no other singular points). If there are $k \geq 2$ nodal singular points on a fiber then the fiber is a union of k spheres, each of which intersects two others transversely (and positively) at one point. In either case, the base of the fibration of the neighborhood is an open disk.

In contrast, the presence of hyperbolic singularities in a fibration causes the base to become non-smoothable and allows the topology of the singular fibers to become complicated. For instance, in dimension two the base for a fibered neighborhood of a hyperbolic singularity is an "X" and recovering the symplectic manifold from the base becomes impossible without extra data: several hyperbolic singular points and the arcs connecting them are typically mapped to one singular point in the base so the topological type of the singular fiber would in general have to be stipulated.

Note that in dimension two the only manifolds that admit a Lagrangian fibration with no hyperbolic singularities are the sphere and the torus. However, all other (orientable) surfaces can be fibered once one allows hyperbolic singularities. This prompts the following:

QUESTION 4.4. What symplectic four-manifolds admit Lagrangian fibrations with topologically stable non-degenerate singularities?

In this paper our focus is on using base diagrams to determine symplectic manifolds. Therefore we henceforth exclude hyperbolic singularities and define

DEFINITION 4.5. A non-degenerate Lagrangian fibration $\pi : (M, \omega) \to B$ of a symplectic four-manifold is an *almost toric fibration* if it is a non-degenerate topologically stable fibration with no hyperbolic singularities. A triple $(B, \mathcal{A}, \mathcal{S})$ is an *almost toric base* if it is the base of such a fibration. A symplectic four-manifold equipped with such a fibration is an *almost toric manifold.*

In the above definition $\mathcal{A}$ is the affine structure on $B - \{s_i\}$ where the s_i are the images of the nodal singularities, and $\mathcal{S}$ is defined so that the s_i belong to the top-dimensional stratum.

The definition clearly generalizes to other dimensions, however we restrict our discussion to dimension four. Notice that an almost toric manifold can fail to be toric in one of two ways: 1) it has nodal fibers or 2) there is no affine immersion of the base into $(\mathbb{R}^2, \mathcal{A}_0)$.

4.2. Neighborhood of a nodal fiber.

DEFINITION 4.6. A *nodal fiber* in a Lagrangian fibered four-manifold is a singular fiber with nodal singularities and a *node* is the image of such a fiber under the fibration map.

Generically, nodal singularities of a Lagrangian fibration of a four-manifold occur in distinct fibers. Furthermore, one can always vary a fibration locally so that each fiber has only one singularity. (See Exercise 6.7.) Therefore, we in general assume a nodal fiber has just one nodal singularity and specify that the nodal fiber and the corresponding node in the base are multiplicity k whenever there are $k \geq 2$ singular points on the fiber.

Topologically, a nodal fiber is the singular fiber of a Lefschetz fibration (in which it appears in a family of symplectic fibers). Specifically, a neighborhood of a nodal fiber is diffeomorphic to $T^2 \times D^2$ with a -1-framed 2-handle attached along a simple closed curve in $T^2 \times \{x\}$ for some $\{x\} \in \partial D^2$. This is explained in [**19**], for instance. It also follows from the techniques of this paper: the core of the 2-handle is the vanishing disk (the disk whose boundary represents the vanishing class, defined in Section 4.3) and the -1-framing is calculated in Lemma 7.11.

An example of a nodal fiber is the image of the zero section of a disk bundle in T^*S^2 under a self-plumbing of the disk bundle. To construct this, first rewrite the normal form for the nodal singularity in terms of complex coordinates $\mathrm{x} = x_1 + ix_2$, $\mathrm{y} = y_1 + iy_2$ with $\omega_0 = \mathrm{Re}\, d\overline{\mathrm{x}} \wedge dy$. Thinking of the cotangent bundle as a complex line bundle with Chern class $c_1 = -2$, we build a disk bundle using two coordinate charts

$$U := \{(\mathrm{x},\mathrm{y}) | \; |\mathrm{xy}| < \epsilon < 1, |\mathrm{x}| < 2, |\mathrm{y}| < \tfrac{1}{4}\} \quad \text{and} \tag{4.6}$$

$$V := \{(\mathrm{u},\mathrm{v}) | \; |\mathrm{uv}| < \epsilon < 1, |\mathrm{u}| < 2, |\mathrm{v}| < \tfrac{1}{4}\} \tag{4.7}$$

with $\mathrm{u} = u_1 + iu_2$ and $\mathrm{v} = v_1 + iv_2$. These are glued together via the map

$$(\mathrm{u},\mathrm{v}) = \varphi(\mathrm{x},\mathrm{y}) = \left(\frac{1}{\mathrm{x}}, \overline{\mathrm{x}}^2 \mathrm{y}\right) \tag{4.8}$$

defined on $U \cap \{\frac{1}{2} < |\mathrm{x}| < 2\}$. This map is clearly a symplectomorphism if we equip the second coordinate chart with the symplectic form $\mathrm{Re}(-d\overline{\mathrm{u}} \wedge d\mathrm{v})$.

Now construct a self-plumbing of the disk bundle by identifying neighborhoods of the origin in each coordinate chart, say $\{(\mathrm{x},\mathrm{y}) | \; |\mathrm{x}|, |\mathrm{y}| < \frac{1}{4}\}$ and $\{(\mathrm{u},\mathrm{v}) | \; |\mathrm{u}|, |\mathrm{v}| < \frac{1}{4}\}$, via the map

$$(\mathrm{u},\mathrm{v}) = \psi(\mathrm{x},\mathrm{y}) = (\overline{\mathrm{y}}, \overline{\mathrm{x}}) \tag{4.9}$$

which again preserves the symplectic form.

The Lagrangian fibration given by $\pi(\mathrm{x},\mathrm{y}) = \bar{\mathrm{x}}\mathrm{y}$, $\pi(\mathrm{u},\mathrm{v}) = \bar{\mathrm{u}}\mathrm{v}$ then has one singular point at $(\mathrm{x},\mathrm{y}) = (\mathrm{u},\mathrm{v}) = (0,0)$.

EXERCISE 4.7. Check that the singular fiber is a sphere with one positive self-intersection. (When checking the positivity of the intersection, note that the manifold is oriented by $\omega^2 = dx_1 \wedge dy_1 \wedge dx_2 \wedge dy_2 = dx_1 \wedge dx_2 \wedge dy_1 \wedge (-dy_2)$.)

PROPOSITION 4.8. *The germ of a neighborhood of an isolated nodal fiber in a Lagrangian fibered symplectic manifold is unique up to symplectomorphism. Furthermore, given any two symplectomorphic fibered neighborhoods of a nodal fiber the symplectomorphism between them can be chosen to be fiber-preserving on the complement of smaller fibered neighborhoods. (See* [**36**].*)*

It is not hard to generalize the above proposition to include multiplicity k nodal fibers (in which case k is of course a topological invariant of the neighborhood).

REMARK 4.9. Note that the germ of a fibered neighborhood of a nodal fiber is not unique up to fiber-preserving symplectomorphism; San [**33**] has identified a non-trivial invariant that classifies the germs of such neighborhoods up to fiberwise symplectomorphism.

4.3. Monodromy around a nodal fiber. We can detect the monodromy around a nodal fiber from the local model in Section 4.2. Indeed, continuing with the complex variables notation, restrict to the torus bundle over a circle of radius $\delta < \epsilon$ in $\mathbb{C} = \mathbb{R}^2$ and let $\theta = \arg(z)$ parameterize the circle in the base, so $\theta = \arg(\bar{\mathrm{x}}\mathrm{y}) = \arg(\bar{\mathrm{u}}\mathrm{v})$. Then the intersection of the torus bundle over the circle with the two coordinate charts U, V gives two cylinder bundles that can be trivialized by coordinates (r,α) and (ρ,β) where $\mathrm{x} = re^{i\alpha}$ and $\mathrm{u} = \rho e^{i\beta}$. The restrictions of the maps φ and ψ show that these bundles are glued together by $\rho e^{i\beta} = \frac{1}{r}e^{-i\alpha}$ when $\frac{1}{2} < r < 2$ and $\rho e^{i\beta} = \frac{\delta}{r}e^{-i(\alpha+\theta)}$ when $4\delta < r < \frac{1}{4}$. Figure 4 shows the two cylinders that make up a fiber torus F_b over $b = z = \delta e^{i\theta}$. On this torus two curves are indicated: γ_1 given by $r = \rho = 1$ and γ_2 given by $\alpha = 0$ in U. As θ varies the curves γ_1, γ_2 give a trivialization of the torus bundle over the interval $0 \leq \theta < 2\pi$. Meanwhile, γ_2 shows the positive Dehn twist in the torus fiber as θ varies from 0 to 2π and reveals that the topological monodromy around a nodal fiber is

$$\Psi(\gamma) = A_{(1,0)} = \begin{pmatrix} 1 & 1 \\ 0 & 1 \end{pmatrix} \tag{4.10}$$

with respect to a basis $\{[\gamma_1],[\gamma_2]\}$ for $H_1(F_b,\mathbb{Z})$.

Changing the basis for $H_1(F_b,\mathbb{Z})$ causes the image of Ψ to be a conjugate of $A_{(1,0)}$. Such matrices are of the form

$$A_{(a,b)} = \begin{pmatrix} 1-ab & a^2 \\ -b^2 & 1+ab \end{pmatrix} \tag{4.11}$$

with eigenvector (a,b).

Consider an almost toric fibration $\pi : (M,\omega) \to B$ that has a node at s. Let η be an embedded curve with endpoints at s and a regular point $b \in B_0$ such that $\eta - \{s\} \subset B_0$ contains no other nodes. Associated to s and η is the (well defined up to sign) *vanishing class* in $H_1(F_b,\mathbb{Z})$, namely the class whose representatives bound a disk in $\pi^{-1}(\eta)$. (With respect to the language for Lefschetz fibrations, the

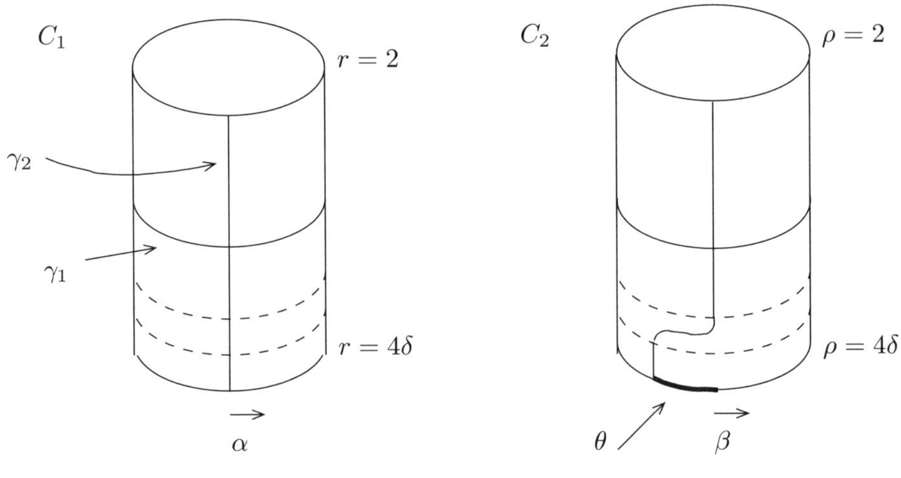

FIGURE 4. Positive Dehn twist as θ goes from 0 to 2π

vanishing class is the homology class of the vanishing cycle.) The *vanishing covector* $w^* \in T_b^*B$ is the primitive covector (defined up to sign) that determines vectors $w(x) \in T_x^{\text{vert}}M$ for each $x \in \pi^{-1}b$ such that the integral curves of this vector field represent vanishing class.

LEMMA 4.10. *Suppose γ is a positively oriented loop based at b that is the boundary of a closed neighborhood of s containing η. Then the vanishing class is the unique class (up to scale) that is preserved by the monodromy along γ. With respect to a basis for $H_1(F_b, \mathbb{Z})$ for which the monodromy matrix is $A_{(a,b)}$, the vanishing class is the class (a, b).*

PROOF. Without loss of generality we work with the model neighborhood of Section 4.2 and with basis $\{[\gamma_1], [\gamma_2]\}$ for $H_1(F_b, \mathbb{Z})$, as above, so the topological monodromy matrix is $A_{(1,0)}$. Now note that γ_1 bounds a disk: the disk $\{(\mathrm{x}, \mathrm{y})|\ \arg(\bar{\mathrm{x}}\mathrm{y}) = 0, |\mathrm{y}| = |\mathrm{x}| \leq \sqrt{\delta}\}$. Thus the vanishing class is $(1, 0)$, the eigenvector of the topological monodromy. □

Because a nodal fiber of multiplicity k (a fiber with k nodal singular points) arises from coalescing k nodal fibers of multiplicity 1, the monodromy around a nodal fiber of multiplicity k is, up to conjugation,

$$\Psi(\gamma) = A_{(1,0)}^k = \begin{pmatrix} 1 & k \\ 0 & 1 \end{pmatrix}. \tag{4.12}$$

4.4. Affine structure near a node. Suppose $\pi : (M, \omega) \to B$ is a Lagrangian fibered neighborhood of a nodal fiber in a four-manifold with $s \in B$ the node. The restriction of the fibration to $B - \{s\}$ is a regular fibration and hence $B - \{s\}$ has an induced affine structure $\mathcal{A}$. Because of the non-trivial affine monodromy around the node (which we call the *nodal monodromy*) there is no immersion of $(B - \{s\}, \mathcal{A})$ into $(\mathbb{R}^2, \mathcal{A}_0)$. However, we can find an affine immersion if we remove a curve from B that has an endpoint at the node, leaving us with a simply connected surface. Therefore, as a base diagram for the neighborhood of

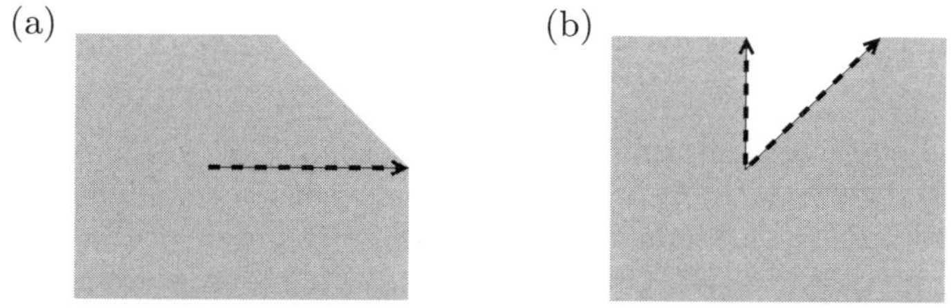

FIGURE 5. Neighborhood of a node

a node, we will use the image of an affine immersion of the complement of such a curve.

For convenience we choose, whenever possible, the curve we remove to be a ray R based at the node s. Then, up to translation, any immersion $\Phi : (B - R, \mathcal{A}) \to (\mathbb{R}^2, \mathcal{A}_0)$ has an image $\Phi(B - R)$ that is a neighborhood of the origin minus a sector. Different choices of the ray R cause the internal angle of the sector to vary in the interval $[0, \pi)$. Figures 5(a) and (b) show base diagrams in which the shaded part of the complement of the heavy dotted lines is the image $\Phi(B - R)$ for two different choices of R.

As we will see below, the sector has angle zero when R belongs to the one well-defined affine linear submanifold through the node s:

DEFINITION 4.11. The *eigenline* L through a node s is the unique maximal affine linear immersed one-manifold through the node for which there is a regular point $b \in L$, arbitrarily close to s, such that the affine monodromy along an arbitrarily small loop around s and based at b preserves $T_bL \subset T_bB$. An *eigenray* is either of the two maximal affine linear submanifolds that has an endpoint at the node and is a subset of the eigenline.

In Figure 5(a) the indicated ray is an eigenray.

We now construct a family of integral affine punctured planes, $(V_u, \mathcal{A}_u)$ where V_u is diffeomorphic to $\mathbb{R}^2 - \{0\}$. Let $(\widetilde{W}, \mathcal{A}_0)$ be the universal cover of $(\mathbb{R}^2 - \{0\}, \mathcal{A}_0)$ with affine projection Ψ. For each vector u representing a point in $\mathbb{R}^2$ let $\widetilde{V}_u$ be a sector in $\widetilde{W}$ with interior angle $> \pi$ and $\leq 2\pi$ and that is bounded by rays R, R' consisting of points x, x' such that $\Psi(x) = \lambda u$ and $\Psi(x') = \lambda A_{(1,0)} u$. (We are ignoring the ambiguity of how $\widetilde{V}_u$ is affinely embedded in $\widetilde{W}$ since this is irrelevant for our construction.) Then define $(V_u, \mathcal{A}_u)$ where $V_u := \widetilde{V}_u / x \sim x'$ for x, x' as above and $\mathcal{A}_u$ is defined by the induced lattice Λ_u on TV_u:

$$(TV_u, \Lambda_u) = (\widetilde{V}_u \times \mathbb{R}^2, \Lambda_0)/(x, v) \sim (x', A_{(1,0)} v). \tag{4.13}$$

Of course, $(\widetilde{V}_u, \mathcal{A}_u)$ and $(\widetilde{V}_{\lambda u}, \mathcal{A}_{\lambda u})$ are isomorphic for any $\lambda > 0$. More importantly,

LEMMA 4.12. *For all $u \in \mathbb{R}^2$, $(V_u, \mathcal{A}_u)$ is isomorphic to $(V_{e_1}, \mathcal{A}_{e_1})$.*

Therefore we simplify notation and call this affine manifold $(V, \mathcal{A})$.

EXERCISE 4.13. Prove Lemma 4.12

If a node has multiplicity k then local models can be constructed exactly as above with $A_{(1,0)}$ replaced by $A^k_{(1,0)}$. We call such affine punctured planes $(V^k, \mathcal{A}^k)$. Note that if R is an eigenray, then the image $\Phi(B - R)$ is independent of k.

PROPOSITION 4.14. *Any node $s \subset B$ of multiplicity $k \geq 1$ has a neighborhood that is affine isomorphic to a neighborhood of the puncture in $(V^k, \mathcal{A}^k)$.*

PROOF. Let U be a neighborhood of s and let $R \subset U$ be a ray based at s. Then $(U - R, \mathcal{A})$ is homomorphic to a disk and hence there is an affine immersion $\Phi : (U - R, \mathcal{A}) \to (\mathbb{R}^2, \mathcal{A}_0)$ such that the origin is the limit of the images of points in B whose limit is s. Corollary 9.16 implies that this is an embedding and hence is isomorphic to a neighborhood of $r = 0$ in the interior of $(\widetilde{V}_u, \mathcal{A}_0)$ for some u. The result then follows from Lemma 4.12 and the remarks about nodes of multiplicity greater than one. □

Furthermore, for each choice of ray $R \subset B$ based at s there is an immersion $\Phi : (B - R, \mathcal{A}) \to (\mathbb{R}^2, \Lambda_0)$ and a u such that $\Phi(B - R)$ is the intersection of $\Psi(\text{Int}\widetilde{V}_u)$ and a neighborhood of the origin in $\mathbb{R}^2$. In particular, when R is an eigenray, $\Phi(B - R)$ is the intersection of $\Psi(\text{Int}\widetilde{V}_{\pm e_1})$ and a neighborhood of the origin in $\mathbb{R}^2$, as in Figure 5(a).

5. Almost toric manifolds

5.1. Classification in terms of base. The base of an almost toric fibration has a geometric structure very similar to a toric base. Indeed, the only difference between an almost toric and a toric base (on a local level) is the presence of nodes which in turn have neighborhoods whose affine structure (on the complement of the node) depends only on the multiplicity of the node.

DEFINITION 5.1. An *integral affine manifold with nodes* $(B, \mathcal{A})$ is a two-manifold B equipped with an integral affine structure on $B - \{s_i\}$ such that each s_i has a neighborhood U_i such that $(U_i - s_i, \mathcal{A})$ is affine isomorphic to a neighborhood of the puncture in $(V^k, \mathcal{A}^k)$ for some $k \geq 1$.

As for toric manifolds, we are interested in using the base of an almost-toric fibration to understand topological and symplectic features of the total space. The stratification $\mathcal{S}$ of a toric base is an important invariant of the fibration and to extend this notion to the almost toric case we declare that nodes belong to the top-dimensional stratum. Then triples $(B, \mathcal{A}, \mathcal{S})$ that can appear as bases of almost toric fibrations can be characterized as follows:

THEOREM 5.2. *Consider a triple $(B, \mathcal{A}, \mathcal{S})$ such that $(B, \mathcal{A})$ is an integral affine manifold with nodes $\{s_i\}_{i=1}^N$. Then $(B, \mathcal{A}, \mathcal{S})$ is an almost toric base if and only if every point in $B - \{s_i\}_{i=1}^N$ has a neighborhood that is a toric base.*

PROOF. An almost toric fibration is by definition locally toric on the complement of its nodal fibers. This implies that the base must, on the complement of the nodes, have the local structure of a toric base.

Conversely, if the base is locally toric on the complement of the nodes, we can construct an almost toric fibration in a similar fashion to a toric fibration: Take the Lagrangian fibration of $((B - \{s_i\}_{i=1}^N) \times T^2, \omega_0)$ given by projection onto the first factor. The local toric structure guarantees that boundary reduction to achieve the stratification $\mathcal{S}$ yields a toric fibered smooth manifold, say $\pi' : (M', \omega') \to (B - \{s_i\}_{i=1}^N, \mathcal{A}, \mathcal{S})$. (Recall that π' does not define a toric fibration unless there is an immersion $(B - \{s_i\}_{i=1}^N, \mathcal{A}) \to (\mathbb{R}^2, \mathcal{A}_0)$). Furthermore, the definition of an affine manifold with nodes implies that each node has a neighborhood that is the base of a model neighborhood of a nodal singular fiber. Theorem 3.19 guarantees that the fibration over the regular points of this neighborhood is unique. Therefore there is a fiber-preserving symplectomorphism of the complement of the singular fiber in its model neighborhood and the restriction of our toric fibration $\pi' : (M', \omega') \to (B - \{s_i\}_{i=1}^N, \mathcal{A}, \mathcal{S})$ to the preimage of a punctured neighborhood of the node. Using such fiber-preserving symplectomorphisms we can glue in nodal fibers and achieve an almost toric fibration with base $(B, \mathcal{A}, \mathcal{S})$. □

Our definition of almost toric manifolds (Definition 4.5) is motivated by a classification of non-degenerate Lagrangian fibrations due to Zung [**39**]. To classify these fibrations up to fiber-preserving symplectomorphism (or symplectic equivalence) he introduced the following notion: two non-degenerate Lagrangian fibrations over the same base $(B, \mathcal{A}, \mathcal{S})$ are *roughly symplectically equivalent* if their singular fibers have fiberwise symplectomorphic neighborhoods. Here, $\mathcal{A}, \mathcal{S}$ are the affine structure and stratification that Zung showed exist on the base of a non-degenerate Lagrangian fibration.

Zung also defined an invariant of a non-degenerate Lagrangian fibration, the *Lagrangian Chern class.* It is an element of the first homology of the base with values in the sheaf of closed basic 1-forms (1-forms that vanish on vectors tangent to fibers) modulo those forms that arise from contracting the vector field for a symplectic fiber-preserving circle action with the symplectic form. (To be precise, this Chern class is actually a relative class in the sense that it is defined relative to a given reference fibration.)

THEOREM 5.3 (Zung [**39**]). *Two non-degenerate Lagrangian fibrations over the same base $(B, \mathcal{A}, \mathcal{S})$ are symplectically equivalent if and only if they are roughly symplectically equivalent and have the same Lagrangian Chern class.*

In dimension four with the absence of hyperbolic singular points, we are led to the following corollary:

COROLLARY 5.4. *Let $(B, \mathcal{A}, \mathcal{S})$ be an almost toric base which is either non-compact or has non-empty boundary. Then $(B, \mathcal{A}, \mathcal{S})$ is the base of an almost toric fibration of a unique symplectic manifold.*

PROOF. The hypothesis that the two-dimensional base be either non-compact or have non-empty boundary implies that it has the homotopy type of a one-dimensional space. Therefore the Lagrangian Chern class vanishes and hence all roughly symplectically equivalent fibrations with base $(B, \mathcal{A}, \mathcal{S})$ are fiberwise symplectomorphic.

Now suppose we have two almost toric fibrations over $(B, \mathcal{A}, \mathcal{S})$. Because of the hypothesis on the topology of B, we know that there is a unique fibration over $(B - \{s_i\}_{i=1}^N, \mathcal{A}, \mathcal{S})$. Completing this to a fibration over $(B, \mathcal{A}, \mathcal{S})$ as in the proof of Theorem 5.2 we are not guaranteed of getting a unique Lagrangian fibration (see San [**33**]). However the proof of Proposition 4.8 given in [**36**] shows that the structure of the fibration can be perturbed via a compactly supported symplectic isotopy, thereby establishing uniqueness up to symplectomorphism of the total space. □

5.2. Base diagrams. Our objective now is to draw base diagrams in $\mathbb{R}^2$ that prescribe the construction of almost toric bases $(B, \mathcal{A}, \mathcal{S})$. We restrict ourselves first to the case when $(B - \{R_i\}_{i=1}^N, \mathcal{A})$ can be immersed in $(\mathbb{R}^2, \mathcal{A}_0)$ where the R_i are a

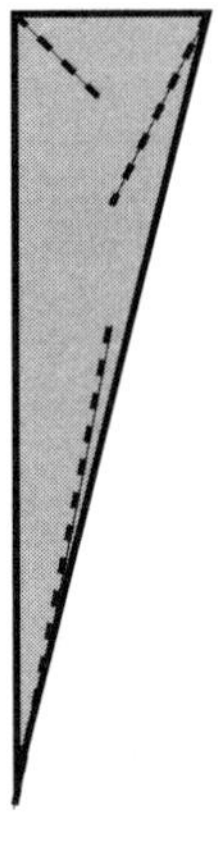

FIGURE 6. Base diagram for an almost-toric fibration of $\mathbb{C}P^2$.

minimal set of disjoint curves, each with at an endpoint at the node s_i. We call the curves R_i *branch curves.* Let P_i be the possible values of Φ if it were extended, as a discontinuous map, to R_i, i.e. let $P_i = \{p \in \mathbb{R}^2 | p = \lim_{x \to y_i} \Phi(x), y_i \in R_i\}$ and call $\cup_{i=1}^N P_i$ the *branch locus.* Take $\Delta = \Phi(B - \{R_i\}_{i=1}^N) \cup \{P_i\}_{i=1}^N$ and indicate each set P_i by a line with heavy dashes. Mark the image of a node s_i with the integer k_i if the node has multiplicity $k_i \geq 2$ and the branch curve R_i based at s_i is an eigenray. (If R_i is an eigenray and the node is not marked we assume it has multiplicity 1.) Now use the conventions of Section 3.2 to indicate the stratification. In the special case that a component of the branch locus emanating from a node intersects the reduced boundary at a point in the 0-stratum rather that the 1-stratum we would mark intersection point with a heavy dot.

EXERCISE 5.5. Check that the notation for the multiplicity of a node is sufficient and not redundant. Specifically, prove that if two vectors $u, v \in \mathbb{R}^2$ satisfy $A^k_{(a,b)} u = v$ and $u \neq v$, then u and v determine a, b and k, while $u = v$ implies the vector (a, b) is parallel to u and k is not determined.

EXAMPLE 5.6. Figures 5(a) and (b) are both base diagrams for the neighborhood of a node. In the notation of Section 4.4 they are $\Psi(\widetilde{V}_{e_1})$ and $\Psi(\widetilde{V}_{e_2})$, thereby representing $(V_{e_1}, \mathcal{A}_{e_1})$ and $(V_{e_2}, \mathcal{A}_{e_2})$. In Figure 5(b) the map Φ extends to a discontinuous map in which the set P (described above) is the boundary of the missing sector. In Figure 5(a) the set P is a ray and the extension of Φ is continuous but not differentiable.

EXAMPLE 5.7. Figure 6 shows a base diagram for an almost toric fibration with three nodal fibers. While at first glance one might think that this is a base diagram for a space with orbifold singularities, the manifold it defines is in fact smooth. Sections 5.3 and 6.1 describe the techniques that allow one to see that the manifold is $\mathbb{C}P^2$.

To reconstruct the total space from a base diagram we can use Observation 3.21 to determine the homology classes of circles that are collapsed at the reduced boundary and the following observation to determine the vanishing class of a nodal fiber.

OBSERVATION 5.8. In a base diagram for an almost toric base, if a primitive integral vector v is perpendicular to the eigenline through a node then $v^* =< v, \cdot >$ is a covector that defines the vanishing class of the fiber (with respect to the eigenline). This is an immediate consequence of Proposition 2.17 and the fact that $v^* u =< v, u >$.

An almost toric fibration can fail to be toric because of the presence of nodal fibers or because of the lack of an affine immersion of the base into $\mathbb{R}^2$. To extend to cases where there is no choice of branch curves $\{R_i\}_{i=1}^N$ such that $B - \{R_i\}_{i=1}^N$ immerses in $\mathbb{R}^2$ we follow the usual convention for representing surfaces in $\mathbb{R}^2$: we cut B along more curves and use light lines with arrows to indicate components that need to be identified (with trivial monodromy). (This same convention could be used to resolve the ambiguities faced when an immersion into $\mathbb{R}^2$ obscures the structure of the base due to multiple covering of portions of the image.) Proposition 3.22 provides criteria for when smooth components of the boundaries of bases can be identified.

Using these conventions we can reconstruct an almost toric base from its base diagram. This is done with the understanding that smooth components of the branch locus are identified according to the unique nodal monodromy transformation (of the correct multiplicity) that maps one onto the other.

EXAMPLE 5.9. The manifold $S^2 \times T^2$ does not admit a Hamiltonian torus action, however it is easy to define an almost toric fibration of it.

Let $S^2 \subset \mathbb{R}^3$ be a sphere of radius 1 with symplectic structure ω_1 being the induced area form. Then the height function gives a Lagrangian fibration $\pi_1 : (S^2, \omega_1) \to [-1, 1]$ with elliptic singularities at the north and south poles. Equip the torus T^2 with angular coordinates (s, t), both of period 2π, symplectic form $\omega_2 = ds \wedge dt$ and Lagrangian fibration $\pi_2(s, t) = s$. Then $\pi_1 \oplus \pi_2 : (S^2 \times T^2, \omega_1 \oplus \omega_2) \to [-1, 1] \times S^1$ is an almost toric fibration.

EXERCISE 5.10. Draw a base diagram for this fibration of $S^2 \times T^2$.

EXAMPLE 5.11. In Section 11 we show examples of base diagrams of almost toric fibrations of the K3 surface. For now we simply give some background, including a source for Lagrangian fibrations of the K3.

The smooth K3 surface is a manifold diffeomorphic to a smooth degree four hypersurface in $\mathbb{C}P^3$, i.e. the zero locus of a non-degenerate degree four homogeneous polynomial in three variables. While there are many complex K3 surfaces (with different complex structures) they are all diffeomorphic.

We start by defining a family of holomorphic fibrations of K3 surfaces. Consider a pair of (distinct) smooth cubic curves in $\mathbb{C}P^2$ defined by degree three polynomials P_1, P_2; these are tori that intersect each other in 9 points $\{x_1, \ldots, x_9\}$. The pencil of cubics given by $\{\alpha P_1 + \beta P_2 = 0\}$ for $[\alpha : \beta] \in \mathbb{C}P^1$ defines a fibration of $\mathbb{C}P^2 - \{x_1, \ldots, x_9\}$. Blowing up those points yields a holomorphic fibration of $\mathbb{C}P^2 \# 9\mathbb{C}P^1$ over $\mathbb{C}P^1$; the generic fiber is a torus (an elliptic curve), and for generic choices of P_1, P_2 the singular fibers each have one node on them (and no other singularities). When equipped with such a holomorphic fibration by elliptic curves, this manifold is called $E(1)$. Now take a double branch cover of $E(1)$, branched over a pair of (disjoint) regular torus fibers. The result is an elliptically fibered K3 surface, also known as $E(2)$.

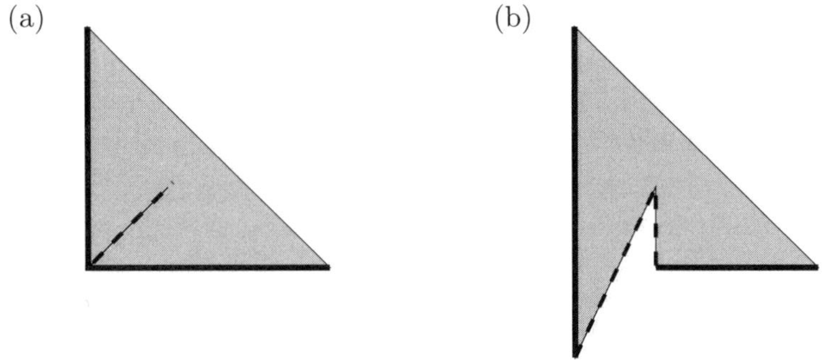

FIGURE 7. Two base diagrams of the same fibration.

Such a complex K3 surface is equipped with a hyper-Kähler structure, namely a sphere's worth of Kähler structures (complex structures $aI+bJ+cK$ and Hermitian metrics whose anti-symmetric parts are symplectic structures $a\omega_I + b\omega_J + c\omega_K$, $a^2+b^2+c^2$). Given a fibration that is holomorphic with respect to J, it is Lagrangian with respect to ω_I (or $a\omega_I + c\omega_K$, $a^2 + c^2 = 1$). (In fact it is special Lagrangian, cf. [**23**].) Generically, a holomorphic fibration has 24 nodal fibers and hence so does a Lagrangian fibration.

5.3. Branch moves. Due to the choices made in the drawing of a base diagram, many base diagrams correspond to the same almost toric fibration. The variations resulting from an application of $AGL(2,\mathbb{Z})$ are quite straightforward. More involved changes occur when we alter how we cut the base before projecting it to $(\mathbb{R}^2, \mathcal{A}_0)$. In particular, a change in a choice of the branch curves R_i constitutes a *branch move*. For instance, Figures 5 (a) and (b) differ by a branch move.

EXERCISE 5.12. Verify that Figure 7 shows two different base diagrams for one almost toric fibration and that they differ by a branch move. In both diagrams the coordinates of the node are $(1,1)$ and the coordinates of the endpoints of the line segment with slope -1 are $(3,0)$ and $(0,3)$.

EXAMPLE 5.13. Figure 8(a) shows a base diagram for $(S^2 \times T^2)\#\overline{\mathbb{C}P}^2$. This follows from Exercise 5.10 and the almost toric blow-up described in Section 5.4. (Filling in the missing wedge gives the answer to Exercise 5.10.) Notice that the eigenline of the node is parallel to the reduced boundary.

EXERCISE 5.14. Verify that Figure 8(b) is a base diagram for a fibration of $(S^2\tilde{\times}T^2)\#\overline{\mathbb{C}P}^2$. Hint: Fill in the missing wedge to blow down (as per Section 5.4) and then convince yourself that this is the base diagram for a sphere bundle over a torus. Use the techniques of Section 7.3 to calculate that the self-intersections of the tori represented by the connected components of the reduced boundary are ± 1.

As an application of branch moves we can use base diagrams to see that

PROPOSITION 5.15. *The manifolds* $(S^2 \times T^2)\#\overline{\mathbb{C}P}^2$ *and* $(S^2\tilde{\times}T^2)\#\overline{\mathbb{C}P}^2$ *are diffeomorphic.*

PROOF. As already noted, Figures 8(a) and (b) are base diagrams for $(T^2 \times S^2)\#\overline{\mathbb{C}P}^2$ and $(S^2\tilde{\times}T^2)\#\overline{\mathbb{C}P}^2$ respectively. In Figure 8(a) the monodromy across

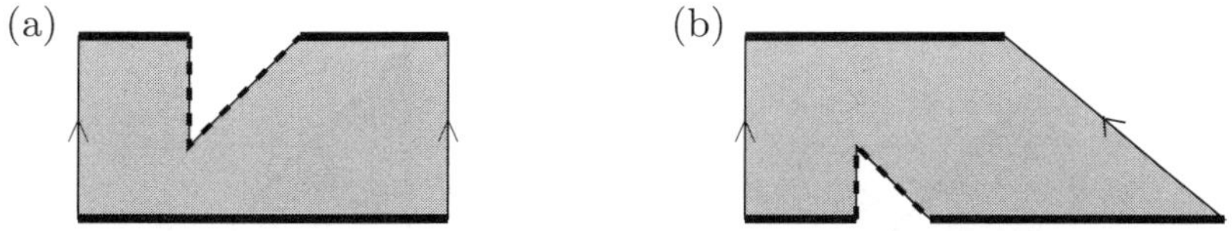

FIGURE 8. Base diagrams for one fibration of $(S^2 \times T^2)\#\overline{\mathbb{C}P}^2 = (S^2 \tilde{\times} T^2)\#\overline{\mathbb{C}P}^2$.

the vertical upward ray is $A_{(1,0)}$ and across the downward vertical ray (which is not drawn) is the identity. To get from the first to the second by a branch move cut Figure 8(a) along the downward vertical and apply the linear map $A_{(1,0)}^{-1}$ to the right half of the image (with the origin at the node); this makes the monodromy across the upward vertical ray the identity and the monodromy across the downward vertical ray $A_{(1,0)}^{-1}$. Reglue along the upward vertical ray, thereby obtaining Figure 8(b). □

EXERCISE 5.16. Adapt this argument to show that $(S^2 \times \Sigma)\#\overline{\mathbb{C}P}^2$ and $(S^2 \tilde{\times} \Sigma)\#\overline{\mathbb{C}P}^2$ are diffeomorphic for any surface Σ. (The case of $\Sigma = S^2$ is discussed in the appendix.) Hint: Blow up inside a toric fibration of $S^2 \times D^2$ where D^2 is a disk in Σ. Choose the symplectic structure and the disk so that the area of the fiber S^2 is smaller than the area of the D^2.

5.4. Blowing up and down. Blowing up is a simple surgery which, on a four-manifold, amounts to replacing a ball with a neighborhood of a sphere of self-intersection -1. It is well known how, in the toric category, to blow up at a point that is itself a fiber (i.e. a rank zero elliptic singular point). After reviewing this we explain how, in the almost toric category, to blow up at a point that lies on a one-dimensional fiber at the expense of introducing a nodal fiber. Blowing down is just the reverse operation to blowing up.

To blow up in the symplectic category one proceeds as follows: find a symplectic embedding $\varphi : (B^4(r), \omega_0) \to (M, \omega)$ of the closed ball of radius r with the standard symplectic structure, consider $M - \mathrm{Int}(\varphi(B^4(r)))$ and perform a boundary reduction along $\varphi(\partial B^4(r))$, thereby collapsing the fibers of the Hopf fibration.

In general the symplectic embedding φ could be very complicated. However, if (M, ω) is equipped with a Lagrangian fibration that has a rank zero elliptic singularity, then for sufficiently small r one can choose the embedding φ so that the origin gets mapped to the rank zero singular point and it is fiber preserving with respect to the standard toric fibration of $B^4(r) \subset (\mathbb{R}^4, \omega_0)$ (as in Example 3.23). Then the surgery can be described in terms of a surgery on the base: remove from the base a triangular neighborhood of the vertex that is the image of the rank zero singular point; do this in such a way that the edges of the triangle meeting at that vertex have length $\frac{1}{2}r^2$ times the length of a primitive integral tangent vector; then connect the two resulting vertices of the base with a heavy line. (See Figure 9(a).) The resulting base reveals the change in topology: the new edge represents a symplectic 2-sphere of area πr^2 whose self-intersection is -1. (See Sections 7.2 and 7.3.)

With the allowance of nodal fibers there is another "fibration compatible" way to blow up, namely an *almost toric blow-up*. We describe the surgery on the base

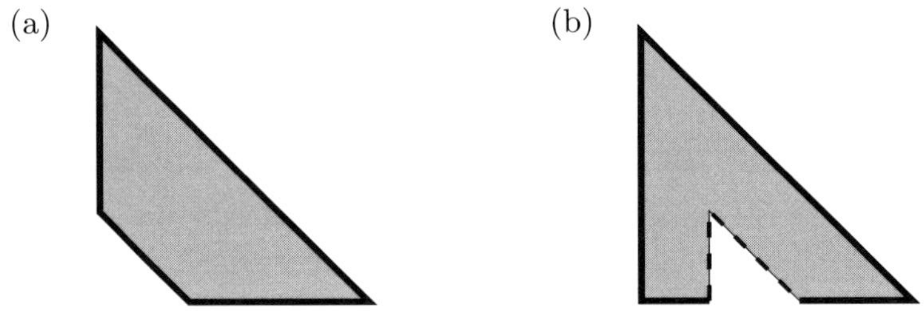

FIGURE 9. Two fibrations of $\mathbb{C}P^2\#\overline{\mathbb{C}P}^2$

in terms of the base diagram and then explain how it corresponds to blowing up a point. (This operation was initially observed by Zung [**39**].)

Consider an almost toric fibration $\pi : (M,\omega) \to (B,\mathcal{A},\mathcal{S})$ such that the 1-stratum of B is non-empty. (Recall that the 1-stratum must belong to $\partial_R B$.) Choose a base diagram for B such that for some $a, \varepsilon > 0$ the set $\{(p_1,p_2)|p_1 > -\varepsilon, p_2 \geq 0, p_1+p_2 < a+\varepsilon\}$ with the boundary components marked by heavy lines represents a fibered neighborhood of a point in the 1-stratum. Remove the triangle with vertices $(0,0)$, $(a,0)$, $(0,a)$ and insert line segments with heavy dashes so that their endpoints are $(0,a),(0,0)$ and $(0,a),(a,0)$. Viewing $(0,a)$ as the origin, $A_{(1,0)}$ maps the vertical dashed segment to the diagonal one and preserves horizontal vectors. Therefore, this is a base diagram for an almost toric fibration and with respect to the induced standard coordinates the nodal fiber has vanishing class $[\frac{\partial}{\partial q_2}]$. Notice that this is the same as the collapsing class of the component of the 1-stratum in question. See Figure 9 for an example.

Theorem 5.4 guarantees that the described operation on the base gives a well-defined surgery on the manifold (M,ω). To see that this surgery amounts to blowing up a point notice that we start with a neighborhood diffeomorphic to $S^1 \times D^3$; on a topological level, the effect of the surgery is to add a vanishing disk (to create the nodal fiber) whose boundary represents the same class as the collapsing class of the almost toric fibered $S^1 \times D^3$, i.e. to adding a -1-framed 2-handle along a contractible circle. But this is the same thing, topologically, as making a connect sum with $\overline{\mathbb{C}P}^2$.

EXERCISE 5.17. Use the techniques of Section 7 to verify that the preimage of either segment marked by heavy dashes is a visible sphere of self-intersection -1. (Both segments yield the same visible sphere.) This is the exceptional sphere introduced by the almost toric blow-up.

REMARK 5.18. In the toric case – and in the complex category – it is clear what point is being blown up; this is not so in either a symplectic blow-up or an almost toric blow-up.

5.5. Closed almost toric manifolds. As mentioned in the introduction, the set of closed manifolds that admit almost toric fibrations is a modest expansion over those that admit toric fibrations. More precisely, we can list them:

THEOREM 5.19 (Leung and Symington [**30**]). *If M is a closed almost toric manifold then M is diffeomorphic to $S^2 \times S^2$, $S^2 \times T^2$, $N\#n\overline{\mathbb{C}P}^2$ with $N = \mathbb{C}P^2$*

or $S^2\tilde{\times}T^2$ *and* $n \geq 0$, *a torus bundle over a torus with monodromy*

$$\left\{\begin{pmatrix}1 & \lambda\\ 0 & 1\end{pmatrix}, Id\right\} \quad \text{with } \lambda \in \mathbb{Z}, \tag{5.1}$$

a torus bundle over the Klein bottle that is double covered by one of the above bundles over a torus, the K3 surface or the Enriques surface.

In this section we give a brief sketch of the proof, referring the reader to [**30**] for details.

There are two parts to the proof of Theorem 5.19: showing that the aforementioned manifolds admit almost toric fibrations and showing that there are no other almost toric manifolds.

As to the first part, notice that $S^2 \times S^2$ and $\mathbb{C}P^2\#n\overline{\mathbb{C}P}^2$ admit toric fibrations and hence almost toric fibrations. Example 5.9 gives an almost toric fibration of $S^2 \times T^2$. A base defining an almost toric fibration of $S^2\tilde{\times}T^2$ was constructed in Exercise 5.14. Applying the almost toric blow-up of Section 5.4 to these examples yields almost toric fibrations of $S^2 \times T^2\#n\overline{\mathbb{C}P}^2$ and $S^2\tilde{\times}T^2\#n\overline{\mathbb{C}P}^2$ (with the case $n = 1$ shown in Figures 8(a) and (b)). Note that we do not need to list separately the blow-ups of these trivial and non-trivial bundles as they are diffeomorphic. (See Section 6.2.) Any fibration over the Moëbius band is a $\mathbb{Z}_2$ quotient of a fibration over the cylinder and hence can be constructed explicitly in terms of a $\mathbb{Z}_2$ quotient of the cylinder with an integral affine structure.

Meanwhile, Geiges' classification of torus bundles over a torus that admit a symplectic structure [**17**] implies that a such a bundle can have Lagrangian fibers if and only if the manifold is diffeomorphic to a bundle with the monodromy given in the statement of Theorem 5.19 . Taking $\mathbb{Z}_2$ quotients of the base, when possible, yields the total spaces of all the regular torus bundles over the Klein bottle that admit Lagrangian fibers.

Explicit almost toric fibrations of the K3 surface are given in Section 11. The first of these, built by taking symplectic sums of eight copies of the fibration defined by Figure 16, has lots of symmetry. Taking a fiber-preserving $\mathbb{Z}_2$ quotient of this fibration yields a fibration with 12 singular fibers and base $\mathbb{R}P^2$. One way to see that this is the Enriques surface is to notice that the resulting manifold is diffeomorphic to the one obtained by symplectic summing four copies of the fibration defined by Figure 16 to give a fibration of $E(1) = \mathbb{C}P^2\#9\overline{\mathbb{C}P}^2$ and then performing two smooth multiplicity two log transforms along tori isotopic to the preimage of the reduced boundary. Since this latter construction is known to yield the Enriques surface (cf. [**19**]) we are done.

The proof that the set of manifolds mentioned in Theorem 5.19 contains all four-manifolds that admit an almost toric fibration is outlined by the following lemmas which are proved in [**30**].

LEMMA 5.20. *If* $(B, \mathcal{A}, \mathcal{S})$ *is a compact almost toric base such that* $\partial_R B = \partial B$, *then* B *must have non-negative Euler characteristic and hence must be homeomorphic to a disk, a cylinder, a Moëbius band, a sphere,* $\mathbb{R}P^2$, *a torus or a Klein bottle. Furthermore, if* B *is a closed manifold then it has* 12χ *nodes (counted with multiplicity) where* χ *is the Euler characteristic of* B.

(This lemma, in a more general form, is stated but not proven in [**39**].) The proof is an application of the Gauss-Bonnet theorem. We separate the arguments for when B has boundary and when it does not. Notice that the hypotheses that B is compact and $\partial_R B = \partial B$ imply that B is the base of a fibration of a closed manifold.

If B is a closed manifold then there must be $12k$ nodes for some integer $k \geq 0$ since any product of matrices conjugate to $A_{(1,0)}$ that equals the identity is a product of $12k$ such matrices (cf. [**19**] or [**15**]). Choosing k disjoint open (topological) disks that each contain 12 nodes around which the monodromy is trivial, put a flat metric on the complement of the union of the disks and then extend the metric over the rest of the base B. The fact that the sphere has an affine structure with 24 nodes then implies that each collection of 12 nodes (with trivial monodromy) contributes 2π to the total curvature, or 1 to the Euler characteristic. Therefore a closed base can only be a sphere with 24 nodes, $\mathbb{R}P^2$ with 12 nodes, or T^2 or the Klein bottle with no nodes.

If B has boundary then heuristically, B has positive Euler characteristic because its boundary is locally convex and each node contributes non-negative curvature. To calculate the Euler characteristic we can appeal to the Gauss-Bonnet formula for polygons – that the sum of the total curvature of the polygon, the geodesic curvature of its boundary and the discontinuities at the vertices equals 2π (cf. [**34**]). Triangulating B and summing over the triangles we find the Euler characteristic as $\frac{1}{2\pi}$ times the sum of the contributions from the triangles. Since an affine manifold with nodes has no metric that is compatible with the affine structure we work instead with a compact manifold $(B', \mathcal{A}')$ that is homeomorphic to $(B, \mathcal{A})$, has no nodes, and such that $(B' - P', \mathcal{A}')$ is isomorphic to $(B - P, \mathcal{A})$ where P' is a subset of $\partial B'$ and P is a union of disjoint curves, each with one endpoint at a node in $(B, \mathcal{A})$. Then the affine structure $\mathcal{A}'$ induces a flat metric on B', allowing us to carry through the Gauss-Bonnet argument.

Now we can consider what manifolds can have an almost toric fibration over each of the bases listed in Lemma 5.20.

LEMMA 5.21. *If there is an almost toric fibration* $\pi : (M, \omega) \to (B, \mathcal{A}, \mathcal{S})$ *where* B *is homeomorphic to a disk and* M *is a closed manifold, then there is a toric fibration* $\pi' : (M, \omega') \to (B, \mathcal{A}')$.

The proof of this lemma relies on branch moves to change the fibration to a new one in which the base is equivalent, after making nodal trades, to a toric base. We do not include the details here but merely note that to make the branch moves it may be necessary to deform the symplectic structure (as in Section 6.3), hence the indicated change in the symplectic structure from ω to ω'.

LEMMA 5.22. *Let* $(B, \mathcal{A}, \mathcal{S})$ *be an almost toric base such that* $\partial_R B = \partial B$. *If* B *is a closed cylinder then two boundary components are parallel. If* B *is either a cylinder or a Moëbius band, then the eigenlines of any nodes must be parallel to* ∂B.

The essence of this lemma can be seen in Figure 8(a) where, if the reduced boundary components were not parallel, there would be no integral affine transformation that could glue the collar neighborhoods of the non-reduced boundary components to obtain a cylinder. Furthermore, the introduction of a node whose eigenline is not parallel to the boundary would have the effect of making the boundary components converge. With these restrictions, the only closed manifolds that have almost toric fibrations over a cylinder or a Moëbius band are sphere bundles over tori and their blow-ups, and we have constructed fibrations of all of them.

We already saw above that Geiges' classification implies that any almost toric fibration over a torus or a Klein bottle is diffeomorphic to one of the manifolds listed in the theorem.

If the base is a sphere then it must have 24 nodes. Then it is the base of an almost toric fibration with 24 nodal fibers. With respect to a different symplectic structure, the manifold is the total space of a Lefschetz fibration with 24 singular fibers and hence is the K3 surface (cf. [**19**]). Finally, an almost toric fibration over $\mathbb{R}P^2$ must be double covered by an almost toric fibration over S^2, namely the K3 surface. Therefore, up to diffeomorphism we have only one such manifold.

6. Modifying fibrations

6.1. Nodal slides and nodal trades. The first operation we introduce in this section is a nodal slide:

DEFINITION 6.1. Two almost toric bases $(B, \mathcal{A}_i, \mathcal{S}_i)$, $i = 1, 2$, are related by a *nodal slide* if there is a curve $\gamma \subset B$ such that $(B - \gamma, \mathcal{A}_1, \mathcal{S}_1)$ and $(B - \gamma, \mathcal{A}_2, \mathcal{S}_2)$ are isomorphic and γ contains one node and belongs to the eigenline L_i through that node.

Any two bases that are related by a nodal slide belong to a family of bases in which the node "slides" along the eigenline. (The affine structures on the complement of the eigenlines are not affected by the position of the node.) These define a family of almost toric fibrations of one manifold and hence a family of symplectic structures on that manifold. Furthermore, the manifolds are (fiberwise) symplectomorphic on the complement of a compact set belonging to the preimage of the eigenline. Therefore we can work in a contractible neighborhood of that preimage where the symplectic structures are exact (since the symplectic structure in a neighborhood of a regular Lagrangian fiber is exact). Invoking Moser's argument [**31**] we can conclude that the there is an isotopy, compactly supported in that neighborhood, that defines a symplectomorphism between the almost toric manifolds fibering over the two bases. Therefore,

PROPOSITION 6.2. *If two bases are related by a nodal slide then they define the same manifold equipped with isotopic symplectic structures.*

Under certain circumstances, sliding a node right into an edge in $\partial_R B$ and replacing the limit point in the edge with a vertex yields an almost toric base. If so, then the new base again determines a manifold symplectomorphic to the original one. For instance, Figure 1(a) and Figure 7(a) are different fibrations of the same symplectic four-ball. Specifically,

LEMMA 6.3. *Given an almost toric base $(B, \mathcal{A}, \mathcal{S})$ let R be an embedded eigenray that connects a node s with a point b in an edge $E \subset \partial_R B$ such that there are no other nodes on R. Let $v^*, w^* \in T_b^* B$ be the vanishing and collapsing covectors. If v^* and w^* span Λ_b^* then there is an almost toric base $(B, \mathcal{A}', \mathcal{S}')$ such that*

(1) *$(B, \mathcal{A}', \mathcal{S}')$ contains one fewer node than $(B, \mathcal{A}, \mathcal{S})$,*
(2) *$(B - R, \mathcal{A}', \mathcal{S}')$ is isomorphic to $(B - R, \mathcal{A}, \mathcal{S})$ and*
(3) *in $(B, \mathcal{A}', \mathcal{S}')$ the intersection of R and $\partial_R B$ is a vertex.*

PROOF. Without loss of generality we can work locally in a neighborhood of R. The condition that v^* and w^* span $\Lambda_b^* B$ implies that we can choose a base diagram in which $v^* = (-1, 1)$ and $w^* = (0, 1)$. Then, possibly after rescaling, a neighborhood of the dashed line in Figure 7(a) provides a base diagram for $(B, \mathcal{A}, \mathcal{S})$ with the dashed line representing R. Simply erasing the dashed line reveals a neighborhood in an appropriately scaled copy of Figure 1(a) that serves as $(B, \mathcal{A}', \mathcal{S}')$, clearly satisfying all of the conditions listed in the lemma. This suffices to prove the lemma. Note that the condition that v^* and w^* span Λ_b^* is required to prevent the vertex from being the image of an orbifold singular point. □

REMARK 6.4. If the edge in B and an eigenray emanating from the node do not intersect inside B, the symplectic manifold defined by B is still a ball, but the ball does not have a toric fibration that agrees with the almost toric fibration near the boundary.

Two almost toric bases $(B, \mathcal{A}, \mathcal{S})$ and $(B, \mathcal{A}', \mathcal{S}')$ are related by a *nodal trade* if they satisfy the hypotheses of Lemma 6.3.

THEOREM 6.5. *Two almost toric bases that are related by a nodal trade are symplectomorphic; in fact their symplectic structures are isotopic.*

PROOF. It suffices to prove that the almost toric bases $(U, \mathcal{A}, \mathcal{S})$ and $(U, \mathcal{A}', \mathcal{S}')$ in the proof of Lemma 6.3 define diffeomorphic four-manifolds with symplectic forms that are isotopic via an isotopy that is compactly supported in a smaller neighborhood. We already know that $(U, \mathcal{A}, \mathcal{S})$ is the base of a fibration of an open four-ball. To see that $(U, \mathcal{A}', \mathcal{S}')$ also defines a four-ball, notice that it can be viewed as D^4 with a 1-handle attached (to get $S^1 \times D^3$) and then a -1-framed 2-handle (a thickening of the vanishing disk) attached along a circle that runs once over the 1-handle (since v^*, w^* span Λ_b^*). Such a 2-handle cancels the 1-handle so the manifold is D^4.

Now repeat the Moser argument proving Proposition 6.2 to complete the proof of the theorem. □

REMARK 6.6. As observed by Zung [**39**], the one parameter family of fibrations corresponding to sliding a node into an edge and achieving a nodal trade is, in the language of integrable systems, a Hamiltonian Hopf bifurcation that occurs in many Hamiltonian systems [**37**].

EXERCISE 6.7. Suppose $(B, \mathcal{A}, \mathcal{S})$ has a multiplicity k node at s and $(B, \mathcal{A}', \mathcal{S}')$ has k distinct multiplicity 1 nodes on a segment L of the eigenline through these nodes. Use the same type of arguments to prove that if $(B - L, \mathcal{A}, \mathcal{S})$ and $(B - L, \mathcal{A}', \mathcal{S}')$ are isomorphic, then $(B, \mathcal{A}, \mathcal{S})$ and $(B, \mathcal{A}', \mathcal{S}')$ define the same four-manifold with isotopic symplectic structures.

6.2. Connecting toric fibrations. It is well known that $(S^2 \times S^2, \omega \oplus \lambda\omega)$ can admit distinct toric fibrations and that the number of such fibrations depends on λ. If, without loss of generality, we assume $\lambda \geq 1$, then the number of such fibrations is $\lceil \lambda \rceil$, the smallest integer greater than or equal to λ. The number of these fibrations is interesting in that it equals the number of non-conjugate tori in the group of Hamiltonian symplectomorphisms of $(S^2 \times S^2, \omega \oplus \lambda\omega)$ [**26**]. Furthermore, each time λ passes an integer the topology of the full symplectomorphism group changes [**21, 1, 2, 3**].

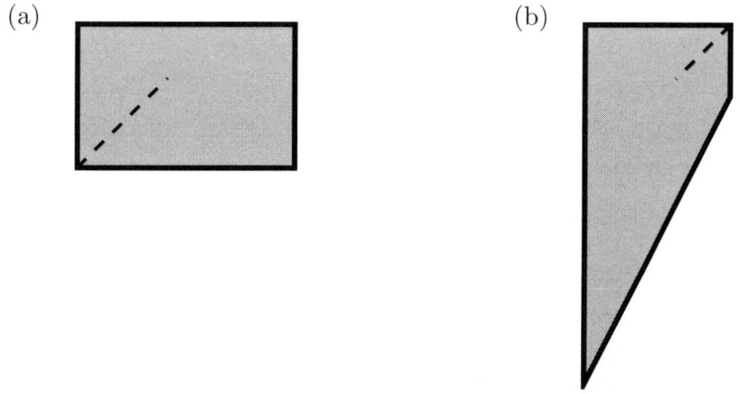

FIGURE 10. Branch move for one fibration of $(S^2 \times S^2, \sigma \oplus \frac{3}{2}\sigma)$

Using nodal slides and branch moves the different toric fibrations of $(S^2 \times S^2, \omega \oplus \lambda\omega)$ can be connected by a path of almost toric fibrations. This provides a simple geometric proof of the the diffeomorphism equivalence of the manifolds defined by the $\lceil \lambda \rceil$ different toric bases and also, invoking a Moser argument, the isotopy equivalence of the induced symplectic forms.

We illustrate this in Figure 10 in which case $\lambda = \frac{3}{2}$. Figures 10(a) and (b) show fibrations that clearly differ from toric ones by a nodal trade and the two figures differ from each other by a branch move.

CONJECTURE 6.8. *Any two toric fibrations of (M, ω) can be connected by a path of almost toric fibrations of (M, ω). (Note that ω is fixed throughout this path.)*

6.3. Deformations. A *deformation* of a symplectic manifold is a one-parameter family of symplectic structures. In general the cohomology class of the symplectic structure varies during the deformation. Any one-parameter family of almost toric bases that defines the same smooth manifold defines a deformation of the symplectic structure.

One obvious deformation arises from rescaling an almost toric base, thereby causing a rescaling of the symplectic structure. Also, if the base is non-compact or has boundary then removing portions of the top stratum of the base or adding to it so as to yield a new base homeomorphic to the first will give a deformation.

Two other useful variations of the base are "thickening" and "thinning" along the reduced boundary. (These deformations were originally introduced in [**32**] where they were used to prove the diffeomorphism equivalence of certain four-manifolds.) Specifically, an edge in the reduced boundary has a collar neighborhood fibered by curves that are affine linear. When the edge contains two vertices the collar neighborhood has a base diagram as shown in Figure 2 for some choice of length a of the heavy horizontal line and slope $\frac{1}{n}$, $n \in \mathbb{Z}$, for the rightmost heavy line. The thickness of the neighborhood is the length of the vertical heavy line, say t. Call such a neighborhood $(N_{a,n,t}, \mathcal{A}_0, \mathcal{S})$. Notice that for fixed lengths a, t and fixed n the base $(N_{a,n,t}, \mathcal{A}_0, \mathcal{S})$ belongs to a family $(N_{a-n\tau,n,t+\tau}, \mathcal{A}_0, \mathcal{S})$ where $\tau \in (-t, \frac{a}{n})$ if $n > 0$ and $\tau \in (-t, \infty)$ if $n \leq 0$. (The upper bound on τ when $n > 0$ merely guarantees that $a - n\tau$ stays positive.) Then to *thicken* or *thin* $(B, \mathcal{A}, \mathcal{S})$ along an edge containing two vertices means to replace a collar neighborhood of it that is isomorphic to $(N_{a,n,t}, \mathcal{A}_0, \mathcal{S})$ with one isomorphic to $(N_{a-n\tau,n,t+\tau}, \mathcal{A}_0, \mathcal{S})$ for some $\tau > 0$ or $\tau < 0$ respectively.

EXERCISE 6.9. Extend this definition of thickening and thinning to the case when the 1-stratum is compact. Hint: Examples 5.9 and 5.7 contain base diagrams in which the 1-stratum is compact.

The effect on the cohomology class of the symplectic structure is evident from the changes in the areas of the spheres (or torus) that are the preimage of the closed edges of B. This is discussed again near the end of Section 7.3.

7. Reading the base I: surfaces

7.1. Visible surfaces. Some surfaces embedded in a toric fibered manifold are easily reconstructed from their images in the base, such as the symplectic spheres that are preimages of edges in the base of a toric fibration of a closed manifold. In this section we explain how a curve in the base of an almost toric fibration and the choice of elements of H_1 of regular fibers of the curve together determine surfaces in the total space. We also show how this data determines whether or not the surface is symplectic or Lagrangian.

Throughout this section, given an immersed curve $\gamma : I \to (B, \mathcal{A}, \mathcal{S})$, we use $\{\gamma_i\}_{i=1}^k$ to denote the continuous components of $\gamma|_{\gamma^{-1}(B-\partial_R B)}$. We assume that if γ approaches and then coincides with a component of the 1-stratum of $\partial_R B$ (on a set with non-empty interior) that the tangency of γ with $\partial_R B$ at a point of first contact is sufficiently high so as to ensure that Σ_γ is non-singular. Specifically, if there are local standard coordinates in which γ is given by $(p_1, f(p_1))$ then at such a point of first tangency $\sqrt{f}$ is differentiable.

DEFINITION 7.1. A circle in a regular fiber F_b of a Lagrangian fibration is an *affine circle* if it has a tangent vector field belonging to Λ^{vert}.

DEFINITION 7.2. A *visible surface* Σ_γ in an almost toric fibered manifold $\pi : (M, \omega) \to (B, \mathcal{A}, \mathcal{S})$ is a an immersed surface whose image is an immersed curve γ with transverse self-intersections such that $\pi|_{\Sigma_\gamma \cap \pi^{-1}(B_0)}$ is a submersion onto $\gamma \cap B_0$, any non-empty intersection of Σ_γ with a regular fiber is a union of affine circles, and no component of $\partial\Sigma_\gamma$ projects to a node.

Notice that a visible surface must be a sphere, a disk, a cylinder or torus.

DEFINITION 7.3. Given an immersed curve $\gamma : I \to (B, \mathcal{A}, \mathcal{S})$, let $\{\gamma_i\}_{i=1}^k$ be the continuous components of $\gamma|_{\gamma^{-1}(B-\partial_R B)}$. A primitive class $\mathbf{a}_i$ in $H_1(\pi^{-1}(\gamma_i), \mathbb{Z})$ (such that $\pi_* \mathbf{a}_i = 0$ if γ_i is a loop) is *compatible* with γ if all of the following are satisfied:

(1) $\mathbf{a}_i$ is the vanishing class of every node in γ_i,
(2) $|\mathbf{a}_i \cdot \mathbf{c}| \in \{0, 1\}$ for each $\mathbf{c}$ that is the collapsing class, with respect to $\overline{\gamma}_i$, for a component of the 1-stratum of $\partial_R B$ that intersects $\overline{\gamma}_i$,
(3) $|\mathbf{a}_i \cdot \mathbf{c}| = 1$ if $\overline{\gamma}_i$ intersects the 1-stratum non-transversely,
(4) $|\mathbf{a}_i \cdot \mathbf{d}| = 1$ for each $\mathbf{d}$ that is one of the two collapsing classes at a vertex contained in the closure of γ_i. (Here $\cdot$ is the intersection pairing in $H_1(\pi^{-1}(\gamma_i), \mathbb{Z})$ and $\overline{\gamma}_i$ is the closure of γ_i.)

Note that we defined the collapsing class to be in the first homology of a regular fiber. But since the collapsing class in $H_1(F_b, \mathbb{Z})$, F_b being any fiber in $\pi^{-1}(\gamma_i)$, has the same image in $H_1(\pi^{-1}(\gamma_i), \mathbb{Z})$ induced by the inclusion map, it makes sense to view the collapsing class as living in $H_1(\pi^{-1}(\gamma_i), \mathbb{Z})$.

THEOREM 7.4. *Suppose $(B, \mathcal{A}, \mathcal{S})$ is an almost toric base such that each node has multiplicity one. An immersed curve $\gamma : I \to (B, \mathcal{A}, \mathcal{S})$ with transverse self-intersections and a set of compatible classes $\{\mathbf{a}_i\}_{i=1}^k$ together determine a visible surface Σ_γ such that for each $b \in \gamma_i$,*

$$\iota_*[\Sigma_\gamma \cap F_b] = \mathbf{a}_i \tag{7.1}$$

where $\iota : F_b \to \pi^{-1}(\gamma_i)$ is the inclusion map. The surface Σ_γ is unique up to isotopy among visible surfaces in the preimage of γ that satisfy Equation 7.1. Furthermore, no such surface exists if the classes $\mathbf{a}_i$ are not compatible with γ.

PROOF. If $\gamma \subset \partial_R B$ then γ defines a symplectic submanifold with an induced almost toric fibration over γ. In this case Σ_γ is a sphere or disk if γ has two endpoints or one endpoint, respectively, in the 0-stratum; it is a torus if γ is compact and without boundary; otherwise Σ_γ is a cylinder. Tori projecting to $\gamma \subset \partial_R B$ can be found in Examples 5.7 and 5.13, among others.

Now suppose $\gamma \cap (B - \partial_R B) \neq \emptyset$. In that case we certainly need to specify classes $\mathbf{a}_i$ in order to obtain a visible surface Σ_γ that is well defined up to isotopy. Since the compatibility conditions for the $\mathbf{a}_i$ deal only with local phenomena (near a given fiber), we assume for simplicity of exposition that γ is embedded.

First suppose that γ_i contains no nodes. Then $\pi^{-1}(\gamma_i)$ is a trivial torus bundle over γ_i. Choose a trivialization $T^2 \times \gamma_i$ of this torus bundle. Then for any $\mathbf{a}_i$ and any section $\sigma : \gamma_i \to \gamma_i \times T^2$ there is a unique visible surface that is a cylinder which contains the image of the section and whose intersection with any fiber represents $\mathbf{a}_i$. Since all sections are isotopic, Σ_{γ_i} is unique up to isotopy among visible surfaces over γ_i.

If γ_i (and hence γ) has a node s as an endpoint, then the requirement that no component of $\partial\Sigma_\gamma$ project to a node implies $\pi^{-1}(s) \cap \Sigma_\gamma$ is the nodal singular point. Furthermore, the smoothness of Σ_γ implies that $\mathbf{a}_i$ is the vanishing class of that node. If a node s is on the interior of γ_i, then because the vanishing class is the only well defined class in a neighborhood of the node, in order for the class of the affine circles in Σ_γ to be well defined as γ passed through a node we must have the $\mathbf{a}_i$ is the vanishing class for that node.

Next suppose $\overline{\gamma}_i$ intersects at b a component of the 1-stratum of $\partial_R B$ that has collapsing class $\mathbf{c}$ with respect to $\overline{\gamma}_i$. If the preimage of b in Σ_γ is a point (in which case b must be an endpoint of γ), then the smoothness of Σ_γ implies that $\mathbf{a}_i$ must be, up to sign, the collapsing class $\mathbf{c}$, i.e. $\mathbf{a}_i \cdot \mathbf{c} = 0$. Furthermore, as can be checked in local coordinates, transversality of the intersection of γ and the reduced boundary at b guarantees smoothness of Σ_γ. If however the preimage of b in Σ_γ is a circle C that represents the generator of $H_1(F_b, \mathbb{Z}) = \mathbb{Z}$, for C to be a limit of representatives of $\mathbf{a}_i$ we must have $\mathbf{a}_i = \mathbf{d} + k\mathbf{c}$ for some $k \in \mathbb{Z}$ where $\mathbf{d} \in H_1(\pi^{-1}(\gamma_i), \mathbb{Z})$ is such that $\mathbf{c}$ and $\mathbf{d}$ generate $H_1(\pi^{-1}(\gamma_i), \mathbb{Z})$ (or a two dimensional subgroup of it if γ_i is a loop). This is equivalent to having $|\mathbf{a}_i \cdot \mathbf{c}| = 1$.

If the closure of γ_i contains a vertex then an affine circle representing a class $\mathbf{a}_i$ is a torus knot in an embedded S^3 bounding a neighborhood of the preimage of the vertex. In fact, it is a torus knot $T(k, l)$ where $\mathbf{a}_i = k\mathbf{b}_1 + l\mathbf{b}_2$ where $\mathbf{b}_1, \mathbf{b}_2$ are the vanishing classes of the two edges that meet at the vertex. But a torus knot

$T(k,l)$ is trivial if and only if k or l equals ± 1 and $|\mathbf{b}_i \cdot \mathbf{b}_j|$ equals 0 if $i = j$ and 1 if $i \neq j$. Therefore k or l equals ± 1 if and only if $\mathbf{a} \cdot \mathbf{b}_i = \pm 1$ for $i = 1$ or 2.

The sufficiency of the compatibility conditions for the existence of Σ_γ can be checked directly in the the local models for neighborhoods of the singular fibers. Because the family of affine circles in a nodal fiber that represent the vanishing class is connected and because the node is the only isolated point that is the limit of affine circles in neighboring regular fibers, the presence of a node in γ does not disrupt the surface Σ_γ being well-defined up to the stated fiberwise isotopy. Furthermore, any fiberwise isotopy of Σ_{γ_i} extends to its closure in an obvious way. □

In light of our goal of reading topological and symplectic properties of a manifold from a base diagram, we want to use Theorem 7.4 to reconstruct a visible surfaces from a curve in a base diagram. Again, for simplicity of exposition we assume that our base is a subset of $(\mathbb{R}^2, \mathcal{A}_0)$.

Recall that a class in the first homology of a regular fiber F_b is determined by a covector in T_b^*B. Given a base diagram we can then use the standard inner product on $TB \subset T\mathbb{R}^2$ to represent the covector by its dual in T_bB. Therefore, to record the choices of compatible classes $\mathbf{a}_i$ we assign a *compatible vector* $v_i \in \mathbb{R}^2$ to γ_i such that the integral curves of the vector field $v_i \frac{\partial}{\partial q} \subset \Lambda^{\text{vert}}$ represent $\mathbf{a}_i$. If γ_i contains a node then v_i is forced on us so we do not need to label γ_i with that vector.

For reading base diagrams, and in particular for determining whether or not vectors v_i are compatible, the following direct calculation is helpful:

PROPOSITION 7.5. *Suppose F_b is a regular fiber in an almost toric fibration defined by $(B, \mathcal{A}_0, \mathcal{S}) \subset (\mathbb{R}^2, \mathcal{A}_0)$. Given two elements $\mathbf{a}, \mathbf{b} \in H_1(F_b, \mathbb{Z})$ and vectors $v, w \in \mathbb{R}^2$ such that $[v\frac{\partial}{\partial q}] = \mathbf{a}$ and $[w\frac{\partial}{\partial q}] = \mathbf{b}$ we have $|\mathbf{a} \cdot \mathbf{b}| = |\det(vw)| = |v \times w|$.*

Also recall Observations 3.21 and 5.8.

Of course, if the image of γ_i in a base diagram intersects the branch locus, the compatible vector $v_i = (k_i, l_i)$ will not stay constant as γ crosses this locus (unless it is the eigenvector of the monodromy for the corresponding node). However, the change in the vector is determined by the base diagram and therefore it suffices to indicate v_i at just one point of γ_i.

REMARK 7.6. The above restrictions that all nodes have multiplicity one and that no boundary component of $\partial\Sigma_\gamma$ project to a node were made only to simplify reconstruction of Σ_γ from a base diagram. With greater willingness to add data to base diagrams these could be eliminated.

7.2. Symplectic area. It is easy to read the (signed) symplectic area of a visible surface off of its image in a base diagram. The toric case is well known: If γ is an edge that is a smooth component of the reduced boundary, then the area of $\pi^{-1}(\gamma)$ is $2\pi|a|$ if the edge is av for some primitive integral vector v. Indeed, by applying an element of $GL(2, \mathbb{Z})$ to the base we can assume that a neighborhood of γ is as in Example 3.26 in which case the area of $\pi^{-1}(\gamma)$ equals $2\pi|a|$ where $|a|$ is the length of γ.

In general we need a way to keep track of orientations.

DEFINITION 7.7. Consider a parameterized curve $\gamma : I \to (B, \mathcal{A}, \mathcal{S})$, its image in a base diagram and a set of compatible vectors $\{v_i\}$ that define a visible surface Σ_γ. Let $b = \gamma(t)$ be a point on the interior of γ_i and $x \in M$ such that $\pi(x) = b$.

Let $\tilde{u}, \tilde{v}_i \in T_x\Sigma_\gamma \subset T_xM$ be as in Theorem 2.6 where $u = \gamma'(t) \in T_b\mathbb{R}^2$. Then the $\{v_i\}_{i=1}^N$ are *co-oriented* if each ordered pair $(\tilde{u}, \tilde{v}_i)$ defines the chosen orientation of Σ_γ.

Then we have:

PROPOSITION 7.8. *Let $\gamma : I \to (B, \mathcal{A}, \mathcal{S})$ be a parameterized immersed curve and $\{v_i\}_{i=1}^N$ a set of co-oriented compatible vectors in a base diagram that define an oriented surface Σ_γ. Then the (signed) symplectic area of Σ_γ is*

$$\text{Area}(\Sigma_\gamma) = 2\pi \int_0^1 \gamma'(t) \cdot v(t)dt \tag{7.2}$$

where $v(t) = v_i$ if $\gamma(t) \in \gamma_i$ and for other values of t (when $\gamma \subset \partial_R B$) $v(t)$ is in integral vector such that $u(t) \times v(t) = 1$ for some integral vector $u(t) = \lambda\gamma'(t)$, $\lambda > 0$.

Notice that when $\gamma(t)$ belongs to a component of $\partial_R B$ that intersects $\overline{\gamma}_i$ we can choose $v(t)$ to be equal to v_i.

PROOF. With respect to standard coordinates we can parameterize the interior of each Σ_{γ_i} as $p = \gamma_i(t)$, $q = v_i s$ for $0 \leq \alpha_i < t < \beta_i \leq 1$ and $0 \leq s \leq 1$. Then since $\omega = dq \wedge dq$ it is immediate that $\text{Area}(\Sigma_{\gamma_i}) = \int_{\alpha_i}^{\beta_i}\int_0^1 \gamma_i'(t) \cdot v_i dtds = \int_{\alpha_i}^{\beta_i} \gamma_i'(t) \cdot v_i dt$. As for the components of γ that lie in a component of the 1-stratum of $\partial_R B$, each one lifts to a visible surface in the boundary recovery $B \times T^2$. Supposing that the collapsing class of this component of the stratum is $\mathbf{b}$, the visible surface is well defined up to isotopy and choice of class $\mathbf{a}$ such that $|\mathbf{a} \cdot \mathbf{b}| = 1$. With $v\frac{\partial}{\partial q}$ being the tangent vector to integral curves representing $\mathbf{a}$, the inner product $\gamma'(t) \cdot v$ is independent of our choice of $\mathbf{a}$ since for two choices $\mathbf{a}, \mathbf{a}'$, $(\mathbf{a}-\mathbf{a}')\cdot\mathbf{b} = (v-v')\cdot\gamma'(t) = 0$. □

COROLLARY 7.9. *If a visible surface Σ_γ is Lagrangian then γ is an affine linear one-dimensional submanifold that contains no open subsets of the 1-stratum of $\partial_R B$.*

PROOF. For Σ_γ to be Lagrangian we must have that $\gamma'(t)\cdot v(t) = 0$ for all $t \in I$. Since $v(t)$ is an integral vector valued function it is piecewise constant, therefore, because γ is smooth we have that γ is linear. The constraint $\gamma'(t) \cdot v(t) = 0$ and the integrality of $v(t)$ then force γ to be affine linear. Furthermore, γ cannot contain an open subset of the 1-stratum of $\partial_R B$ since for such points we would have $\gamma'(t) \cdot v(t) > 0$. □

7.3. Intersections numbers.

In this section we explain how to calculate the intersections and self-intersections of visible surfaces. The latter are of particular importance from the symplectic standpoint because the symplectic neighborhood theorem (cf. [**31**]) implies that the germ of a symplectic neighborhood of a symplectic surface in a symplectic four-manifold is determined by its self-intersection and area.

We start by calculating, separately, the signed transverse geometric intersection of visible surfaces in the preimage of a regular point or a node.

LEMMA 7.10. *Consider two oriented visible surfaces Σ_{γ_i}, $i = 1, 2$, defined by parameterized arcs $\gamma_i : I \to B_0 \subset (B, \mathcal{A}, \mathcal{S})$ and co-oriented compatible vectors v_i in a base diagram. Suppose γ_1 and γ_2 intersect transversely at one point $b \in B_0$ and* *assume that* Σ_{γ_1} *and* Σ_{γ_2} *intersect transversely in* F_b. *Let* u_i *be a tangent vector at* b *that is positively oriented with respect to the parameterization of* γ_i. *Then* Σ_{γ_1} *intersects* Σ_{γ_2} *in* $|\det(v_1v_2)|$ *points in* F_b *and the sign of all of the intersections equals the sign of the product* $\det(u_1u_2)\det(v_1v_2)$.

PROOF. The vectors v_i determine homology classes $[v_i\frac{\partial}{\partial q}]$ in $H_1(F_b,\mathbb{Z})$ whose algebraic intersection in F_b is $[v_1\frac{\partial}{\partial q}]\cdot[v_2\frac{\partial}{\partial q}] = v_1 \times v_2 = \det(v_1v_2)$. Since the representatives of these classes are affine circles the algebraic intersection equals the geometric intersection. To figure the sign of the intersection of the two surfaces in the four-manifold from the intersection of curves in the fiber torus, note that the sign of the intersection equals the sign of $\det(\tilde{u}_1\tilde{v}_1\tilde{u}_2\tilde{v}_2) = \det(u_1u_2)\det(v_1v_2)$. $\square$

Notice that two visible surfaces can have a transverse intersection at a node only if they share an endpoint at that node. Indeed, as we saw in the proof of Theorem 7.4, if γ has a node on its interior then the preimage of that point in Σ_γ is an affine circle representing the vanishing class. The only transverse intersection two such circles, or one such and the node, can have is an empty intersection.

LEMMA 7.11. *The signed intersection of a pair of vanishing disks that intersect once transversely at a nodal singular point is* -1.

PROOF. To calculate this intersection we refer to the local model in Section 4.2. There we have used complex coordinates $\mathbf{x} = x_1 + ix_2$, $\mathbf{y} = y_1 + iy_2$ on a neighborhood of the nodal singularity (at the origin) and the total space has the orientation given by $(\mathbf{x}, \overline{\mathbf{y}})$. Since the neighborhood of the nodal fiber supports the S^1 action defined by $t\cdot(\mathbf{x},\mathbf{y}) = (e^{2\pi it}\mathbf{x}, e^{2\pi it}\mathbf{y})$(and its continuation into the $(\mathbf{u},\mathbf{v})$ coordinate chart), the vanishing cycles can be chosen to be $T(1,1)$ torus knots in an S^3 that is the boundary of a neighborhood of the nodal singularity. These intersect each other positively once with respect to the usual orientation on $\mathbb{C}^2$, but since we have the reverse orientation the intersection is -1. $\square$

EXERCISE 7.12. Check that the only transverse intersections of visible surfaces that lie in the preimage of the 1-stratum are empty.

We now consider how to calculate the self-intersection of a visible sphere or torus Σ_γ from its image γ in a base diagram. First, if possible, find an isotopic visible surface Σ_η such that the intersection of Σ_γ and Σ_η (hence the self-intersection of Σ_γ) can be calculated using Lemmas 7.10 and 7.11. If such a Σ_η does not exist (as when $\gamma = \partial B = \partial_R B$) then calculate the self-intersection by applying the above lemmas to the intersections of two surfaces Σ_{γ_1}, Σ_{γ_2} that are isotopic to Σ_γ.

We make the calculation explicit for the two most useful cases.

PROPOSITION 7.13. *Suppose* Σ_γ *is a sphere such that* $\gamma \subset \partial_R B$ *(so* γ *is an edge). Let* v_1, v_2 *be the inward pointing collapsing vectors for the two components of the* 1*-stratum whose closures intersect the endpoints of* γ. *Let* v_γ *be the inward pointing collapsing vectors for the component of the* 1*-stratum contained in* γ. *Then the self-intersection of the sphere* Σ_γ *equals* $v_1 \times v_2$ *where the vectors are indexed so that* $v_1 \times v_\gamma > 0$.

PROOF. Figure 11 shows two surfaces Σ_{γ_i}, $i = 1, 2$ that are isotopic to Σ_γ. For each the compatible vector is the collapsing vector for the 1-stratum that the curve intersects transversely. Applying Lemma 7.10 then gives the formula. $\square$

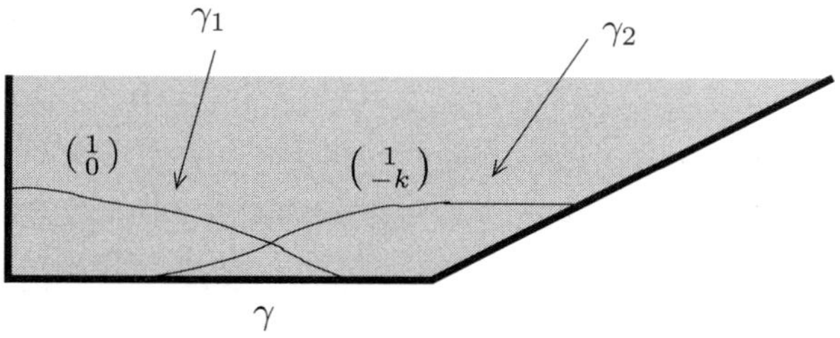

FIGURE 11. Isotopic spheres Σ_{γ_1} and Σ_{γ_2}.

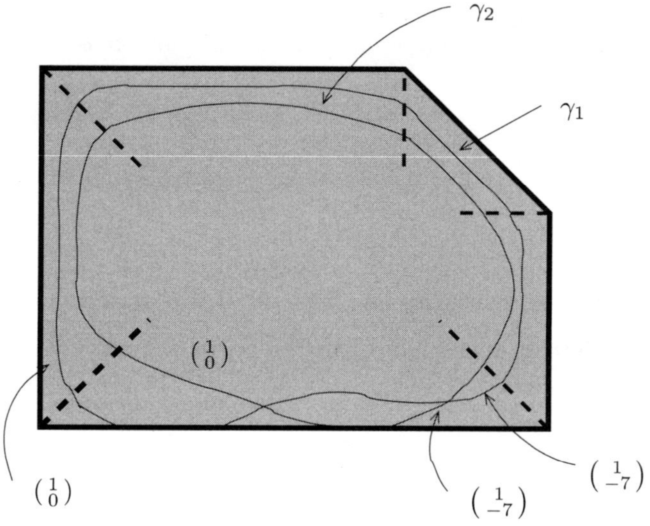

FIGURE 12. Isotopic tori Σ_{γ_1}, Σ_{γ_2} in $S^2 \times S^2 \# \overline{\mathbb{C}P}^2$.

PROPOSITION 7.14. *Suppose Σ_γ is a torus with $\gamma \subset \partial_R B$, so $\gamma = \partial_R B = \partial B$. Let v be an inward pointing collapsing vector for the reduced boundary. Then the self-intersection of Σ_γ equals $v \times v'$ where v' is the parallel transport of v clockwise along γ.*

PROOF. In a base diagram γ appears as a union of $k \geq 3$ line segments with oriented integral tangent vectors $\{u_1, \ldots, u_k\}$ such that $u_{j+1} = A_j u_j$ where A_j is conjugate to $A_{(1,0)}$ and the vectors are indexed so that j increases as one moves clockwise along the boundary. Without loss of generality let v be a primitive integral vector normal to the edge with tangent vector u_1. Now choose two curves γ_1, γ_2 isotopic to γ in a small collar neighborhood of γ that contains no nodes. Choose them so that $\gamma_i \cap \gamma$ for $i = 1, 2$ each consists of a closed subset of "first edge" – the segment with tangent vector u_1. Choose compatible vectors v_1, v_2 for $\Sigma_{\gamma_1}, \Sigma_{\gamma_2}$ to equal u_1 just to the left of the intersection of γ_i with the first edge. (See Figure 12 for an example.) Then on the other side of the intersection the compatible vector representing the same homology class is $A_k^{-T} \cdots A_2^{-T} A_1^{-T} u_1$. This equals $A_k \cdots A_2 A_1 v$, which is the parallel transport of v along the curves γ_i, because $v = J u_1$ and $A_j J = J A_j^{-T}$. Applying Lemma 7.10 we get the result. □

When the image γ of a visible sphere or torus belongs to $\partial_R B$ the surface is necessarily symplectic. A collar neighborhood of γ is fibered by parallel affine linear submanifolds η_t and the rate of change of the affine length of these submanifolds measures the self-intersection of the surface, i.e. the first Chern class of the normal bundle. Indeed, the pre-image of η_t is a circle bundle over the surface whose first Chern class equals the self-intersection of Σ_γ. Furthermore, the rate of change of the affine length of η_t measures the Chern class of the normal bundle of the surface (or self-intersection of the surface).

Specifically, let u be a primitive integral vector tangent to γ and w be an inward pointing integral vector such that $|n \times w| = 1$. Choose the parameter t so that η_t passes through points $b + tw$ for $b \in \gamma$. Letting $l(t)$ be the affine length of γ, $\frac{dl}{dt}$ is a constant and is equal to the self-intersection of Σ_γ. We leave it to the reader to verify this alternative formula for the self-intersection. It is a special case of the following:

PROPOSITION 7.15. *(cf.* **[31]***) Given a circle bundle $\pi : P \to X$ with first Chern class c_1, then there is an S^1 invariant symplectic form ω on the manifold $P \times I$ with Hamiltonian function H equal to the projection $P \times I \to I$ and with reduced spaces (X, τ_λ) where $[\tau_\lambda] = [\tau_\mu] + (\lambda - \mu)c_1$.*

8. Reading the base II: total space

8.1. Euler characteristic and fundamental group. Given an almost toric fibration we can calculate the Euler characteristic of the total space M very easily from the base:

$$\chi(M) = V + \sum_i k_i \tag{8.1}$$

where V is the number of vertices and k_i is the multiplicity of each node s_i.

This follows from the additivity property of the Euler characteristic and the Euler characteristics of almost toric fibered neighborhoods of the different types of fibers. Each neighborhood has the homotopy type of the singular fiber so the Euler characteristic of the neighborhood of a point, circle, or torus fiber is $1, 0$ and 0 respectively. Accordingly, the neighborhood of a multiplicity k nodal fiber has the homotopy type of either a pinched torus (if $k = 1$) or a cyclic chain of k spheres, each of which intersects two others once. Such a fiber has Euler characteristic k. We can choose to cover the manifold with fibered neighborhoods so that the intersections of the neighborhoods are all either empty or diffeomorphic to a torus bundle or $S^1 \times D^3$ and hence always have Euler characteristic equal to 0.

EXERCISE 8.1. Note that by encoding the collapsing and vanishing classes, the base of an almost toric fibration reveals the fundamental group of the total space. Verify that any closed toric manifold is simply connected.

8.2. First Chern class. Given a symplectic manifold (M, ω) its first Chern class $c_1(M)$ is $c_1(TM)$ where TM is a complex vector bundle with respect to a compatible almost complex structure on M. (An almost complex structure J is *compatible* if $\omega(u, Ju) > 0$ for all $u \neq 0$ and $\omega(Ju, Jv) = \omega(u, v)$ for all u, v.) Recall that the set of compatible almost complex structures is contractible, so in particular $c_1(M)$ is well-defined.

One way to define the first Chern class of a complex vector bundle of rank k is as the Poincaré dual to the homology class of the set D of points where a generic k-frame becomes degenerate (i.e. drops in rank); in doing so the sign of the homology class is chosen with regard to orientations. For a Lagrangian fibered symplectic manifold we bypass choosing an almost complex structure: a representative of the homology class Poincaré dual to $c_1(M)$, up to sign, is given by the set of points where a generic Lagrangian k-frame of TM degenerates.

To get the orientations right we can proceed as follows: Choose a generic n-tuple of sections $\eta_1, \ldots, \eta_n$ that span a Lagrangian subspace of each tangent plane in which they are linearly independent. Let $D \subset M$ be the set of points where they are linearly dependent, and $D' \subset D$ the set of points where they span an $n-1$-dimensional subspace. In a neighborhood of a point $x \in D'$ choose two other sections μ_1, μ_2 such that $\omega(\mu_1, \mu_2) > 0$ and $\{\eta_j\}_{j<n} \cup \{\mu_i\}$ defines a Lagrangian frame for either $i = 1$ or 2. Then $\eta_n = \sum_{j<n} f_j \eta_j + g\mu_1 + h\mu_2$ for smooth functions f_j, g, h. Now let Ω be an orientation for the component of D' containing x so that $\Omega \wedge dg \wedge dh$ gives the orientation of M. Then the homology class represented by D with the orientations given by Ω_x for $x \in D'$ gives the Poincaré dual to $c_1(M)$. This calculation follows from using an almost complex structure to complete the Lagrangian frame to a complex frame and then carrying out the usual calculation of the orientations (cf. [**20**]). The answer is independent of whatever local compatible almost complex structure we choose.

PROPOSITION 8.2. *Consider an almost toric fibration $\pi : (M, \omega) \to (B, \mathcal{A}, \mathcal{S})$ of a closed manifold M. The Poincaré dual to $c_1(M)$ is given by the homology class of $\pi^{-1}(\partial_R B)$.*

PROOF. This is especially easy for a toric manifold: we have standard coordinates (p, q) defined on $\pi^{-1}(B_0)$. There, we can take sections $\eta_i = \frac{\partial}{\partial q_i}$, $i = 1, \ldots, n$. These become linearly dependent exactly on the preimage of the reduced boundary and span an $n-1$-dimensional subspace on the preimage of the $n-1$-stratum of the reduced boundary. It suffices to consider the case when the collapsing class is $[\frac{\partial}{\partial q_n}]$ in which case we have local coordinates $(p_1, q_1, \ldots p_{n-1}, q_{n-1}, x_n, y_n)$. Taking $\mu_1 = \frac{\partial}{\partial x_n}$, $\mu_2 = \frac{\partial}{\partial y_n}$ we have $\frac{\partial}{\partial q_n} = -y_n \frac{\partial}{\partial x_n} + x_n \frac{\partial}{\partial y_n}$ so $dg \wedge dh = dx_n \wedge dy_n$. Therefore the smooth components of the preimage of the reduced boundary contribute with their orientations as symplectic manifolds.

Restricting to dimension four, we do not necessarily have global sections $\frac{\partial}{\partial q_i}$ in an almost toric fibration. However, on the complement of nodal singularities $\frac{\partial}{\partial q_1} \wedge \frac{\partial}{\partial q_2}$ is well defined since it is invariant under the action of $GL(2, \mathbb{Z})$. Therefore the Chern class can be calculated using local sections that give this same section of $\wedge^2 TM$. Since the nodal singularities are codimension four they do not contribute to the Chern class calculation. □

8.3. Volume. Given an almost toric fibration $\pi : (M, \omega) \to (B, \mathcal{A}, \mathcal{S})$, the symplectic volume of M is equal to the symplectic volume of $\pi^{-1}(B_0)$ where B_0 is the set of regular values. This is a regular Lagrangian fibration with $\omega = dp \wedge dq$ with respect to standard local coordinates (which are compatible with the fibration). Therefore the symplectic volume of M, $\int_M \omega^n$, equals $n! \int_{\pi^{-1}(B_0)} dp_1 \wedge dq_1 \wedge \cdots \wedge dp_n \wedge dq_n = n!(2\pi)^n \int_{B_0} dp_1 \wedge \cdots \wedge dp_n$ since q is defined mod 2π. (Note that $dp_1 \wedge \cdots \wedge dp^n$ gives a well defined volume form on the base because elements of $AGL(n, \mathbb{Z})$ preserve volume.) For instance, the symplectic volume of $(\mathbb{C}P^2, \omega)$ where ω is determined by an almost toric fibration over the base in Figure 1(b), is $4\pi^2$.

9. Lens spaces and manifolds they bound

9.1. The cone on a lens space. Two common definitions of the lens space $L(n, m)$ where n, m are relatively prime integers, $n \geq 1$, are:

- The quotient of the sphere $S^3 = \{(z_1, z_2) | |z_1|^2 + |z_2|^2 = 1\} \subset C^2$ by the free $\mathbb{Z}_n$ action generated by $\alpha \cdot (z_1, z_2) = (\alpha^m z_1, \alpha z_2)$ where $\alpha = e^{\frac{2\pi i}{n}}$.
- The space obtained by gluing two solid tori along their boundaries T_1, T_2 as follows: choose generators (a_i, b_i), $i = 1, 2$, for $H_1(T_i, \mathbb{Z})$ such that a_i is the class of a meridinal circle in T_i and $a_i \cdot b_i = 1$; then glue the two solid tori together by a diffeomorphism $\phi : T_1 \to T_2$ such that $\phi_* a_1 = -m a_2 + n b_2$.

For the first definition to make sense, one certainly must assume that $n \neq 0$, hence $L(0, 1)$ is not a lens space. However, $n = 0$ does make sense in the second definition and one can easily see that $L(0, 1) = S^2 \times S^1$. The orientation of the lens space $L(n, m)$ is the one it inherits from S^3 in the first definition and from the solid tori oriented by the frame $\{\frac{\partial}{\partial r_i}, \frac{\partial}{\partial \theta_i}, \frac{\partial}{\partial \phi_i}\}$ where r_i is a radial coordinate and θ_i, ϕ_i are cyclic coordinates such that the meridians a_i are represented by curves in the boundaries T_i given by $\phi_i = \text{const}$.

For our purposes, the following rephrasing of the second definition is most natural:

DEFINITION 9.1. Let (q_1, q_2) be coordinates on T^2, t a coordinate on I and orient $T^2 \times I$ by the frame $\{\frac{\partial}{\partial t}, \frac{\partial}{\partial q_1}, \frac{\partial}{\partial q_2}\}$. The *lens space* $L(n, m)$ is the quotient of $T^2 \times I / \sim$ where $\sim$ collapses the integral curves of $\frac{\partial}{\partial q_1}$ on $T^2 \times \{0\}$ and collapses the integral curves of $-m\frac{\partial}{\partial q_1} + n\frac{\partial}{\partial q_2}$ on $T^2 \times \{1\}$.

Using Definition 9.1 it is very easy to construct a toric fibration of the cone on a lens space:

EXAMPLE 9.2. Let $u = (n, m)$ and $v = (0, 1)$ be two vectors in $(\mathbb{R}^2, \mathcal{A}_0)$ and define the domain $V_{n,m} := \{xu \pm yv \in \mathbb{R}^2 | x, y \geq 0\} - \{(0, 0)\}$ and its closure $\overline{V_{n,m}}$. Then $(V_{n,m}, \mathcal{A}_0, \mathcal{S})$ where $\mathcal{S}$ is specified by heavy lines on the two edges of $V_{n,m}$, is the base or a toric fibration of $(L(n, m) \times \mathbb{R}_+, \omega_{n,m})$. To see this, let p_2 be the coordinate on the vertical axis in $\mathbb{R}^2$; then Definition 9.1 implies immediately that $\pi^{-1}(V_{n,m} \cap \{p_2 = s\})$, $s > 0$ is the lens space $L(n, m)$. Notice that given any embedded curve $\gamma \subset V_{n,m}$ which intersects the boundary of $V_{n,m}$ once transversely on each edge, $\pi^{-1}(\gamma) = L(n, m)$.

REMARK 9.3. Notice that in general the four-dimensional space defined by $(\overline{V_{n,m}}, \mathcal{A}_0, \mathcal{S})$ is not smooth. Indeed, the cone on a lens space $L(n, m)$ is smooth only when $n = 1$, i.e. when the lens space is S^3. Otherwise it has an *orbifold singularity.*

The following theorem, which records the redundancy in the above definitions is well known (cf. [**25**]).

THEOREM 9.4. *Two lens spaces $L(n,m)$ and $L(n,m')$ are orientation preserving diffeomorphic if and only if*

$$m \equiv m' \ (\mathrm{mod}\ n) \quad \text{or} \tag{9.1}$$

$$mm' \equiv 1 \ (\mathrm{mod}\ n). \tag{9.2}$$

Note that since the fundamental group of $L(n,m)$ is $\mathbb{Z}_n$, $L(n,m)$ and $L(n',m')$ can only be diffeomorphic if $n = n'$.

We can use base diagrams to "see" these diffeomorphism equivalences. We merely need to apply linear maps $A \in GL(2,\mathbb{Z})$ to the base diagram for $(V_{n,m}, \mathcal{A}_0, \mathcal{S})$ with the vertex at the origin (e.g. Example 9.2). If we choose A so that it leaves $v = (0,1)$ fixed then $A = \left(\begin{smallmatrix} 1 & 0 \\ k & 1 \end{smallmatrix}\right)$. so the image of $V_{n,m}$ is $V_{n,m'}$ where $m' = kn + m$. To see the second equivalence, choose A so that it sends (n,m) to $(0,1)$, i.e. $A = \left(\begin{smallmatrix} -m & n \\ c & m' \end{smallmatrix}\right)$ with $\det A = -1$.

EXERCISE 9.5. Two lens spaces $L(n,m)$ and $L(n,m')$ are orientation reversing diffeomorphic if and only if $m \equiv -m'$ (mod n) or $mm' \equiv -1$ (mod n). Use base diagrams to verify that such lens spaces are diffeomorphic. How does the reversal of the orientations show up?

Toric fibrations also give a way to see the equivalence of the two definitions of the lens space $L(n,m)$, (1) as a quotient of S^3 and (2) as a union of solid tori.

To explain this, we present a symplectic n-fold covering of $L(n,m) \times \mathbb{R}_+$ by $L(1,0) \times \mathbb{R}_+ = S^3 \times \mathbb{R}_+$ from which we can extract the quotient construction. Indeed, define a linear map $B : \mathbb{R}^2 \to \mathbb{R}^2$ by

$$B = \begin{pmatrix} 1 & 0 \\ \frac{m}{n} & \frac{1}{n} \end{pmatrix} \tag{9.3}$$

which maps $V_{1,0}$ onto $V_{n,m}$. The map $(B, B^{-T}) : (V_{1,0} \times T^2, \omega_0) \to (V_{n,m} \times T^2, \omega_0)$ is then a symplectic n-fold cover that descends under boundary reduction to a symplectic n-fold cover of $L(n,m) \times \mathbb{R}_+$ by $L(1,0) \times \mathbb{R}_+$ that preserves the almost toric fibrations defined by $V_{1,0}$ and $V_{n,m}$. Restricting this n-fold cover to a sphere S^3 that projects to an embedded curve $\gamma \subset V_{1,0}$, we see $L(n,m)$ as the stated quotient of S^3. Because Definition 9.1 is just a rewording of (2), we are done.

9.2. An induced contact structure. A *contact structure* ξ on a three-manifold N is a completely non-integrable plane field. Locally (and often globally) such a plane field can be prescribed as the kernel of a 1-form α such that

$$\alpha \wedge d\alpha \neq 0. \tag{9.4}$$

Conversely, given any 1-form α satisfying Equation 9.4, $\xi = \ker\, \alpha$ is a contact structure.

Contact structures are the cousins of symplectic structures that live on odd dimensional manifolds. In particular, contact manifolds arise naturally as hypersurfaces in symplectic manifolds.

DEFINITION 9.6. A vector field X on a symplectic manifold (M,ω) is *expanding* if it satisfies $\mathcal{L}_X\omega = \omega$.

Recall that $\mathcal{L}_X\omega := d(\iota_X\omega) + \iota_X d\omega$. In our case, since $d\omega = 0$ only the first term is non-zero.

A basic result from contact geometry is:

PROPOSITION 9.7. *Let $Y \subset (M, \omega)$ be a hypersurface. If there is an expanding vector field X defined on a neighborhood of Y that is transverse to Y, then $\alpha := \iota_X \omega$ defines a contact structure on Y.*

An almost toric fibration of a symplectic manifold allows for a simple criterion to determine the existence of an induced contact structure on certain lens space hypersurfaces.

DEFINITION 9.8. A *visible lens space* in a symplectic manifold with almost toric fibration $\pi : (M, \omega) \to (B, \mathcal{A}, \mathcal{S})$ is the preimage of an embedded curve $\gamma \subset (B, \mathcal{A}, \mathcal{S})$ that is disjoint from nodes and the 0-stratum and intersects the 1-stratum of ∂B transversely at its endpoints.

That $\pi^{-1}(\gamma)$ is a lens space follows directly from Definition 9.1. Notice that a neighborhood of γ, chosen small enough to contain no nodes, necessarily admits an integral affine immersion to $\mathbb{R}^2$ (since it is simply connected).

PROPOSITION 9.9. *Let $\gamma \subset (B, \mathcal{A}, \mathcal{S})$ be an embedded curve that defines a visible lens space and let $\mathcal{N}(\gamma)$ be a neighborhood of γ. Consider an immersion $\Psi : (\mathcal{N}(\gamma), \mathcal{A}) \to (\mathbb{R}^2, \mathcal{A}_0)$. If there is a point $x_0 \in \mathbb{R}^2$ such that $\Psi(\gamma)$ is transverse to all lines through x_0 and those lines through x_0 that intersect the endpoints of $\Psi(\gamma)$ are tangent to the image of $\partial_R B$ at those intersections, then the lens space has an induced contact structure.*

PROOF. Let $\Phi_n : (\widetilde{W}_n, \mathcal{A}_n) \to (\mathbb{R}^2 - \{0\}, \mathcal{A}_0)$ be an n-fold cover equipped with local coordinates $p = (p_1, p_2)$ induced from $\mathbb{R}^2$. This defines a Lagrangian fibration $\pi_n : (\mathcal{N}(\gamma) \times T^2, \omega_n) \to (W_n, \mathcal{A}_n)$ where $\omega = dp \wedge dq$ with respect to local standard coordinates. Without loss of generality take x_0 to be the origin in $\mathbb{R}^2$. The transversality assumption on γ implies that if $\mathcal{N}(\gamma)$ is chosen small enough, the immersion Ψ factors through an embedding $\Psi_n : (\mathcal{N}(\gamma), \mathcal{A}) \to (\widetilde{W}_n, \mathcal{A}_n)$ so $\Psi = \Phi_n \circ \Psi_n$. Then the vector field $p\frac{\partial}{\partial p}$ on $(\mathbb{R}^2 - \{0\}, \mathcal{A}_0)$ defines a vector field of the same form on $(\widetilde{W}_n, \mathcal{A}_n)$, and likewise a vector field $X = p\frac{\partial}{\partial p}$ on $\mathcal{N}(\gamma) \times T^2 \subset (\widetilde{W}_n \times T^2, \omega_n)$. Note that X is transverse to $\pi_n^{-1}(\Psi_n(\gamma))$. Therefore, since $d(\iota_X dp \wedge dq) = d(pdq) = dp \wedge dq$, X defines a contact structure on $\gamma \times T^2 \subset (N(\gamma) \times T^2, \omega_n)$. The assumed tangency at the endpoints of γ implies this vector field extends to a well defined vector field on the boundary reduction of $\mathcal{N}(\gamma) \times T^2$ along $\pi_n^{-1}\Psi_n(\partial_R B \cap \mathcal{N}(\gamma))$. □

In contact geometry on three-manifolds there is a dichotomy of fundamental importance: a contact structure can be tight or overtwisted, and from a topological standpoint the tight structures are the interesting ones.

DEFINITION 9.10. A contact structure ξ on a three-manifold Y is *overtwisted* if there is a an embedded (two-dimensional) disk $D \subset Y$ such ξ is tangent to D at all points of ∂D. It is *tight* if there is no such disk.

For visible lens spaces, there is an easy way detect overtwistedness of a contact structure. The following construction of overtwisted contact structures was first made by Lerman [**28**] to illustrate the existence of a countable number of non-conjugate tori of dimension two in the group of contactomorphisms of these contact structures on lens spaces.

PROPOSITION 9.11. *Suppose $\gamma \subset (B, \mathcal{A}, \mathcal{S})$ determines a visible lens space. Consider the image of γ in $(\mathbb{R}^2, \mathcal{A}_0)$ and a point x_0 in $\mathbb{R}^2$ as in Proposition 9.9. If the image of* $\mathrm{Int}(\gamma)$ *intersects any line through x_0 more than once, then the induced contact structure on the lens space is overtwisted.*

PROOF. View γ as an immersion $\gamma : [0,1] \to (\mathbb{R}^2, \mathcal{A}_0)$ with tangent vectors $\gamma'(t)$. We can assume, without loss of generality, that x_0 is the origin in $\mathbb{R}^2$, that $\gamma(0)$ is on the positive vertical axis and $\gamma'(0)$ has a positive horizontal component. Proposition 9.9 implies the existence of a contact structure ξ that in terms of local coordinates can be given as $\xi = \ker pdq$.

Consider the visible disk $D = \Sigma_{\gamma([0,\tau])}$, $0 < \tau < 1$, with compatible vector $(1,0)$. (See Section 7.1 for definitions.) The boundary of the disk consists of points that project to $\gamma(\tau)$. In terms of local coordinates, the tangent plane to the disk at these points is spanned by the vectors $\gamma'(\tau) = v\frac{\partial}{\partial p}$, $v \in \mathbb{R}^2$, and $\frac{\partial}{\partial q_1}$. Meanwhile, the contact planes are spanned by $\gamma'(\tau)$ and $(Jp)\frac{\partial}{\partial q}$. Here, $J = \left(\begin{smallmatrix} 0 & -1 \\ 1 & 0 \end{smallmatrix}\right)$. Thus tangency of D and the contact planes occurs all along the boundary if we can choose τ so that $\gamma(\tau)$ lies on the vertical axis. Definition 9.10 implies that under these conditions the induced contact structure is overtwisted. □

Note that in the above construction it does not suffice for the endpoints of γ to lie on a line. If they do, and $\mathrm{Int}(\gamma)$ belongs to an open half-space, then $\gamma([0,\tau]) = \gamma([0,1])$ defines a sphere – and hence is not a candidate for an overtwisted disk.

EXERCISE 9.12. Check that the characteristic foliation (defined in [**11**]) on the overtwisted disk D in the preceding proof is exactly the degenerate characteristic foliation shown in Figure 3 of [**11**] in this volume.

Determining tightness of a contact structure is a difficult problem in general. One of the most useful criteria is that of fillability:

DEFINITION 9.13. Consider a symplectic manifold (M, ω) with boundary $Y = \partial M$. The boundary is *ω-convex* if a collar neighborhood of Y admits an outward pointing expanding vector field transverse to N. Then the induced contact structure ξ on Y is *(strongly symplectically) fillable* and (M, ω) is a *(strong symplectic) filling* of (Y, ξ).

THEOREM 9.14 (Gromov, Eliashberg; cf. [**12**]). *If a contact three-manifold (Y, ξ) is fillable then the contact structure ξ is tight.*

There are several notions of fillability of contact manifolds; any one of them suffices in the above theorem. For more details see [**13**].

COROLLARY 9.15. *If the boundary of a symplectic manifold (M, ω) is ω-convex, then the induced contact structure is tight.*

COROLLARY 9.16. *Let $\pi : (M, \omega) \to (B, \mathcal{A}, \mathcal{S})$ be an almost toric fibration of a neighborhood of a nodal fiber, $R \subset B$ an eigenray based at the node, and $\Phi : (B - R, \mathcal{A}) \to (\mathbb{R}^2, \mathcal{A}_0)$ an immersion. Then Φ is an embedding.*

PROOF. Without loss of generality suppose that the image (under Φ) of the node is the origin in $\mathbb{R}^2$. Let $\gamma \subset (B, \mathcal{A}, \mathcal{S})$ be a loop around the node, the image of $\partial M'$ where $M' \subset M$ is a closed almost toric neighborhood of the nodal fiber.

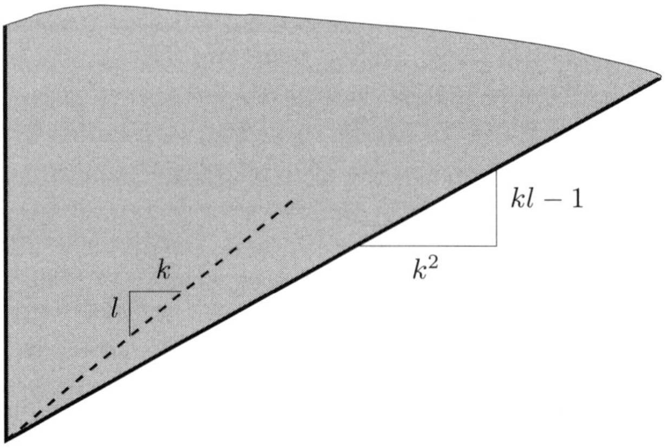

FIGURE 13. Rational ball with boundary $L(k^2, kl-1)$.

Suppose without loss of generality that the image of $\gamma \cap (B-R)$ in $(\mathbb{R}^2, \mathcal{A}_0)$ is transverse to the vector field $p\frac{\partial}{\partial p}$. With respect to standard coordinates on the preimage of $B-R$, $p\frac{\partial}{\partial p}$ is a vector field that extends to an expanding vector field X on a full collar neighborhood of $\partial M'$. Since X is transverse to $\partial M'$, Theorem 9.14 implies the induced contact structure $\xi = \ker(\iota_X \omega)$ is tight. But Proposition 9.11 implies this can happen only if Φ is one to one. $\square$

9.3. Rational balls. The base $\overline{V}_{n,m}$ defines a toric fibration of the cone on $L(n,m)$ and has an orbifold singularity whenever $n \neq 1$. When $n=1$ we have seen in Section 6.1 that an another almost-toric fibration (of $\mathbb{R}^4$, the cone on S^3) can be obtained by trading the vertex for a node. We can in fact mimic this construction whenever $n=k^2, m=kl-1$ for relatively prime integers k,l because the linear map $A_{l,k}$ (defined in Equation (4.11)) maps the vector $(k^2, kl-1)$ to $(0,-1)$. Specifically, modify the base diagram for $\overline{V}_{n,m}$ by removing the vertex and including a ray R marked by heavy dashes, based at (tk, tl) for some $t>0$ and passing through the origin. The resulting base $\widetilde{V}_{n,m}$ is the base of an almost toric fibration of a smooth manifold. (See Figure 13.)

What is the manifold? The structure of the base implies that this manifold, say M, is obtained from $T^2 \times D^2$ by forcing $(-l,k)$ to be a vanishing class (thereby introducing a nodal fiber) and having $(1,0)$ be a collapsing class in a boundary reduction. Since these two vectors span $\mathbb{Q}^2$, we have $H_*(M,\mathbb{Q}) = H_*(\mathbb{R}^4,\mathbb{Q})$. The construction also reveals that $H_1(M,\mathbb{Z}) = \mathbb{Z}_k$.

Now consider any embedded arc γ with vertices on the two edges of the base diagram for $\widetilde{V}_{n,m}$. If γ does not intersect the ray R then it defines a visible lens space $L(k^2, kl-1)$ that bounds a compact manifold with boundary whose homology with rational coefficients is that of a ball; this manifold with boundary, denoted here as $B_{k,l}$, is called a *rational ball.* Invoking Proposition 9.9, the boundary is equipped with an induced tight contact structure if γ is transverse to the radial vector field $p\frac{\partial}{\partial p}$.

EXERCISE 9.17. Check that if γ intersects the ray R once (away from the node) then it defines a visible surface that is diffeomorphic to $S^1 \times S^2$. Does the transversality of γ with the vector field $p\frac{\partial}{\partial p}$ ensure an induced tight contact structure?

9.4. Chains of spheres. The rational balls of the previous section are almost toric manifolds bounded by the lens spaces $L(k^2, kl-1)$. In fact, every lens space bounds a smooth manifold that can be equipped with a symplectic structure and a toric fibration.

DEFINITION 9.18. A *chain of spheres* in a smooth four-manifold is a collection of embedded spheres S_i, $i = 1, \dots k$ such that $S_i \cdot S_{i+1} = 1$ for $i = 1, \dots, k-1$, $S_i \cdot S_j = 0$ for $j \neq i, i+1$, and all intersections are transverse.

A model for a neighborhood of a chain of spheres in a four-manifold is a linear plumbing of disk bundles:

DEFINITION 9.19. A *plumbing* of two disk bundles over spheres $\pi_i : X_i \to \Sigma_i$ is the space $(X_1 \cup X_2)/\varphi$ where $\varphi : \pi_2^{-1}(D_2) \to \pi_1^{-1}(D_1)$ is an orientation preserving diffeomorphism that maps the disk $D_2 \subset \Sigma_2$ to a fiber of $\pi_1^{-1}(D_1)$ and maps a fiber of $\pi_2^{-1}(D_2)$ to the disk $D_1 \subset \Sigma_1$. A *linear plumbing* of disk bundles $\pi_i : X_i \to \Sigma_i$, $i = 1, \dots k$, is the space $\sqcup_i X_i/\varphi_1, \dots \varphi_{k-1}$ where $\varphi_i : \pi_{i+1}^{-1}(D_{i+1}) \to \pi_i^{-1}(D_i')$ is as above and for each $i \in \{2, \dots, k-1\}$, $\pi_i^{-1}(D_i)$ and $\pi_i^{-1}(D_i')$ are disjoint.

The following theorem is well known (cf. [**25**]). Here we give a proof in terms of base diagrams.

THEOREM 9.20. *Let N be a closed neighborhood of a chain of spheres $\{S_i\}_{i=1,\dots k}$ in a four-manifold such that the spheres S_i have self-intersections $-b_i$. Then $\partial N = L(n,m)$ where*

$$\frac{n}{m} = [b_1, b_2, \dots b_k] := b_1 - \cfrac{1}{b_2 - \cdots \frac{1}{b_k}}. \tag{9.5}$$

PROOF. Let U_i be the quadrilateral in $\mathbb{R}^2$ that is the intersection of the open half-plane $\{p_2 < \varepsilon\}$ and $U_{n,1}$ of Example 3.26 with $n = b_i$ and $s = a_i$. Thus the base diagram U_i defines a model neighborhood of a sphere with self-intersection $-b_i$. Now "plumb" the base diagrams U_i, $i = 1, \dots k$. Specifically, let $u_i = (a_i, 0)$ and $A_i = \left(\begin{smallmatrix} b_i & -1 \\ 1 & 0 \end{smallmatrix}\right)$. Let D_i, D_i' be two dimensional polydisks such as shown in Figure 14 and then construct $B = \sqcup_{i=1}^k U_i/\phi_1, \dots \phi_{k-1}$ where $\phi_i : D_{i+1} \to D_i'$ is defined by $\phi_i(x) = A_i x + u_i$. The parameter ε is assumed to be chosen small enough that the D_i, D_i' are disjoint. Then Proposition 3.22 implies we have constructed a base that defines a plumbing of the disk bundles defined by the U_i. An example is shown in Figure 15 with $k = 3$, $b_1 = 2$, $b_2 = 1$, and $b_3 = 3$. Finally, an embedded smooth curve $\gamma \subset B$ that projects to a curve with endpoints on the "initial" and "final" edges (as in Figure 15) defines a visible lens space that is the boundary of a closed neighborhood of the spheres.

To calculate which lens space bounds this neighborhood, choose an immersion $\Phi : (B, \mathcal{A}) \to (\mathbb{R}^2, \mathcal{A}_0)$ such that $\Phi(U_1) = U_{b_1,1}$. Then notice that the image (under Φ) of the final vector is

$$A_1 \cdots A_{k-1} \begin{pmatrix} b_k \\ 1 \end{pmatrix}. \tag{9.6}$$

FIGURE 14. Polydisks D_i, D'_i.

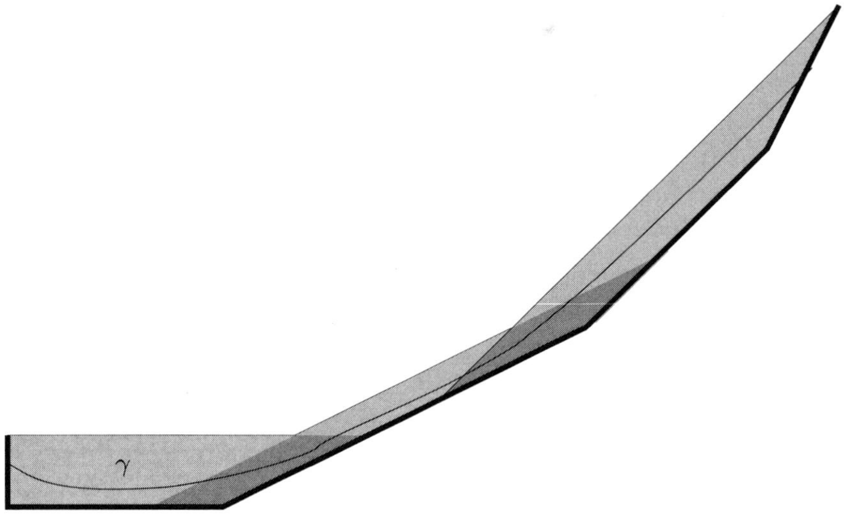

FIGURE 15. Chain of spheres in a linear plumbing.

Now,

$$A_j \begin{pmatrix} x \\ y \end{pmatrix} = \begin{pmatrix} b_j x - y \\ x \end{pmatrix} \tag{9.7}$$

which changes the ratio $\frac{x}{y}$ to $b_j - 1/\frac{x}{y}$. The conclusion of the theorem follows immediately. □

EXERCISE 9.21. Use base diagrams to check that if $b_i = 1$ for some $1 \leq i \leq k$ then $[b_1, \ldots, b_{i-1}, b_i, b_{i+1}, \ldots, b_k] = [b_1, \ldots, b_{i-1} - 1, b_{i+1} - 1, \ldots, b_k]$.

EXERCISE 9.22. Consider a linear plumbing that is a model neighborhood for a chain of spheres $\{S_i\}_{i=1,\ldots k}$ with self-intersections $b_i \geq 2$. Show that inside the linear plumbing there is an arbitrarily small closed toric neighborhood of the chain of spheres whose boundary is ω-convex. (See Definition 9.13.)

10. Surgeries

10.1. Rational blowdown. Almost toric fibrations of open manifolds and manifolds with boundary are very useful for understanding surgeries of symplectic manifolds. The key is that the gluing locus for such surgeries often admit an almost toric fibration.

Blowing up and down are the simplest examples of this. One need not have an almost toric fibration of a symplectic manifold (M, ω) in order to "see" a blow-up or blow-down; one only needs a toric fibration of an embedded symplectic ball or a neighborhood of a symplectic -1-sphere. In the first case one can push forward the fibration of the standard ball that is being embedded; in the second case the symplectic neighborhood theorem guarantees that the -1 sphere has a neighborhood symplectomorphic to a toric fibered model. The fact that balls and spheres of the appropriate sizes have fibrations whose bases are equivalent near their boundaries gives a "toric proof" that this surgery can be achieved.

More general than blowing down, in which one replaces a neighborhood of a -1-sphere by a ball, one can replace certain chains of spheres (Section 9.4) by a rational ball. This surgery is called a (generalized) rational blowdown:

DEFINITION 10.1. If a smooth four-manifold M contains a chain of spheres such that the boundary of a closed neighborhood N of the spheres is the lens space $L(n^2, nm-1)$, then the *generalized rational blowdown* of M along the spheres is

$$\widetilde{M} := (M - \mathrm{Int}(N)) \cup_{\varphi} B_{n,m} \tag{10.1}$$

where φ is an orientation preserving diffeomorphism from the boundary of N to the boundary of the rational ball $B_{n,m}$.

This smooth surgery (in the case $m = 1$) was used by Fintushel and Stern to prove the diffeomorphism classification of elliptic surfaces and to construct interesting smooth four-manifolds including an infinite family of simply connected four-manifolds not homotopic to a complex surface. They also used it to give an alterative construction of the exotic K3 surfaces of Gompf and Mrowka (manifolds homeomorphic but not diffeomorphic to the K3 surface).

Almost toric fibrations yield a virtually pictorial proof of

THEOREM 10.2 (Symington [**36**]). *If a symplectic four-manifold (M, ω) contains a chain of spheres, each of which is symplectic, such that the boundary of a closed neighborhood of the spheres is the lens space $L(k^2, kl-1)$, $k \geq 2$ and $l \geq 1$, then the generalized rational blowdown of (M, ω) along the spheres admits a symplectic structure induced by ω.*

Indeed, we constructed in Section 9.4 a toric neighborhood of a chain of spheres whose boundary is $L(n, m)$. Taking $n = k^2$, $m = kl - 1$, a slight generalization of the symplectic neighborhood theorem implies that the chain of spheres in the theorem have a neighborhood symplectomorphic to our model toric one (for some curve γ). We also constructed, in Section 9.3, an almost toric fibration of a rational ball. The structure of the bases of these fibrations makes it clear that whenever a rational ball with boundary $L(k^2, kl-1)$ can replace a neighborhood of a chain of spheres smoothly then this can be done symplectically.

As a consequence of Theorem 10.2, the smooth manifolds mentioned above (constructed by Fintushel and Stern) all admit symplectic structures.

10.2. Symplectic sum. The symplectic sum was defined in [**22**] by Gromov and used to great effect (mostly in dimension four) by Gompf [**18**] and Fintushel and Stern [**14**].

Definition 10.3. Consider two symplectic manifolds (M_i, ω_i), $i = 1, 2$, that contain symplectomorphic codimension two submanifolds Σ_i with anti-isomorphic normal bundles. Let $\overline{M_i - \Sigma_i}$ be the manifold with boundary such that $\overline{M_i - \Sigma_i} - Y_i$ is symplectomorphic to $(M_i - \Sigma_i, \omega)$ with $Y_i \subset \partial \overline{M_i - \Sigma_i}$ being diffeomorphic to a circle bundle over Σ_i. The *fiber sum* $M_1 \#_{\Sigma_1 = \Sigma_2} M_2$ is then $\overline{M_1 - \Sigma_1} \cup_\varphi \overline{M_2 - \Sigma_2}$ with the induced symplectic structure, where $\varphi : Y_1 \to Y_2$ is an orientation reversing diffeomorphism.

In dimension four, the normal bundles of surfaces Σ_i are anti-isomorphic when their self-intersections sum to zero and the surfaces themselves are symplectomorphic if they have the same genus and area.

When the surfaces are spheres base diagrams give a way to picture the symplectic sum. Specifically, if the surfaces are spheres of self-intersection n and $-n$, invoking the symplectic neighborhood theorem we know that we can find neighborhoods of the Σ_i that admit toric fibrations with bases $U_{n,1}$, $U_{-n,1}$ of Example 3.26 for some choice of s. Change the stratification in each base by making the interior of the edge that is the image of Σ_i part of the non-reduced boundary. This gives base diagrams that define toric fibrations of collar neighborhoods of the Y_i in $(\overline{M_i - \Sigma_i}, \omega)$. Under the hypotheses of the theorem, Proposition 3.22 implies the bases for the collar neighborhoods of the Y_i can be joined along the images of the Y_i to yield a base for a toric fibration of the gluing locus in the summed manifold.

Remark 10.4. When performing the symplectic sum of symplectic four-manifolds it is not necessary for the surfaces Σ_1, Σ_2 to be symplectomorphic. Unless they have self-intersection zero, it suffices that they have the same genus and opposite self-intersection numbers, and that one be willing to thicken or thin one of the (M_i, ω_i), $i = 1$ or 2, along Σ_i.

If Σ_1, Σ_2 are symplectic tori then they have almost toric neighborhoods. Modifying $U_{n,1}, U_{-n,1}$ by replacing the left and right edges by light lines marked with arrows to indicate that they are identified yields base diagrams for such neighborhoods. Then, as above, carrying out the symplectic sum along these surfaces corresponds to replacing the reduced boundary component in each base by a non-reduced boundary component and then identifying these two boundary components by an affine isomorphism.

Of course, if an almost toric fibration is given and the Σ_i are spheres or tori belonging to the preimage of the reduced boundary in each M_i, then the above procedure shows the symplectic sum can be performed within the category of almost toric fibered manifolds.

Exercise 10.5. If the surfaces Σ_i have genus ≥ 2 then they do not have an almost toric neighborhood. Nonetheless, one can still draw base diagrams that encode the sum. We leave it to the interested reader to sort this out. Hint: Decompose each surface as $\Sigma_i = D_i \cup \Sigma_i'$ where $\Sigma_i' = \Sigma_i - D_i'$ and $D_i' \subset D_i$ are disks such that the areas of $\Sigma_i - D_i'$ are equal and so are the areas of the $D_i - D_i'$. Find disk bundle neighborhoods of the Σ_i and local coordinates that trivialize the bundles over the $\Sigma - D_i'$ with compatible almost toric fibrations of the bundles over the $D_i - D_i'$ and the D_i. The base diagrams for the fibrations of the disk bundles over the D_i then reveal the essence of the symplectic sum.

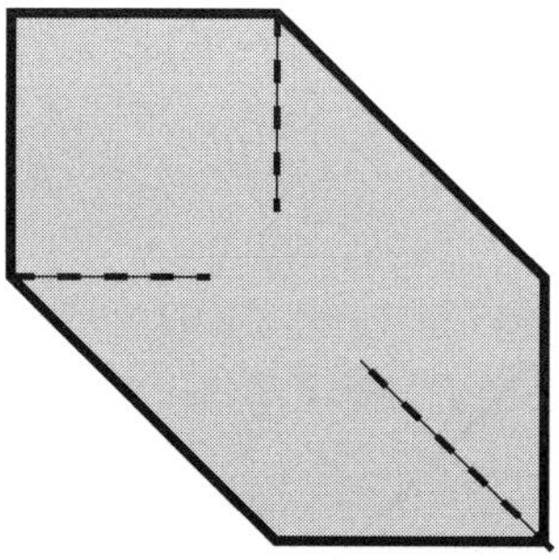

FIGURE 16. Fibration of $\mathbb{C}P^2\#3\overline{\mathbb{C}P}^2$.

11. Images of the K3 surface

The base of a Lagrangian fibered K3 surface is a sphere so there are no base diagrams per se. However, using symplectic sums we can give presentations of the K3 as a union of fibrations over topological disks for which we can draw base diagrams.

Our first construction was initially observed by Zung [**39**]. Consider the base diagram shown in Figure 16 that defines a Lagrangian fibration $\pi : (M, \omega) \to (B, \mathcal{A}, \mathcal{S})$ where $M = \mathbb{C}P^2\#3\overline{\mathbb{C}P}^2$ The essence is to perform successive symplectic sums of eight copies of this manifold by summing eight copies of $(B, \mathcal{A}, \mathcal{S})$, i.e. by performing symplectic sums that are compatible with the Lagrangian fibrations. Specifically, the reduced boundary shown in Figure 16 is a union of three rational line segments that are the images of three symplectic spheres of self-intersection zero, all having equal area. If we sum two copies of the base we get a second base that defined a fibered manifold with six nodal fibers and two spheres of self-intersection zero that project to the boundary of the base. (The new spheres are connect sums of two pairs of the original ones.) Performing a symplectic sum of two copies of this second base yields a third; it defines a fibration over a disk with 12 nodes and one smooth component to the reduced boundary. (As a shortcut, one can sum four copies of the original manifold at once using the four-fold sum [**32**], also in terms of gluing the bases together.) Lemma 5.21 and Theorem A.1 together imply that the resulting manifold is diffeomorphic to E(1). The preimage of the boundary is a symplectic torus of self-intersection zero so we can perform a symplectic sum of two copies to get a fibration over the sphere with 24 nodal fibers; by the proof of Theorem 5.19 this is a K3 surface.

A standard fact about the K3 surface is that its intersection form, with respect to an appropriate basis for $H_2(K3, \mathbb{Z})$, is $-E_8 \oplus -E_8 \oplus 3\left(\begin{smallmatrix} 0 & 1 \\ 1 & -2 \end{smallmatrix}\right)$. Our goal now is to present an almost toric fibration of the K3 surface so that, via base diagrams, we can identify surfaces that represent such a basis for $H_2(K3, \mathbb{Z})$.

We do this by presenting the K3 surface via an almost toric fibration of E(1)= $\mathbb{C}P^2\#9\overline{\mathbb{C}P}^2$ with two key properties:

- the reduced boundary has one smooth component that is the whole boundary and

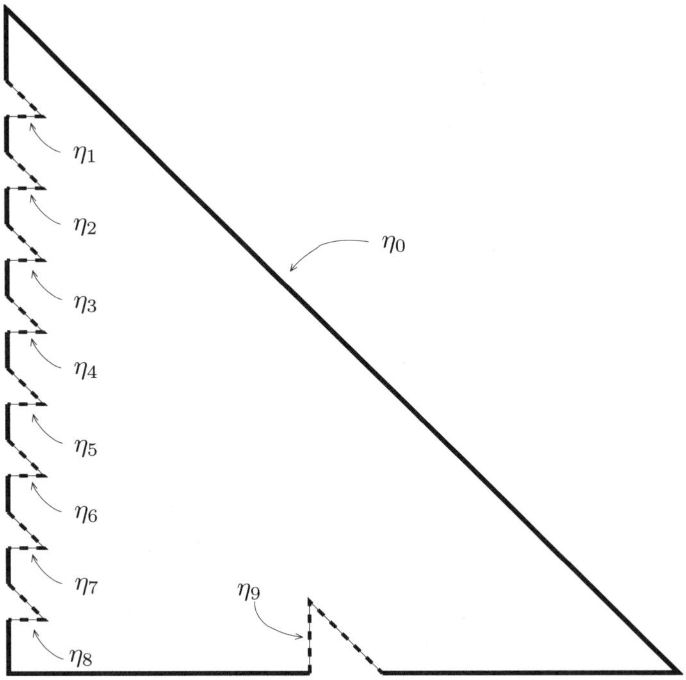

FIGURE 17. Visible surfaces showing basis for $H_2(E(1), \mathbb{Z})$.

- we can identify visible surfaces, disjoint form the preimage of the reduced boundary, that represent a basis for $H_2(E(1), \mathbb{Z})$ with respect to which the intersection form is $-E_8 \oplus \left(\begin{smallmatrix} 0 & 1 \\ 1 & -1 \end{smallmatrix}\right)$.

The first property allows us to define an almost toric fibration of the K3 surface by identifying two copies of this base diagram along their boundaries. This amounts to performing a symplectic sum of two copies of E(1) along the tori that are the preimage of the reduced boundary. The second property allows us to immediately see most of the surfaces we are looking for in the K3 surface. We explain how to identify all 22 of them.

As preparation we start with a base diagram (Figure 17) for an almost toric fibration of E(1)= $\mathbb{C}P^2 \# 9\overline{\mathbb{C}P}^2$ in which we can see visible surfaces $\Sigma_{\eta_0}, \Sigma_{\eta_1}, \ldots \Sigma_{\eta_9}$ where each Σ_{η_i}, $i \geq 1$ is a -1-sphere coming from a blow-up. Letting $h = [\Sigma_{\eta_0}]$ and $e_i = [\Sigma_{\eta_i}]$ for $i \geq 1$ we get a basis for the second homology in which the intersection form is $< 1 > \oplus 9 < -1 >$.

Now making nodal trades at each vertex we get the fibration defined by the base diagram in Figure 18. The visible surfaces Σ_{γ_i}, $i = 1, \ldots 10$ represent a basis for the second homology with respect to which we get the intersection form $-E_8 \oplus \left(\begin{smallmatrix} 0 & 1 \\ 1 & -1 \end{smallmatrix}\right)$. Indeed, $[\Sigma_{\gamma_j}] = e_j - e_{j+1}$, $j = 1, \ldots 7$, $[\Sigma_{\gamma_8}] = e_6 + e_7 + e_8 - h$, $[\Sigma_{\gamma_9}] = e_9$, and $[\Sigma_{\gamma_{10}}] = 3h - \sum_{i=1}^{9} e_i$ when the appropriate orientations are chosen. Notice that Σ_{γ_8} does define a smooth sphere: With respect to the induced standard coordinates the compatible vector must be the vanishing covector at each endpoint, namely $(-1, 1)$ and $(2, 1)$; each time γ_8 crosses a wedge corresponding to a blow-up the

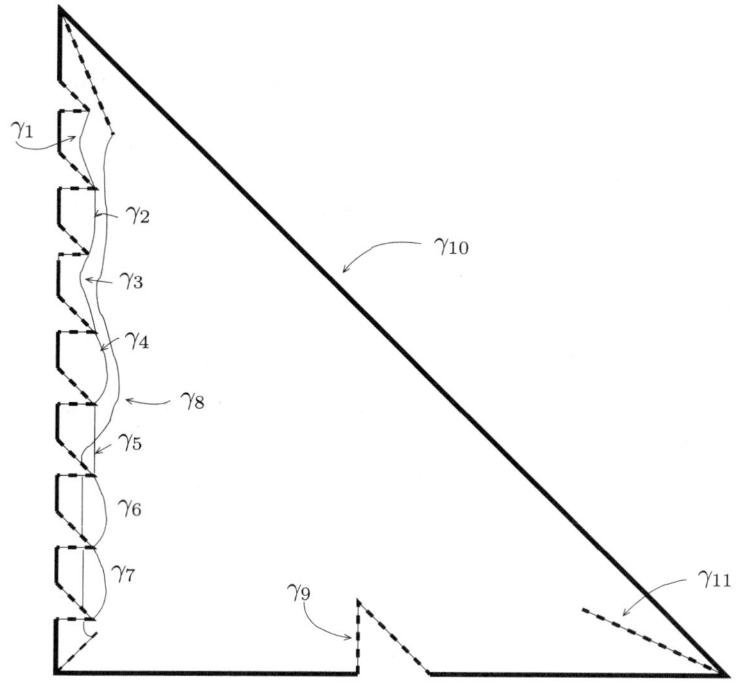

FIGURE 18. Another basis for $H_2(E(1), \mathbb{Z})$.

coordinates of the vector change by an application of $\left(\begin{smallmatrix}0&1\\1&1\end{smallmatrix}\right)$; since $\left(\begin{smallmatrix}0&1\\1&1\end{smallmatrix}\right)^3\left(\begin{smallmatrix}-1\\1\end{smallmatrix}\right) = \left(\begin{smallmatrix}2\\1\end{smallmatrix}\right)$ the compatible vector is well defined.

Using the techniques developed in Section 7.1 the homology classes represented by the surfaces Σ_{γ_i} can be read off the base diagram (for which the previous fibration of E(1) may be helpful) and it is straightforward to check the intersection pairing of the surfaces.

Now perform the symplectic sum of the two copies of E(1): remove the torus $\Sigma_{\gamma_{10}}$ from each E(1) and replace it with a 3-torus (i.e. replace the almost toric fibered E(1) with its boundary recovery, Definition 3.14) and then identify the two resulting four-manifolds M_1, M_2 along their boundaries. In doing so, we can lift $\Sigma_{\gamma_{10}}$ to a torus inside the boundary of each M_i and can identify $\partial M_1 = \partial M_2$ so that the lifted two-tori in the two boundaries are identified. Call the homology class of this torus in K3 surface T_1.

Because each γ_i, $i = 1, \dots 8$ is disjoint from the reduced boundary, the spheres Σ_{γ_i}, $i = 1, \dots, 8$ embed into the K3 surface after we have performed the symplectic sum. Furthermore, the induced map on homology is injective on the subspace spanned by the corresponding homology classes so the images of the $[\Sigma_{\gamma_j}]$, $j = 1, \dots, 7$ and of $[\Sigma_{\gamma_8}]$ represent distinct homology classes that we name V_j^i and W^i respectively, where i indicates which copy of E(1) the surface came from.

Meanwhile, the surfaces Σ_{γ_9} in each copy of E(1) lift to closed disks in M_1, M_2 and the visible disks $\Sigma_{\gamma_{11}}$ in each $E(1)$ remain visible disks in M_1, M_2. It is easy to perform the symplectic sum in such a way that the boundaries of these corresponding visible disks in each M_i are identified. The result will be two spheres of self-intersection -2, the first of which intersects T_1 transversely in one point. Call the homology classes of these spheres S_1 and S_2 respectively. Let S_3 be the homology class of a section of the almost toric fibration of the K3.

Inside the 3-torus boundary of M_1 we can find two other tori that are products of affine circles so that together with the lift of $\Sigma_{\gamma_{10}}$ they span $H_2(\partial M_1, \mathbb{Z})$. Let T_2, T_3 be the homology classes of the images in the K3 surface (via inclusion) of these tori, with T_3 corresponding to the one that is a Lagrangian fiber in M_1.

Then it is straightforward to check that V_j^i, W^i, T_k, S_k, $i = 1, 2$, $j = 1, \dots 7$, $k = 1, 2, 3$ is a basis for the second homology of the K3 with respect to which the intersection form is $-E_8 \oplus -E_8 \oplus 3\left(\begin{smallmatrix} 0 & 1 \\ 1 & -2 \end{smallmatrix}\right)$.

Appendix A. Closed toric four-manifolds

The following basic theorem is well known but direct proofs in the symplectic category are difficult to find.

THEOREM A.1. *Suppose M is a closed four-manifold that admits a symplectic structure and toric fibration $\pi : (M, \omega) \to (B, \mathcal{A}, \mathcal{S})$. Let k be the number of vertices of the base. Then M is diffeomorphic to $\mathbb{C}P^2 \# (k-3)\overline{\mathbb{C}P}^2$ if $k \neq 4$ and is diffeomorphic to $\mathbb{C}P^2 \# \overline{\mathbb{C}P}^2$ or $S^2 \times S^2$ if $k = 4$. The latter two cases can be distinguished by whether or not (respectively) a pair of non-adjacent edges has odd cross product (thereby revealing a sphere with odd self-intersection).*

Delzant [8] proved that a closed symplectic toric manifold is diffeomorphic to a smooth compact toric variety. (See also the appendix of [6] for the dimension four case.) Then one can appeal to the classification of smooth compact toric complex surfaces, cf. [16]. Here we provide a simple proof that is in keeping with the perspective of this paper, namely we classify the bases B that can define a toric fibration of a closed four-manifold. When there are few vertices in the base (three or four) it is easy to determine the possible bases and the corresponding manifolds. The main issue is to prove that if there are more than four vertices then at least one of the edges must be the image of a -1-sphere that can be blown down (Proposition A.5).

First we set some notation to be used throughout this section. Since the base of a toric fibration of a closed manifold is isomorphic to the image of the moment map for a torus action $(B, \mathcal{A})$ must be isomorphic to a convex polygon $(B, \mathcal{A}_0) \subset (\mathbb{R}^2, \mathcal{A}_0)$. Label vertices of the polygon in $\mathbb{R}^2$ clockwise as x_i, $i = 1, \dots n$. Let e_i be the primitive integral vector that points from x_{i+1} to x_i (where $x_{n+1} = x_1$) and let S_i be the sphere whose image is the edge with vertices x_i, x_{i+1}.

The fact that M is a smooth closed manifold implies that at each vertex x_i we have $|e_{i+1} \times e_i| = 1$. The clockwise numbering of the vertices and the convexity condition then imply the refinement

$$e_{i+1} \times e_i = 1. \tag{A.1}$$

Furthermore,

LEMMA A.2. *The sphere S_j has self-intersection $e_{j-1} \times e_{j+1}$.*

EXERCISE A.3. Prove this as a corollary of Proposition 7.13

LEMMA A.4. *Let $\pi : (M, \omega) \to (B, \mathcal{A}_0, \mathcal{S})$ be a toric fibration of a closed manifold M. If the base has three vertices then M must be diffeomorphic to $\mathbb{C}P^2$ and* *there is a unique fibration (up to scaling). If the base has four vertices then M is diffeomorphic to $S^2 \times S^2$ or the non-trivial S^2 bundle over S^2 (or, equivalently, $\mathbb{C}P^2 \# \overline{\mathbb{C}P}^2$).*

PROOF. Assume without loss of generality that $x_1 = (a_1, 0)$ for some $a_1 > 0$, $x_2 = (0,0)$ and $x_3 = (0, a_2)$. This implies $e_1 = (1,0)$, $a_2 > 0$ and $e_2 = (0,-1)$. Suppose $(B, \mathcal{A}_0, \mathcal{S})$ has three vertices. The smoothness and convexity constraint (Equation A.1) at x_1 and x_3 forces $e_3 = (-1,1)$ and $a_1 = a_2$. This shows that up to scaling and integral affine isomorphism there is only one base $(B, \mathcal{A}_0, \mathcal{S})$ with three vertices. Consulting Figure 1(b) we see that this base defines $\mathbb{C}P^2$ and that, up to scaling of the total volume (Section 8.3) all Lagrangian fibrations of $\mathbb{C}P^2$ are equivalent.

If $(B, \mathcal{A}_0, \mathcal{S})$ has four vertices, the smoothness and convexity constraint Equation A.1 at x_3 and x_1 forces $e_3 = (-1, n_3)$ and $e_4 = (m_4, 1)$. Applying the constraint at x_4 implies either $n_3 = 0$ or $m_4 = 0$. Our choice of which edge to put on the horizontal axis allows us to stipulate, without loss of generality, that $n_3 = 0$ and $m_4 = n \geq 0$. Appealing to Example 3.26 and our calculations of self-intersections in Section 7.3 we see that M is the union of two disk bundles over the sphere glued along their boundaries and hence is a sphere bundle over a sphere. Because $\pi_1(\mathrm{Diff}^+ S^2) = \pi_1(SO(3)) = \mathbb{Z}_2$ there are only two possibilities: a trivial bundle $S^2 \times S^2$ and a non-trivial bundle. Taking S_1, S_2 as a basis for the second homology the intersection forms is $Q_M = \left(\begin{smallmatrix} -n & 1 \\ 1 & 0 \end{smallmatrix}\right)$. This implies M is diffeomorphic to $M_n = S^2 \times S^2$ when n is even and $M_n = S^2 \tilde{\times} S^2$ when n is odd. To see that the non-trivial sphere bundle is diffeomorphic to $\mathbb{C}P^2 \# \overline{\mathbb{C}}P^2$, notice that when $n = 1$, S_1 is the image of a -1-sphere and that blowing this sphere down (Section 10.1) yields $\mathbb{C}P^2$. □

It is not so convenient to enumerate the possible bases $(B, \mathcal{A}_0, \mathcal{S})$ that have five or more vertices (and satisfy $\partial_R B = \partial B$ so that they define a closed manifold). However,

PROPOSITION A.5. *If $\pi : (M, \omega) \to (B, \mathcal{A}_0, \mathcal{S})$ is a toric fibration of a closed manifold M such that the base has at least five vertices, then some edge in ∂B is the image of a sphere with self-intersection -1.*

LEMMA A.6. *If $(B, \mathcal{A}_0, \mathcal{S})$ has at least five vertices, then there are at most two edges in ∂B that are the image of a sphere with non-negative self-intersection and these two edges must be adjacent. Furthermore, only one of these two spheres can have positive self-intersection.*

PROOF. Position the vertices e_i so that their bases are all at the origin and let α_i be the angle between e_i and e_{i+1} and observe that $\sum_i \alpha_i = 2\pi$. (Note that having done so one sees the *fan* corresponding to the complex toric variety diffeomorphic to the manifold defined by $(B, \mathcal{A}_0, \mathcal{S})$). Lemma A.2 implies that an edge of B with tangent vector e_i is the image of a sphere S_i of non-negative self-intersection if and only if $\alpha_{i-1} + \alpha_i \geq \pi$.

Suppose that three of the spheres have non-negative self-intersection. Then the condition on the angles implies the spheres are adjacent and there are at most four angles, and hence at most four edges contradicting our assumption of at least five edges.

Knowing now that there are at most two such spheres, suppose there are two and that they are not adjacent. Let S_2 be one of the two spheres with non-negative self-intersection and, without loss of generality, let $e_1 = (1,0)$. Then $\alpha_1 + \alpha_2 \geq \pi$ implies $e_3 = (a,b)$ with $b \leq 0$. Since $\sum_i \alpha_i = 2\pi$, the spheres of non-negative self-intersection being adjacent forces the other such sphere to be S_4, $e_3 = (-1,0)$. Hence both S_2, S_4 must have self-intersection zero and there are only four spheres, again a contradiction.

Finally, suppose there is a pair of adjacent spheres having positive self-intersection. Refer again to the base in $(\mathbb{R}^2, \mathcal{A}_0)$ with the vectors e_1, e_2 chosen as in the proof of Lemma A.4. Smoothness at x_1 and x_3 implies that $e_n = (-k,1)$ and $e_3 = (-1,l)$ for some $k, l > 0$. But then convexity of the image implies that $k = l$ and $e_3 = e_n$ so there are only three edges, a contradiction. □

Proof of Proposition A.5. Assume that no S_i is a -1-sphere.

By Lemma A.6 we can assume that either

(1) $S_i \cdot S_i \leq -2$ for all i, or
(2) $S_n \cdot S_n \geq 0$, $S_{n-1} \cdot S_{n-1} = 0$ and $S_i \cdot S_i \leq -2$ for all other i, or
(3) $S_n \cdot S_n \geq 0$ and $S_i \cdot S_i \leq -2$ for all other i.

Appealing to Exercise 9.22 we see right away that the first case is impossible since $(B, \mathcal{A}_0)$ is a closed polygon. In the second case, since $S_{n-1} \cdot S_{n-1} = 0$, we can assume that $e_{n-2} = (1,0)$, $e_{n-1} = (0,-1)$ and $e_n = (-1,0)$. Then $S_n \cdot S_n \geq 0$ implies $e_1 = (m,1)$ with $m \geq 0$ while $S_i \cdot S_i \leq -2$ for $1 \leq i \leq n-2$ implies $e_1 = (k,l)$ for some relatively prime $k, l > 0$ such that $\frac{k}{l} = [b_{n-2}, \ldots, b_2]$ with $b_j \geq 2$ for all $2 \leq j \leq n-2$. For these constraints on e_1 to agree we must have $l = 1$ which forces $n - 2 = 2$ in which case there are only four edges, a contradiction. In the third case we can assume that $e_n = (0,-1)$ and $e_{n-1} = (1,0)$. Then $S_n \cdot S_n \geq 0$ implies $e_1 = (-1,k)$, $k \geq 0$ and $S_i \cdot S_i \leq -2$ for $2 \leq i \leq n-1$ implies $e_1 = (k,l)$ as in the previous case; but this is a contradiction since $k, l > 0$. □

Proof of Theorem A.1. We now know that we can blow down any closed toric manifold fibering over a polygon with at least five sides to yield a toric fibered closed manifold whose base has one less side. It follows that any such manifold is the blow-up of $S^2 \times S^2$ or $\mathbb{C}P^2 \# \overline{\mathbb{C}P}^2$. But we can refine this statement a bit thanks to the following fact:

Exercise A.7. Use base diagrams (moment map images) for toric fibrations of $S^2 \times S^2$ and $\mathbb{C}P^2 \# \overline{\mathbb{C}P}^2$ to show that $(S^2 \times S^2) \# \overline{\mathbb{C}P}^2$ and $\mathbb{C}P^2 \# 2\overline{\mathbb{C}P}^2$ are diffeomorphic.

This completes the proof of Theorem A.1.

□

References

1. Miguel Abreu, *Topology of symplectomorphism groups of $S^2 \times S^2$*, Invent. Math. **131** (1998), no. 1, 1–23.
2. Miguel Abreu and Dusa McDuff, *Topology of symplectomorphism groups of rational ruled surfaces*, J. Amer. Math. Soc. **13** (2000), no. 4, 971–1009 (electronic).
3. Silvia Anjos, *Homotopy type of symplectomorphism groups of S^2XS^2*, Geom. Topol. **6** (2002), 195–218.

4. V. I. Arnold, *Mathematical methods of classical mechanics*, Springer-Verlag, New York, 1978, Translated from the Russian by K. Vogtmann and A. Weinstein, Graduate Texts in Mathematics, 60.
5. M. F. Atiyah, *Convexity and commuting Hamiltonians*, Bull. London Math. Soc. **14** (1982), no. 1, 1–15.
6. Michèle Audin, *The topology of torus actions on symplectic manifolds*, Birkhäuser Verlag, Basel, 1991.
7. Mohamed Boucetta and Pierre Molino, *Géométrie globale des systèmes hamiltoniens complètement intégrables: fibrations lagrangiennes singulières et coordonnées action-angle à singularités*, C. R. Acad. Sci. Paris Sér. I Math. **308** (1989), no. 13, 421–424.
8. Thomas Delzant, *Hamiltoniens périodiques et images convexes de l'application moment*, Bull. Soc. Math. France **116** (1988), no. 3, 315–339.
9. J. J. Duistermaat, *On global action-angle coordinates*, Comm. Pure Appl. Math. **33** (1980), no. 6, 687–706.
10. L. H. Eliasson, *Normal forms for Hamiltonian systems with Poisson commuting integrals—elliptic case*, Comment. Math. Helv. **65** (1990), no. 1, 4–35.
11. John B. Etnyre, *Introductory lectures on contact geometry*, This volume.
12. ———, *Symplectic convexity in low-dimensional topology*, Topology Appl. **88** (1998), no. 1-2, 3–25, Symplectic, contact and low-dimensional topology (Athens, GA, 1996).
13. John B. Etnyre and Ko Honda, *Tight contact structures with no symplectic fillings*, Invent. Math. **148** (2002), no. 3, 609–626.
14. Ronald Fintushel and Ronald J. Stern, *Knots, links, and 4-manifolds*, Invent. Math. **134** (1998), no. 2, 363–400.
15. Robert Friedman and John W. Morgan, *Smooth four-manifolds and complex surfaces*, Springer-Verlag, Berlin, 1994.
16. William Fulton, *Introduction to toric varieties*, Princeton University Press, Princeton, NJ, 1993, The William H. Roever Lectures in Geometry.
17. Hansjörg Geiges, *Symplectic structures on T^2-bundles over T^2*, Duke Math. J. **67** (1992), no. 3, 539–555.
18. Robert E. Gompf, *A new construction of symplectic manifolds*, Ann. of Math. (2) **142** (1995), no. 3, 527–595.
19. Robert E. Gompf and András I. Stipsicz, *4-manifolds and Kirby calculus*, American Mathematical Society, Providence, RI, 1999.
20. Phillip Griffiths and Joseph Harris, *Principles of algebraic geometry*, Wiley-Interscience [John Wiley & Sons], New York, 1978, Pure and Applied Mathematics.
21. M. Gromov, *Pseudoholomorphic curves in symplectic manifolds*, Invent. Math. **82** (1985), no. 2, 307–347.
22. Mikhael Gromov, *Partial differential relations*, Springer-Verlag, Berlin, 1986.
23. Mark Gross and P. M. H. Wilson, *Mirror symmetry via 3-tori for a class of Calabi-Yau threefolds*, Math. Ann. **309** (1997), no. 3, 505–531.
24. V. Guillemin and S. Sternberg, *Convexity properties of the moment mapping*, Invent. Math. **67** (1982), no. 3, 491–513.
25. F. Hirzebruch, W. D. Neumann, and S. S. Koh, *Differentiable manifolds and quadratic forms*, Marcel Dekker Inc., New York, 1971, Appendix II by W. Scharlau, Lecture Notes in Pure and Applied Mathematics, Vol. 4.
26. Yael Karshon, *Maximal tori in the symplectomorphism groups of Hirzebruch surfaces*, Preprint.
27. Eugene Lerman, *Symplectic cuts*, Math. Res. Lett. **2** (1995), no. 3, 247–258.
28. ———, *Contact cuts*, Israel J. Math. **124** (2001), 77–92.
29. Eugene Lerman and Susan Tolman, *Hamiltonian torus actions on symplectic orbifolds and toric varieties*, Trans. Amer. Math. Soc. **349** (1997), no. 10, 4201–4230.
30. Nai-Chung Leung and Margaret Symington, *Almost toric symplectic four-manifolds*, In preparation.
31. Dusa McDuff and Dietmar Salamon, *Introduction to symplectic topology*, second ed., The Clarendon Press Oxford University Press, New York, 1998.
32. Dusa McDuff and Margaret Symington, *Associativity properties of the symplectic sum*, Math. Res. Lett. **3** (1996), no. 5, 591–608.

33. Vu Ngoc San, *On semi-global invariants for focus-focus singularities*, To appear in Topology, 2002.
34. Michael Spivak, *A comprehensive introduction to differential geometry. Vol. III*, second ed., Publish or Perish Inc., Wilmington, Del., 1979.
35. Margaret Symington, *Symplectic rational blowdowns*, J. Differential Geom. **50** (1998), no. 3, 505–518.
36. ______, *Generalized symplectic rational blowdowns*, Algebr. Geom. Topol. **1** (2001), 503–518 (electronic).
37. Jan-Cees van der Meer, *The Hamiltonian Hopf bifurcation*, Springer-Verlag, Berlin, 1985.
38. J. Williamson, *On the algegraic problem concerning the normal forms of linear dynamical systems*, Amer. J. Math. **58** (1936), no. 1, 141–163.
39. Nguyen Tien Zung, *Symplectic topology of integrable Hamiltonian systems, II*, Preprint.

School of Mathematics, Georgia Inst. of Technology, Atlanta, GA 30332
E-mail address: msyming@math.gatech.edu

Proceedings of Symposia in Pure Mathematics
Volume **71**, 2003

On the integer valued SU(3) Casson invariant

Hans U. Boden, Christopher M. Herald, and Paul Kirk

In 1986, Andrew Casson constructed a new invariant $\lambda_{SU(2)}(X)$ for oriented, integral homology 3-spheres X by counting conjugacy classes of irreducible $SU(2)$ representations $\alpha\colon \pi_1 X \to SU(2)$ with sign. The homology restriction on the 3-manifold X guarantees that the set of conjugacy classes of irreducible $SU(2)$ representations is compact, and Casson described a method for perturbing the representation variety to obtain an oriented manifold of dimension zero and used compactness of the irreducible stratum to prove that the resulting algebraic count of points is independent of perturbation.

This article is a survey of the results in [**8, 9**] generalizing the Casson invariant to the $SU(3)$ setting. The main challenge is that the irreducible stratum is no longer compact, and consequently a naive count of conjugacy classes of irreducible representations $\alpha\colon \pi_1 X \to SU(3)$ does not produce a well defined invariant of integral homology 3-spheres.

A similar problem was encountered by Walker in generalizing Casson's original invariant from integral to rational homology 3-spheres [**24**]. The homology condition on a rational homology 3-sphere X does not guarantee compactness of the set of irreducible representations $\alpha\colon \pi_1 X \to SU(2)$. As a consequence a direct count (with sign) of the conjugacy classes of irreducible $SU(2)$ representations ends up depending on the choice of perturbation. Walker solved this problem by introducing a correction term defined entirely in terms of the stratum of reducible (abelian) representations.

In generalizing the Casson invariant from $SU(2)$ to $SU(3)$, the direct count of conjugacy classes of irreducible representations $\alpha\colon \pi_1 X \to SU(3)$ depends on the choice of perturbation, and so again one needs to introduce a correction term defined in terms of the stratum of reducible representations. Correction terms of several different types have been proposed (cf. [**5, 8, 13**]). All three proposals adopt a gauge theoretic point of view, treating the Chern-Simons function or a perturbation of it as a Morse function on the space of connections and viewing the Casson invariant as an Euler characteristic for the space of connections modulo gauge, a framework first developed for $SU(2)$ by Floer and Taubes (cf. [**16, 23**]).

In this article, we describe the complications involved in defining an invariant in the $SU(3)$ gauge theory setting. We discuss several alternatives for resolving the difficulties, following [**5, 8**]. We also present some calculations of the resulting invariant $\tau_{SU(3)}$, following [**9**]. Our approach to the integer valued $SU(3)$ Casson invariant differs from that given in [**8**] in that here, $\tau_{SU(3)}(X)$ is defined as the average of two $SU(3)$ invariants, $\tau_+(X)$ and $\tau_-(X)$. Both of these invariants are a priori integer valued, and they differ only in the choice of basepoints used to determine the correction term.

1. The Casson invariant as an intersection number

We begin with a review of Casson's approach to the transversality and perturbation issues. First, we establish some notation. Given a space X and a Lie group G, let

$$R(X,G) = \operatorname{Hom}(\pi_1 X, G)/\text{conjugation}$$

denote the (real-algebraic) variety of conjugacy classes of representations $\alpha\colon \pi_1 X \to G$. Let $R^*(X,G) \subset R(X,G)$ denote the subset of conjugacy classes of representations $\alpha\colon \pi_1 X \to G$ whose stabilizer $\{g \in G \mid g\alpha(x)g^{-1} = \alpha(x) \text{ for all } x \in \pi_1 X\}$ coincides with the center of G. Such representations are called *irreducible representations*, and the complement $R(X,G) - R^*(X,G)$ is the subvariety of *reducible representations*.

If G is a subgroup of the general linear group $GL(n,\mathbb{C})$ (resp. $GL(n,\mathbb{R})$), then a representation $\alpha : \pi_1 X \to G$ induces a representation of $\pi_1 X$ on the vector space $\mathbb{C}^n$ (resp. $\mathbb{R}^n$). The above definition of reducibility is equivalent to the existence of a proper invariant complex (real) subspace.

A continuous map $f\colon X \to Y$ induces an algebraic map $f^*\colon R(Y,G) \to R(X,G)$. This map is injective if $f_*\colon \pi_1 X \to \pi_1 Y$ is surjective. Notice that if α is a reducible representation, then $f^*(\alpha)$ is also reducible.

Suppose X is a closed 3-manifold, and choose a Heegaard splitting (H_1, H_2, Σ) for X; i.e. H_1 and H_2 are solid handlebodies of genus g with boundary a closed surface Σ and

$$X = H_1 \cup_\Sigma H_2.$$

The inclusions $\pi_1\Sigma \to \pi_1 H_1$ and $\pi_1\Sigma \to \pi_1 H_2$ are surjective, and so the Seifert-Van Kampen theorem implies that $\pi_1 H_1 \to \pi_1 X$ and $\pi_1 H_2 \to \pi_1 X$ are surjective. Thus every morphism in the diagram

$$\begin{array}{ccccc} & & R(H_1, SU(2)) & & \\ & \nearrow & & \searrow & \\ R(X, SU(2)) & & & & R(\Sigma, SU(2)) \\ & \searrow & & \nearrow & \\ & & R(H_2, SU(2)) & & \end{array}$$

is injective. This identifies $R(X, SU(2))$ as the intersection of $R(H_1, SU(2))$ with $R(H_2, SU(2))$ in $R(\Sigma, SU(2))$.

The representation varieties $R(H_1, SU(2))$, $R(H_2, SU(2))$ and $R(\Sigma, SU(2))$ are singular along the reducible representations (those conjugacy classes whose stabilizer is larger than the center $\mathbb{Z}_2 \subset SU(2)$), and if $g \geq 2$ the irreducible points $R^*(\Sigma, SU(2))$, $R^*(H_1, SU(2))$, and $R^*(H_2, SU(2))$ are smooth (non-compact) manifolds of dimension $6g-6$, $3g-3$, $3g-3$, respectively.

When X is an integral homology 3-sphere, every intersection point except for the trivial representation lies in the intersection of the irreducible strata

$$R^*(X, SU(2)) = R^*(H_1, SU(2)) \cap R^*(H_2, SU(2)).$$

This is because a reducible $SU(2)$ representation has abelian image, but $\pi_1 X$ has no nontrivial abelian representations if $H_1(X;\mathbb{Z}) = 0$. (There are nontrivial reducible representations if one replaces $SU(2)$ by $SU(3)$, or if $H_1(X;\mathbb{Z})$ is non-zero; this is the source of the difficulty in extending Casson's invariant to $SU(3)$ or to rational homology 3-spheres.)

An orientation on X induces orientations on each of the three representation varieties $R^*(H_1, SU(2)), R^*(H_2, SU(2))$ and $R^*(\Sigma, SU(2))$ in a natural way, and Casson defined $\lambda_{SU(2)}(X)$ to be one half times the algebraic intersection number of these two $(3g-3)$-dimensional submanifolds of $R^*(\Sigma, SU(2))$:

$$\lambda_{SU(2)}(X) := \tfrac{1}{2} R^*(H_1, SU(2)) \cdot R^*(H_2, SU(2)).$$

To make sense of this intersection number Casson proved that one can make $R^*(H_1, SU(2))$ and $R^*(H_2, SU(2))$ transverse using a compactly supported isotopy of $R^*(H_2, SU(2))$ in $R^*(\Sigma, SU(2))$.

That one can choose the isotopy to be compactly supported in this context is a crucial point. Even though neither $R^*(H_1, SU(2))$ nor $R^*(H_2, SU(2))$ are themselves compact, their intersection $R^*(X, SU(2)) = R^*(H_1, SU(2)) \cap R^*(H_2, SU(2))$ is compact. This follows from the fact that, up to conjugacy, there is one only reducible representation (the trivial representation), and its conjugacy class is an isolated point in $R(X, SU(2))$.

If X is not a homology 3-sphere, then $R^*(X, SU(2))$ is not generally compact. This leads to problems in trying to use the algebraic intersection number of $R^*(H_1, SU(2))$ and $R^*(H_2, SU(2))$ to produce a well defined invariant. For example, if X is a rational homology 3-sphere but not an integral homology 3-sphere (so $H_1(X;\mathbb{Z})$ is nontrivial and finite), Walker observed that as one perturbs the representation variety $R(H_1, SU(2))$ relative to $R(H_2, SU(2))$ in $R(\Sigma, SU(2))$, an intersection point of $R^*(H_1, SU(2))$ and $R^*(H_2, SU(2))$ can suddenly appear out of (or disappear into) the stratum of reducible representations. Coincident with such a birth (or death) is a simultaneous change in a certain Maslov index determined by the varieties $R(H_1, U(1))$ and $R(H_2, U(1))$ in $R(\Sigma, U(1))$ and the normal bundles of these strata in the full representation space.

Using these Maslov indices, Walker derived a correction term involving the reducible representations which cancels out the effect of these births and deaths when X is a rational homology 3-sphere. Thus Walker was able to extend Casson's invariant to rational homology 3-spheres by adding this correction term to the intersection number of $R^*(H_1, SU(2))$ and $R^*(H_2, SU(2))$ [**24**].

The behavior of Casson's invariant and Walker's extension under various cut-and-paste operations is well understood. The most important such result is the Dehn surgery formula for the Casson-Walker invariant. Motivated by the surgery formula, Lescop gave a combinatorial definition of an invariant for all 3-manifolds which agrees with the Casson-Walker invariant on rational homology 3-spheres [**22**].

We now briefly review the behavior of the Casson-Walker-Lescop invariant $\lambda_{SU(2)}$ under change of orientation, connected sum, and Dehn surgery operations (cf. §1.5, [**22**]). If $-X$ denotes the 3-manifold X with the opposite orientation, then

$$\lambda_{SU(2)}(-X) = (-1)^{b_1(X)+1} \lambda_{SU(2)}(X),$$

where $b_1(X) = \operatorname{rank} H_1(X;\mathbb{Z})$. Furthermore,

$$\lambda_{SU(2)}(X_1 \# X_2) = |H_1(X_2;\mathbb{Z})| \lambda_{SU(2)}(X_1) + |H_1(X_1;\mathbb{Z})| \lambda_{SU(2)}(X_2),$$

where

$$|H_1(X;\mathbb{Z})| = \begin{cases} \text{the order of } H_1(X;\mathbb{Z}) & \text{if } b_1(X)=0, \\ 0 & \text{if } b_1(X)>0. \end{cases}$$

Finally, Walker and Lescop have generalized Casson's original surgery formula, which states: Given a knot $K \subset X$ in an integral homology 3-sphere X, if $\Delta_K(t)$ is the (symmetric) Alexander polynomial of K and X_n is the integral homology 3-sphere obtained by performing $1/n$-Dehn surgery on K, then

$$\lambda_{SU(2)}(X_n) = \lambda_{SU(2)}(X) + \frac{n}{2}\Delta_K''(1).$$

There is a similar formula for surgeries on a link which effectively computes the Casson-Walker-Lescop invariant for all 3-manifolds, since every 3-manifold is surgery on some link L in S^3.

In the next section, we describe Taubes' reinterpretation of the Casson invariant as an Euler characteristic [**23**], using gauge theory. This is the approach we take in generalizing the Casson invariant to the $SU(3)$ context.

2. The Casson invariant as an Euler characteristic

Throughout the remainder of the paper, we assume that X is an oriented integral homology 3-sphere. We now recall the gauge theoretic description of the Casson invariant provided by Taubes [**23**]. For the purposes of expediency, we develop notation in the general $SU(n)$ context here even though a treatment of the $SU(2)$ case would suffice for this section.

Let $P_n = X \times SU(n)$ be the trivial(ized) principal $SU(n)$ bundle over X, and denote by $\mathscr{A}_n$ and $\mathscr{G}_n$ the space of connections and gauge transformations on P_n. Since P_n is trivialized, any smooth $SU(n)$ connection can be described as a differential operator $d_A : C^\infty(X, \mathbb{C}^n) \to C^\infty(T^*X \otimes \mathbb{C}^n)$ of the form $d_A = d + A$, where d denotes the de Rham exterior derivative and $A \in \Gamma(T^*X \otimes su(n))$, i.e. A is a 1-form with values in the Lie algebra $su(n)$ (where we view an $su(n)$ element as an endomorphism of $\mathbb{C}^n$). The group $\mathscr{G}_n$ is $C^\infty(X, SU(n))$, the set of functions from X to $SU(n)$. There is a group action of $\mathscr{G}_n$ on $\mathscr{A}_n$ defined by

$$d_{g\cdot A} = g \circ d_A \circ g^{-1}.$$

A connection is called *irreducible* if its stabilizer is just the center of $\mathscr{G}_n$ (which consists precisely of constant functions on X into the center of $SU(n)$). If a connection has a larger stabilizer, it is called *reducible.*

For analytical reasons, one must use the completions of the space of connections and gauge transformations with respect to certain Sobolev norms: $\mathscr{A}_n$ is the Banach manifold obtained by completing the space of smooth connections on P_n in the L^2_1 norm, and $\mathscr{G}_n$ is the Banach Lie group obtained by completing the group of smooth bundle automorphisms of P_n in the L^2_2 norm. We will not stress this point here.

We adopt the usual notational convention $d_A \leftrightarrow A$ which identifies connections with their associated connection 1-form and denote by θ the canonical trivial connection (i.e. $d \leftrightarrow \theta$). In this notation the $\mathscr{G}_n$ action on 1-forms is written

$$g \cdot A = gAg^{-1} - (dg)g^{-1}.$$

We say A and $g \cdot A$ are *gauge equivalent.*

Let $\Omega(X, x_0)$ denote the monoid of piecewise smooth loops in X based at x_0. Associated to each smooth connection A is its *holonomy representation*

$$hol_A \colon \Omega(X, x_0) \longrightarrow SU(n)$$

obtained from parallel translation with respect to A. The connection is called *flat* if its curvature 2-form $F(A) = dA + A \wedge A$ vanishes. By an elliptic regularity argument, flat connections are (up to gauge) smooth, and for a flat smooth connection the holonomy depends only on the homotopy class of a loop. Thus the holonomy representation of a flat connection A factors through the fundamental group $\pi_1(X)$. Gauge equivalent flat connections have conjugate holonomy representations. This establishes a homeomorphism between the space of gauge equivalence classes of flat $SU(n)$ connections on X and conjugacy classes of $SU(n)$ representations of $\pi_1 X$. This correspondence preserves the notions of reducibility.

The Chern-Simons function $cs \colon \mathscr{A}_n \longrightarrow \mathbb{R}$ is given by the formula (viewing a connection as a matrix valued 1-form):

$$cs(A) = \frac{1}{8\pi^2} \int_X tr\left(A \wedge dA + \tfrac{2}{3} A \wedge A \wedge A\right).$$

This is a smooth function with critical point set equal to the set of flat connections in $\mathscr{A}_n$. As we next explain, this suggests that the Casson invariant, which is an algebraic count of flat connections up to gauge, is the analogue of the Euler characteristic (for the quotient $\mathscr{A}_n / \mathscr{G}_n$), assuming that cs satisfies a Morse condition adapted to this equivariant, infinite dimensional setting.

To explain this properly, we first recall the analogous results in finite dimensions, namely the Poincaré-Hopf theorem. Suppose M is a smooth compact manifold and $f \colon M \to \mathbb{R}$ a smooth function. A point $p \in M$ is called a *critical point* of f if $df_p = 0$.

The *Hessian* of f at a critical point p is the symmetric bilinear form

$$\operatorname{Hess} f_p : T_pM \times T_pM \to \mathbb{R}$$

defined by $\operatorname{Hess} f_p(X, Y)) = \tilde{X}(\tilde{Y}(f))$, where $\tilde{X}$ and $\tilde{Y}$ are vector fields which extend X and Y, and vector fields act on functions in the usual way, i.e. $X(f) = df(X)$. A function is called *Morse* if its Hessian is nondegenerate at each critical point.

If we endow M with a Riemannian metric, the gradient vector field of f is defined by the condition

$$X(f) = \langle \operatorname{grad} f_p, X \rangle_p$$

for all tangent vectors $X \in T_pM$. In other words $\operatorname{grad} f$ is dual to the differential df with respect to the Riemannian metric. A point $p \in M$ is a critical point of f if $\operatorname{grad} f_p = 0$, and the function f is Morse if $\operatorname{grad} f$ is transverse to the zero section.

At a critical point $p \in M$ of f, the *Morse index* $\mu(p)$ is defined to be the dimension of the negative eigenspace of the Hessian of f at p. Then the Poincaré-Hopf theorem states that the number of critical points, counted with sign determined by the Morse index, equals the Euler characteristic:

$$\sum_{p \in \operatorname{Crit}(f)} (-1)^{\mu(p)} = \chi(M). \tag{2.1}$$

There is a very useful way to describe the difference in Morse indices at critical points which will be used to extend the discussion to the infinite dimensional case. In general the Hessian of f is only defined at a critical point. The following construction provides a suitable extension of the Hessian to all points in M.

Let $\nabla : C^\infty(TM) \times C^\infty(TM) \to C^\infty(TM)$ denote the Levi-Civita connection on M. Then for *any* point $p \in M$, the linear map

$$H_p : T_pM \to T_pM \tag{2.2}$$

defined by $H_p(X) = \nabla_X(\operatorname{grad} f)_p$ satisfies

$$\langle H_p(X), Y\rangle = \langle X, H_p(Y)\rangle$$

for $X, Y \in T_pM$, hence is self-adjoint with respect to the Riemannian metric. Moreover, at a critical point p, H_p is related to the Hessian by the equation

$$\operatorname{Hess}(f)_p(X,Y) = \langle H_p(X), Y\rangle_p.$$

Thus, if p_0 and p_1 are two critical points of f, the difference $\mu(p_1) - \mu(p_0)$ equals the spectral flow of the family $-H_{p_t}$ for any path p_t in M from p_0 to p_1. (Given a continuous path D_t, $t \in [0,1]$ of self-adjoint matrices, or more generally of self-adjoint Fredholm operators, the spectral flow of the path $SF(D_t)$ is the difference between the number of eigenvalues that change from negative to non-negative minus the number that change from non-negative to negative for the path of operators D_t. This is the so-called $(-\varepsilon, -\varepsilon)$ *convention.* Spectral flow is a homotopy invariant of the path of operators, rel endpoints; in addition, with this convention, spectral flow is additive under composition of paths. A careful construction of the spectral flow can be found in [**11**].)

Thus the symmetric linear map H_p extends the notion of the Hessian to all points in M. We remark that the preceding discussion applies with no change to circle valued functions $f : M \to S^1$.

We now explain how Taubes adapted these ideas to the infinite dimensional setting of the Chern-Simons function $cs : \mathscr{A}_n \to \mathbb{R}$. The infinite dimensional manifold $\mathscr{A}_n$ is an affine space modeled on $\Omega^1_X \otimes su(n)$. We choose a Riemannian metric on X, and then this space of forms inherits an L^2 inner product defined by

$$\langle a, b\rangle_{L^2} = -\int_X tr(a \wedge *b)$$

for $a, b \in T_A\mathscr{A}_n = \Omega^1_X \otimes su(n)$, where $* : \Omega^p \to \Omega^{3-p}$ denotes the Hodge star operator. This gives an inner product on the tangent space $T_A\mathscr{A}_n = L^2_1(\Omega^1_X \otimes su(n))$ for each $A \in \mathscr{A}_n$. Alternatively, we can view it as a Riemannian metric on $\mathscr{A}_n$. Computing the derivative of the Chern-Simons function, we see that if $A \in \mathscr{A}_n$ and $a \in \Omega^1(X; su(n))$, then

$$\begin{aligned}
\left.\frac{d}{dt}\right|_{t=0} cs(A + ta) &= \frac{1}{8\pi^2}\int_X tr(A \wedge da + a \wedge dA + 2a \wedge A \wedge A) \\
&= \frac{1}{8\pi^2}\int_X tr(-dA \wedge a + a \wedge dA + 2a \wedge A \wedge A) \\
&= \frac{1}{4\pi^2}\int_X tr(a \wedge (dA + A \wedge A)) \\
&= -\frac{1}{4\pi^2}\langle a, *F(A)\rangle_{L^2},
\end{aligned}$$

where $F(A) = dA + A \wedge A \in \Omega^2 \otimes su(n)$ denotes the curvature of the connection A. Hence, the gradient of the Chern-Simons function is given by

$$\operatorname{grad} cs_A = -\frac{1}{4\pi^2} * F(A),$$

and the critical set of the Chern-Simons function consists of all flat connections, namely all $A \in \mathscr{A}_n$ for which $F(A) \equiv 0$.

In order to extend the constructions described above in the finite-dimensional case to this new setting, one must deal with several issues. First, the critical set of the Chern-Simons function is infinite dimensional. This is because $F(g \cdot A) = ad_g(F(A))$. Hence if A is a critical point so is $g \cdot A$ for any g in the infinite dimensional group $\mathscr{G}_n$.

This difficulty is resolved by passing to the quotient

$$\mathscr{B}_n := \mathscr{A}_n / \mathscr{G}_n.$$

As is typical in the context of group actions, one describes the tangent space of an orbit space using the slice theorem. In the present context, the tangent space to $\mathscr{G}_n$ at the identity is just $\Omega^0_X \otimes su(n)$. For fixed $A \in \mathscr{A}_n$, the differential of the action map

$$\mathscr{G}_n \to \mathscr{A}_n, \ g \mapsto g \cdot A$$

at the identity is given by the covariant derivative:

$$d_A : \Omega^0_X \otimes su(n) = T_1\mathscr{G}_n \to \Omega^1_X \otimes su(n) = T_A\mathscr{A}_n.$$

The action of $\mathscr{G}_n$ is not free, but as long as one restricts to the open subset $\mathscr{A}^*_n$ of irreducible connections, $\mathscr{G}_n$ acts freely (up to its center) and so $\mathscr{B}^*_n = \mathscr{A}^*_n/\mathscr{G}_n$ is a (Banach) manifold. The slice theorem then identifies the tangent space of $\mathscr{B}^*_n$ at $[A]$ with the cokernel of $d_A : \Omega^0_X \otimes su(n) \to \Omega^1_X \otimes su(n)$. The L^2 inner product then identifies this cokernel with the kernel of $d^*_A : \Omega^1_X \otimes su(n) \to \Omega^0_X \otimes su(n)$.

The gauge invariance property of cs (invariance under the action of the identity component of $\mathscr{G}$) implies that $\operatorname{grad} cs$ is orthogonal to the orbit tangent space, and may be viewed either as a $\mathscr{G}$ equivariant vector field on $\mathscr{A}$ or as a tangent vector field on the quotient (at least on $\mathscr{B}^*$, where this makes sense).

In particular, let $\mathscr{V}_n \to \mathscr{A}^*_n$ be the subbundle of the tangent bundle $T\mathscr{A}^*_n$ whose fiber at $A \in \mathscr{A}^*_n$ is $\ker d^*_A$. Using the trivial connection on $T\mathscr{A}^*_n \to \mathscr{A}^*_n$ and the fiberwise L^2 projection onto the subbundle $\mathscr{V}_n$, we construct a connection ∇ on $T\mathscr{B}^*_n \to \mathscr{B}^*_n$ defined for vector fields $X, Y : \mathscr{B}^*_n \to T\mathscr{B}^*_n$ as follows. First lift X and Y to equivariant sections $\tilde{X}, \tilde{Y} : \mathscr{A}^*_n \to \mathscr{V}_n$. Then let $\nabla_X(Y)$ be the vector field on $\mathscr{B}^*_n$ whose lift to $\mathscr{V}_n$ at A equals $\operatorname{proj}_{\ker d^*_A}(d_X(Y))$, where d_X denotes the trivial connection in the product bundle $T\mathscr{A}^*_n \to \mathscr{A}^*_n$. It is typical to (and we will) abuse notation and blur the distinction between sections of $T\mathscr{B}^*_n \to \mathscr{B}^*_n$ and equivariant sections of $\mathscr{V}_n \to \mathscr{A}^*_n$.

Having described the tangent space to $\mathscr{B}^*_n$, we continue our discussion of the Morse theory of the Chern-Simons function. The Chern-Simons function is not invariant under the action of $\mathscr{G}_n$. In fact, $cs(g \cdot A) = cs(A) + \deg g$. (The *degree* $\deg g$ of a gauge transformation $g : X \to SU(n)$ is defined by the equation $g_*([X]) = \deg g \cdot [SU(2)]$, where $[X] \in H_3(X; \mathbb{Z})$ denotes the fundamental class of X, and $[SU(2)] \in H_3(SU(n); \mathbb{Z}) \cong \mathbb{Z}$ denotes the class represented by the inclusion $SU(2) \subset SU(n)$.) Since deg: $\pi_0\mathscr{G}_n \to \mathbb{Z}$ is an isomorphism, the Chern-Simons function descends to a circle valued function:

$$cs : \mathscr{B}_n \to \mathbb{R}/\mathbb{Z}.$$

The gradient of the Chern-Simons function on $\mathscr{B}^*_n$ (or rather, its lift to $\mathscr{V}_n$) again satisfies $\operatorname{grad} cs_A = -\frac{1}{4\pi^2} * F(A)$. Thus the critical set is the set of flat connections modulo gauge transformations, a space homeomorphic (via the holonomy map) to $R(X, SU(n))$, the finite dimensional real-analytic variety of conjugation classes of $SU(n)$ representations of $\pi_1 X$.

The second issue concerns the Morse index $\mu(p)$. Mimicking the construction in the finite dimensional case, we define the Hessian, or rather its substitute H (as in equation (2.2)), using the connection ∇ on $T\mathscr{B}^*_n$. (For convenience, we adjust by a factor of $4\pi^2$.) Thus we define

$$H_A : T_A\mathscr{B}^*_n \to T_A\mathscr{B}^*_n, \ b \mapsto 4\pi^2 \nabla_b (\operatorname{grad} cs)_A,$$

which we compute using the definition of ∇:

$$\begin{aligned} H_A(b) &= 4\pi^2 \nabla_b (\operatorname{grad} cs)_A \\ &= -\operatorname{proj}_{\ker d^*_A} \left(\tfrac{d}{dt}\big|_{t=0} * F(A+tb)\right) \\ &= -\operatorname{proj}_{\ker d^*_A} (*d_A(b)). \end{aligned} \tag{2.3}$$

The operator H_A is closed, unbounded, and self-adjoint. It has infinitely many positive and negative eigenvalues and so the traditional definition of the Morse index $\mu(A)$ at a critical (i.e. flat) connection does not make sense. Taubes overcame this obstacle by using the spectral flow in place of the Morse index (as we described above in the finite dimensional context). This gives a relative sign to all critical points, and, assuming cs is Morse, determines an invariant up to an overall sign. Specifying the sign then amounts to deciding on a basepoint from which to measure spectral flow. We will return to the basepoint question shortly.

An interesting wrinkle that appears is that the spectral flow of $-H_{A_t}$ changes by a multiple of $4n$ if we replace A_0 or A_1 by gauge equivalent connections, but in any case the parity of $SF(-H_{A_t})$ is well defined, and this is enough to make sense of equation (2.1) in this context assuming the Chern-Simons function is Morse. We will sometimes be sloppy and write $SF(A_0, A_1)$ for $SF(-H_{A_t})$.

If $cs : \mathscr{B}^*_n \to \mathbb{R}/\mathbb{Z}$ is not Morse, we wish to perturb it by adding a gauge invariant function $h\colon \mathscr{A}_n \to \mathbb{R}$ so that $cs + h$ has only isolated nondegenerate critical points. For the remainder of the section, we restrict to the case $n = 2$. Taubes and Floer described a class of functions $\mathscr{F}$ satisfying:

(i) $\operatorname{grad}(cs + h)$ is a compact perturbation of $\operatorname{grad}(cs)$ for all $h \in \mathscr{F}$, and
(ii) $cs + h$ is a Morse function for generic $h \in \mathscr{F}$.

Let $\mathscr{M}^*_h \subset \mathscr{B}^*_2$ denote the irreducible gauge orbits in the critical set of the perturbed Chern-Simons function $cs + h$.

The main result in [**23**] states that for a generic small perturbation h,

$$\lambda_{SU(2)}(X) = \tfrac{1}{2} \sum_{[A]\in\mathscr{M}^*_h} (-1)^{SF(\theta,A)}. \tag{2.4}$$

This is the infinite dimensional analogue of the Poincaré-Hopf Theorem (2.1) for the manifold $\mathscr{B}^*_2$ and the Chern-Simons function.

The right hand side of equation (2.4) has the virtue that it gives an invariant of homology 3-spheres without making use of a Heegaard splitting. Indeed, Taubes's proof that the right hand side of equation (2.4) is well defined avoids a sticky issue that comes up in Casson's original proof, namely stabilization of Heegaard diagrams. Thus in trying to define an SU(3) version of Casson's invariant one is motivated to generalize Taubes's construction, rather than Casson's, and this is what has been done in the articles [**5, 8, 13**].

We end this section with a more detailed discussion of exactly what is meant by "$SF(\theta, A)$, $[A] \in \mathscr{M}_h$" in equation (2.4). There are two issues: The first is that we described H_A above as the Hessian of the Chern-Simons function, but one must extend this to the perturbed Chern-Simons function $cs+h$. The second issue is that the base point θ (the trivial connection) is not irreducible and therefore its gauge orbit $[\theta]$ is a singular point of $\mathscr{B}_2$. So H_θ is not well-defined. We will replace the self-adjoint operator H_A with an operator that makes sense over reducible connections as well as irreducibles.

We first describe the space $\mathscr{F}$ of admissible perturbations. This space consists of gauge invariant functions $h : \mathscr{A}_2 \to \mathbb{R}$ constructed as follows. Fix a collection of thickened loops $\{\gamma_i \colon S^1 \times D^2 \to X \mid i = 1, \ldots, n\}$ whose cores generate $\pi_1 X$ and with a common normal disk at one point, i.e. $\gamma_i(s_0, x) = \gamma_j(s_0, x)$ for all i, j. For any connection A, denote by $hol_i(x, A)$ the holonomy of A around the ith loop $\gamma_i(S^1 \times \{x\})$ from some fixed basepoint. Choose $\eta(x)$ to be a bump function on the 2-disk and define the space $\mathscr{F}$ of admissible perturbations $h \colon \mathscr{A}_2 \to \mathbb{R}$ to be functions of the form

$$h(A) = \int_{D_2} f(hol_1(x, A), \ldots, hol_n(x, A))\eta(x) d^2x,$$

where $f \colon SU(2)^n \to \mathbb{R}$ is a C^3 function invariant under the adjoint action. For example, for every word $\omega = \omega(g_1, \ldots, g_n)$ in the free group, one obtains an admissible perturbation by taking $f = tr(\omega)$.

The gradient $\operatorname{grad}(cs+h)_A = -\frac{1}{4\pi^2} * F(A) + \operatorname{grad} h_A$ can be viewed either as an equivariant vector field on $\mathscr{A}^*$ or as a vector field on $\mathscr{B}^*$. A connection A is called *h-perturbed flat* if $[A]$ is a critical point of $cs+h$ (i.e. if $\left(*F(A) - 4\pi^2 \operatorname{grad}(h)_A\right) = 0$).

Arguing as before we define the self-adjoint operator $H_{A,h} : \ker d_A^* \to \ker d_A^*$ for any A and h by

$$H_{A,h}(b) = 4\pi^2 \nabla_b(\operatorname{grad}(cs + h)) = \operatorname{proj}_{\ker d_A^*} \left(- * d_A(b) + 4\pi^2 d \operatorname{grad}(h)_A(b)\right).$$

Abusing notation slightly, write

$$\operatorname{Hess} h_A(b) = d \operatorname{grad}(h)_A(b).$$

If A_0, A_1 are two h-perturbed flat irreducible connections, $SF(A_0, A_1)$ means the spectral flow of the path of operators $-H_{A_t,h}$, where A_t is a path of irreducible connections joining A_0 to A_1.

In order to simplify and extend the definition of $SF(A_0, A_1)$ to reducible connections, Taubes introduced the following trick. For notational convenience write Ω^i for $\Omega^i_X \otimes su(2)$. Given an arbitrary connection $A \in \mathscr{A}_2$ and an admissible perturbation h, write

$$d_{A,h} = d_A - *4\pi^2 (\operatorname{Hess} h)_A : \Omega^1 \to \Omega^2$$

and consider the operator

$$K_{A,h} : \Omega^0 \oplus \Omega^1 \to \Omega^0 \oplus \Omega^1$$

defined by $K_{A,h}(\xi, a) = (d_A^* a, d_A\xi + *d_{A,h}(a))$. Then $K_{A,h}$ is self-adjoint (with respect to the L^2 inner product). Using the Hodge decomposition

$$\Omega^0 \oplus \Omega^1 = \Omega^0 \oplus \text{image } d_A \oplus \ker d_A^*,$$

one can check that if A is an irreducible connection, the difference

$$K_{A,h} - \begin{pmatrix} 0 & d_A^* & 0 \\ d_A & 0 & 0 \\ 0 & 0 & -H_{A,h} \end{pmatrix}$$

is a compact operator. (The compactness properties of the perturbation enter at this point.)

Moreover, if A is h-perturbed flat then the two maps $d_A \circ d_{A,h} : \Omega^1 \to \Omega^3$ and $d_{A,h} \circ d_A : \Omega^0 \to \Omega^2$ are both zero. It follows that for A an h-perturbed flat connection,

$$K_{A,h} = \begin{pmatrix} 0 & d_A^* & 0 \\ d_A & 0 & 0 \\ 0 & 0 & -H_{A,h} \end{pmatrix}.$$

Thus at an h-perturbed flat connection A, the $K_{A,h}$ is obtained from $-H_{A,h}$ by adding an operator with symmetric spectrum.

These facts, together with the invariance of spectral flow with respect to homotopy of paths of self-adjoint operators rel endpoints, imply that if A_0, A_1 are any two h-perturbed irreducible flat connections, and A_t is any path of irreducible connections joining them, the spectral flow of the family $K_{A_t,h}$ equals the spectral flow of the family $-H_{A_t,h}$.

The advantage is that $K_{A,h}$ is a smooth family of self-adjoint operators *for any perturbation h and any connection A*. Thus we can define the spectral flow $SF((A_0, h_0), (A_1, h_1))$ for any pair of connections and admissible perturbations to be the spectral flow of the family K_{A_t,h_t} for any path A_t joining A_0 to A_1 and any path h_t joining h_0 to h_1. This is well defined since the space of connections and the space of perturbations is contractible. It agrees with the previous definition of $SF(A_0, A_1)$ in terms of the Hessian $H_{A,h}$ when A_0 and A_1 are irreducible h-perturbed flat connections for a fixed perturbation h.

If A is an h-perturbed flat connection, we use the shorthand $SF(\theta, A)$ for $SF((\theta, 0), (A, h))$. This, finally, is the meaning of this term in equation (2.4). We call the operator $K_{A,h}$ the *perturbed odd signature operator*, since when A is a flat connection, $K_{A,0}$ is exactly the odd signature operator coupled to A as described in [**3**]. The reader will note that this discussion applies equally well to $SU(n)$ connections. In this general case we have $SF(A_0, g \cdot A_1) = SF(A_0, A_1) + 4n \deg g$.

3. Reducible connections, singularities and bifurcations

As explained in the previous section, Taubes interpreted the Casson invariant as an infinite dimensional Morse theoretic Euler characteristic for the space $\mathscr{B}_2^*$ of $SU(2)$ connections modulo gauge. This quotient space is singular along the reducible connections, but since X is a homology 3-sphere the critical set of the function cs avoids the singularities (except for the orbit of the trivial connection, which is isolated from other critical orbits), and so the singularities do not cause any serious difficulties.

In the setting of $SU(3)$ gauge theory, a reducible flat connection need not be gauge equivalent to the trivial connection, even on a homology 3-sphere. Restricting a gauge transformation to a fiber of the principal bundle determines an isomorphism from the stabilizer subgroup of a connection A to a subgroup of $SU(3)$. Up to conjugation, there are five possibilities for this subgroup:

(1) The center $\mathbb{Z}_3$ (so A is irreducible),
(2) The 1-dimensional torus $\{\operatorname{diag}(\lambda, \lambda, \lambda^{-2}) \mid \lambda \in U(1)\}$,
(3) The maximal torus $\{\operatorname{diag}(\lambda_1, \lambda_2, (\lambda_1\lambda_2)^{-1}) \mid \lambda_i \in U(1)\}$,
(4) $S(U(2) \times U(1)) = \left\{ \begin{pmatrix} A & 0 \\ 0 & \det A^{-1} \end{pmatrix} \middle| A \in U(2) \right\}$, and
(5) $SU(3)$.

Since $H_1(X;\mathbb{Z}) = 0$, all abelian representations of $\pi_1 X$ are trivial, so the third and fourth possibilities do not occur as stabilizers of representations (or of flat connections). The three types of flat connections on a homology 3-sphere, then, are the following:

(1) Irreducible connections A, with $\operatorname{stab}(A) = \mathbb{Z}_3$.
(2) Connections A with $\operatorname{stab}(A)$ equal to a 1-dimensional torus. For these connections, the image of the holonomy representation $hol_A \colon \pi_1 X \to SU(3)$ lies in $SU(2) \subset S(U(2) \times U(1))$ up to conjugation.
(3) Connections which are gauge equivalent to the trivial connection. For these connections, $\operatorname{stab}(A) \cong SU(3)$.

The presence of $S(U(2) \times U(1))$ flat connections means that, unlike the $SU(2)$ case, the $SU(3)$ situation requires analysis of reducible connections to define a perturbation independent quantity, because the singular stratum is not disjoint from the critical set of the Chern-Simons function. (This is similar to what Walker faced in extending the $SU(2)$ invariant to rational homology 3-spheres; in that case there were also three strata to consider.) As in the $SU(2)$ setting, the gauge orbit of the trivial connection is isolated in the space of flat connections modulo gauge. Thus the difficulties are focused on the singular stratum in the quotient $\mathscr{B}_3 = \mathscr{A}_3/\mathscr{G}_3$ consisting of connections whose stabilizer is a 1-dimensional torus. Notice that the reducible flat $SU(3)$ connections may be viewed as irreducible flat $SU(2)$ connections, i.e. those considered by Taubes.

The irreducible portion $\mathscr{B}_3^* \subset \mathscr{B}_3$ is a manifold. In addition, the stratum

$$\mathscr{B}_3^r = \{A \in \mathscr{A}_3 \mid \operatorname{stab}(A) \cong U(1)\}/\mathscr{G}_3$$

is also smooth (and in fact has a cone bundle neighborhood in $\mathscr{B}_3$). The strata of $\mathscr{B}_3$ corresponding to the third and fourth orbit type listed above are disjoint from the critical set of the Chern-Simons function, so we will not dwell on them.

Again, $\operatorname{grad} cs$ and $\operatorname{grad}(cs + h)$ may be viewed as either equivariant vector fields on $\mathscr{A}_3$ or as tangent vector fields on each stratum. (Invariance implies that at a reducible connection A, the gradient is tangent to the space of reducible connections.) Before describing the $SU(3)$ invariants, we illustrate the difficulties in defining an invariant by counting critical points (zeros of a gradient vector field) in such a stratified context.

For the purpose of illustration, we consider a finite dimensional manifold M with a semifree $U(1)$ action. Recall that an action of a group G on a space M is called *semifree* if the stabilizer subgroup of every point in M is either the trivial subgroup or G.

Let $L \subset M$ denote the set of fixed points of the group action. Like $\mathscr{B}^* \cup \mathscr{B}^r$, the quotient space $M/U(1)$ consists of two strata, $L/U(1) \cong L$ and $(M - L)/U(1)$, and the fixed point stratum has a normal bundle similar to that of $\mathscr{B}^r$ in $\mathscr{B}$. Namely, the normal bundle of $L/U(1)$ has fiber homeomorphic to $c(\mathbb{C}P^k)$, whereas the normal bundle of $\mathscr{B}^r$ has fiber $c(\mathbb{C}P^\infty)$. In other words, in this illustration, the non-fixed points play the role of the irreducible $SU(3)$ connections, and the fixed points play the role of the (nontrivial) reducible connections.

Let $f\colon M \to \mathbb{R}$ be a $U(1)$ invariant smooth function. If $m \in M$ is a critical point for f, then so is every point in the orbit of m. Thus the critical set of f consists of a collection of fixed points together with a union of circle orbits of critical points in $M - L$. Fix an invariant Riemannian metric on M so that we can identify the critical set of f with the zero set of the $U(1)$-equivariant gradient vector field $\operatorname{grad} f$. Because $U(1)$ acts with no nonzero fixed vectors on the normal bundle to L, if $p \in L$ then $\operatorname{grad} f(p) \in T_pL$. It follows that the critical fixed points are exactly the critical points of $f|_L$.

On a compact manifold M without group actions, standard techniques in differential topology show that generic functions are Morse (i.e. have gradient vector fields transverse to the zero section of TM). This nondegeneracy condition implies that the critical set of the function (the set of zeros of the vector field) is compact. Moreover, for a generic path of functions connecting two Morse functions, the only topological changes in the critical set are births and deaths of pairs of critical points of Morse indices differing by one. This fact can be used to give a Cerf theoretic proof that the signed count of critical points is independent of the Morse function, for the algebraic contribution from such a pair will be zero. (In fact, as mentioned in Section 2, the signed count can be identified with the Euler characteristic, so there are other ways of seeing that it is an invariant, too.)

Things are more complicated in the equivariant setting. We may view $M/U(1)$ as a union of two manifolds, $(M - L)/U(1)$ and $L/U(1) \cong L$. The singular nature of the quotient space has to do with the normal bundle structure of the second in the first. A theorem of Wasserman **[25]** asserts in this setting that a generic invariant function on M will give a Morse function with finite critical set on each of these strata. Since $(M - L)/U(1)$ is not compact, finiteness here is less straightforward. But Wasserman's theorem is stronger than the statement that the function restricted to each stratum is Morse. In addition, at a critical point $p \in L$, $\operatorname{Hess}(f)_p$ is also nondegenerate on the normal bundle fiber to N_pL. This prevents critical orbits in $M - L$ from accumulating at p.

For a generic path of invariant functions on M connecting two such generic functions, three types of bifurcations can occur in the critical set. The first two correspond to standard Morse births and deaths of cancelling pairs of critical points, in L, and in the (smooth) quotient space $(M - L)/U(1)$. In other words, a pair of critical points in L of Morse index difference one can be born or can annihilate one another, and similarly for a pair of circle orbits of critical points of index difference one.

The third type of bifurcation involves a critical circle orbit popping out of a critical fixed point (or this process in reverse). That is to say, there is a critical point in L, with no other critical points nearby for $t \leq t_0$, and then for $t > t_0$ there is in addition a circle orbit of critical points nearby. Figure 1 illustrates the topology of the critical set (in a neighborhood nearby). This phenomenon is illustrated with the path of functions $f_t\colon \mathbb{C} \to \mathbb{R}$, $f_t(z) = -t|z|^2 + |z|^4$.

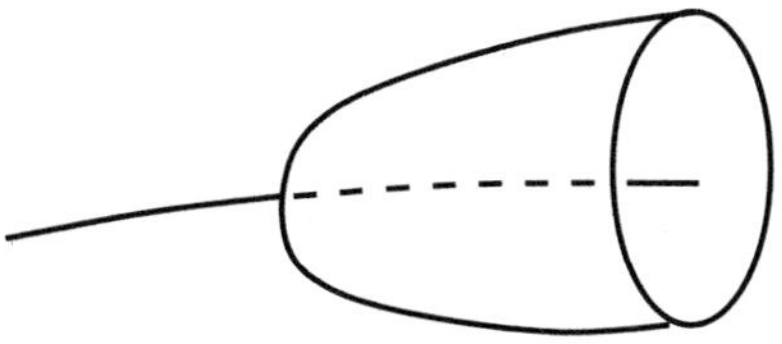

FIGURE 1. A PICTURE OF THE PARAMETERIZED CRITICAL SET NEAR t_0. For $t \leq t_0$, the critical set consists of only one fixed point, but for $t > t_0$, it also includes a circle orbit.

As a consequence, for a generic path f_t of functions joining two generic functions, the *parameterized critical set*

$$\{([m], t) \in M/U(1) \times [0,1] \mid m \text{ is a critical point of } f_t \}$$

has a description as a compact 1-dimensional singular bordism with 'T'-intersections from the critical set of f_0 to the critical set of f_1. More precisely, it is the union of

$$C_{\text{fixed}} := \{([p], t) \in L/U(1) \times [0,1] \mid p \in L \text{ is a critical fixed point of } f_t \}$$

and

$$C_{\text{free}} := \{([m], t) \in M/U(1) \times [0,1] \mid m \in M - L \text{ is a critical point of } f_t \}.$$

C_{fixed} is a compact 1-dimensional manifold with boundary, and the boundary lies in the "ends" $L/U(1) \times \{0,1\}$. On the other hand, C_{free} is a 1-manifold (generally not compact). All of its endpoints lie in $M/U(1) \times \{0,1\}$, but its noncompact ends limit to points in the interior of C_{fixed}. An example of a possible parameterized critical set is illustrated in Figure 2.

To obtain an analogue of the Euler characteristic in this $U(1)$ manifold setting, we must modify the usual signed count of critical points (or critical orbits) to obtain a formula which is invariant under all three types of bifurcations. The crucial observation is that, simultaneous with a bifurcation of the third type described above, there is a change in the normal Morse index of the function at the critical fixed point.

To describe the change in detail, we must refine our notion of Morse index at the critical points in L. The fact that the Hessian of an invariant function at a

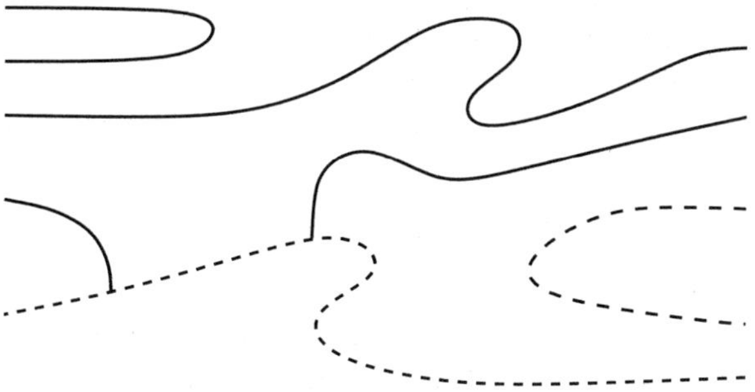

FIGURE 2. A COBORDISM WITH 'T'-INTERSECTIONS. The dashed curves represent the varying critical set in L, and the solid curves represent the varying critical set (modulo $U(1)$) in $M - L$.

critical point $p \in L$ is invariant under the (linearized) group action on the tangent space T_pM implies that the Hessian decomposes into a direct sum of operators on the two summands $T_pL \oplus N_pL \cong T_pM$. We define the tangential and normal components of the Morse index to be the number of negative eigenvalues of these summands, and denote them by $\mu_t(p)$ and $\mu_n(p)$. On N_pL, the Hessian commutes with the $U(1)$ action, so its eigenspaces are $U(1)$ invariant subspaces. The $U(1)$ action on N_pL has no trivial subrepresentations, therefore $\mu_n(p)$ is always even.

A close examination of the relationship between the Morse indices of the critical points involved in the bifurcations demonstrates that the quantity

$$\sum_{q\in((M-L)\cap crit(f))/U(1)} (-1)^{\mu(q)} - \sum_{p\in L\cap crit(f)} (-1)^{\mu_t(p)} \left(\frac{\mu_n(p)}{2}\right) \tag{3.1}$$

is invariant under all three bifurcations and hence is independent of the invariant Morse function. In fact, one can see by choosing the function appropriately that this invariant of the $U(1)$ manifold M is the relative Euler characteristic of the quotient space relative to the singular set (i.e. the Euler characteristic of the relative cohomology).

4. Correction terms in the SU(3) setting

In this section, we consider the $SU(3)$ gauge theory setting (and so we now drop the subscript on $\mathscr{A}$). There is a classification of the types of bifurcations that occur in the critical set for a generic one parameter family of admissible perturbations of the Chern-Simons function analogous to that given in the previous section. That is to say, the bifurcations involve standard Morse births/deaths in the irreducible stratum of $\mathscr{B}$, standard Morse births/deaths in the reducible (i.e. $S(U(2) \times U(1))$) stratum of $\mathscr{B}$, and irreducible orbits of critical points popping out of reducible critical points.

Motivated by the finite dimensional model discussed in the previous section and Taubes' gauge theoretic description of the Casson invariant, it is natural to try to define an $SU(3)$ Casson invariant by replacing the Morse indices in (3.1) with the spectral flow analogues. We begin by describing the spectral flow analogue of the decomposition of Morse index $\mu(p) = \mu_t(p) + \mu_n(p)$ at critical points with circle stabilizer.

We first identify the tangent space of the space $\mathscr{A}$ of $SU(3)$ connections with $\Omega^1(X) \otimes su(3)$, the space of differential 1-forms on X with values in the Lie algebra $su(3)$. Every reducible connection is gauge equivalent to an $S(U(2) \times U(1))$ connection. Under the adjoint action of $S(U(2) \times U(1))$ on the Lie algebra $su(3)$, we have the decomposition

$$su(3) = s(u(2) \times u(1)) \oplus \mathbb{C}^2, \tag{4.1}$$

where an element $A \oplus (\det A)^{-1} \in S(U(2) \times U(1))$ acts by the adjoint action on the first factor and by matrix multiplication by the $U(2)$ matrix $\det(A)A$ on the second factor. Any reducible $SU(3)$ connection can be gauge transformed into the path connected space of $S(U(2) \times U(1))$ connections. Thus one can decompose the spectral flow of the odd signature operator K_{A_t} along a path A_t of reducible connections into $s(u(2) \times u(1))$ and $\mathbb{C}^2$ components according to the decomposition of equation (4.1). We denote the spectral flow in the $s(u(2) \times u(1))$ (resp. $\mathbb{C}^2$)

subspace suggestively by SF_t (resp. SF_n), where 't' stands for tangential and 'n' for normal.

The observation which underlies the construction of the $SU(3)$ Casson invariant is the following: *At a bifurcation where a new irreducible critical point pops out of a reducible one, an eigenvalue of multiplicity two in the* $\mathbb{C}^2$ *component crosses zero.* More precisely, if $(A_t, h_t)_{t\in(-\varepsilon,\varepsilon)}$ is a generic path of reducible connections such that A_t is h_t-perturbed flat, and so that a bifurcation occurs at $t = 0$, then $SF_n(A_t) = \pm 2$.

Let $\mathscr{M}_h \subset \mathscr{A}/\mathscr{G}$ be the h-perturbed flat moduli space, which consists of the gauge orbits of critical points of $cs + h$ for an admissible perturbation h. We can split $\mathscr{M}_h = \mathscr{M}_h^* \cup \mathscr{M}_h^{red} \cup \{[\theta]\}$ as a disjoint union of irreducible and reducible orbits and the trivial orbit.

Replacing the Morse indices of formula (3.1) with the corresponding spectral flows, we obtain, when $\mathscr{M}_h$ is generic,

$$\sum_{[A]\in\mathscr{M}_h^*} (-1)^{SF(\theta,A)} - \sum_{[B]\in\mathscr{M}_h^{red}} (-1)^{SF_t(\theta,B)} \left(\frac{SF_n(\theta,B)}{2}\right). \tag{4.2}$$

Equation (4.2) is the expression which generalizes to the gauge theory setting the discussion we carried out above in the finite-dimensional setting, and which presents the natural candidate for the definition of an $SU(3)$ Casson invariant following the approach of Taubes.

But there is a hitch, arising from the fact that the spectral flow is not well defined on gauge orbits. For the exponents of (-1), this is not a problem, since $SF(\theta,[A])$ is well defined modulo 12 and $SF_t(\theta,[B])$ is well defined modulo 8. But the gauge ambiguity of the normal spectral flow, which is only well defined modulo 4, makes formula (4.2) dependent on the gauge representatives chosen for the reducible perturbed flat orbits.

Several approaches have been proposed to correct this gauge ambiguity. In [**5**], the perturbation h is taken to be small and the correction term is defined by replacing $\frac{1}{2}SF_n(\theta,B)$ with $\frac{1}{2}(SF_n(\theta,B) - 4cs(\widehat{B}) + 2)$ in the second sum of (4.2). Here, $\widehat{B}$ is a genuine flat connection near B. This quantity does not depend on the choice of gauge representative B for the orbit $[B]$ because

$$SF_n(\theta, gB) - 4cs(g\widehat{B}) = SF_n(\theta,B) - 4cs(\widehat{B}).$$

This choice for the $SU(3)$ correction term is motivated by Walker's correction term for rational homology 3-spheres.

The resulting real valued invariant,

$$\lambda_{SU(3)} := \sum_{[A]\in\mathscr{M}_h^*} (-1)^{SF(\theta,A)} - \tfrac{1}{2} \sum_{[B]\in\mathscr{M}_h^{red}} (-1)^{SF_t(\theta,B)} \left(SF_n(\theta,B) - 4cs(\widehat{B}) + 2\right),$$

was shown in [**6**] to satisfy:

$$\begin{aligned} &\lambda_{SU(3)}(-X) = \lambda_{SU(3)}(X) \\ &\lambda_{SU(3)}(X\#Y) = \lambda_{SU(3)}(X) + \lambda_{SU(3)}(Y) + 16\lambda_{SU(2)}(X)\lambda_{SU(2)}(Y). \end{aligned} \tag{4.3}$$

The behavior of $\lambda_{SU(3)}$ under Dehn surgery appears to be very complicated. A rather delicate analysis was used to compute the values of $\lambda_{SU(3)}$ for surgeries on $(2,q)$-torus knots in [**7**], using analytical surgery methods to compute spectral flow. The data obtained from these calculations did not suggest any obvious surgery formula for $\lambda_{SU(3)}$. Moreover, the calculations show that $\lambda_{SU(3)}$ is not a finite-type invariant.

The value of $\lambda_{SU(3)}$ for Seifert fibered homology spheres is rational since the Chern-Simons invariant of flat connections on Seifert fibered spaces are known to be rational. But it is not generally known whether the value of $\lambda_{SU(3)}$ on homology 3-spheres is always a rational number. This is closely related to the conjectured rationality of the Chern-Simons invariants, which remains an open problem.

Cappell, Lee, and Miller presented an alternative method for correcting the gauge ambiguities and derived a new $SU(3)$ Casson invariant [**13**]. Their result is based on the observation that one can correct for the gauge ambiguities of (4.2) using the tangential spectral flow instead of the Chern-Simons invariant: the quantity $SF_n(\theta, B) - SF_t(\theta, B)/2$ is independent of the choice of gauge representative of $[B]$. Thus Cappell, Lee, and Miller solved the problem of gauge ambiguities by replacing $SF_n(\theta, B)$ in the second sum in (4.2) by $SF_n(\theta, B) - SF_t(\theta, B)/2 + 5/4$. This restores gauge invariance but destroys the independence of perturbation: one still needs to correct for births and deaths of pairs of *reducible* critical orbits. To handle this phenomenon, they add to

$$\sum_{[A]\in\mathscr{M}_h^*} (-1)^{SF(\theta,A)} - \tfrac{1}{2} \sum_{[B]\in\mathscr{M}_h^{red}} (-1)^{SF_t(\theta,B)} \left(SF_n(\theta, B) - SF_t(\theta, B)/2 + 5/4\right)$$

a third correction term involving the ranks of the boundary operators for the Floer instanton homology complex (on $S(U(2) \times U(1))$ connections). That the resulting expression is an invariant follows from the fact that Floer homology is itself independent of the perturbation, and hence a birth and death of a pair of reducibles coincides with a change (of ± 1) in the rank of some boundary operator.

The Cappell-Lee-Miller invariant λ_{CLM} takes values in $\frac{1}{4}\mathbb{Z}$ and also satisfies $\lambda_{CLM}(-X) = \lambda_{CLM}(X)$, but its behavior under connected sum is obscured by complications arising from the third correction term. These complications are closely related to the subtle problem of determining the Floer homology for connected sums.

In [**8**], the present authors developed a different method for defining a correction term, which we now describe in detail. Let $\mathscr{G}_0$ denote the identity component of $\mathscr{G}$, which consists of degree zero gauge transformations. Then $\mathscr{A}/\mathscr{G}_0 \longrightarrow \mathscr{A}/\mathscr{G}$ is a nontrivial $\mathbb{Z}$ covering and is classified by the Chern-Simons function. Thus spectral flow is well defined on $\mathscr{A}/\mathscr{G}_0$. This has the consequence that if $[A]$ and $[B]$ are close enough in $\mathscr{A}/\mathscr{G}$ then they have close lifts in $\mathscr{A}/\mathscr{G}_0$ and the spectral flow between these two lifts is well defined and independent of the lift (provided they are close). If $[A]$ and $[B]$ happen to be reducible and close, then we can also define the normal spectral flow between them and it is also independent of the choice of close lifts. In fact, for any connected subset $\mathscr{U} \subset \mathscr{A}^{red}/\mathscr{G}$ over which $\widetilde{\mathscr{U}} \longrightarrow \mathscr{U}$ is the trivial $\mathbb{Z}$ cover, we define the normal spectral flow $SF_n([A], [B])$ between any two gauge orbits $[A], [B] \in \mathscr{U}$ by simply choosing lifts of $[A]$ and $[B]$ and a path between them in the same connected component of $\widetilde{\mathscr{U}}$. For the remainder of this article, we adopt this convention for defining the normal spectral flow between reducible gauge orbits.

We apply this to the path components of $\mathscr{M}^{red}$. Let $\widetilde{\mathscr{M}}^{red} \subset \mathscr{A}/\mathscr{G}_0$ be the lift (i.e. inverse image) of the moduli space $\mathscr{M}^{red}$ of reducible flat $SU(3)$ connections. Then since

(i) $cs(g \cdot A) = cs(A) + \deg(g)$, and
(ii) $cs\colon \mathscr{A} \longrightarrow \mathbb{R}$ is constant on components of flat connections,

the cover $\widetilde{\mathscr{M}}^{red} \longrightarrow \mathscr{M}^{red}$ is trivial and so the normal spectral flow is well defined between gauge orbits lying in the same connected component of $\widetilde{\mathscr{M}}^{red}$ and, more generally, the normal spectral flow is defined between gauge orbits lying in a small enough neighborhood of a path component of $\mathscr{M}^{red}$.

The new correction term is defined in terms of $SF_n([B_i^\pm],[A])$ for fixed basepoints $[B_i^\pm]$ chosen as follows. Let X be a homology 3-sphere and index the components of the reducible flat moduli space $C_1,\ldots,C_n$. Each component C_i is compact, hence the normal spectral flow is bounded and one can choose orbits $[B_i^-]$ and $[B_i^+]$ in C_i for which the normal spectral flow $SF_n([B_i^-],[B_i^+])$ is maximal. If h is a small perturbation, then every reducible h-perturbed flat connection $[B]$ is close to one of the components C_i, and so there is an unambiguous notion of normal spectral flow from the basepoints $[B_i^\pm]$ to $[B]$. Although the basepoints need not be unique, the normal spectral flows are well defined since different choices $[\widehat{B}_i^\pm]$ satisfy

$$SF_n([B_i^-],[\widehat{B}_i^-]) = 0 = SF_n([B_i^+],[\widehat{B}_i^+]).$$

Given a small, nondegenerate perturbation h, we define $\tau_\pm(X)$ by

$$\tau_+(X) = \sum_{[A]}(-1)^{SF(\theta,A)} - \tfrac{1}{2}\sum_{i=1}^{n}\sum_{[B]}(-1)^{SF_t(\theta,B)} SF_n([B_i^+],[B])$$

$$\tau_-(X) = \sum_{[A]}(-1)^{SF(\theta,A)} - \tfrac{1}{2}\sum_{i=1}^{n}\sum_{[B]}(-1)^{SF_t(\theta,B)} \Big(SF_n([B_i^-],[B]) + \dim H^1_{B_i^-}(X;\mathbb{C}^2)\Big),$$

where the first and second sums in $\tau_\pm(X)$ are over $[A] \in \mathscr{M}_h^*$ and $[B] \in \mathscr{M}_h^{red} \cap \mathscr{U}_i$, and $\mathscr{U}_i \subset \mathscr{B}^{red}$ is a small open neighborhood of C_i.

Note that the normal spectral flow $SF_n([B_i^\pm],[B])$ is always even because it counts the (real) eigenvalues of a complex operator. This observation shows that

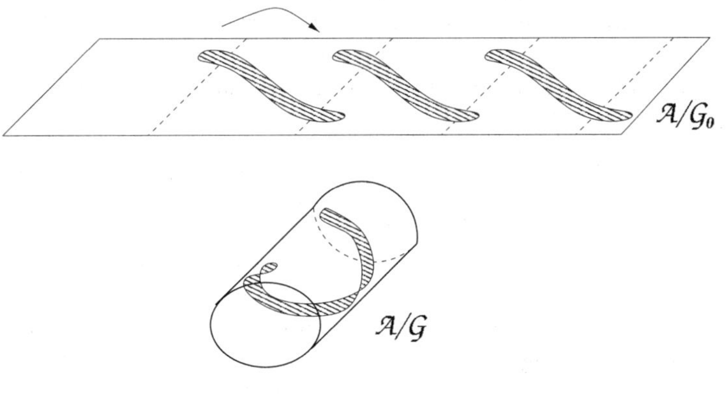

FIGURE 3. A SET $\mathscr{U}$ WITH TRIVIAL COVER. Given $[A],[B] \in \mathscr{U}$, compute the spectral flow by choosing lifts in the same component of $\widetilde{\mathscr{U}}$.

both $\tau_+(X)$ and $\tau_-(X)$ are integer valued. Under change in orientations, one can easily verify that $\tau_+(-X) = \tau_-(X)$. (The appearance of the $\dim H^1$ term in the definition of τ_- comes from our choice of convention for spectral flow.) Hence it is natural to define an orientation-independent invariant by averaging them. Thus we set

$$\begin{aligned}\tau_{SU(3)}(X) =& \tfrac{1}{2}(\tau_+(X) + \tau_-(X)) \\ =& \sum_{[A]} (-1)^{SF(\theta,A)} - \tfrac{1}{4} \sum_{i=1}^{n} \sum_{[B]} (-1)^{SF_t(\theta,B)} \Big(SF_n([B_i^+],[B]) \\ & \qquad + SF_n([B_i^-],[B]) + \dim H^1_{B_i^-}(X;\mathbb{C}^2)\Big),\end{aligned}$$

where the sums are defined as before.

We claim that $\tau_{SU(3)}(X)$ is integer valued. To prove this claim, it is sufficient to show that, for any $B \in \mathscr{M}_h^*$, the quantity

$$SF_n([B_i^+],[B]) + SF_n([B_i^-],[B]) + \dim H^1_{B_i^-}(X;\mathbb{C}^2)$$

is divisible by 4.

Identifying $\mathbb{C}^2$ with the quaternions and $SU(2)$ with the unit quaternions, the natural action of $SU(2)$ on $\mathbb{C}^2$ becomes left quaternionic multiplication. Since B_i^- is a flat $SU(2)$ connection, right quaternionic multiplication endows the cohomology group $H^1_{B_i^-}(X;\mathbb{C}^2)$ with a quaternionic structure, which implies that the dimension of this cohomology is a multiple of 4. Furthermore, for any $SU(2)$ connection, the twisted signature operator on $\mathbb{C}^2$ valued forms commutes with this right multiplication, so its eigenspaces are all quaternionic.

This leaves the two normal spectral flow terms. Additivity of the spectral flow under composition of paths gives that

$$SF_n([B_i^+],[B]) + SF_n([B_i^-],[B]) = 2SF_n([B_i^+],[B]) + SF_n([B_i^-],[B_i^+]).$$

The complex structure shows that $SF_n([B_i^+],[B])$ is even. Furthermore, since both B_i^+ and B_i^- are flat $SU(2)$ connections, they can be connected by a path of $SU(2)$ connections. As noted above, the corresponding path of twisted signature operators will have quaternionic eigenspaces, which implies that $SF_n([B_i^-],[B_i^+])$ is divisible by 4. This completes the proof of the integrality claim.

Thus $\tau_{SU(3)}$ is integer valued, invariant under change of orientation, and moreover Theorem 4 of [**8**] shows that

$$\tau_{SU(3)}(X\#Y) = \tau_{SU(3)}(X) + \tau_{SU(3)}(Y) + 16\lambda_{SU(2)}(X)\lambda_{SU(2)}(Y).$$

(The difference in coefficients here and in [**8**] is the result of our normalizing $\lambda_{SU(2)}(X)$ as Casson originally did here, whereas it was normalized according to Walker's convention [**24**] in [**8**].)

The invariant $\tau_{SU(3)}$ is vastly easier to compute than either $\lambda_{SU(3)}$ or λ_{CLM}. To see why, consider a homology 3-sphere X whose cohomology $H^1_\alpha(X;\mathbb{C}^2)$ vanishes for every $SU(2)$ representation $\alpha\colon X \to SU(2)$. Since the twisted signature operator at each reducible flat connection has no zero modes on $\mathbb{C}^2$ valued forms, all normal spectral flows $SF_n(B_i^\pm, B)$ vanish for all sufficiently small perturbations. This implies that the correction term for $\tau_{SU(3)}(X)$ vanishes for small perturbations, and thus $\tau_{SU(3)}(X)$ simply equals the quantity

$$\sum_{[A]\in\mathscr{M}_h^*} (-1)^{SF(\theta,A)}$$

obtained by perturbing the $SU(3)$ moduli space and counting only irreducible, perturbed flat gauge orbits. In this sense, the advantage of $\tau_{SU(3)}$ over $\lambda_{SU(3)}$ and λ_{CLM} is that the correction term for $\tau_{SU(3)}$ is nontrivial only when absolutely necessary. For example, the correction term vanishes for Brieskorn spheres of the form $\Sigma(2,p,q)$. This was used in the computations of $\tau_{SU(3)}(\Sigma(2,p,q))$ in [**8**]. In Section 5, we present computations of $\tau_{SU(3)}$ in cases where the correction term does not vanish.

5. Computations for Brieskorn spheres $\Sigma(p,q,r)$

In this section, we outline a method for computing $\tau_{SU(3)}$ on Brieskorn homology 3-spheres [**9**].

5.1. SU(3) representations of Brieskorn 3-spheres. Fix pairwise relatively prime numbers p,q,r and consider the Brieskorn homology 3-sphere

$$\Sigma(p,q,r) = \{(x,y,z)\in\mathbb{C}^3 \mid x^p+y^q+z^r=0\}\cap S^5_\epsilon.$$

Taking a,b,c with

$$aqr+bpr+cpq=1,$$

the fundamental group of $\Sigma(p,q,r)$ has a presentation

$$\pi_1(\Sigma(p,q,r)) = \langle x,y,z,h \mid h \text{ central}, x^p=h^a, y^q=h^b, z^r=h^c, xyz=1\rangle.$$

The next result is not difficult to prove; details can be found in [**9**].

PROPOSITION 5.1. *If $\alpha\colon \pi_1\Sigma(p,q,r)\to SU(3)$ is irreducible, then $\alpha(h) = e^{2\pi ik/3}I$ is central. If $\alpha\colon \pi_1\Sigma(p,q,r)\to SU(3)$ is reducible, then up to conjugation* image$(\alpha)\subset SU(2)\times\{1\}$ *and*

$$\alpha(h) = \begin{bmatrix} \pm1 & 0 & 0\\ 0 & \pm1 & 0\\ 0 & 0 & 1\end{bmatrix}.$$

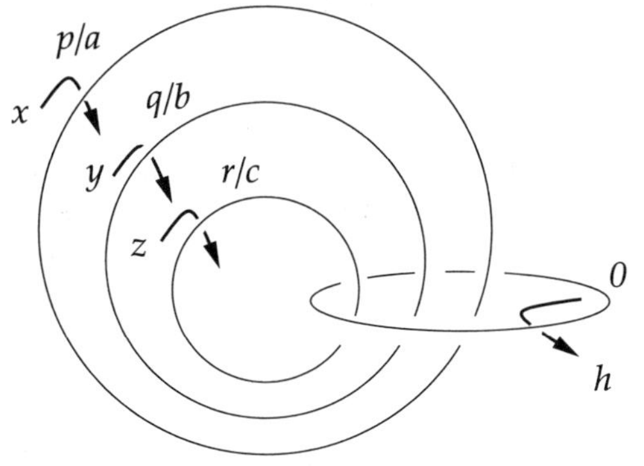

FIGURE 4. THE BRIESKORN SPHERE $\Sigma(p,q,r)$ AS SURGERY ON A LINK.

FIGURE 5. COMPONENTS OF TYPES III AND IV IN $R(\Sigma, SU(3))$.

Path components of $R(\Sigma(p,q,r), SU(3))$ are indexed by fixing $\alpha(h)$ and choosing $\alpha(x), \alpha(y)$ and $\alpha(z)$ to be p-th, q-th and r-th roots of $\alpha(h^a)$, $\alpha(h^b)$ and $\alpha(h^c)$, respectively.

Using the Seifert fibration $\Sigma(p,q,r) \to S^2$, one can interpret $R(\Sigma(p,q,r), SU(3))$ in terms of moduli spaces of parabolic bundles over S^2 with 3 marked points. This allows us to make use of the description of these moduli spaces given in [**4**] and [**19**]. In these articles it is shown that each path component of $R(\Sigma(p,q,r), SU(3))$ is homeomorphic to either an isolated point or a 2-sphere. These can be further broken down into 4 types:

(1) *Type I* consist of isolated irreducible $SU(3)$ representations,
(2) *Type II* are isolated reducible $SU(3)$ representations,
(3) *Type III* are smooth 2-spheres consisting entirely of irreducible $SU(3)$ representations, and
(4) *Type IV* are 2-spheres which have one non-smooth point: the smooth points are irreducible $SU(3)$ representations and the non-smooth point is a reducible representation. These components cause the most trouble: we call them *pointed 2-spheres.*

When $p = 2$, there are no 2-sphere components. In this case, $\tau_{SU(3)}(\Sigma(p,q,r))$ is simply a count of the Type I points; the Type II points do not contribute to $\tau_{SU(3)}$ since $H^1_\alpha(\Sigma; \mathbb{C}^2) = 0$ at these points (see the last paragraph of Section 4). The count of the Type I points is carried out in [**4**], where it is shown that each Type I point contributes positively to $\tau_{SU(3)}$. Table 1 gives some calculations of $\tau_{SU(3)}(\Sigma(2,q,r))$ for various q and r.

5.2. Twisting perturbations. If $p,q,r > 2$, then $R(\Sigma(p,q,r), SU(3))$ always contains pointed 2-spheres. Thus perturbations are needed to resolve them. In [**9**] we use a preliminary step to improve the moduli space $\mathscr{M}$: we show that a certain class of *twisting* perturbations resolve each Type IV component into the union of one Type II point and one Type III smooth 2-sphere. Rather than to introduce the twisting perturbations explicitly here, we describe their effect.

The starting point is that pertubations alter the flatness equations in the neighborhood of a knot. A perturbation function $h\colon \mathscr{A} \to \mathbb{R}$ takes a connection A to a number determined by the holonomies of the A along curves $S^1 \times \{z\}$ in a solid torus $S^1 \times D^2$ contained in the homology 3-sphere. Floer proved that for such a perturbation h, an h-perturbed flat connection is in fact flat outside the solid torus [**16**].

Let $Y = S^1 \times D^2$ be a neighborhood of the singular r-fiber in $\Sigma = \Sigma(p,q,r)$ and set $Z = \Sigma - Y$. Then

$$\Sigma = Y \cup_T Z.$$

We perturb the flatness equations in the solid torus Y and study the effect on a pointed 2-sphere. To better visualize this situation, we regard $\mathscr{M}(\Sigma)$ and $\mathscr{M}_h(\Sigma)$ as the pullback of an intersection in the 4-dimensional representation space of the torus

$$\mathscr{M}(T) = T^2 \times T^2/S_3,$$

where S_3 acts diagonally. Let h denote the twisting perturbation and let h_t be a path from 0 to h. Consider the diagram

$$\begin{array}{ccccc} & & \mathscr{M}_{h_t}(Y) & & \\ & \nearrow & & \searrow^{i^*} & \\ \mathscr{M}_{h_t}(\Sigma) & & & & \mathscr{M}(T) \\ & \searrow & & \nearrow_{j^*} & \\ & & \mathscr{M}(Z) & & \end{array}$$

The flat moduli space of Z near an intersection giving rise to a Type IV component has the form of a union of a 2-dimensional reducible stratum and a 2-parameter family of 2-spheres. This second piece consists of all irreducible points except for a single point on each of the spheres along a curve in the 2-parameter family, and this arc of reducible points is identified with a curve in the 2-dimensional reducible stratum. The restriction map $j^* : \mathscr{M}(Z) \to \mathscr{M}(T)$ collapses 2-spheres to points. See Figure 6.

When $t = 0$, $\mathscr{M}_0(\Sigma)$ contains a pointed 2-sphere, corresponding to an intersection point of $i^*\mathscr{M}_0(Y)$ and $j^*\mathscr{M}(Z)$. The important property of the twisting perturbation is that as t increases, this intersection point moves transversely to the image of $\mathscr{M}^*(Z)$ under j^*. Thus, for $t > 0$, $j^*|_{\mathscr{M}^*(Z)}$ is a submersion with nondegenerate 2-sphere fibers over the intersection point of $i^*\mathscr{M}_{h_t}(Y)$ and $j^*\mathscr{M}(Z)$.

Suppose $[B_0]$ is the reducible flat connection on the pointed 2-sphere. We claim that this point persists under the twisting perturbation. To see why, consider the intersection of $\mathscr{M}_{h_t}(Y)$ and $\mathscr{M}^{red}(Z)$ in $\mathscr{M}(T)$. At $t = 0$, $\mathscr{M}_0(Y) \cap \mathscr{M}^{red}(Z) =$

Σ	$\tau_{SU(3)}(\Sigma)$
$\Sigma(2,3,6k \pm 1)$	$3k^2 \pm k$
$\Sigma(2,5,10k \pm 1)$	$33k^2 \pm 9k$
$\Sigma(2,5,10k \pm 3)$	$33k^2 \pm 19k + 2$
$\Sigma(2,7,14k \pm 1)$	$138k^2 \pm 26k$
$\Sigma(2,7,14k \pm 3)$	$138k^2 \pm 62k + 4$
$\Sigma(2,7,14k \pm 5)$	$138k^2 \pm 102k + 16$
$\Sigma(2,9,18k \pm 1)$	$390k^2 \pm 58k$
$\Sigma(2,9,18k \pm 5)$	$390k^2 \pm 210k + 24$
$\Sigma(2,9,18k \pm 7)$	$390k^2 \pm 298k + 52$

TABLE 1. CALCULATIONS OF THE INTEGER VALUED SU(3) CASSON INVARIANT FOR BRIESKORN SPHERES $\Sigma(2,q,r)$.

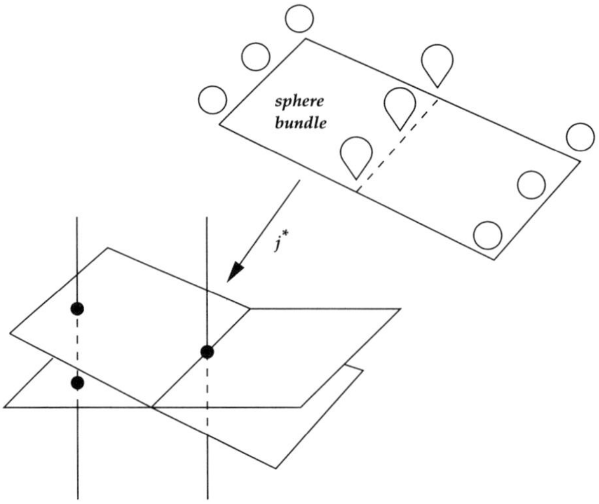

FIGURE 6. THE EFFECT OF A TWISTING PERTURBATION. The two intersecting 2-planes here represent the images of $\mathscr{M}^{red}(Z)$ and $\mathscr{M}^*(Z)$ in $\mathscr{M}(T)$. The two vertical lines, from left to right, represent the images of $\mathscr{M}_h(Y)$ and $\mathscr{M}(Y)$, which are 2-dimensional. Also depicted is the 2-sphere bundle $\mathscr{M}^*(Z)$ being mapped under j^* to the slanted 2-plane.

$\{[B_0]\}$, but the map $\mathscr{M}^{red}(Z) \to \mathscr{M}(T)$ is transverse to $\mathscr{M}_0(Y)$ at $[B_0]$ along the reducible stratum. The situation is illustrated in Figure 6. Thus, after perturbing, the intersection $\mathscr{M}_h(\Sigma) = \mathscr{M}(Z) \cap \mathscr{M}_h(Y)$ near the pointed sphere consists of one isolated reducible orbit and a smooth nondegenerate 2-sphere of irreducible orbits (see Figure 7).

Note that the perturbed Chern-Simons function is not generic, since there are critical 2-spheres. However, the 2-sphere is a Bott-Morse critical submanifold, and it is a general fact that such submanifolds contribute their Euler characteristic to the $SU(3)$ Casson invariant, up to sign. Precisely, a Bott-Morse smooth critical submanifold $M \subset \mathscr{M}_h^*$ contributes $\chi(M) \cdot (-1)^{SF(\theta,A)}$ where A is any connection whose gauge orbit lies on M. In the present case, $\chi(S^2) = 2$ and $SF(\theta, A)$ was shown to be even in [**4**]. Thus this 2-sphere contributes 2 to $\tau_{SU(3)}(\Sigma)$. A similar comment applies to compute the contribution from the Type III smooth 2-sphere components, which remain smooth 2-spheres for small enough perturbations. The Type I isolated irreducible points remain isolated irreducible for sufficiently small perturbations.

The results of [**4**] then apply to calculate the contribution of the Type I, Type III, and the resolved smooth 2-spheres coming from the Type IV components. The Type II components do not contribute to $\tau_{SU(3)}(\Sigma)$. To complete the calculation it remains to compute the contribution of the reducible point in $\mathscr{M}_h(\Sigma)$ to the second term of $\tau_{SU(3)}(\Sigma)$ coming from the resolution of the Type IV components.

5.3. Normal spectral flow along the path of twisted perturbed flat reducible connections. For $0 \leq t \leq \varepsilon$, let B_t be the path of h_t-perturbed flat reducible connections near a pointed 2-sphere, i.e. B_t corresponds to the intersection

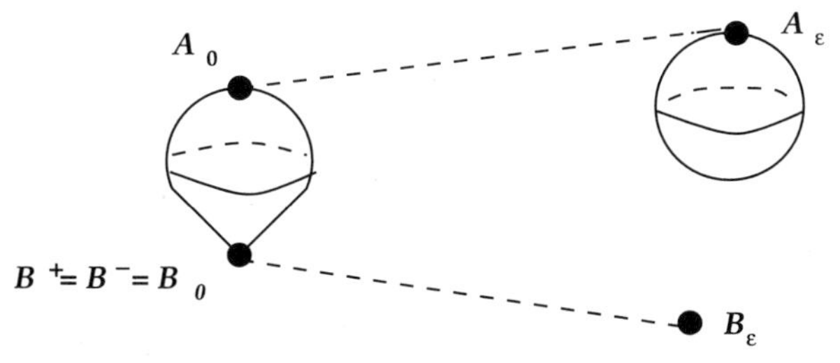

FIGURE 7. A POINTED 2-SPHERE RESOLVES INTO A SMOOTH 2-SPHERE AND A REDUCIBLE POINT.

of $\mathscr{M}_{h_t}(Y)$ with $\mathscr{M}(Z)$ in $\mathscr{M}(T)$ near the singular point of the pointed 2-sphere. In light of the definition of the second term of $\tau_{SU(3)}$, we need to calculate the normal spectral flow $SF_n(B^{\pm}, B_{\varepsilon})$. Notice that B_0 is isolated in the reducible flat moduli space, and so we take $B^+ = B^- = B_0$.

To calculate $SF_n(B_0, B_\epsilon)$, we split the spectral flow according to the manifold decomposition

$$\Sigma = Y \cup_T Z.$$

We use Atiyah-Patodi-Singer boundary conditions on Y and Z to get self-adjoint operators. Consider first a connection A on Y whose restriction to a collar of the boundary is flat, and let $a = A|_T$. The tangential operator S_a of the odd signature operator K_A is

$$S_a \colon \Omega^{0+1+2}(T; \mathbb{C}^2) \to \Omega^{0+1+2}(T; \mathbb{C}^2)$$

$$S_a(\alpha, \beta, \gamma) = (*d_a\beta, - * d_a\alpha - d_a * \gamma, d_a * \beta)\,.$$

(In this formula, $d_a \colon \Omega^i(T;\mathbb{C}^2) \to \Omega^{i+1}(T;\mathbb{C}^2)$ is the exterior derivative associated to the connection a.) After a gauge transformation, if needed, the formula $K_A = \sigma(du)(S_a + \frac{\partial}{\partial u})$ holds in a collar $T \times I$ of the boundary, where σ denotes the symbol of K_A and u is the inward normal coordinate.

In this context Atiyah, Patodi and Singer in [**2**] showed that the operator K_A acting on those forms in $\Omega^{0+1}(Y;\mathbb{C}^2)$ whose restriction to the boundary lie in the positive eigenspan P_a^+ of S_a is Fredholm, and self-adjoint if in addition $\ker S_a = 0$. When a is flat, $\ker S_a$ is isomorphic via the Hodge theorem to $H_a^{0+1+2}(T;\mathbb{C}^2)$. This cohomology group vanishes for nontrivial flat connections a.

Applying this to the path B_t restricted to Y gives a family of self-adjoint operators whose (normal) spectral flow is denoted by $SF_n(B_t; P^+; Y; 0 \leq t \leq \varepsilon)$. Similar considerations on Z (taking into account the change in orientation of the inward normal) gives the spectral flow $SF_n(B_t; P^-; Z; 0 \leq t \leq \varepsilon)$. A theorem of Bunke ([**12**], see [**14**] for general splitting theorems for spectral flow) implies that

$$SF_n(B_t;\Sigma;0 \leq t \leq \varepsilon) = SF_n(B_t;P^+;Y;0 \leq t \leq \varepsilon) + SF_n(B_t;P^-;Z;0 \leq t \leq \varepsilon).$$

THEOREM 5.2. *Let B_t be the path of reducible perturbed flat connections near a pointed 2-sphere described above. Then*

(i) $SF_n(B_t, P^+; Y; 0 \leq t \leq \varepsilon) = 0$,
(ii) $SF_n(B_t, P^-; Z; 0 \leq t \leq \varepsilon) = -2$.

Hence each pointed 2-sphere contributes $+2$ to $\tau_{SU(3)}(\Sigma(p,q,r))$.

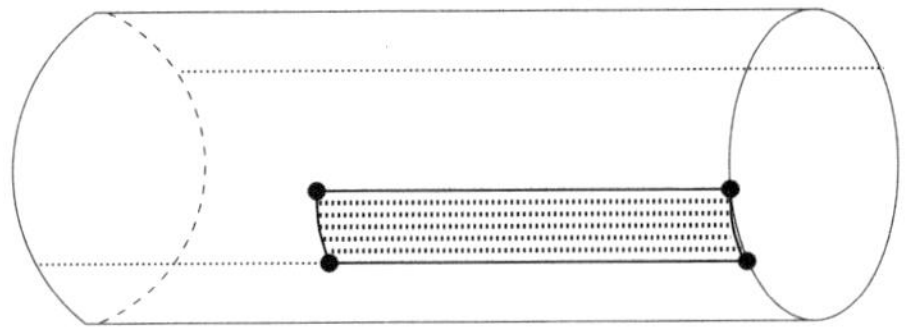

FIGURE 8. WHEN p AND q ARE BOTH ODD, THE COMPONENTS OF REDUCIBLE $SU(3)$ REPRESENTATIONS OF $\pi_1 Z$ ARE HOMEOMORPHIC TO CYLINDERS $S^1 \times [0,1]$.

Sketch of proof. Part (i) follows from vanishing of cohomology $H^{0+1}_{B_0}(Y;\mathbb{C}^2)$ and is left to the reader. Instead, we outline the argument for part (ii), which uses additivity of the spectral flow under composition of paths.

The space of conjugacy classes of $SU(2)$ representations of $\pi_1 Z$ is a union of arcs $\rho_s, s \in [0,1]$. The endpoints of these arcs are abelian $SU(2)$ representations. (These facts are proved in [**21**].) Since $H_1(Z;\mathbb{Z}) = \mathbb{Z}$, we can twist each $SU(2)$ representation ρ_s to get a representation $\pi_1 Z \to U(2)$ by simply multiplying ρ_s by the (unique) abelian representation $\chi_t \colon \pi_1 Z \to U(1)$ with $\chi_t(\gamma) = e^{2\pi i t}$ for γ generator of $H_1(Z;\mathbb{Z})$.

Identifying $U(2)$ with the subgroup $S(U(2) \times U(1)) \subset SU(3)$, the family $\alpha_{s,t} = \chi_t \rho_s$, parameterizes a family of reducible $SU(3)$ representations where $(s,t) \in [0,1] \times (-\epsilon, \epsilon)$ and $\epsilon > 0$ is chosen small.

The restriction to Z of the path B_t is a path of flat connections whose holonomy is essentially of the form $\alpha_{s_0,t}$ for some fixed $s_0 \in (0,1)$. Thus there is a two parameter family $B_{s,t}$ of flat connections on Z defined for $(s,t) \in [0,1] \times [0,\varepsilon]$ with $B_{s_0,t} = B_t$ and $B_{1,t}$ abelian.

Using the invariance of spectral flow with respect to homotopy rel endpoints and the additivity with respect to composition of paths, we have:

$$\begin{aligned} SF(B_t; P^-; Z; 0 \le t \le \varepsilon) &= SF(B_{s_0,t}; P^-; Z; 0 \le t \le \varepsilon), \\ &= SF(B_{s,0}; P^-; Z; s_0 \le s \le 1) \\ &\quad + SF(B_{1,t}; P^-; Z; 0 \le t \le \varepsilon) \\ &\quad - SF(B_{s,1}; P^-; Z; s_0 \le s \le 1), \\ &= SF(B_{1,t}; P^-; Z; 0 \le t \le \varepsilon). \end{aligned}$$

To see the last step, note that $\dim H^1_{B_{s,0}}(Z;\mathbb{C}^2) = 4$ and $\dim H^1_{B_{s,1}}(Z;\mathbb{C}^2) = 0$ are both constant in s, hence $SF(B_{s,0}; P^-; Z; s_0 \le s \le 1)$ and $SF(B_{s,1}; P^-; Z; s_0 \le s \le 1)$ vanish.

Now $\{[B_{1,t}] \mid 0 \le t \le \varepsilon\} \subset \mathcal{M}^{ab}(Z)$ is a path of connections with abelian holonomy. The operator $K_{B_{1,t}}$ on Z with P^- boundary conditions has four zero modes at $t = 0$ and no zero modes for $0 < t \le \varepsilon$. Working with abelian connections is simpler and one can show (using for example Levine-Tristam signatures, as in [**20**]) that as t increases from $t = 0$, two of the zero modes go up and the other two go down, hence (with our convention for spectral flow)

$$SF_n(B_t; Z; P^-; 0 \le t \le \varepsilon) = SF(B_{1,t}; P^-; Z; 0 \le t \le \varepsilon) = -2.$$

This establishes (ii).

Σ	$\tau_{SU(3)}(\Sigma)$
$\Sigma(3,4,12k\pm 1)$	$105k^2 \pm 21k$
$\Sigma(3,4,12k\pm 5)$	$105k^2 \pm 87k+16$
$\Sigma(3,5,15k\pm 1)$	$276k^2 \pm 40k$
$\Sigma(3,5,15k\pm 2)$	$276k^2 \pm 74k+2$
$\Sigma(3,5,15k\pm 4)$	$276k^2 \pm 148k+16$
$\Sigma(3,5,15k\pm 7)$	$276k^2 \pm 254k+56$

TABLE 2. CALCULATIONS OF THE INTEGER VALUED SU(3) CASSON INVARIANT FOR BRIESKORN SPHERES $\Sigma(p,q,r)$ WITH $p>2$.

Using Bunke's theorem, the fact that $B^+ = B^- = B_0$ in this case, the computation that $\dim H^1_{B_0}(\Sigma;\mathbb{C}^2) = 4$, and the definition of the correction term for $\tau_{SU(3)}$, it follows that the perturbed flat reducible gauge orbit $[B_\varepsilon]$ contributes

$$\begin{aligned}-\tfrac{1}{4}(-1)^{SF(\theta,B_\varepsilon)}&\left(2SF_n(B_0,B_\varepsilon)+\dim H^1_{B_0}(\Sigma;\mathbb{C}^2)\right)\\ &=-\tfrac{1}{4}(-1)^{2SF(\theta,B_\varepsilon)}\left(2(-2)+4\right)=0\end{aligned}$$

to $\tau_{SU(3)}(\Sigma)$. Thus the only contribution of a pointed 2-sphere to $\tau_{SU(3)}(\Sigma)$ comes from the smooth 2-sphere of irreducibles obtained after perturbing which contributes $\chi(S^2)=2$ to (the first term in the definition of) $\tau_{SU(3)}(\Sigma)$. □

Table 2 gives some sample computations of the integer valued Casson invariant. Let $K_{p,q}$ = the (p,q) torus knot and set $X_n = 1/n$ Dehn surgery on $K_{p,q}$. Then $X_n = \pm\Sigma(p,q,r)$ for $r=|pqn-1|$. Table 3 gives the value of $\tau_{SU(3)}(X_n)$ for various p,q. These calculations suggest several conjectures. First of all, in all these computations, $\tau_{SU(3)}(\Sigma)$ is even, which lends evidence to the following conjecture of [**8**].

CONJECTURE 5.3. *$\tau_{SU(3)}(X)$ is even for all integral homology 3-spheres.*

One reason for believing Conjecture 5.3 (apart from the empirical evidence) is the existence of a natural involution on $\mathscr{M}_{SU(3)}$ induced by complex conjugation on $SU(3)$ and $su(3)$. For example, one can use the involution to prove the conjecture under the hypothesis that the flat moduli space $\mathscr{M}_{SU(3)}$ is regular, which is the condition that $H^1_\alpha(X;su(3))=0$ for every nontrivial representation $\alpha:\pi_1 X\to SU(3)$).

A second pattern observed in the data is quadratic growth in the surgery coefficient for successive surgeries on a fixed knot.

CONJECTURE 5.4. *If $K\subset X$ is a knot in a homology 3-sphere and X_n is the result of $1/n$ Dehn surgery on K, then the limit*

$$\lim_{n\to\infty}\tau_{SU(3)}(X_n)/n^2 = A(K)$$

exists and depends only on the knot K.

$p=2$	$\tau_{SU(3)}(X_n)$	$p=3$	$\tau_{SU(3)}(X_n)$	$p=4$	$\tau_{SU(3)}(X_n)$
$K_{2,3}$	$3n^2-n$	$K_{3,4}$	$105n^2-21n$	$K_{4,5}$	$1011n^2-111n$
$K_{2,5}$	$33n^2-9n$	$K_{3,5}$	$276n^2-40n$	$K_{4,7}$	$4110n^2-320n$
$K_{2,7}$	$138n^2-26n$	$K_{3,7}$	$1128n^2-124n$	$K_{4,9}$	$11\,490n^2-712n$
$K_{2,9}$	$390n^2-58n$	$K_{3,8}$	$1953n^2-179n$	$K_{4,11}$	$25\,935n^2-1297n$
$K_{2,11}$	$885n^2-107n$	$K_{3,10}$	$4851n^2-367n$	$K_{4,13}$	$50\,925n^2-2171n$
$K_{2,13}$	$1743n^2-179n$	$K_{3,11}$	$7140n^2-476n$	$K_{4,15}$	$90\,636n^2-3320n$
$K_{2,15}$	$3108n^2-276n$	$K_{3,13}$	$14\,028n^2-812n$	$K_{4,17}$	$149\,940n^2-4888n$
$K_{2,17}$	$5148n^2-404n$	$K_{3,14}$	$18\,915n^2-993n$	$K_{4,19}$	$234\,405n^2-6789n$
$K_{2,19}$	$8055n^2-565n$	$K_{3,16}$	$32\,385n^2-1517n$	$K_{4,21}$	$350\,295n^2-9231n$
$K_{2,21}$	$12\,045n^2-765n$	$K_{3,17}$	$41\,328n^2-1788n$	$K_{4,23}$	$504\,570n^2-12\,072n$
$K_{2,23}$	$17\,358n^2-1006n$	$K_{3,19}$	$64\,620n^2-2544n$	$K_{4,25}$	$704\,886n^2-15\,600n$
$K_{2,25}$	$24\,258n^2-1294n$	$K_{3,20}$	$79\,401n^2-2923n$	$K_{4,27}$	$959\,595n^2-19\,569n$
$K_{2,27}$	$33\,033n^2-1631n$	$K_{3,22}$	$116\,403n^2-3951n$	$K_{4,29}$	$1\,277\,745n^2-24\,363n$

TABLE 3. THE INTEGER VALUED SU(3) CASSON INVARIANT FOR HOMOLOGY 3-SPHERES X_n OBTAINED BY $1/n$ SURGERY ON $K_{p,q}$

The data from Table 3 even gives a precise (conjectural) formula for $A(K)$ for torus knots:

$$A(K_{p,q}) = \frac{(p^2-1)(q^2-1)(2p^2q^2-3p^2-3q^2-3)}{240},$$

which is consistent with the more general formula (also conjectural) that

$$A(K) = 6c_4(K) + 3c_2(K)^2, \tag{5.1}$$

where $\Delta_K(z) = \sum_{i\geq 0} c_{2i}(K)z^{2i}$ is the Conway polynomial of K. Based on Frohman's work [**17**], formula (5.1) is essentially equivalent to the conjecture that $A(K)$ equals is six times the Frohman-Nicas $SU(3)$ knot invariant, at least for fibered knots (cf. [**18, 10**]).

Table 3 suggests a stronger result, namely that

$$\tau_{SU(3)}(X_n) = A(K)n^2 - B(K)n + \tau_{SU(3)}(X) \tag{5.2}$$

is a quadratic polynomial. Here, the assumptions are as before: $K \subset X$ is a knot in a homology 3-sphere and X_n is the result of performing $1/n$ Dehn surgery on K. For example, interpolating the data from Table 3,we get the formula

$$B(K_{2,q}) = \begin{cases} \frac{1}{12}(q^3-4q+3) & \text{if } q\equiv 1 \mod 4, \\ \frac{1}{12}(q^3-4q-3) & \text{if } q\equiv 3 \mod 4, \end{cases}$$

for $p=2$. Further, for $p=3$ and $p=4$, we get the formulas

$$B(K_{3,q}) = \begin{cases} \frac{1}{54}(20q^3+3q^2-48q+25) & \text{if } q\equiv 1 \mod 6, \\ \frac{1}{54}(20q^3-3q^2-48q+2) & \text{if } q\equiv 2 \mod 6, \\ \frac{1}{54}(20q^3+3q^2-48q-2) & \text{if } q\equiv 4 \mod 6, \\ \frac{1}{54}(20q^3-3q^2-48q-25) & \text{if } q\equiv 5 \mod 6, \end{cases}$$

and

$$B(K_{4,q}) = \begin{cases} \frac{1}{16}(16q^3 + q^2 - 42q + 25) & \text{if } q \equiv 1 \mod 8, \\ \frac{1}{16}(16q^3 - q^2 - 42q + 39) & \text{if } q \equiv 3 \mod 8, \\ \frac{1}{16}(16q^3 + q^2 - 42q - 39) & \text{if } q \equiv 5 \mod 8, \\ \frac{1}{16}(16q^3 - q^2 - 42q - 25) & \text{if } q \equiv 7 \mod 8. \end{cases}$$

The complexity of these formulas makes it difficult to guess a general formula for $B(K)$ in terms of classical invariants of the knot.

Notice that the conjectured formula (5.1) for $A(K)$ is not consistent with the conjecture (5.2) that $\tau_{SU(3)}(X_n)$ is a quadratic polynomial in the surgery coefficient. We explain this point using the well known fact that, for any amphicheiral knot K in S^3, $X_{-n} = -X_n$. Since $\tau_{SU(3)}(-X) = \tau_{SU(3)}(X)$, the conjectured formula (5.2) would imply that $B(K) = 0$ and that $\tau_{SU(3)}(X_n) = A(K)n^2$ for amphicheiral knots in S^3.

The figure eight knot K is amphicheiral, and ± 1 surgery on K gives the Brieskorn sphere $\pm\Sigma(2,3,7)$. Hence $\tau_{SU(3)}(X_{\pm 1}) = 4$. Assuming Conjecture 5.4 is true (so there is a quadratic growth coefficient $A(K)$), the formula (5.1) predicts that $A(K) = 3$ (because the Conway polynomial of K equals $1 - z^2$). But $B(K) = 0$, $\tau_{SU(3)}(S^3) = 0$ and the value of $\tau_{SU(3)}(X_{\pm 1})$ for ± 1 surgery on K equals 4, hence $A(K) = 4$. So if $\tau_{SU(3)}(X_n)$ is a quadratic polynomial as in (5.2), then formula (5.1) for $A(K)$ is incorrect.

Finding a direct method for computing $\tau_{SU(3)}$ for surgeries on the figure eight knot is an important step that promises to help determine the correct surgery formula for the integer valued $SU(3)$ Casson invariant.

Acknowledgements The authors gratefully acknowledge support from the National Science Foundation grants DMS-9971578, DMS-0072564 and DMS-9971020.

References

[1] S. Akbulut and J. McCarthy, Casson's invariant for oriented homology 3-spheres, an exposition, Math. Notes, Vol. **36**, Princeton University Press, Princeton, NJ, 1990.

[2] M. F. Atiyah, V. K. Patodi, and I. M. Singer, *Spectral asymmetry and Riemannian geometry. I.* Math. Proc. Cambridge Philos. Soc. **77** (1975), no. 1 43–69.

[3] M. F. Atiyah, V. K. Patodi, and I. M. Singer, *Spectral asymmetry and Riemannian geometry. II.* Math. Proc. Cambridge Philos. Soc. **78** (1975), no. 3 405–432.

[4] H. U. Boden, *Unitary representations of Brieskorn spheres,* Duke J. Math. **75** (1994) 193–220.

[5] H. U. Boden and C. M. Herald, *The SU(3) Casson invariant for integral homology 3-spheres,* J. Differential Geom. **50** (1998) 147–206.

[6] H. U. Boden and C. M. Herald, *A connected sum formula for the SU(3) Casson invariant,* J. Differential Geom. **53** (1999) 443–465.

[7] H. U. Boden, C. M. Herald, P. Kirk, and E. P. Klassen, *Gauge theoretic invariants of Dehn surgeries on knots,* Geom. Topology **5** (2001) 143–226.

[8] H. U. Boden, C. M. Herald and P. Kirk, *An integer valued* $SU(3)$ *Casson invariant,* Math. Research Letters **8** (2001) 589–602.

[9] H. U. Boden, C. M. Herald and P. Kirk, *The integer valued* $SU(3)$ *Casson invariant for Brieskorn homology spheres,* in preparation.

[10] H. U. Boden and A. Nicas, *Universal formulae for SU(n) Casson invariants of knots,* Trans. Amer. Math. Soc. **352** (2000) 3149–3187.

[11] B. Booss-Bavnbek, M. Lesch, and J. Phillips, *Unbounded Fredholm Operators and Spectral Flow,* 2001 preprint, math.FA/0108014.

[12] U. Bunke, *On the gluing problem for the η-invariant,* J. Differential Geom. **41** (1995), no. 2, 397–448.

[13] S. Cappell, R. Lee, and E. Miller, *A perturbative $SU(3)$ Casson invariant,* math.DG/0006018, Comment. Math. Helv. **77** (2002), no. 3, 491–523.

[14] A. M. Daniel and P. Kirk, *A general splitting formula for the spectral flow.* With an appendix by K. P. Wojciechowski. Michigan Math. J. **46** (1999), no. 3, 589–617.

[15] S. Donaldson and P. Kronheimer, *The Geometry of Four-Manifolds,* Oxford Mathematical Monographs, Clarendon Press, Oxford, 1990.

[16] A. Floer, *An instanton invariant for 3-manifolds,* Comm. Math. Phys. **118** (1989) 215–240.

[17] C. Frohman, *Unitary representations of knot groups*, Topology **32** (1993) 121–144.

[18] C. Frohman and A. Nicas, *An intersection homology invariant for knots in a rational homology 3-sphere*, Topology **33** (1994) 123–158.

[19] M. Hayashi, *The moduli space of SU(3)-Flat connections and the fusion rules,* Proc. Amer. Math. Soc., **127**, no. 5 (1999), 1545–1555.

[20] P. Kirk, E. Klassen, and D. Ruberman, *Splitting the spectral flow and the Alexander matrix,* Comment. Math. Helv. **69** (1994), no. 3, 375–416.

[21] E. Klassen, *Representations of knot groups in* SU(2), Trans. Amer. Math. Soc. **326** (1991), no. 2, 795–828.

[22] C. Lescop, Global surgery formula for the Casson-Walker invariant, Annals of Math Studies **140**, Princeton University Press, Princeton, NJ, 1996.

[23] C. Taubes, *Casson's invariant and gauge theory,* J. Differential Geom. **31** (1990) 547–599.

[24] K. Walker, An extension of Casson's invariant, Annals of Math Studies **126**, Princeton University Press, Princeton, NJ, 1992.

[25] A. Wasserman, *Equivariant differential topology,* Topology **8** (1969) 127–150.

Department of Mathematics & Statistics, McMaster University, Hamilton, Ontario L8S-4K1, Canada

E-mail address: `boden@mcmaster.ca`

Department of Mathematics/084, University of Nevada, Reno NV 89557-0045, USA

E-mail address: `herald@unr.edu`

Department of Mathematics, Indiana University, Bloomington IN 47405, USA

E-mail address: `pkirk@indiana.edu`

Proceedings of Symposia in Pure Mathematics
Volume **71**, 2003

On Donaldson and Seiberg-Witten invariants

Paul M. N. Feehan and Thomas G. Leness

ABSTRACT. We sketch a proof of Witten's formula relating the Donaldson and Seiberg-Witten series modulo powers of degree $c+2$, with $c = -\frac{1}{4}(7\chi + 11\sigma)$, for four-manifolds obeying some mild conditions, where χ and σ are their Euler characteristic and signature. We use the moduli space of SO(3) monopoles as a cobordism between a link of the Donaldson moduli space of anti-self-dual SO(3) connections and links of the moduli spaces of Seiberg-Witten monopoles. Gluing techniques allow us to compute contributions from Seiberg-Witten moduli spaces lying in the first (or 'one-bubble') level of the Uhlenbeck compactification of the moduli space of SO(3) monopoles.

1. Introduction

This article consists of lightly edited notes for a lecture by the first author at the International Georgia Topology Conference 2001. Although we shall only briefly mention technical details and qualifications appropriate for more complete accounts published elsewhere [**10**], [**11**], [**15**], we hope that these notes provide a convenient survey of our recent work on the SO(3)-monopole program.

1.1. Witten's conjecture. Two kinds of invariants can be used to explore the classification problem for compact, smooth 4-manifolds:

- *Donaldson invariants*, defined using an SO(3) Yang-Mills gauge theory (discovered in 1986).
- *Seiberg-Witten invariants*, defined using a U(1) monopole gauge theory (1994).

We shall restrict our attention throughout to the case of closed, oriented 4-manifolds with $b_1 = 0$ and odd $b_2^+ > 1$. The conjectured relationship between these gauge theory invariants is described below:

CONJECTURE 1.1. [**30**] A 4-manifold X has KM-simple type if and only if it has SW-simple type. If X has simple type, the KM and SW basic classes coincide, and the Donaldson and Seiberg-Witten series obey

$$(1.1) \qquad \mathbf{D}^w_X(h) = 2^{2-c(X)} e^{\frac{1}{2}Q_X(h,h)} \mathbf{SW}^w_X(h), \quad h \in H_2(X;\mathbb{R}).$$

PMNF was supported in part by NSF grants DMS-9704174 and DMS-0125170.
TGL was supported in part by NSF grant DMS-0103677.

Here, the 4-manifold X has intersection form

$$Q_X : H_2(X;\mathbb{Z}) \times H_2(X;\mathbb{Z}) \to \mathbb{Z},$$

Euler characteristic χ, signature σ and

$$c(X) = -\frac{1}{4}(7\chi + 11\sigma).$$

We shall recall the definitions of the Donaldson and Seiberg-Witten series shortly.

1.2. Remarks on the problem. Before proceeding to discuss our work on Witten's conjecture, it is interesting to compare the mathematicians' and physicists' approaches to establishing (1.1).

Witten employs a certain $N = 2$ *supersymmetric quantum Yang-Mills theory*. He uses rescaling, $g_t = t^2 g$, of the Riemannian metric g on X and metric independence of the correlation functions to relate the Donaldson invariants ($t \to 0$) with the Seiberg-Witten invariants ($t \to \infty$).

The mathematical approach to a proof of Witten's formula proposed by Pidstrigatch and Tyurin [**27**] instead employs an SO(3) monopole gauge theory which generalizes both the instanton and U(1) monopole gauge theories. All three gauge theories are classical field theories and their solutions are invariant under metric rescaling, whereas Witten's quantum field theory is sensitive to metric rescaling.

Apparently, the SO(3) monopole gauge theory provides a purely classical field theory alternative to Witten's quantum field theory method. The problem of determining the relationship between these two approaches is surely an important one worth exploring further.

2. SO(3) monopoles

2.1. Clifford modules and spin structures. Given a Riemannian metric g on X, let $V \to X$ be a Hermitian bundle with a linear Clifford map $\rho : T^*X \to \mathrm{End}_{\mathbb{C}}(V)$,

$$\rho(\alpha)^\dagger = -\rho(\alpha), \quad \text{and} \quad \rho(\alpha)^2 = -g(\alpha,\alpha), \quad \alpha \in \Omega^1(X,\mathbb{R}).$$

Then (ρ, V) defines a Clifford or $\mathbb{C}\ell(T^*X)$ module structure on V. If $W \to X$ is a complex-rank four Hermitian bundle, then $\mathfrak{s} = (\rho, W)$ is a *spin^c structure*, familiar from Seiberg-Witten theory [**28**]. If $V \to X$ is a complex-rank eight Hermitian bundle, then we call $\mathfrak{t} = (\rho, V)$ a *spin^u structure*.

2.2. From spin^u structures to SO(3) bundles. Given $\mathfrak{t} = (\rho, V)$ on X, one obtains

- An SO(3) *subbundle*,
$$\mathfrak{g}_{\mathfrak{t}} \subset \mathfrak{su}(V),$$
characterized as the span of the sections ξ of $\mathfrak{su}(V)$ such that $[\xi, \rho(\omega)] = 0$, for all $\omega \in \Omega^\bullet(X,\mathbb{R})$.
- A complex line bundle,
$$\det{}^{\frac{1}{2}}(V^+),$$
where $V = V^+ \oplus V^-$ and $V^\pm$ are the ∓ 1 eigenspaces of $\rho(\mathrm{vol})$ on V.

- Splittings, if $\Lambda^2(T^*X) = \Lambda^+ \oplus \Lambda^-$ and $\Lambda^\pm$ are the ± 1 eigenspaces of $*_g$ on $\Lambda^2(T^*X)$, and $\rho : \Lambda^\pm \cong \mathfrak{su}(W^\pm)$ are the usual isomorphisms of SO(3) bundles,
$$\mathfrak{su}(V^\pm) \cong \rho(\Lambda^\pm) \oplus \rho(\Lambda^\pm) \otimes \mathfrak{g}_{\mathfrak{t}} \oplus \mathfrak{g}_{\mathfrak{t}}.$$

Moreover, for any choice of spin^c structure $\mathfrak{s} = (\rho, W)$, one further obtains

- A complex-rank two Hermitian bundle,
$$E = \mathrm{Hom}_{\mathbb{C}\ell(T^*X)}(W, V).$$
- A Clifford module isomorphism,
$$V \cong W \otimes E.$$
- An isomorphism of SO(3) bundles,
$$\mathfrak{su}(E) \cong \mathfrak{g}_{\mathfrak{t}}.$$
- An isomorphism of complex line bundles,
$$\det{}^{\frac{1}{2}}(V^+) \cong \det(W^+) \otimes \det(E).$$

2.3. SO(3)-monopole equations. We call a pair (A, Φ) an SO(3) *monopole* if
$$\begin{aligned} \mathrm{ad}^{-1}(F^+_{\hat{A}}) - \rho^{-1}(\Phi \otimes \Phi^*)_{00} &= 0, \\ D_A \Phi &= 0, \end{aligned}$$
where A is a spin connection on V, inducing a fixed connection $A|_{\det(V^+)} = 2A_\Lambda$ on $\det(V^+)$ and Φ is a section of V^+; $\hat{A}$ is the induced connection on the SO(3) bundle $\mathfrak{g}_{\mathfrak{t}} \subset \mathfrak{su}(V)$; $F^+_{\hat{A}}$ is the self-dual component of the curvature of $\hat{A}$; the term $(\Phi \otimes \Phi^*)_{00}$ is the component of $\Phi \otimes \Phi^*$ lying in $\rho(\Lambda^+) \otimes \mathfrak{g}_{\mathfrak{t}}$; $D_A : C^\infty(V^+) \to C^\infty(V^-)$ is the Dirac operator. We let $\mathcal{M}_{\mathfrak{t}}$ be the space of SO(3) monopoles for $\mathfrak{t} = (\rho, V)$, modulo gauge transformations.

2.4. Singularities in SO(3)-monopole space. We now classify the fixed points of the circle action on $\mathcal{M}_{\mathfrak{t}}$ given by complex multiplication on the spinor components.

An SO(3) monopole (A, Φ) is a *Yang-Mills* or *instanton* solution if
$$\Phi \equiv 0 \quad \text{and} \quad F^+_{\hat{A}} = 0.$$
Hence, there is a moduli subspace of SO(3) instantons,
$$M^w_\kappa \hookrightarrow \mathcal{M}_{\mathfrak{t}},$$
where $\kappa = -\frac{1}{4}p_1(\mathfrak{g}_{\mathfrak{t}})$ and $w \in H^2(X; \mathbb{Z})$ lifts $w_2(\mathfrak{g}_{\mathfrak{t}}) \in H^2(X; \mathbb{Z}/2\mathbb{Z})$.

An SO(3) monopole (A, Φ) is a *Seiberg-Witten* or *reducible* solution if
$$A = B \oplus B \otimes A_L \quad \text{on} \quad V = W \oplus W \otimes L,$$
for some Hermitian line bundle L, a unitary connection A_L on L, and $\Phi = \Psi \oplus 0$ with Ψ a section of W^+ obeying
$$\begin{aligned} \mathrm{Tr}(F^+_B) - \rho^{-1}(\Psi \otimes \Psi^*)_0 - F^+_{A_\Lambda} &= 0, \\ D_B \Psi &= 0. \end{aligned}$$
Hence, there are moduli subspaces of Seiberg-Witten monopoles for $\mathfrak{s} = (\rho, W)$,
$$M_{\mathfrak{s}} \hookrightarrow \mathcal{M}_{\mathfrak{t}},$$

whenever $V = W \oplus W \otimes L$.

3. Invariants of smooth 4-manifolds

We sketch definitions of the Donaldson series [**22**] and Seiberg-Witten series [**30**].

3.1. Donaldson invariants. Set $\mathbb{A}(X) = \mathrm{Sym}(H_0(X;\mathbb{R}) \oplus H_2(X;\mathbb{R}))$, so $z \in \mathbb{A}(X)$ is a linear combination of monomials,

$$x^m \beta_1 \beta_2 \cdots \beta_{\delta-2m},$$

with $x \in H_0(X;\mathbb{Z})$ being the positive generator and $\beta_i \in H_2(X;\mathbb{R})$. Cohomology classes on M^w_κ can be defined via a map [**2**], [**5**],

$$\mu_p : H_i(X;\mathbb{R}) \to H^{4-i}(M^w_\kappa;\mathbb{R}).$$

The Donaldson invariant is then a linear function,

$$D^w_X : \mathbb{A}(X) \to \mathbb{R},$$

where, for a monomial z with $\deg(z) = 2\delta$,

$$D^w_X(z) = \langle \mu_p(z), [M^w_\kappa] \rangle$$

with $\mu_p(z) = \mu_p(x)^m \smile \mu_p(\beta_1) \smile \cdots \smile \mu_p(\beta_{\delta-2m})$.

3.2. Kronheimer-Mrowka structure theorem. One says that a 4-manifold X has *KM-simple type* if for some w and all $z \in \mathbb{A}(X)$,

$$D^w_X(x^2 z) = 4D^w_X(z).$$

One defines the *Donaldson series* by setting

$$\mathbf{D}^w_X(h) = D^w_X((1 + \tfrac{1}{2}x)e^h), \quad h \in H_2(X;\mathbb{R}). \tag{3.1}$$

We recall the celebrated

THEOREM 3.1. [**22**] *If X has KM-simple type, then there exist $a_r \in \mathbb{Q}$ and $K_r \in H^2(X;\mathbb{Z})$, the KM-basic classes, such that*

$$\mathbf{D}^w_X(h) = e^{\frac{1}{2}Q_X(h,h)} \sum_{r=1}^{s} (-1)^{\frac{1}{2}(w^2 + w\cdot K_r)} a_r e^{\langle K_r, h\rangle}, \quad h \in H_2(X;\mathbb{R}). \tag{3.2}$$

See also [**17**], for a similar result and independent proof by different methods.

3.3. Seiberg-Witten invariants. The *Seiberg-Witten invariants* comprise a function,

$$SW_X : \mathrm{Spin}^c(X) \to \mathbb{Z},$$

where

$$SW_X(\mathfrak{s}) = \langle \mu_{\mathfrak{s}}(x)^{\max}, [M_{\mathfrak{s}}] \rangle,$$

and $\mu_{\mathfrak{s}}(x) \in H^2(M_{\mathfrak{s}};\mathbb{Z})$ is a cohomology class associated to a circle action. One says that a 4-manifold X has *SW-simple type* if for all $\mathfrak{s}$ for which $M_{\mathfrak{s}}$ has positive dimension one has that

$$SW_X(\mathfrak{s}) = 0,$$

and calls $c_1(\mathfrak{s}) = c_1(W^+)$ an *SW-basic class* if $SW_X(\mathfrak{s}) \neq 0$. We define the *Seiberg-Witten series* by

$$\mathbf{SW}^w_X(h) = \sum_{\mathfrak{s} \in \mathrm{Spin}^c(X)} (-1)^{\frac{1}{2}(w^2 + c_1(\mathfrak{s})\cdot w)} SW_X(\mathfrak{s}) e^{\langle c_1(\mathfrak{s}), h\rangle}, \tag{3.3}$$

for all $h \in H_2(X;\mathbb{R})$. Witten's prediction then takes the form stated in Conjecture 1.1.

4. SO(3) monopole cobordism

4.1. Bubbling and Uhlenbeck compactness. In order to apply the moduli space $\mathcal{M}_{\mathfrak{t}}$ of SO(3) monopoles as a cobordism, we must use a compactification. If $\{[A_\alpha,\Phi_\alpha]\}_{\alpha\in\mathbb{N}} \subset \mathcal{M}_{\mathfrak{t}}$, then the sequence converges to an *ideal* SO(3) *monopole* $([A_\infty,\Phi_\infty],\mathbf{x})$ in $M_{\mathfrak{t}_\ell} \times \mathrm{Sym}^\ell(X)$ if

- $(A_\alpha,\Phi_\alpha) \to (A_\infty,\Phi_\infty)$ in C^∞ on $X \setminus \{\mathbf{x}\}$, modulo gauge transformations,
- $|F_{A_\alpha}|^2 \to |F_{A_\infty}|^2 + 8\pi^2\sum_{x\in\mathbf{x}}\delta(x)$, as measures, where $\delta(x)$ denotes the Dirac measure centered at x.

We let $\bar{\mathcal{M}}_{\mathfrak{t}}$ be the closure of $\mathcal{M}_{\mathfrak{t}}$ with respect to the Uhlenbeck topology, implicit above, in the space of ideal SO(3) monopoles,

$$\bigsqcup_{\ell=0}^{N}\left(M_{\mathfrak{t}_\ell} \times \mathrm{Sym}^\ell(X)\right),$$

where $\mathfrak{t}_\ell = (\rho, V_\ell)$ is a spin^u structure with characteristic classes

$$p_1(\mathfrak{g}_{\mathfrak{t}_\ell}) = p_1(\mathfrak{g}_{\mathfrak{t}}) + 4\ell, \quad w_2(\mathfrak{g}_{\mathfrak{t}_\ell}) = w_2(\mathfrak{g}_{\mathfrak{t}}), \text{ and } c_1(\mathfrak{t}_\ell) = c_1(\mathfrak{t}),$$

and $c_1(\mathfrak{t}) := c_1(\det^{\frac{1}{2}}(V^+))$. The space $\bar{\mathcal{M}}_{\mathfrak{t}}$ is smoothly stratified, with *top* or *zeroth level* $\mathcal{M}_{\mathfrak{t}}$, and *lower levels* $\mathcal{M}_{\mathfrak{t}_\ell} \times \mathrm{Sym}^\ell(X)$, $\ell \geq 1$.

4.2. Stratification of the space of SO(3) monopoles. For the top level, one has a stratification

$$\mathcal{M}_{\mathfrak{t}} = M_\kappa^w \sqcup \mathcal{M}_{\mathfrak{t}}^{*,0} \sqcup \bigcup_{\mathfrak{t}=\mathfrak{s}\oplus\mathfrak{s}'} M_{\mathfrak{s}}. \tag{4.1}$$

The complement $\mathcal{M}_{\mathfrak{t}}^{*,0}$ in $\mathcal{M}_{\mathfrak{t}}$ of the Yang-Mills and Seiberg-Witten solutions is a smooth manifold, cut out transversely by the SO(3) monopole equations [**8**].

A stratification of the form (4.1) arises in each level, $\mathcal{M}_{\mathfrak{t}_\ell} \times \mathrm{Sym}^\ell(X)$, of the compactification $\bar{\mathcal{M}}_{\mathfrak{t}}$. Though dimension-counting arguments rule out contributions from the instanton moduli subspace $M_{\kappa-\ell}^w$ of $\mathcal{M}_{\mathfrak{t}_\ell}$ to pairings with the cohomology classes appearing in equation (4.2), Seiberg-Witten moduli subspaces of $\mathcal{M}_{\mathfrak{t}_\ell}$ can contribute to Donaldson invariants computed using $M_\kappa^w \subset \mathcal{M}_{\mathfrak{t}}$. To apply the cobordism, we

- define a link $\bar{\mathbf{L}}_{\mathfrak{t},\kappa}^w \subset \bar{\mathcal{M}}_{\mathfrak{t}}/S^1$ of the instanton moduli subspace,
$$\bar{M}_\kappa^w \subset \bar{\mathcal{M}}_{\mathfrak{t}},$$
by restricting to spinors with L^2 norm equal to a small positive constant, and
- use gluing theory [**12**], [**13**] to construct links $\bar{\mathbf{L}}_{\mathfrak{t},\mathfrak{s}} \subset \bar{\mathcal{M}}_{\mathfrak{t}}/S^1$ of ideal Seiberg-Witten moduli subspaces,
$$M_{\mathfrak{s}} \times \mathrm{Sym}^\ell(X) \subset \bar{\mathcal{M}}_{\mathfrak{t}}.$$

The links $\bar{\mathbf{L}}_{\mathfrak{t},\mathfrak{s}}$ are considerably more difficult to construct than $\bar{\mathbf{L}}_{\mathfrak{t},\kappa}^w$, especially when ℓ is large.

4.3. SO(3)-monopole cobordism formula. The cobordism $\bar{\mathcal{M}}_{\mathfrak{t}}/S^1$ now yields the raw identity,

$$\left\langle \mu_p(z) \smile \mu_c^{\delta_c-1}, [\bar{\mathbf{L}}^w_{\mathfrak{t},\kappa}] \right\rangle = - \sum_{\mathfrak{s}\in \mathrm{Spin}^c(X)} \left\langle \mu_p(z) \smile \mu_c^{\delta_c-1}, [\bar{\mathbf{L}}_{\mathfrak{t},\mathfrak{s}}] \right\rangle, \tag{4.2}$$

where $\mu_c \in H^2(\mathcal{M}^{*,0}_{\mathfrak{t}};\mathbb{Z})$ is a class associated to a circle action on $\mathcal{M}^{*,0}_{\mathfrak{t}}$. One finds that

$$\langle \mu_p(z) \smile \mu_c^{\delta_c-1}, [\bar{\mathbf{L}}^w_{\mathfrak{t},\kappa}] \rangle$$

is a multiple of the Donaldson invariant, $D^w_X(z)$. The sum in (4.2) is over all $\mathfrak{s} \in \mathrm{Spin}^c(X)$, with $\bar{\mathbf{L}}_{\mathfrak{t},\mathfrak{s}}$ empty unless $M_{\mathfrak{s}} \times \mathrm{Sym}^\ell(X) \subset \bar{\mathcal{M}}_{\mathfrak{t}}$, for some $\ell(\mathfrak{t},\mathfrak{s}) \geq 0$.

The difficult aspect of using (4.2) to derive Witten's formula (1.1) is to show that

$$\langle \mu_p(z) \smile \mu_c^{\delta_c-1}, [\bar{\mathbf{L}}_{\mathfrak{t},\mathfrak{s}}] \rangle \tag{4.3}$$

is the correct multiple of the Seiberg-Witten invariant $SW_X(\mathfrak{s})$; the degree of difficulty grows rapidly with $\ell \geq 0$. The assertion that the pairing (4.3) is a multiple of $SW_X(\mathfrak{s})$ is refered to as the multiplicity conjecture. As this conjecture follows from the work in [**14**], we shall assume it for the rest of this note.

5. Application of the cobordism

We may consider following situations, arranged in increasing order of complexity:

- There are no Seiberg-Witten moduli spaces with non-zero invariants in $\bar{\mathcal{M}}_{\mathfrak{t}}$, so the intersection $(M_{\mathfrak{s}} \times \mathrm{Sym}^\ell(X)) \cap \bar{\mathcal{M}}_{\mathfrak{t}}$ is empty for all $\ell \geq 0$ and $SW_X(\mathfrak{s}) \neq 0$. The Donaldson invariants defined by $\bar{M}^w_\kappa \subset \bar{\mathcal{M}}_{\mathfrak{t}}$ are then zero and, eventually, this observation leads to a vanishing result:
$$\mathbf{D}^w_X(h) \equiv 0 \equiv \mathbf{SW}^w_X(h) \pmod{h^{c(X)-2}}, \quad h \in H_2(X;\mathbb{R}).$$
- Calculation of contributions from $M_{\mathfrak{s}} \subset \mathcal{M}_{\mathfrak{t}}$ leads to Witten's formula, mod $h^{c(X)}$.
- Calculation of contributions from $M_{\mathfrak{s}} \times \mathrm{Sym}^\ell(X) \subset \bar{\mathcal{M}}_{\mathfrak{t}}$ for $\ell = 0, 1$ leads to Witten's formula, mod $h^{c(X)+2}$.
- Calculation of contributions from $M_{\mathfrak{s}} \times \mathrm{Sym}^\ell(X) \subset \bar{\mathcal{M}}_{\mathfrak{t}}$ for $\ell = 0, 1, 2$ leads to Witten's formula, mod $h^{c(X)+4}$.
- Calculation of contributions from $M_{\mathfrak{s}} \times \mathrm{Sym}^\ell(X) \subset \bar{\mathcal{M}}_{\mathfrak{t}}$ for $\ell \geq 3$ should lead to a verification of Witten's formula (1.1).

We have considered the cases $\ell = 0$ and $\ell = 1$ in detail [**10**], [**11**], [**15**] and we would expect the case $\ell = 2$ to follow in a similar manner, by exploiting work of Leness [**23**] on the wall-crossing formula for Donaldson invariants.

At present, we can compute the general shape of the contributions for $\ell \geq 3$ (see [**14**]); complete, direct computations of those contributions appear to be difficult, though we expect indirect methods will yield the desired result [**16**].

5.1. Level-zero Seiberg-Witten contributions. Let $B \subset H^2(X;\mathbb{Z})$ be the set of Seiberg-Witten basic classes and let $B^\perp \subset H^2(X;\mathbb{Z})$ be the Q_X-orthogonal complement of B. We call a 4-manifold X *abundant* if $Q_X|_{B^\perp}$ has a hyperbolic sublattice. Every compact, complex algebraic, simply-connected surface with $b_2^+ \geq 3$ is abundant [**10**].

THEOREM 5.1. [**11**] *Assume X is abundant, has $b_1 = 0$, odd $b_2^+ \geq 3$, and SW-simple type. Suppose $\Lambda \in B^\perp$ exists with $\Lambda^2 = 2 - (\chi + \sigma)$. For such Λ and $w \in H^2(X;\mathbb{Z})$ with $w - \Lambda \equiv w_2(X) \pmod 2$ one has, for all $h \in H_2(X;\mathbb{R})$,*

$$\mathbf{D}_X^w(h) \equiv 0 \equiv \mathbf{SW}_X^w(h) \pmod{h^{c(X)-2}}, \tag{5.1}$$

$$\mathbf{D}_X^w(h) \equiv 2^{2-c(X)} e^{\frac{1}{2}Q_X(h,h)} \mathbf{SW}_X^w(h) \pmod{h^{c(X)}}. \tag{5.2}$$

The vanishing assertion (5.1) for the Seiberg-Witten series is a statement that the Moore-Mariño-Peradze conjecture holds for (abundant) 4-manifolds of SW-simple type [**9**], [**24**], [**25**].

5.2. Level-one Seiberg-Witten contributions. With more sophisticated analytical tools, specifically gluing theory, we can compute contributions from $M_{\mathfrak{s}} \times X \subset \bar{\mathcal{M}}_{\mathfrak{t}}$, and these computations lead to the

THEOREM 5.2. [**15**] *Same hypotheses as Theorem 5.1, but now suppose $\Lambda \in B^\perp$ exists with $\Lambda^2 = 4 - (\chi + \sigma)$. For such Λ and $w \in H^2(X;\mathbb{Z})$ with $w - \Lambda \equiv w_2(X)$ $\pmod 2$ one has, for all $h \in H_2(X;\mathbb{R})$,*

$$\mathbf{D}_X^w(h) \equiv 0 \equiv \mathbf{SW}_X^w(h) \pmod{h^{c(X)-2}}, \tag{5.3}$$

$$\mathbf{D}_X^w(h) \equiv 2^{2-c(X)} e^{\frac{1}{2}Q_X(h,h)} \mathbf{SW}_X^w(h) \pmod{h^{c(X)+2}}. \tag{5.4}$$

We expect an identity similar to (5.4), but mod $h^{c(X)+4}$, by computing contributions for $\ell = 2$, when $\Lambda^2 = 6 - (\chi + \sigma)$. The restrictive hypotheses on existence of classes Λ with prescribed even squares can be dropped if one can consider contributions for arbitrary $\ell \geq 0$.

5.3. Seiberg-Witten contributions from arbitrary levels. More generally, we establish the following in [**14**]:

THEOREM 5.3. *Let X be a closed, connected, oriented smooth four-manifold with $b_1(X) = 0$ and odd $b_2^+(X) > 1$. Let $\Lambda, w \in H^2(X;\mathbb{Z})$ obey $w - \Lambda \equiv w_2(X)$ $\pmod 2$. Let δ, m be non-negative integers for which $m \leq [\delta/2]$, where $[\,\cdot\,]$ denotes the greatest integer function, and $\delta \equiv -w^2 - \frac{3}{4}(\chi+\sigma) \pmod 4$, with Λ and δ obeying $\delta < i(\Lambda)$, where $i(\Lambda) = \Lambda^2 - \frac{1}{4}(\chi + \sigma)$. Then for any $h \in H_2(X;\mathbb{R})$ and generator $x \in H_0(X;\mathbb{Z})$, we have the following expression for the Donaldson invariant:*

$$\begin{aligned} D_X^w(h^{\delta-2m}x^m) = &\sum_{\mathfrak{s}\in \mathrm{Spin}^c(X)} (-1)^{\frac{1}{2}(w^2+w\cdot c_1(\mathfrak{s}))} SW_X(\mathfrak{s}) \\ &\times \sum_{i=0}^{\min(\ell,[\delta/2]-m)} \left(p_{\delta,\ell,m,i}(c_1(\mathfrak{s}) - \Lambda, \Lambda) Q_X^i\right)(h), \end{aligned} \tag{5.5}$$

where Q_X is the intersection form on $H_2(X;\mathbb{R})$, $\ell = \frac{1}{4}(\delta + (c_1(\mathfrak{s}) - \Lambda)^2 + \frac{3}{4}(\chi+\sigma))$ and $p_{\delta,\ell,m,i}(\cdot,\cdot)$ is a homogeneous polynomial of degree $\delta - 2m - 2i$ with coefficients which are universal functions of

$$\chi,\ \sigma,\ c_1(\mathfrak{s})^2,\ \Lambda^2,\ c_1(\mathfrak{s})\cdot\Lambda,\ \delta,\ m,\ \ell.$$

Although Theorem 5.3 does not immediately yield Witten's formula (1.1), it is still powerful enough to prove that the Seiberg-Witten invariants determine the Donaldson invariants. Furthermore, Witten's formula (1.1) itself should follow from Theorem 5.3 by indirect calculations of the remaining unknown coefficients [**16**].

6. Outline of the proofs of Theorems 5.1, 5.2, and 5.3

We shall first sketch how to compute the rough form of the pairings,

$$\langle \mu_p(z) \smile \mu_c^{\delta_c - 1}, [\bar{\mathbf{L}}_{\mathfrak{t},\mathfrak{s}}] \rangle, \tag{6.1}$$

or, at least why these pairings have the form

$$SW_X(\mathfrak{s}) \times (\text{Factors depending only on topology}).$$

We use our gluing theory [**12**], [**13**] to construct a topological model for a neighborhood in $\bar{\mathcal{M}}_{\mathfrak{t}}$ and hence a link, $\bar{\mathbf{L}}_{\mathfrak{t},\mathfrak{s}} \subset \bar{\mathcal{M}}_{\mathfrak{t}}/S^1$, of the 'stratum'

$$M_{\mathfrak{s}} \times \mathrm{Sym}^{\ell}(X) \subset \bar{\mathcal{M}}_{\mathfrak{t}}.$$

Given this topological model for $\bar{\mathbf{L}}_{\mathfrak{t},\mathfrak{s}}$, we can then apply intersection theory methods to partly compute the pairings (6.1).

This suffices to prove the 'rough version' (5.5) of Witten's formula (1.1), which is enough to show that the Seiberg-Witten invariants determine the Donaldson invariants. We shall illustrate the method below, often assuming $\ell = 1$ for the sake of simplicity [**15**].

The passage from this stage to Witten's formula (1.1) requires us to compute the many universal, but unknown coefficients in the rough version (5.5) of Witten's formula.

6.1. Neighborhood of a Seiberg-Witten stratum.

Our gluing theory [**12**], [**13**] allows us to construct a model for a neighborhood of the level $M_{\mathfrak{s}} \times \mathrm{Sym}^{\ell}(X)$ in a local, 'virtual' moduli space,

$$\bar{\mathcal{M}}^{\mathrm{vir}}_{\mathfrak{t},\mathfrak{s}} := \tilde{N}_{\mathfrak{t}_\ell,\mathfrak{s}}(\varepsilon) \times_{\mathcal{G}_{\mathfrak{s}}} \bar{\mathrm{Gl}}_{\mathfrak{t}_\ell}(\delta), \tag{6.2}$$

where $\bar{\mathrm{Gl}}_{\mathfrak{t}_\ell}(\delta)$ is a space of instanton gluing data and $\tilde{N}_{\mathfrak{t}_\ell,\mathfrak{s}}(\varepsilon)$ is a radius-ε disk subbundle of the vector bundle (6.3).

A neighborhood of $M_{\mathfrak{s}} \times \mathrm{Sym}^{\ell}(X)$ in the true moduli space $\bar{\mathcal{M}}_{\mathfrak{t}}$ then takes the shape

$$\boldsymbol{\gamma}\left(\boldsymbol{\chi}^{-1}(0) \cap \bar{\mathcal{M}}^{\mathrm{vir}}_{\mathfrak{t},\mathfrak{s}}\right) = \bar{\mathcal{M}}_{\mathfrak{t}} \cap \boldsymbol{\gamma}(\bar{\mathcal{M}}^{\mathrm{vir}}_{\mathfrak{t},\mathfrak{s}}),$$

where $\boldsymbol{\gamma}$ and $\boldsymbol{\chi}$ are described below. The stratum $M_{\mathfrak{s}}$ in $\mathcal{M}_{\mathfrak{t}_\ell}$ has a 'virtual' normal bundle,

$$N_{\mathfrak{t},\mathfrak{s}} \to M_{\mathfrak{s}}, \tag{6.3}$$

and an obstruction bundle with section $\boldsymbol{\chi}$,

$$\Xi_{\mathfrak{t},\mathfrak{s}} \to \mathcal{M}_{\mathfrak{t}_\ell},$$

while the gluing map,

$$\boldsymbol{\gamma} : \bar{\mathcal{M}}^{\mathrm{vir}}_{\mathfrak{t},\mathfrak{s}} \to \text{Configuration space of ideal pairs containing } \bar{\mathcal{M}}_{\mathfrak{t}},$$

gives a homeomorphism from $\boldsymbol{\chi}^{-1}(0) \cap \bar{\mathcal{M}}^{\mathrm{vir}}_{\mathfrak{t},\mathfrak{s}}$ onto a neighborhood of $M_{\mathfrak{s}} \times \mathrm{Sym}^{\ell}(X)$ in $\bar{\mathcal{M}}_{\mathfrak{t}}$.

6.2. Instanton component of the topological model. When $\ell = 1$, the instanton gluing-data component is given by

$$\bar{\mathrm{Gl}}_{\mathfrak{t}_\ell}(\delta) = \left(\mathrm{Fr}(\mathfrak{g}_{\mathfrak{t}_1}) \times_X \mathrm{Fr}(T^*X) \times \bar{M}_1^{s,\natural}(S^4,\delta)\right) / (\mathrm{SO}(3) \times \mathrm{SO}(4)). \tag{6.4}$$

Here, $M_k^{s,\natural}(S^4,\delta)$ is the moduli space of k-instantons on S^4, framed at the south pole s, with mass center at the north pole, and scale $\leq \delta$. For $k = 1$ there are homeomorphisms,

$$M_1^{s,\natural}(S^4,\delta) \cong (0,\delta] \times \mathrm{SO}(3),$$
$$\bar{M}_1^{s,\natural}(S^4,\delta) \cong c(\mathrm{SO}(3)),$$

where $c(\mathrm{SO}(3))$ is the cone on $\mathrm{SO}(3)$.

Our gluing-model (6.2) admits an Uhlenbeck stratification, given below when $\ell = 1$:

$$\bar{\mathcal{M}}_{\mathfrak{t},\mathfrak{s}}^{\mathrm{vir}} = \mathcal{M}_{\mathfrak{t},\mathfrak{s}}^{\mathrm{vir}} \sqcup (N_{\mathfrak{t},\mathfrak{s}}(\varepsilon) - M_{\mathfrak{s}}) \times X \sqcup M_{\mathfrak{s}} \times X.$$

When $\ell \geq 2$, the symmetric product has its usual stratification and one can also construct bundles $\mathrm{Gl}_{\mathfrak{t}_\ell}(\delta,\Sigma) \to \Sigma$ with instanton moduli space fibers, for each stratum $\Sigma \subset \mathrm{Sym}^\ell(X)$, following the prescription of Friedman and Morgan [**19**] and developing the idea of Kotschick and Morgan [**21**] and Mrowka [**26**]. The difficult part is to assemble these local gluing data bundles into a space of global gluing data, $\bar{\mathrm{Gl}}_{\mathfrak{t}_\ell}(\delta)$. One essentially has,

$$\bar{\mathrm{Gl}}_{\mathfrak{t}_\ell}(\delta) = \bigcup_{\Sigma \subset \mathrm{Sym}^\ell(X)} \mathrm{Gl}_{\mathfrak{t}_\ell}(\delta,\Sigma),$$

but some modifications of the spaces $\mathrm{Gl}_{\mathfrak{t}_\ell}(\delta,\Sigma)$ are needed to carry out this construction. Unlike the analogous problem for the anti-self-dual moduli space described in [**21**], the gluing maps do not define transition maps because the images of the gluing maps for the SO(3) monopole equations intersect only at the zero-locus of the obstruction map χ. In [**14**], we define a deformation of the moduli space of anti-self-dual connections on S^4 and deformations of the *splicing* maps so that the intersections of the images of the deformed splicing maps define explicitly understood transition maps. The space $\bar{\mathrm{Gl}}_{\mathfrak{t}_\ell}(\delta)$ is defined to be the union of the images of these deformed splicing maps.

6.3. Link of the Seiberg-Witten stratum. We define the link in two steps, first considering the Seiberg-Witten component of the link of $M_{\mathfrak{s}} \times \mathrm{Sym}^\ell(X) \subset \bar{\mathcal{M}}_{\mathfrak{t},\mathfrak{s}}^{\mathrm{vir}}/S^1$,

$$\bar{\mathbf{L}}_{\mathfrak{t},\mathfrak{s}}^{\mathrm{vir},s} := \left(\partial\tilde{N}_{\mathfrak{t}_\ell,\mathfrak{s}}(\varepsilon) \times_{\mathcal{G}_{\mathfrak{s}}} \bar{\mathrm{Gl}}_{\mathfrak{t}_\ell}(\delta)\right)/S^1,$$

and, second, defining the instanton component of the link of $M_{\mathfrak{s}} \times \mathrm{Sym}^\ell(X) \subset \bar{\mathcal{M}}_{\mathfrak{t},\mathfrak{s}}^{\mathrm{vir}}/S^1$,

$$\bar{\mathbf{L}}_{\mathfrak{t},\mathfrak{s}}^{\mathrm{vir},i} := \left(\tilde{N}_{\mathfrak{t}_\ell,\mathfrak{s}}(\varepsilon) \times_{\mathcal{G}_{\mathfrak{s}}} \partial\bar{\mathrm{Gl}}_{\mathfrak{t}_\ell}(\delta)\right)/S^1,$$

a complex disk bundle over $M_{\mathfrak{s}} \times \partial\bar{\mathrm{Gl}}_{\mathfrak{t}_\ell}(\delta)/S^1$, when $\ell = 1$, and a more general fiber bundle (that is, not a product bundle) when $\ell > 1$. When $\ell = 1$, the boundary $\partial\bar{\mathrm{Gl}}_{\mathfrak{t}_\ell}(\delta)$ is defined by restricting to instantons on S^4 with scale δ. We then define the link of $M_{\mathfrak{s}} \times \mathrm{Sym}^\ell(X)$ in the virtual moduli space, $\bar{\mathcal{M}}_{\mathfrak{t},\mathfrak{s}}^{\mathrm{vir}}/S^1$,

$$\bar{\mathbf{L}}_{\mathfrak{t},\mathfrak{s}}^{\mathrm{vir}} := \bar{\mathbf{L}}_{\mathfrak{t},\mathfrak{s}}^{\mathrm{vir},s} \cup \bar{\mathbf{L}}_{\mathfrak{t},\mathfrak{s}}^{\mathrm{vir},i}.$$

Finally, we obtain

$$\bar{\mathbf{L}}^{\text{vir}}_{\mathfrak{t},\mathfrak{s}} := \gamma\left(\chi^{-1}(0)\cap\bar{\mathbf{L}}^{\text{vir}}_{\mathfrak{t},\mathfrak{s}}\right) = (\bar{\mathcal{M}}_{\mathfrak{t}}/S^1)\cap\gamma(\bar{\mathbf{L}}_{\mathfrak{t},\mathfrak{s}}),$$

the link of $M_{\mathfrak{s}}\times\operatorname{Sym}^{\ell}(X)$ in the true moduli space $\bar{\mathcal{M}}_{\mathfrak{t}}/S^1$.

6.4. Fiber bundle structure in Seiberg-Witten link pairings. The virtual moduli space method reduces the computation of the link pairing (6.1) to that of the pairing on the right-hand side below,

$$\langle\mu_p(z)\smile\mu_c^{\delta_c},[\bar{\mathbf{L}}_{\mathfrak{t},\mathfrak{s}}]\rangle = \langle\mu_p(z)\smile\mu_c^{\delta_c}\smile e,[\bar{\mathbf{L}}^{\text{vir}}_{\mathfrak{t},\mathfrak{s}}]\rangle, \tag{6.5}$$

where e is Euler class of the total obstruction bundle over $\bar{\mathcal{M}}^{\text{vir}}_{\mathfrak{t},\mathfrak{s}}$, with section χ, and $\bar{\mathbf{L}}_{\mathfrak{t},\mathfrak{s}}\cong\chi^{-1}(0)\cap\bar{\mathbf{L}}^{\text{vir}}_{\mathfrak{t},\mathfrak{s}}$.

When $\ell=1$, we show in [**15**] that the pairing (6.5) with $[\bar{\mathbf{L}}^{\text{vir}}_{\mathfrak{t},\mathfrak{s}}]$ is expressible in terms of pairings with

$$[M_{\mathfrak{s}}\times\partial\bar{\text{Gl}}_{\mathfrak{t}_\ell}(\delta)/S^1].$$

In particular, we can show that the latter pairings are in turn products of

- Pairings with $[M_{\mathfrak{s}}]$, giving multiples of the Seiberg-Witten invariant, $SW_X(\mathfrak{s})$, and
- Pairings with $[\partial\bar{\text{Gl}}_{\mathfrak{t}_\ell}(\delta)/S^1]$.

The pairings with $[\partial\bar{\text{Gl}}_{\mathfrak{t}_\ell}(\delta)/S^1]$ depend only the topology of the 4-manifold, X, and universal data.

When $\ell>1$, the construction of the space

$$\tilde{N}_{\mathfrak{t}_\ell,\mathfrak{s}}(\delta)\times_{\mathcal{G}_{\mathfrak{s}}\times S^1}\bar{\text{Gl}}_{\mathfrak{t}_\ell}(\delta),$$

shows that the virtual link $\bar{\mathbf{L}}^{\text{vir}}_{\mathfrak{t},\mathfrak{s}}$ admits a fiber bundle structure

$$\bar{\mathbf{L}}^{\text{vir}}_{\mathfrak{t},\mathfrak{s}}\to M_{\mathfrak{s}},$$

with fiber given by $\mathbb{C}^n\times_{S^1}\bar{\text{Gl}}_{\mathfrak{t}_\ell}(\delta)$ and structure group given by $\operatorname{Map}(X,S^1)$. Hence, the pairing (6.5) can be written as products of

- Pairings with $[M_{\mathfrak{s}}]$, giving multiples of the Seiberg-Witten invariant, $SW_X(\mathfrak{s})$,
- Pairings with the fiber $\mathbb{C}^n\times_{S^1}\bar{\text{Gl}}_{\mathfrak{t}_\ell}(\delta)$.

In [**14**], we show how the pairings with the fiber $\mathbb{C}^n\times_{S^1}\bar{\text{Gl}}_{\mathfrak{t}_\ell}(\delta)$ can be qualitatively understood in terms of homotopy data of the spin^u structure $\mathfrak{t}_\ell$, of the spin^c structure $\mathfrak{s}$, and of the manifold X.

The passage from the rough version (5.5) to Witten's formula (1.1) to the exact formula is discussed in [**16**].

7. Witten's conjecture and symplectic 4-manifolds

7.1. Gauge theory and Lefschetz fibrations. Away from finitely many critical points, a *Lefschetz fibration* $\pi : X\to S$ is a smooth fiber bundle over a connected base, whose fibers are closed Riemann surfaces, Σ, of given genus. Possibly after blowing up, all symplectic 4-manifolds admit Lefschetz fibrations [**1**], [**3**], [**4**].

One can ask what is the relationship between gauge theoretic invariants and the Lefschetz fibration structure. As a first step, one could use the product, $X=\Sigma\times S$, as a toy model. In this situation, the relationship between the gauge theory moduli spaces and holomorphic maps can be explored via the following techniques:

- Adiabatic limit analysis, by the scaling metric $g = g_\Sigma \oplus g_S$ as $g_\varepsilon = \varepsilon^2 g_\Sigma \oplus g_S$, with $\varepsilon \to 0$, or
- Restriction of stable, holomorphic bundles over $\Sigma \times S$ to $\Sigma \times \{z\}$, as $z \in S$ varies.

This leads to the following identifications:

(1) SO(3)-instantons over $\Sigma \times S$ are identified with holomorphic maps, $S \to M_\Sigma$, where M_Σ is the space of flat SO(3) connections over Σ. This identification goes back to Dostoglou and Salamon [**7**].

(2) Seiberg-Witten U(1)-monopoles over $\Sigma \times S$ are identified with holomorphic maps, $S \to M_{\Sigma,d}$, where $M_{\Sigma,d}$ is the space of vortices on the line bundle $L \to \Sigma$ with $d = \langle c_1(L), [\Sigma]\rangle$. This identification has been outlined by Salamon [**29**].

(3) SO(3)-monopoles over $\Sigma \times S$ are identified with holomorphic maps from S to a space of non-abelian vortices over Σ. Non-abelian vortices of this kind have been studied by Bradlow, Garcia-Prada, and others.

Of course, there are many unanswered questions:

- What is the relationship between the gauge theory compactifications and the compactifications of spaces of holomorphic maps?
- What is the relationship between the gauge theory invariants and Gromov-Witten invariants of spaces of flat connections or vortices over Riemann surfaces?
- Can one extend the analysis for the toy model, $\Sigma \times S$, to the case of non-trivial surface bundles or Lefschetz fibrations?
- Can symplectic 4-manifolds provide additional data we can use to help determine Witten's formula?

We note that the relationship between Seiberg-Witten invariants and Lefschetz fibration data bears on a closely related program due to Simon Donaldson and Ivan Smith whose aims include giving Lefschetz-fibration proofs of results for symplectic 4-manifolds derived via Seiberg-Witten theory [**6**].

References

[1] D. Auroux, *Asymptotically holomorphic families of symplectic submanifolds*, Geom. Funct. Anal. **7** (1997), 971–995.

[2] S. K. Donaldson, *Polynomial invariants for smooth four-manifolds*, Topology **29** (1990), 257–315.

[3] S. K. Donaldson, *Symplectic submanifolds and almost-complex geometry*, J. Differential Geom. **44** (1996), 666–705.

[4] ———, *Lefschetz fibrations in symplectic geometry*, Proceedings of the International Congress of Mathematicians, Vol. II (Berlin, 1998), 1998, pp. 309–314 (electronic).

[5] S. K. Donaldson and P. B. Kronheimer, *The geometry of four-manifolds*, Oxford Univ. Press, Oxford, 1990.

[6] S. K. Donaldson and I. Smith, *Lefschetz pencils and the canonical class for symplectic 4-manifolds*, math.SG/0012067.

[7] S. Dostoglou and D. A. Salamon, *Self-dual instantons and holomorphic curves*, Ann. of Math. (2) **139** (1994), 581–640.

[8] P. M. N. Feehan, *Generic metrics, irreducible rank-one* PU(2) *monopoles, and transversality*, Comm. Anal. Geom. **8** (2000), 905–967, math.DG/9809001.

[9] P. M. N. Feehan, P. B. Kronheimer, T. G. Leness, and T. S. Mrowka, *PU(2) monopoles and a conjecture of Mariño, Moore, and Peradze*, Math. Res. Lett. **6** (1999), 169–182, arXiv:math.DG/9812125.

[10] P. M. N. Feehan and T. G. Leness, *PU(2) monopoles and links of top-level Seiberg-Witten moduli spaces*, J. Reine Angew. Math. **538** (2001), 57–133, arXiv:math.DG/0007190.
[11] ———, *PU(2) monopoles. II: Top-level Seiberg-Witten moduli spaces and Witten's conjecture in low degrees*, J. Reine Angew. Math. **538** (2001), 135–212, arXiv:dg-ga/9712005.
[12] ———, *PU(2) monopoles. III: Existence of gluing and obstruction maps*, submitted to a print journal, arXiv:math.DG/9907107.
[13] ———, *PU(2) monopoles. IV: Surjectivity of gluing maps*, in preparation.
[14] , ———, *A general SO(3)-monopole cobordism formula relating Donaldson and Seiberg-Witten invariants*, arXiv:math.DG/0203047.
[15] ———, *PU(2) monopoles, level-one Seiberg-Witten moduli spaces, and Witten's conjecture in low degrees*, Topology Appl., to appear, arXiv:math.DG/0106238.
[16] , ———, *Witten's conjecture for four-manifolds of simple type*, in preparation.
[17] R. Fintushel and R. Stern, *Donaldson invariants of 4-manifolds with simple type*, J. Differential Geom. **42** (1995), 577–633.
[18] ———, *Rational blowdowns of smooth* 4*-manifolds*, J. Differential Geom. **46** (1997), 181–235, arXiv:alg-geom/9505018.
[19] R. Friedman and J. W. Morgan, *Smooth four-manifolds and complex surfaces*, Springer, Berlin, 1994.
[20] L. Göttsche, *Modular forms and Donaldson invariants for 4-manifolds with* $b^+ = 1$, J. Amer. Math. Soc. **9** (1996), 827–843, arXiv:alg-geom/9506018.
[21] D. Kotschick and J. W. Morgan, *SO(3) invariants for four-manifolds with* $b^+ = 1$, *II*, J. Differential Geom. **39** (1994), 433–456.
[22] P. B. Kronheimer and T. S. Mrowka, *Embedded surfaces and the structure of Donaldson's polynomial invariants*, J. Differential Geom. **43** (1995), 573–734.
[23] T. G. Leness, *Donaldson wall-crossing formulas via topology*, Forum Math. **11** (1999), 417–457, dg-ga/9603016.
[24] M. Mariño, G. Moore, and G. Peradze, *Four-manifold geography and superconformal symmetry*, Math. Res. Lett. **6** (1999), 429–437, arXiv:math.DG/9812042.
[25] ———, *Superconformal invariance and the geography of four-manifolds*, Comm. Math. Phys. **205** (1999), 691–735, arXiv:hep-th/9812055.
[26] T. S. Mrowka, private communication.
[27] V. Y. Pidstrigatch and A. N. Tyurin, *Localisation of Donaldson invariants along the Seiberg-Witten classes*, arXiv:dg-ga/9507004.
[28] D. Salamon, *Spin geometry and Seiberg-Witten invariants*, Birkhäuser, Boston, to appear.
[29] D. A. Salamon, *Seiberg-Witten invariants of mapping tori, symplectic fixed points, and Lefschetz numbers*, Proceedings of 6th Gökova Geometry-Topology Conference, vol. 23, 1999, pp. 117–143.
[30] E. Witten, *Monopoles and four-manifolds*, Math. Res. Lett. **1** (1994), 769–796, arXiv:hep-th/9411102.

Department of Mathematics, Rutgers University, Piscataway, NJ 08854-8019
E-mail address: `feehan@math.rutgers.edu`
URL: `http://www.math.rutgers.edu/~feehan`

Department of Mathematics, Florida International University, Miami, FL 33199
E-mail address: `lenesst@fiu.edu`
URL: `http://www.fiu.edu/~lenesst`

Proceedings of Symposia in Pure Mathematics
Volume **71**, 2003

Quandle Homology Theory and Cocycle Knot Invariants

J. Scott Carter and Masahico Saito

ABSTRACT. This paper is a survey of several papers in quandle homology theory and cocycle knot invariants that have been published recently. Here we describe cocycle knot invariants that are defined in a state-sum form, quandle homology, and methods of constructing non-trivial cohomology classes.

1. Prologue

We start with an example and its history. Figure 1 is an illustration of the knotted surface diagram for an embedded 2-sphere in the 4-sphere, S^4. The 2-sphere is obtained by doubling a slice disk of the stevedore's knot. The diagram is a *broken surface diagram* that is obtained from a generic projection of the surface into 3-space by indicating over/under crossing information in a way similar to the classical case. Specifically, the portion of the surface that is closest to the hyperplane of projection is depicted as an unbroken sheet while the sheet that is further away is broken locally into two sheets. See [**14**] for details.

Figure 2 indicates the three local pictures at double, triple, and branch points of the projection. A diagram can have branch and triple points in general, although the diagram in Fig. 1 does not. At a triple point, we have a notion of *top, middle and bottom* sheets. The adjectives describe the relative proximity to the hyperplane into which the knotted surface has been projected.

The sphere that is illustrated first appeared in the manuscript [**22**] by Fox and Milnor, and later as Example 10 in Fox's "Quick Trip" [**21**], described in a motion picture form. This knotted sphere is not obtained by the spinning construction [**1**]. This can be seen as follows. The Alexander polynomial of a spun knot agrees with that of the underlying classical knot since their fundamental groups are isomorphic. The first homology $H_1(\tilde{X})$ (called the knot module), of the infinite cyclic cover $\tilde{X}$ of the complement X of the sphere in S^4 in question, is $\mathbb{Z}[T,T^{-1}]/(2-T)$ as a $\Lambda = \mathbb{Z}[T,T^{-1}]$-module, thus the Alexander polynomial is not symmetric.

Fox's Example 11 can be recognized as the same sphere as Example 10 with its orientation reversed. Its Alexander polynomial is $(1-2T)$. Thus the sphere

2000 *Mathematics Subject Classification.* Primary 57M25, 57Q45; Secondary 55N99, 18G99.
Key words and phrases. Knots, links, knotted surfaces, quandle, quandle cohomology, state-sum invariants, non-invertibility.
The first author was supported in part by NSF Grant DMS #9988107.
The second author was supported in part by NSF Grant DMS #9988101.

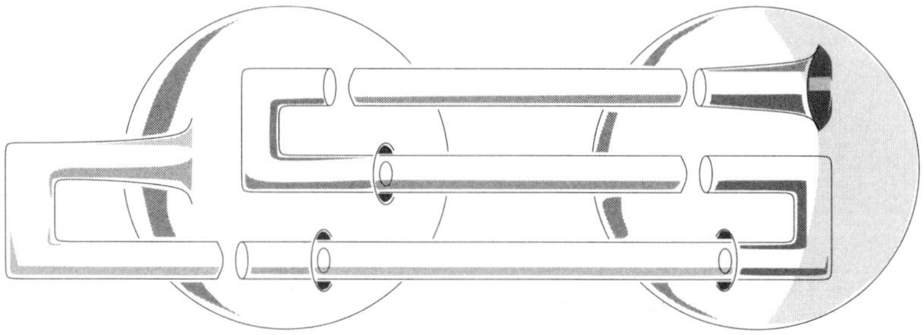

FIGURE 1. Example 10 in "Quick Trip"

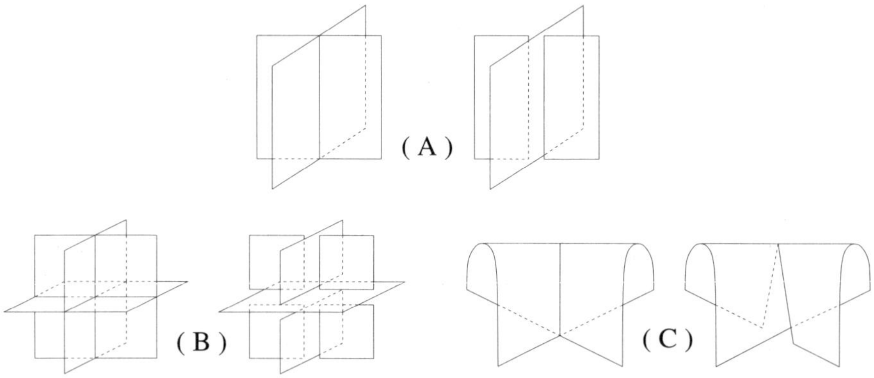

FIGURE 2. Broken surface diagrams

illustrated in Fig. 1 is *non-invertible*: It is not ambiently isotopic to the same surface with its orientation reversed.

Example 12 of "Quick Trip" has as its knot module $\Lambda/(2 - T, 1 - 2T)$. It is obtained from the previous two examples by combining some of their portions. The fact that this ideal is not principal also illustrates the difference between classical knot theory and knotted surfaces. Note that the argument of asymmetric ideals no longer applies to Example 12. It is also interesting to note that this example is in fact the 2-twist spun trefoil [**33**], although Zeeman's twist spin construction appeared later in 1965 [**44**].

Hillman [**24**] showed that this knotted sphere was non-invertible using the Farber-Levine pairing. Ruberman [**40**] used Casson-Gordon invariants to prove the same result, with other new examples of non-invertible knotted spheres. Neither technique applies directly to the same knot with a trivial 1-handle attached. Kawauchi [**31, 32**] has generalized the Farber-Levine pairing to higher genus surfaces, showing that such a torus is also non-invertible. The method we survey in this article shows this fact [**7**] using an invariant defined in a state-sum form from quandle cohomology theory, called the cocycle knot invariant. The cocycle knot invariant has also been used to prove new geometric results [**41**].

We asked Ruberman if he had proved non-invertibility of the 2-twist-spun trefoil on his first excursion to the Georgia Topology Conference in 1982. (Incidentally, the first named author also had his topology debut at GTC1982. The second named

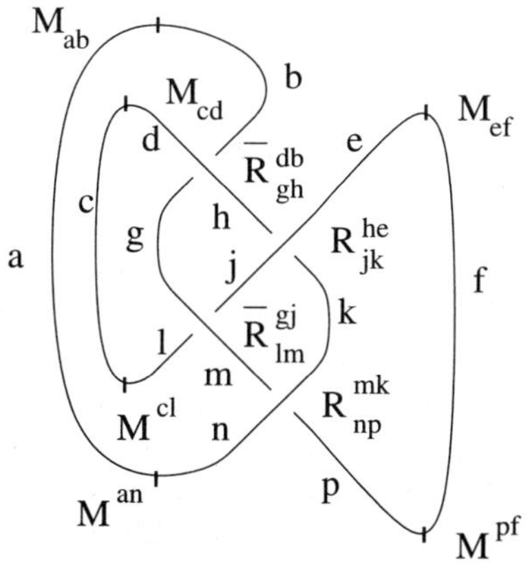

$$\sum M_{ab} M_{cd} M_{ef} \overline{R}^{db}_{gh} R^{he}_{jk} \overline{R}^{gj}_{lm} R^{mk}_{np} M^{cl} M^{an} M^{pf}$$

FIGURE 3. Bracket polynomial of knots

author debuted at GTC1990.) Ruberman told us that the era was correct, although he did not present the result then. His dissertation, however, was inspired by the paper by Sumners [**42**], which showed, in particular, that any 2-sphere in 4-space that contains the Stevedore's knot as a cross-section is knotted, such as the above examples in "Quick Trip."

In Section 4 below, we will give the definition of the cocycle invariant for classical knots and for knotted surfaces in 4-space. Our motivation came from the Jones polynomial and quantum invariants of 3-manifolds. A common feature of the quantum invariants is the state-sum definition, and it has been asked since their discovery whether such invariants exist in higher dimensions (see [**13, 15**] for such attempts). We briefly review the state-sum definition of Jones polynomial and a related invariant for triangulated 3-manifolds — the Dijkgraaf-Witten invariant.

The Bracket Model. The bracket polynomial of a classical knot or link is obtained as follows. The knot is projected generically into the plane, and a height function on the plane is chosen. Let an index set $S = \{1, 2\}$ (in general a finite set), whose elements are called spins, be given and fixed. Let $\mathcal{A}$ denote the set of arcs obtained from the given knot diagram by deleting local maxima, minima, and crossing points. The *coloring* $\mathcal{C}$ is a map $\mathcal{C} : \mathcal{A} \to S$.

Boltzmann weights $B(\tau, \mathcal{C})$ are assigned at minima, maxima, and crossings as follows: Local minima are assigned M^{ab}, local maxima are assigned M_{ab}, crossings are assigned R^{ab}_{cd} if the over crossing arc has positive slope, or $\overline{R}^{ab}_{cd}$ if the overcrossing arc has negative slope, where each weight is defined with a variable A and $i = \sqrt{-1}$ by

$$M^{ab} = M_{ab} = \begin{cases} iA & \text{if} \quad a = 0, b = 1 \\ (iA)^{-1} & \text{if} \quad a = 1, b = 0 \\ 0 & \text{otherwise} \end{cases},$$

$$\begin{aligned} R^{ab}_{cd} &= A\delta^a_c \delta^b_d + A^{-1} M^{ab} M_{cd}, \\ \overline{R}^{ab}_{cd} &= A^{-1}\delta^a_c \delta^b_d + A M^{ab} M_{cd}. \end{aligned}$$

Here, δ denotes Kronecker's delta. The bracket polynomial, as a polynomial in A, is defined by the *state-sum*

$$\langle K \rangle = \sum_{\mathcal{C}} \prod_{\tau} B(\tau, \mathcal{C}),$$

where the product is taken over all crossings, and the sum is taken over all colorings. The Jones polynomial is obtained from the bracket by normalizing and substituting. Specifically, the quantity $\mathcal{L}_K(A) = (-A)^{-3w}\langle K \rangle$ is a knot invariant, where the exponent w is the *writhe* of the diagram K, and $V(t) = \mathcal{L}_K(t^{-1/4})$ is the Jones polynomial. In Fig. 3, a colored knot diagram and its Boltzmann weights are depicted. For a given coloring (denoted by lower case letters a through p excluding i) on arcs, the product of the Boltzmann weights are given at the bottom of the figure. The sum is taken over all colorings, see [**29**] for details.

The Dijkgraaf-Witten Invariant. Similar state-sum invariants were defined for 3-manifolds in [**16**] using group cocycles and the state-sum concept as follows. A combinatorial definition for Chern-Simons invariants with finite gauge groups was given using 3-cocycles of group cohomology. We follow Wakui's description, see [**43**] for a more detailed treatment. Let T be a triangulation of an oriented closed 3-manifold M, with a vertices and n tetrahedra. Give an ordering to the set of vertices. Let G be a finite group. Let $\mathcal{C}$: $\{$ oriented edges $\} \rightarrow G$ be a map such that (1) for any triangle with vertices v_0, v_1, v_2 of T, $\mathcal{C}(\langle v_0, v_2 \rangle) = \mathcal{C}(\langle v_1, v_2 \rangle)\mathcal{C}(\langle v_0, v_1 \rangle)$, where $\langle v_i, v_j \rangle$ denotes the oriented edge with endpoints v_1 and v_2, and (2) $\mathcal{C}(-e) = \mathcal{C}(e)^{-1}$. Such a map $\mathcal{C}$ is called a (group) coloring. Let $\alpha : G \times G \times G \rightarrow A$, $(g, h, k) \mapsto \alpha[g|h|k] \in A$, be a 3-cocycle with values in a multiplicative abelian group A, $\alpha \in Z^3(G; A)$. The 3-cocycle condition is written as

$$\alpha[h|k|l]\alpha[gh|k|l]^{-1}\alpha[g|hk|l]\alpha[g|h|kl]^{-1}\alpha[g|h|k] = 1.$$

Then the Dijkgraaf-Witten invariant is defined by

$$Z_M = \frac{1}{|G|^a} \sum_{\mathcal{C}} \prod_{i=1}^{n} W(\sigma, \mathcal{C})^{\epsilon_i}.$$

Here the Boltzmann weight is given as $W(\sigma, \mathcal{C}) = \alpha[g|h|k]$ where $\mathcal{C}(\langle v_0, v_1 \rangle) = g$, $\mathcal{C}(\langle v_1, v_2 \rangle) = h$, $\mathcal{C}(\langle v_2, v_3 \rangle) = k$, for the tetrahedron $\sigma = |v_0 v_1 v_2 v_3|$ with the ordering $v_0 < v_1 < v_2 < v_3$, and $\epsilon = \pm 1$ according to whether or not the orientation of σ with respect to the vertex ordering matches the orientation of M, see Fig. 4.

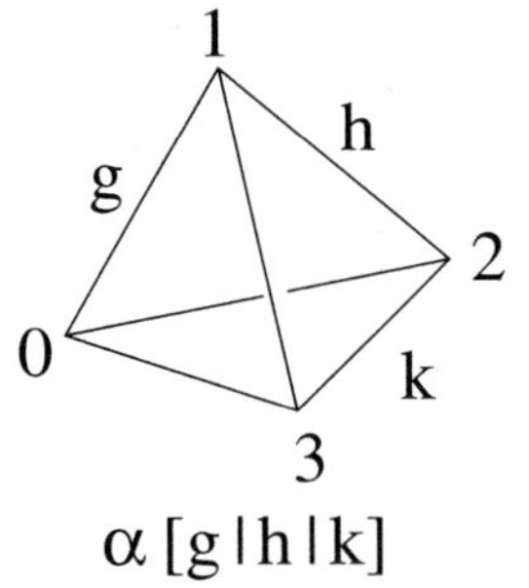

FIGURE 4. A 3-cocycle assigned at a triangle

2. Quandles and Quandle Colorings

In this section we define quandles, quandle colorings, and illustrate that counting quandle colorings can be formulated as a state-sum. This definition will help motivate the definition of the quandle cocycle invariants that we will define in Section 4.

A *quandle*, X, is a set with a binary operation $(a, b) \mapsto a * b$ such that

(I) For any $a \in X$, $a * a = a$.

(II) For any $a, b \in X$, there is a unique $c \in X$ such that $a = c * b$.

(III) For any $a, b, c \in X$, we have $(a * b) * c = (a * c) * (b * c)$.

A *rack* is a set with a binary operation that satisfies (II) and (III).

Racks and quandles have been studied in, for example, [**2, 18, 26, 29, 36**]. The axioms for a quandle correspond respectively to the Reidemeister moves of type I, II, and III (see Fig. 5 and [**18, 29**], for example). A function $f : X \to Y$ between quandles or racks is a *homomorphism* if $f(a * b) = f(a) * f(b)$ for any $a, b \in X$. The following are typical examples of quandles.

- A group $X = G$ with n-fold conjugation as the quandle operation: $a * b = b^{-n} a b^n$.
- Any set X with the operation $x * y = x$ for any $x, y \in X$ is a quandle called the *trivial* quandle. The trivial quandle of n elements is denoted by T_n.
- Let n be a positive integer. For elements $i, j \in \{0, 1, \ldots, n-1\}$, define $i * j \equiv 2j - i \pmod{n}$. Then $*$ defines a quandle structure called the *dihedral quandle*, R_n. This set can be identified with the set of reflections of a regular n-gon with conjugation as the quandle operation.
- Any $\Lambda(= \mathbb{Z}[T, T^{-1}])$-module M is a quandle with $a * b = Ta + (1 - T)b$, $a, b \in M$, called an *Alexander quandle*. Furthermore for a positive integer n, a *mod-n Alexander quandle* $\mathbb{Z}_n[T, T^{-1}]/(h(T))$ is a quandle for a Laurent polynomial $h(T)$. It is finite if the coefficients of the highest and lowest degree terms of h are units in $\mathbb{Z}_n$.

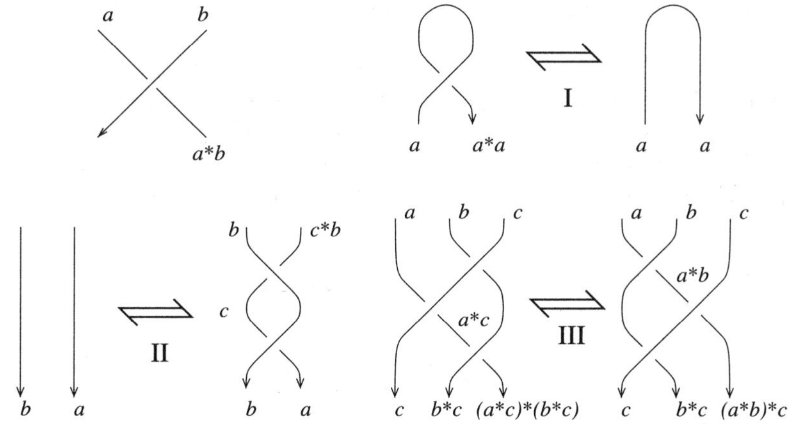

FIGURE 5. Reidemeister moves and quandle conditions

Let X be a fixed quandle. Let K be a given oriented classical knot or link diagram, and let $\mathcal{R}$ be the set of (over-)arcs. The normals are given in such a way that (tangent, normal) matches the orientation of the plane, see Fig. 6. A (quandle) *coloring* $\mathcal{C}$ is a map $\mathcal{C} : \mathcal{R} \to X$ such that at every crossing, the relation depicted in Fig. 6 holds. More specifically, let β be the over-arc at a crossing, and α, β be under-arcs such that the normal of the over-arc points from α to β. Then it is required that $\mathcal{C}(\gamma) = \mathcal{C}(\alpha) * \mathcal{C}(\beta)$.

Alternatively, a coloring can be described as a quandle homomorphism as follows. Classical knots have fundamental quandles that are defined via generators and relations. The theory of quandle presentations is given a complete treatment in [**18**]. Specifically, the generators of the fundamental quandle correspond to the arcs in a diagram. The quandle relation $a * b = c$ holds where a is the generator that corresponds to the underarc away from which the normal to the overarc points, b is the generator that corresponds to the overarc, and c corresponds to the underarc towards which the transversal's normal points, see Fig. 6. A coloring of a classical knot diagram by a quandle X gives rise to a quandle homomorphism from the fundamental quandle to the quandle X.

The number $\mathrm{Col}_X(K)$ of colorings of a knot diagram K by a fixed finite quandle X is a knot invariant, and has a description as a state-sum as follows.

For a finite quandle X, consider the set of maps $\{\mathcal{D} : \mathcal{R} \to X\}$ (without the requirement of a quandle coloring). For a given such a map $\mathcal{D}$, define the Boltzmann weight at a crossing τ, with over-arc β whose normal points from the under-arc α to the under-arc γ, by

$$B(\tau, \mathcal{D}) = \begin{cases} 1 & \text{if } \mathcal{D}(\alpha) * \mathcal{D}(\beta) = \mathcal{D}(\gamma) \\ 0 & \text{otherwise.} \end{cases}$$

Then the number of quandle colorings is written by a state-sum $\mathrm{Col}_X(K) = \sum_{\mathcal{D}} \prod_\tau B(\tau, \mathcal{D})$. We could also use colorings similar to those used in the bracket, or we could write $\mathrm{Col}_X(K) = \sum_{\mathcal{C}} \prod_\tau B_1(\tau, \mathcal{C})$, where $\mathcal{C}$ ranges over only quandle colorings $\mathcal{C}$, and $B_1(\tau, \mathcal{C}) \equiv 1$ is a constant function. Either way, it is natural to ask whether we can modify the weights 1 to a general function.

Fox's n-coloring is a quandle coloring by the dihedral quandle R_n. The classical result that a knot is non-trivially Fox p-colorable (for a prime p) if and only if $p | \Delta(-1)$ (where $\Delta(t)$ denotes the Alexander polynomial) has been generalized by Inoue [**25**] to the following:

THEOREM 2.1. [**25**] *Let $\Delta_K^{(i)}(T)$ denote the greatest common divisor of all $(n-i-1)$ minor determinants of the presentation matrix for the knot module obtained via the Fox calculus. Let p be a prime number, J an ideal of the ring $\Lambda_p = \mathbb{Z}_p[T, T^{-1}]$ and let $Q(K)$ denote a knot quandle. For each $i \geq 0$, put*

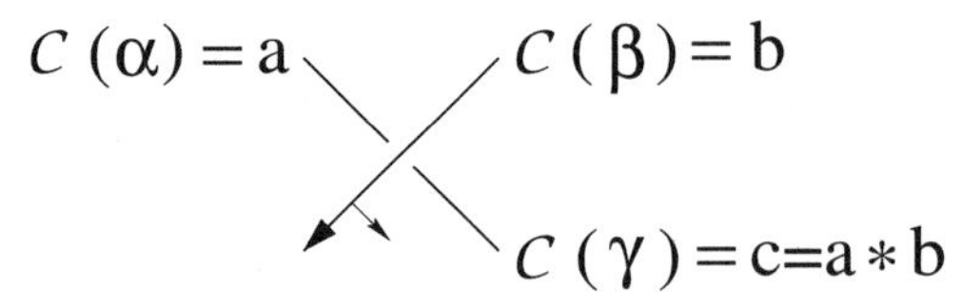

FIGURE 6. Quandle relation at a crossing

$e_i(T) = \Delta_K^{(i)}(T)/\Delta_K^{(i+1)}(T)$. *Then the number of all quandle homomorphisms of the knot quandle $Q(K)$ to the Alexander quandle Λ_p/J is equal to the cardinality of the module $(\Lambda_p/J) \oplus \oplus_{i=0}^{n-2}\{\Lambda_p/(e_i(T), J)\}$*

EXAMPLE 2.2. The Alexander quandle $S_4 = \mathbb{Z}_2[T, T^{-1}]/(T^2 + T + 1)$ has four elements that are represented as $0, 1, T$, and $T + 1$. This quandle colors both the trefoil (3_1) and the figure 8 knot (4_1) as one can easily see directly or by considering the mod-2 reduction of the Alexander polynomials. In either case, the order of S_4 is 4 but the determinants are 3 and 5, for 3_1 and 4_1, respectively. Thus quandle colorings are more general than Fox colorings.

EXAMPLE 2.3. The quandle QS_6 consists of the 4-cycles $a = (1234)$, $A = (1432)$, $b = (1243)$, $B = (1342)$, $c = (1324)$, and $C = (1423)$ with group conjugation as the quandle operation. Figure 7 illustrates a coloring of the knot 6_1 by QS_6. This quandle has R_3 as a quotient quandle. The map $f(a) = f(A) = 0$, $f(b) = f(B) = 1$, and $f(c) = f(C) = 2$ is a quandle homomorphism. The equalizers $(E_y = \{x : f(x) = f(y)\})$ are all the two element trivial quandle. Recently, Angela Harris has shown that QS_6 is not an Alexander quandle of the form $\Lambda_n/(h)$ where h is a polynomial.

3. Quandle Homology and Cohomology Theories

In this section, we present twisted quandle homology, which was discussed in [**4**], and specialize it to the untwisted theory subsequently. Originally, rack homology and homotopy theory were defined and studied in [**19**], and a modification to quandle homology theory was given in [**7**] to define a knot invariant in a state-sum form. Then they were generalized to a twisted theory in [**4**]. Computations are found in [**8, 9**] and also in [**34, 37**] by other authors.

Let $\Lambda = \mathbb{Z}[T, T^{-1}]$, and let $C_n^{\mathrm{TR}}(X) = C_n^{\mathrm{TR}}(X; \Lambda)$ be the free module over Λ generated by n-tuples $(x_1, \ldots, x_n)$ of elements of a quandle X. Define a homomorphism $\partial = \partial_n^T : C_n^{\mathrm{TR}}(X) \to C_{n-1}^{\mathrm{TR}}(X)$ by

$$\begin{aligned}\partial_n^T&(x_1, x_2, \ldots, x_n)\\ &= \sum_{i=1}^{n} (-1)^i \, [T(x_1, x_2, \ldots, x_{i-1}, x_{i+1}, \ldots, x_n)\\ &\quad - (x_1 * x_i, x_2 * x_i, \ldots, x_{i-1} * x_i, x_{i+1}, \ldots, x_n)]\end{aligned}$$

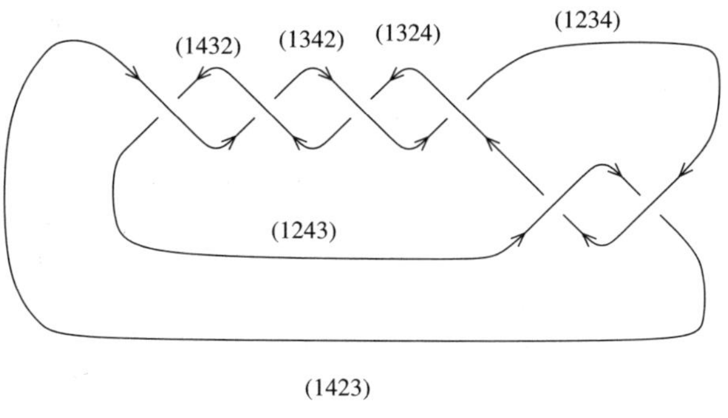

FIGURE 7. A coloring of 6_1 by QS_6

for $n \geq 2$ and $\partial_n^T = 0$ for $n \leq 1$. We regard that the $i = 1$ terms contribute $(1-T)(x_2, \dots, x_n)$. Then $C_*^{\mathrm{TR}}(X) = \{C_n^{\mathrm{TR}}(X), \partial_n^T\}$ is a chain complex. For any Λ-module A, let $C_*^{\mathrm{TR}}(X;A) = \{C_n^{\mathrm{TR}}(X) \otimes_\Lambda A, \partial_n^T\}$ be the induced chain complex, where the induced boundary operator is represented by the same notation. Let $C_{\mathrm{TR}}^n(X;A) = \mathrm{Hom}_\Lambda(C_n^{\mathrm{TQ}}(X), A)$ and define the coboundary operator $\delta = \delta_{\mathrm{TR}}^n : C_{\mathrm{TR}}^n(X;A) \to C_{\mathrm{TR}}^{n+1}(X;A)$ by $(\delta f)(c) = (-1)^n f(\partial c)$ for any $c \in C_n^{\mathrm{TQ}}(X)$ and $f \in C_{\mathrm{TR}}^n(X;A)$. Then $C_{\mathrm{TR}}^*(X;A) = \{C_{\mathrm{TR}}^n(X;A), \delta_{\mathrm{TR}}^n\}$ is a cochain complex. The n-th homology and cohomology groups of these complexes are called *twisted rack homology group* and *cohomology group*, and are denoted by $H_n^{\mathrm{TR}}(X;A)$ and $H_{\mathrm{TR}}^n(X;A)$, respectively.

Let $C_n^{\mathrm{TD}}(X;A)$ be the subset of $C_n^{\mathrm{TR}}(X;A)$ generated by n-tuples $(x_1, \dots, x_n)$ with $x_i = x_{i+1}$ for some $i \in \{1, \dots, n-1\}$ if $n \geq 2$; otherwise let $C_n^{\mathrm{TD}}(X;A) = 0$. If X is a quandle, then $\partial_n^T(C_n^{\mathrm{TD}}(X;A)) \subset C_{n-1}^{\mathrm{TD}}(X;A)$ and $C_*^{\mathrm{TD}}(X;A) = \{C_n^{\mathrm{TD}}(X;A), \partial_n^T\}$ is a sub-complex of $C_*^{\mathrm{TR}}(X;A)$. Similar subcomplexes $C_{\mathrm{TD}}^*(X;A) = \{C_{\mathrm{TD}}^n(X;A), \delta_T^n\}$ are defined for cochain complexes. The n-th homology and cohomology groups of these complexes are called *twisted degeneracy homology group* and *cohomology group*, and are denoted by $H_n^{\mathrm{TD}}(X;A)$ and $H_{\mathrm{TD}}^n(X;A)$, respectively.

Put $C_n^{\mathrm{TQ}}(X;A) = C_n^{\mathrm{TR}}(X;A)/C_n^{\mathrm{TD}}(X;A)$ and $C_*^{\mathrm{TQ}}(X;A) = \{C_n^{\mathrm{TQ}}(X;A), \partial_n^T\}$, where all the induced boundary and coboundary operators are denoted by $\partial = \partial_n^T$ and $\delta = \delta_T^n$, respectively. A cochain complex $C_{\mathrm{TQ}}^*(X;A) = \{C_{\mathrm{TQ}}^n(X;A), \delta_T^n\}$ is similarly defined. The n-th homology and cohomology groups of these complexes are called *twisted homology group* and *cohomology group*, and are denoted by

$$H_n^{\mathrm{TQ}}(X;A) = H_n(C_*^{\mathrm{TQ}}(X;A)), \quad H_{\mathrm{TQ}}^n(X;A) = H^n(C_{\mathrm{TQ}}^*(X;A)).$$

The groups of (co)cycles and (co)boundaries are denoted using similar notations.

EXAMPLE 3.1. The 1-cocycle condition is written for $\eta \in Z_{TQ}^1(X;A)$ as

$$\begin{aligned} -T\eta(x_2) + T\eta(x_1) + \eta(x_2) - \eta(x_1 * x_2) = 0, \quad \text{or} \\ T\eta(x_1) + (1-T)\eta(x_2) = \eta(x_1 * x_2). \end{aligned}$$

Note that this means that $\eta : X \to A$ is a quandle homomorphism.

The 2-cocycle condition is written for $\phi \in Z_{TQ}^2(X;A)$ as

$$\begin{aligned} & T[-\phi(x_2,x_3) + \phi(x_1,x_3) - \phi(x_1,x_2)] \\ & \quad + \ [\phi(x_2,x_3) - \phi(x_1 * x_2, x_3) + \phi(x_1 * x_3, x_2 * x_3)] = 0 \quad \text{or} \\ & T\phi(x_1,x_2) + \phi(x_1 * x_2, x_3) \\ & \quad = \ T\phi(x_1,x_3) + (1-T)\phi(x_2,x_3) + \phi(x_1 * x_3, x_2 * x_3). \end{aligned}$$

The geometric meaning of this condition will become clear in Section 4.

The original untwisted quandle homology is described as a specification of $T = 1$. Specifically, in the definition of the boundary homomorphism ∂_n^T, set $T = 1$, and define all the cycle, boundary, homology groups similarly. Then use $\mathrm{Hom}(\ - \ ; A)$ to define cohomology theory. Thus we assume that the coefficients A simply form an abelian group. We obtain degenerate, rack, and quandle homology groups denoted by $H_*^{\mathrm{W}}(X;A)$ for $W = D, R, Q$, respectively. Similarly, $H_{\mathrm{W}}^n(X;A)$ denotes the corresponding cohomology groups. The cohomology theory H_{R}^* was defined in [**19**]. It was seen in [**9**] that the short exact sequence:

$$0 \to C_n^{\mathrm{D}}(X) \xrightarrow{i} C_n^{\mathrm{R}}(X) \xrightarrow{j} C_n^{\mathrm{Q}}(X) \to 0$$

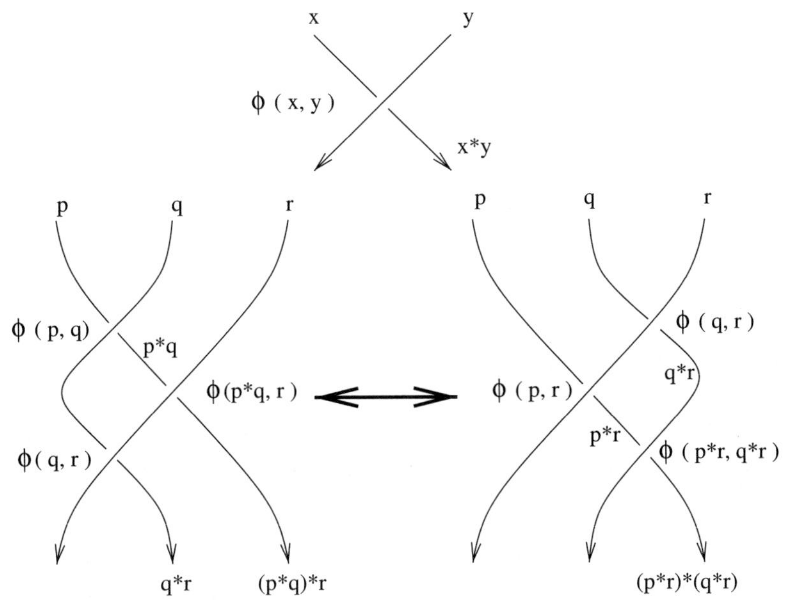

FIGURE 8. The untwisted 2-cocycle condition and type III move

gives rise to the following homology long exact sequence:

$$\cdots \xrightarrow{\partial_*} H_n^{\mathrm{D}}(X;A) \xrightarrow{i_*} H_n^{\mathrm{R}}(X;A) \xrightarrow{j_*} H_n^{\mathrm{Q}}(X;A) \xrightarrow{\partial_*} H_{n-1}^{\mathrm{D}}(X;A) \to \cdots$$

and it was shown in [**11**] by geometric arguments that the sequence splits in low dimensions. This result was improved upon in [**34**] by Litherland and Nelson where they showed the following:

THEOREM 3.2. [**34**] *The above long exact sequence splits into short exact sequences*

$$0 \to H_n^{\mathrm{D}}(X;A) \to H_n^{\mathrm{R}}(X;A) \to H_n^{\mathrm{Q}}(X;A) \to 0.$$

In fact, they construct a projection $p : C_n^{\mathrm{R}}(X) \to C_n^{\mathrm{D}}(X)$ thereby splitting the short exact sequence of chain complexes.

4. Cocycle Knot Invariants

Untwisted Cocycle Invariants. Let K be a classical knot or link diagram. Let a finite quandle X, and an (untwisted) quandle 2-cocycle $\phi \in Z_{\mathrm{Q}}^2(X;A)$ be given. A *(Boltzmann) weight*, $B(\tau, \mathcal{C})$ (that depends on ϕ), at a crossing τ is defined as follows. Let $\mathcal{C}$ denote a coloring $\mathcal{C} : \mathcal{R} \to X$. Let β be the over-arc at τ, and α, γ be under-arcs such that the normal to β points from α to γ, see Fig. 6. Let $x = \mathcal{C}(\alpha)$ and $y = \mathcal{C}(\beta)$. Then define $B(\tau, \mathcal{C}) = \phi(x,y)^{\epsilon(\tau)}$, where $\epsilon(\tau) = 1$ or -1, if (the sign of) the crossing τ is positive or negative, respectively. By convention, the crossing in Fig. 6 is positive if the orientation of the under-arc points downward.

The *(quandle) cocycle knot invariant* is defined by the state-sum expression

$$\Phi(K) = \sum_{\mathcal{C}} \prod_{\tau} B(\tau, \mathcal{C}).$$

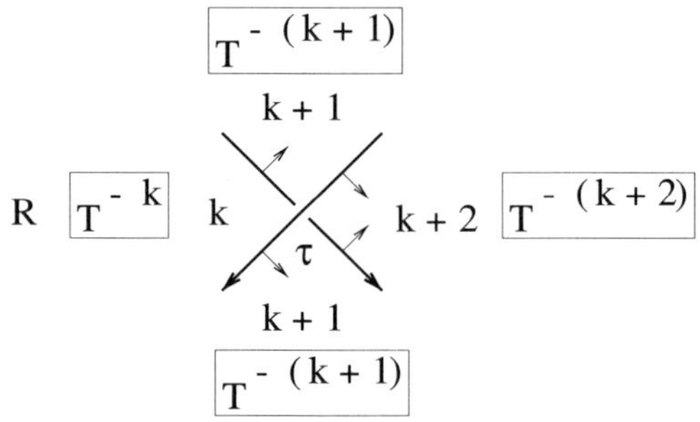

FIGURE 9. Alexander numbering of a crossing

The product is taken over all crossings of the given diagram K, and the sum is taken over all possible colorings. The values of the partition function are taken to be in the group ring $\mathbb{Z}[A]$ where A is the coefficient group written multiplicatively. The state-sum depends on the choice of 2-cocycle ϕ. This is proved [**7**] to be a knot invariant. Figure 8 shows the invariance of the state-sum under the Reidemeister type III move. The sums of cocycles, equated before and after the move, is the 2-cocycle condition given in Example 3.1 with the evaluation $T = 1$.

The following variations have been considered.

- Lopes [**35**] observed that the family $\{\prod_\tau B(\tau, \mathcal{C})\}_\mathcal{C}$ is a knot invariant, without taking summation. In particular, infinite quandles can be used for coloring in this case.
- For a link $L = K_1 \cup \ldots \cup K_n$, let $\mathcal{T}_i$, $i = 1, \ldots, n$, be the set of crossings at which the under-arcs belong to the component K_i. Then it was observed [**4**] that $\{\sum_\mathcal{C} \prod_{\tau \in \mathcal{T}_i} B(\tau, \mathcal{C})\}_{i=1}^n$ is a link invariant, strictly stronger than the single state-sum.

Twisted Cocycle Invariants. Let K be an oriented knot diagram with normals. The (underlying) diagram divides the plane into regions. Take an arc ℓ from the region at infinity to a region H such that ℓ intersects the arcs (missing crossings) of the diagram transversely in finitely many points. The *Alexander numbering* $\mathcal{L}(H)$ of a region H is the number of such intersections counted with signs. This does not depend on the choice of an arc ℓ.

Let τ be a crossing. There are four regions near τ, and the unique region from which normals of over- and under-arcs point is called the *source region* of τ. The *Alexander numbering* $\mathcal{L}(\tau)$ of a crossing τ is defined to be $\mathcal{L}(R)$ where R is the source region of τ. Compare with [**10**]. In other words, $\mathcal{L}(\tau)$ is the number of intersections, counted with signs, between an arc ℓ from the region at infinity to τ approaching from the source region of τ. In Fig. 9, the source region R is the left-most region, and the Alexander numbering of R is k, and so is the Alexander numbering of the crossing τ.

Let a classical knot (or link) diagram K, a finite quandle X, a finite Alexander quandle A be given. A coloring of K by X also is given and is denoted by $\mathcal{C}$. A *twisted (Boltzmann) weight*, $B_T(\tau, \mathcal{C})$, at a crossing τ is defined as follows. Let $\mathcal{C}$ denote a coloring. Let β be the over-arc at τ, and α, γ be under-arcs such that the normal to β points from α to γ. Let $x = \mathcal{C}(\alpha)$ and $y = \mathcal{C}(\beta)$. Pick a twisted quandle 2-cocycle $\phi \in Z^2_{\rm TQ}(X; A)$. Then define $B_T(\tau, \mathcal{C}) = [\phi(x, y)^{\epsilon(\tau)}]^{T^{-\mathcal{L}(\tau)}}$, where $\epsilon(\tau) = 1$ or -1, if the sign of τ is positive or negative, respectively. Here, we use the multiplicative notation of elements of A, so that $\phi(x, y)^{-1}$ denotes the inverse of $\phi(x, y)$. Recall that A admits an action by $\mathbb{Z} = \{T^n\}$, and for $a \in A$, the action of T on a is denoted by a^T. To specify the action by $T^{-\mathcal{L}(\tau)}$ in the figures, each region R with Alexander numbering $\mathcal{L}(R) = k$ is labeled by the power T^{-k} framed with a square, as depicted in Fig. 9.

The *state-sum*, or a *partition function*, is the expression

$$\Phi_T(K) = \sum_{\mathcal{C}} \prod_{\tau} B_T(\tau, \mathcal{C}).$$

The product is taken over all crossings of the given diagram, and the sum is taken over all possible colorings. The value of the weight $B_T(\tau, \mathcal{C})$ is in the coefficient group A written multiplicatively. Hence the value of the state-sum is in the group ring $\mathbb{Z}[A]$.

It was proved in [**5**] that $\Phi_T(K)$ is a knot invariant, called the *(quandle) twisted cocycle invariant.* Figure 10 depicts the invariance under the type III move, where the left-most region is assumed to have Alexander numbering -1. The sum over all these cocycles, equated before and after the move, gives the 2-cocycle condition written in Example 3.1.

Cocycle Invariants for Knotted Surfaces. The state-sum invariant is defined in an analogous way for oriented knotted surfaces in 4-space using their projections and diagrams in 3-space. Specifically, the above steps can be repeated as follows, for a fixed finite quandle X and a knotted surface diagram K.

- The diagrams consist of double curves and isolated branch and triple points [**14**]. Along the double curves, the coloring rule is defined using normals in the same way as classical case, as depicted in the left of Fig. 11.
- The Alexander numbering $\mathcal{L}$ of regions divided by a given diagram is defined similarly.
- The source region R and the Alexander numbering $\mathcal{L}(\tau) = \mathcal{L}(R)$ are defined for a triple point τ using orientation normals.

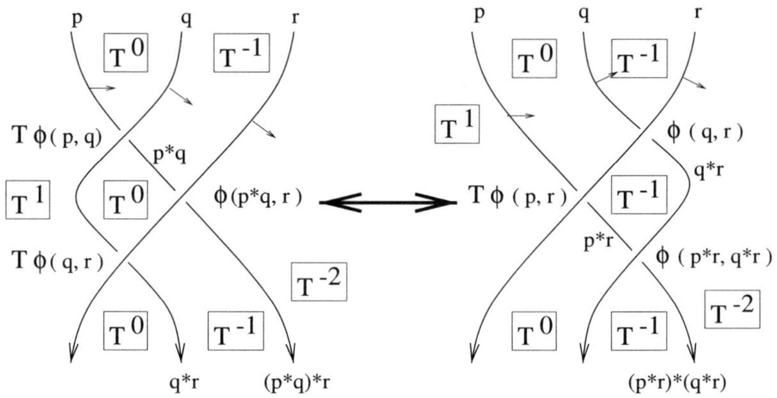

FIGURE 10. The twisted 2-cocycle condition and type III move

- The sign $\epsilon(\tau)$ of a triple point τ is defined [**14**] in such a way that it is positive if and only if the normals to top, middle, bottom sheets, in this order, match the orientation of 3-space.
- For a coloring $\mathcal{C}$, the Boltzmann weight at a triple point τ is defined by $B_T(\tau, \mathcal{C}) = [\theta(x,y,z)^{\epsilon(\tau)}]^{T^{-\mathcal{L}(\tau)}}$, where θ is a 3-cocycle, $\theta \in Z^3_{\mathrm{TQ}}(X;A)$. In the right of Fig. 11, the triple point τ is positive, and $\mathcal{L}(\tau) = 0$, so that $B_T(\tau, \mathcal{C}) = \theta(p,q,r)$.
- The state-sum is defined by $\Phi_T(K) = \sum_{\mathcal{C}} \prod_\tau B_T(\tau, \mathcal{C})$.

By checking the analogues of Reidemeister moves for knotted surface diagrams, called Roseman moves, it was shown in [**5**] that $\Phi_T(K)$ is an invariant, called the *(twisted quandle) cocycle invariant* of knotted surfaces.

Similarly, the state-sum invariant in the untwisted case was defined earlier in [**6**] and [**7**]. In the untwisted case, there is no Alexander numbering, and the Boltzmann weight at a triple point is simply the quantity $B(\tau, \mathcal{C}) = \theta(x,y,z)^{\epsilon(\tau)}$ where x, y, z are the colors on the source regions of the bottom, middle, and top sheets at the triple points.

In all of these cases, the value of the state-sum invariant depends only on the cohomology class represented by the defining cocycle. In particular, a coboundary will simply count the number of colorings of a knot or knotted surface by the quandle X.

Applications. Two important topological applications have been obtained using the cocycle invariants.

- The 2-twist spun trefoil K and its orientation-reversed counterpart $-K$ have shown to have distinct cocycle invariants using a cocycle in $Z^3_Q(R_3;\mathbb{Z}_3)$, providing a proof that K is non-invertible [**7**].

  The higher genus surfaces obtained from K by adding arbitrary number of trivial 1-handles are also non-invertible, since such handle additions do not alter the cocycle invariant.

  We note, again, that this result in higher genus cases is not immediately obtained from [**24, 40**], although higher genus generalizations of the Farber-Levine pairing [**32**] can be used.

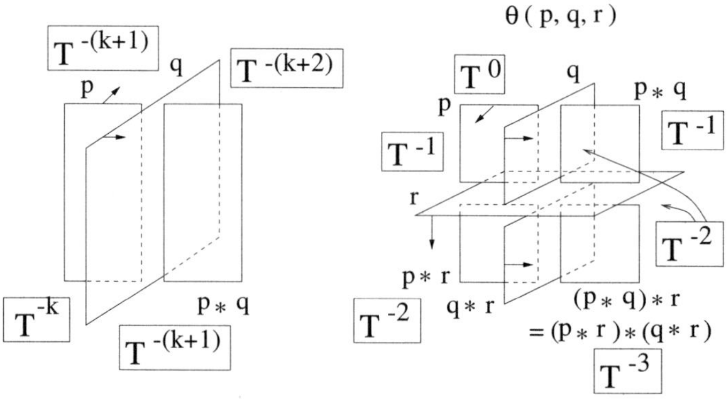

FIGURE 11. Colors at double curves and 3-cocycle at a triple point

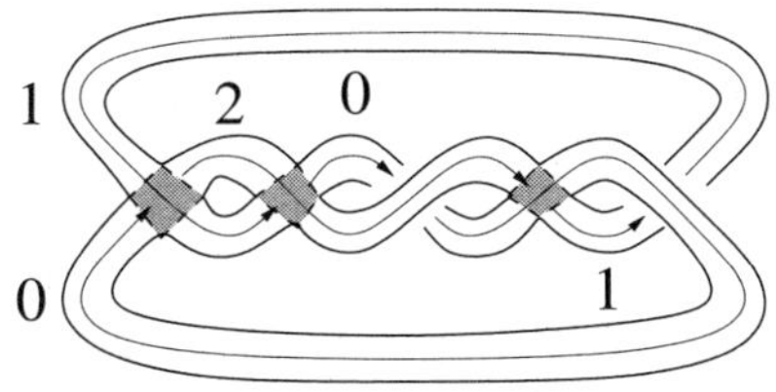

FIGURE 12. A colored virtual knot

- The projection of the 2-twist spun trefoil was shown to have at least four triple points [**41**].

  The same cocycle group $Z^3_Q(R_3; \mathbb{Z}_3)$, but a different cocycle found in [**37**] was used.

5. Virtual Knots and Quandle Homology

In this section, we describe 2-dimensional quandle homology classes as cobordism classes of quandle colored virtual knot diagrams. See [**11, 19, 20, 23**] for more general geometric descriptions of homology classes.

Consider an untwisted quandle homology class of a quandle X, and represent the class by $\eta \in Z^{\mathrm{Q}}_2(X)$. Write η as a sum of 2-chains $\eta = \sum_j \epsilon_j(a_j, b_j)$ where $\epsilon_j = \pm 1$. For each j with $\epsilon_j = 1$, consider a positive crossing diagram in which the over-arc is colored b_j and the under-arc away from which the normal to the overarc points is colored a_j. Similarly, when $\epsilon_j = -1$ we consider a negative crossing of the same form. The boundary of the chain is $\pm(a_j - a_j * b_j)$ which is the difference in the colors on the under-arcs. Since η is a cycle these boundary terms cancel over the sum of the crossings.

Thus to represent the 2-cycle, we take a disjoint union of colored crossings, and join the end-point arcs together when they have the correct orientation and the same color. The arcs are joined together formally, and the joining need not occur on a planar diagram, obtaining a colored "virtual knot diagram." Virtual knots have been popularized by L.H. Kauffman who has found, for example, that the diagram in Fig. 12 has trivial Jones polynomial. A virtual knot can be regarded as a knot on a surface [**27**].

Conversely, a colored virtual knot diagram represents a 2-cycle. In Fig. 12, such a diagram colored with $R_3 = \{0, 1, 2\}$ is depicted. The colored crossings in shaded squares, from left to right, represent 2-chains $(0, 1)$, $(1, 2)$, and $-(1, 0)$, respectively, and therefore, this diagram shows that the 2-chain $(0, 1)+(1, 2)-(1, 0)$ is a 2-cocycle. The unshaded crossings between bands can be regarded as virtual crossings. These bands connecting shaded squares correspond to identifying matching boundaries in the above construction. Some remarks are in order.

- There is a one-to-one correspondence [**11**] between (1) quandle colored virtual knot diagrams modulo the virtual Reidemeister moves and colored cobordisms, and (2) 2-dimensional quandle homology classes.
- This geometric representations were used to estimate the rank of rack homology groups in [**20, 23**] for some racks and quandles.
- This was also used to show that a certain long exact sequence splits [**11**] at low dimensions, as mentioned in Section 4.

- The 2-dimensional regions near crossings in classical diagrams can also be colored to represent 3-cycles. Such colorings were used in [**39**] to give an alternate proof that left- and right handed trefoils are not equivalent.
- The cocycle invariants can be interpreted as a formal sum of the Kronecker product between a fixed cocycle and such cycles constructed above represented by colored diagrams. Such an interpretation was used in [**12**] to evaluate the cocycle invariants.
- Twisted cycles have a similar interpretation, but the consistency of Alexander numbering requires care.
- The untwisted homology group $H_2^{\mathrm{Q}}(R_3)$ is trivial. Thus the cycle $(0,1)+(1,2)-(1,0)$ is a boundary. Meanwhile, the fundamental quandle [**30**] of the virtual knot in Fig. 12 can be computed to be R_3. Thus we have the interesting situation in which a knot, with any coloring by its fundamental quandle elements, is null-homologous in the 2-dimensional cycle group of its fundamental quandle.

6. Constructions of Cocycles from Extension Theory of Quandles

The first constructions of quandle cocycles were a combination of hand and computer calculations [**7, 8**]. Here we summarize two important cases. To describe cocycles, denote the characteristic function by

$$\chi_a(x) = \begin{cases} 1 & \text{if } a = x \\ 0 & \text{if } a \neq x \end{cases},$$

where a, x are n-tuples of elements of a quandle X.

- For the Alexander quandle $S_4 = \mathbb{Z}_2[T, T^{-1}]/(T^2+T+1)$, the $\mathbb{Z}_2$-valued function

$$\phi = \sum_{a \neq b \ a,b \neq T} \chi_{a,b}$$

represents a non-trivial cohomology class in $H^2_{\mathrm{Q}}(S_4;\mathbb{Z}_2) \cong \mathbb{Z}_2$.

- It was computed that $H^3_{\mathrm{Q}}(R_3;\mathbb{Z}_3) \cong \mathbb{Z}_3$ and a generator is given by

$$\theta = -\chi_{(0,1,0)} + \chi_{(0,2,0)} - \chi_{(0,2,1)} + \chi_{(1,0,1)} + \chi_{(1,0,2)} + \chi_{(2,0,2)} + \chi_{(2,1,2)} \in Z^3_{\mathrm{Q}}(R_3;\mathbb{Z}_3).$$

In [**7**] it was mentioned that the trefoil (3_1) and the figure-eight knot (4_1) have non-trivial cocycle invariants with the cocycle ϕ. It was also shown that the 2-twist spun trefoil is not invertible using the 3-cocycle θ. This was proven using similar techniques in [**39**]. Recently, Satoh and Shima [**41**] have shown that any diagram for the 2-twist spun trefoil has at least 4 triple points using a 3-cocycle in $Z^3_{\mathrm{Q}}(R_3;\mathbb{Z}_3)$ discovered by Mochizuki [**37**]. Mochizuki [**37**], Litherland and Nelson [**34**] have developed more techniques for computing quandle homology and cohomology.

For quantum invariants, solutions (R-matrices) to the Yang-Baxter equations were discovered by calculations first, and then Drinfeld [**17**] developed a theory of quantum groups whose representations gave rise to R-matrices. This construction is seen as an obstruction to co-commutativity satisfying the next order (the Yang-Baxter) relation, or, deformation theory of an algebraic structure giving rise to a solution to a higher order relation. Considering analogies between group and quandle cohomology theories, it is, then, natural to seek such methods of finding cocycles in deformation and extension theories of quandles. An extension theory of quandles was developed in [**5**] for the twisted case as follows (see also [**4, 12**]), in analogy with the group cohomology theory (one sees that the following is in parallel to Chapter IV of [**3**]).

- Let X be a quandle and A be an Alexander quandle, so that A admits an action by $\mathbb{Z}$ whose generator is denoted by T. Let $\phi \in Z^2_{\rm TQ}(X;A)$. Let $AE(X,A,\phi)$ be the quandle defined on the set $A \times X$ by the operation $(a_1,x_1) * (a_2,x_2) = (a_1 * a_2 + \phi(x_1,x_2), x_1 * x_2)$.
- The above defined operation $*$ on $A \times X$ indeed defines a quandle $AE(X,A,\phi) = (A \times X, *)$, which is called an *Alexander extension* of X by (A,ϕ).
- Let X be a quandle and A be an Alexander quandle. Recall that $\eta \in Z^1_{\rm TQ}(X;A)$ implies that $\eta : X \to A$ is a quandle homomorphism. Let $0 \to N \xrightarrow{i} G \xrightarrow{p} A \to 0$ be an exact sequence of $\mathbb{Z}[T,T^{-1}]$-module homomorphisms among Alexander quandles. Let $s : A \to G$ be a set-theoretic section (i.e., $ps = {\rm id}_A$) with the "normalization condition" $s(0)=0$. Then $s\eta : X \to G$ is a mapping, which is not necessarily a quandle homomorphism. We measure the failure by 2-cocycles. Since $p[Ts\eta(x_1) + (1-T)s\eta(x_2)] = p[s\eta(x_1 * x_2)]$ for any $x_1, x_2 \in A$, there is $\phi(x_1,x_2) \in N$ such that
$$Ts\eta(x_1) + s\eta(x_2) = i\phi(x_1,x_2) + [Ts\eta(x_2) + s\eta(x_1 * x_2)].$$
This defines a function $\phi \in C^2_{\rm TQ}(X;N)$. Then it was shown that $\phi \in Z^2_{\rm TQ}(X;N)$.
- Let $s' : A \to G$ be another section, and $\phi' \in Z^2_{\rm TQ}(X;N)$ be a 2-cocycle determined by
$$Ts'\eta(x_1) + s'\eta(x_2) = i\phi'(x_1,x_2) + [Ts'\eta(x_2) + s'\eta(x_1 * x_2)].$$
Then it was shown that $[\phi] = [\phi'] \in H^2_{\rm TQ}(X;N)$.
- It was shown that if $[\phi] = 0 \in H^2_{\rm TQ}(X;N)$, then ϕ extends to a quandle homomorphism to G, i.e., there is a quandle homomorphism $\eta' : X \to G$ such that $p\eta' = \eta$.

The above results were summarized as

THEOREM 6.1. [**5**] *The obstruction to extending $\eta : X \to A$ to a quandle homomorphism $X \to G$ lies in $H^2_{\rm TQ}(X;N)$.*

Conversely, we have the following.

LEMMA 6.2. [**5**] *Let X, E be quandles, and A be an Alexander quandle. Suppose there exists a bijection $f : E \to A \times X$ with the following property. There exists a function $\phi : X \times X \to A$ such that for any $e_i \in E$ $(i = 1,2)$, if $f(e_i) = (a_i, x_i)$, then $f(e_1 * e_2) = (a_1 * a_2 + \phi(x_1,x_2), x_1 * x_2)$. Then $\phi \in Z^2_{\rm TQ}(X;A)$.*

This lemma implies that under the same assumption we have $E = AE(X,A,\phi)$, where $\phi \in Z^2_{\rm TQ}(X;A)$, and by identifying such quandles, we obtain cocycles as desired. We identify such examples, and include a proof, as it provides explicit formulas of cocycles.

Let $\Lambda_p = \mathbb{Z}_p[T,T^{-1}]$ for a positive integer p (or $p=0$, in which case Λ_p is understood to be $\Lambda = \mathbb{Z}[T,T^{-1}]$). Note that since T is a unit in Λ_p, $\Lambda_p/(h)$ for a Laurent polynomial $h \in \Lambda_p$ is isomorphic to $\Lambda_p/(T^n h)$ for any integer n, so that we may assume that h is a polynomial with a non-zero constant (without negative exponents of T).

LEMMA 6.3. [**5**] *Let $h \in \Lambda_{p^m}$ be a polynomial with leading and constant coefficients invertible, or $h = 0$. Let $\bar{h} \in \Lambda_{p^{m-1}}$ and $\tilde{h} \in \Lambda_p$ be such that $\bar{h} \equiv h \bmod (p^{m-1})$ and $\tilde{h} \equiv h \bmod (p)$, respectively (in other words, $\bar{h}$ is h with its coefficients reduced modulo p^{m-1}, and $\tilde{h}$ is h with its coefficients reduced modulo p). Then the quandle $E = \Lambda_{p^m}/(h)$ satisfies the conditions in Lemma 6.2 with $X = \Lambda_{p^{m-1}}/(\bar{h})$ and $A = \Lambda_p/(\tilde{h})$.*

In particular, $\Lambda_{p^m}/(h)$ is an Alexander extension of $\Lambda_{p^{m-1}}/(\bar{h})$ by $\Lambda_p/(\tilde{h})$:

$$\Lambda_{p^m}/(h) = AE(\Lambda_{p^{m-1}}/(\bar{h}),\ \Lambda_p/(\tilde{h}),\ \phi),$$

for some $\phi \in Z^2_{\mathrm{TQ}}(\Lambda_{p^{m-1}}/(\bar{h}); \Lambda_p/(\tilde{h}))$.

Proof. Let $A \in \mathbb{Z}_{p^m}$. Represent A in p^m-ary notation as

$$A = \sum_{i=0}^{m-1} A_i p^i$$

where $A_i \in \{0, \dots, p-1\}$. Since p is fixed throughout, we represent A by the sequence

$$[A_{m-1}, A_{m-2}, A_{m-3}, \dots, A_0].$$

Define $\overline{A} = [A_{m-2}, \dots, A_0]$. Observe that $A \equiv \overline{A} \pmod{p^{m-1}}$, and $A \equiv A_0 \pmod{p}$.

Let $\hat{\pi} : \mathbb{Z}_{p^m} \to \mathbb{Z}_{p^{m-1}}$ be the map defined by $\hat{\pi}(A) = \overline{A}$. We obtain a short exact sequence:

$$0 \to \mathbb{Z}_p \xrightarrow{\hat{\imath}} \mathbb{Z}_{p^m} \xrightarrow{\hat{\pi}} \mathbb{Z}_{p^{m-1}} \to 0$$

where $\hat{\imath}(A) = [A, 0, \dots, 0]$. There is a set-theoretic section $\mathbb{Z}_{p^m} \xleftarrow{\hat{s}} \mathbb{Z}_{p^{m-1}}$ defined by $\hat{s}[A_{m-2}, \dots, A_0] = [0, A_{m-2}, \dots, A_0]$. The map $\hat{s}$ satisfies $\hat{\pi}\hat{s} = \mathrm{id}$ and $\hat{s}(0) = 0$.

For a polynomial $L(T) \in \Lambda_{p^m} = \mathbb{Z}_{p^m}[T, T^{-1}]$, write

$$L(T) = \sum_{j=-n}^{k} [A_{j,m-1}, A_{j,m-2}, \dots, A_{j,0}] T^j.$$

Define

$$\overline{L}(T) = \sum_{j=-n}^{k} [A_{j,m-2}, \dots, A_{j,0}] T^j \in \Lambda_{p^{m-1}},$$

and

$$\tilde{L}(T) = \sum_{j=-n}^{k} A_{j,m-1} T^j \in \Lambda_p.$$

There is a one-to-one correspondence $f : \Lambda_{p^m} \to \Lambda_p \times \Lambda_{p^{m-1}}$ given by $f(L) = (\tilde{L}, \overline{L})$. We have a short exact sequence of rings:

$$0 \to \mathbb{Z}_p[T, T^{-1}] \xrightarrow{i} \mathbb{Z}_{p^m}[T, T^{-1}] \xrightarrow{\pi} \mathbb{Z}_{p^{m-1}}[T, T^{-1}] \to 0$$

with a set theoretic section $\mathbb{Z}_{p^m}[T, T^{-1}] \xleftarrow{s} \mathbb{Z}_{p^{m-1}}[T, T^{-1}]$ where i, π and s are the natural maps induced by $\hat{\imath}$, $\hat{\pi}$ and $\hat{s}$, respectively. Note that for $L \in \Lambda_{p^m} =$

$\mathbb{Z}_{p^m}[T,T^{-1}]$ we have $\overline{L} = \pi(L)$, and the section $s : \Lambda_{p^{m-1}} \to \Lambda_{p^m}$ is defined by the formula

$$s\left(\sum_{j=-n}^{k} [A_{j,m-2}, \ldots, A_{j,0}]T^j\right) = \sum_{j=-n}^{k} [0, A_{j,m-2}, \ldots, A_{j,0}]T^j.$$

For $L, M \in \Lambda_{p^m}$, let

$$s(\overline{L}) * s(\overline{M}) = \sum_j [F_{j,m-1}, \ldots, F_{j,0}]T^j \in \Lambda_{p^{m-1}}.$$

If $L = \sum_j A_j T^j$, and $M = \sum_j B_j T^j$, then

$$L * M = B_{-n}T^{-n} + \sum_{j=-n+1}^{k+1} (A_{j-1} - B_{j-1} + B_j)T^j = \sum_{j=-n}^{k} C_j T^j.$$

Furthermore,

$$\begin{aligned}\overline{L} * \overline{M} &= [B_{-n,m-2}, \ldots, B_{-n,0}]T^{-n} \\ &+ \sum_{j=-n+1}^{k+1} ([A_{j-1,m-2}, \ldots, A_{j-1,0}] \\ &\quad -[B_{j-1,m-2}, \ldots, B_{j-1,0}] + [B_{j,m-2}, \ldots, B_{j,0}])\, T^j\end{aligned}$$

and write the right-hand side by $\sum_{j=-n}^{k} D_j T^j$. Note that D_j's are well-defined integers, not only elements of $\mathbb{Z}_{p^{m-2}}$. If D_j is positive, then $F_{j,m-1} = 0$, and if D_j is negative, then $F_{j,m-1} = p - 1$. Hence

$$f(L * M) = (\tilde{L} * \tilde{M} + \phi(\overline{L}, \overline{M}), \overline{L} * \overline{M}),$$

where

$$\phi(\overline{L}, \overline{M}) = \sum_{j=-n}^{k} F_{j,m-1}.$$

This concludes the case $h = 0$.

Now let $h(T) \in \mathbb{Z}_{p^m}[T]$ be a polynomial with leading and constant coefficients being invertible in $\mathbb{Z}_p$. Let (h) denote the ideal generated by h. Since $i(\tilde{h}) \subset (h)$, we obtain a short exact sequence of quotients:

$$0 \to \mathbb{Z}_p[T,T^{-1}]/(\tilde{h}) \xrightarrow{\bar{i}} \mathbb{Z}_{p^m}[T,T^{-1}]/(h) \xrightarrow{\overline{\pi}} \mathbb{Z}_{p^{m-1}}[T,T^{-1}]/(\overline{h}) \to 0$$

with a set-theoretic section $\mathbb{Z}_{p^m}[T,T^{-1}]/(h) \xleftarrow{\bar{s}} \mathbb{Z}_{p^{m-1}}[T,T^{-1}]/(\overline{h})$. Thus we obtain a twisted cocycle

$$\phi : \mathbb{Z}_{p^{m-1}}[T,T^{-1}]/(\overline{h}) \times \mathbb{Z}_{p^{m-1}}[T,T^{-1}]/(\overline{h}) \to \mathbb{Z}_p[T,T^{-1}]/(\tilde{h}) \qquad \square$$

Since $R_n = \Lambda_n/(T+1)$, we have the following.

COROLLARY 6.4. *The dihedral quandle $E = R_{p^m}$, where p, m are positive integers with $m > 1$, satisfies the conditions in Lemma 6.2 with $X = R_{p^{m-1}}$ and $A = R_p$.*

In particular, R_{p^m} is an Alexander extension of $R_{p^{m-1}}$ by R_p: $R_{p^m} = AE(R_{p^{m-1}},\ R_p,\ \phi)$, for some $\phi \in Z^2_{\mathrm{TQ}}(R_{p^{m-1}}; R_p)$.

EXAMPLE 6.5. Let $X = R_3$ and $A = R_3$, then the proof of Lemma 6.3 gives an explicit 2-cocycle ϕ as follows. For $\phi(r_1, r_2) = \phi(1, 2)$, for example, one computes

$$r_1 * r_2 = [0,1] * [0,2] = 2[0,2] - [0,1] = 3 = 3 \cdot 1 + 0 = [1,0],$$

Hence $\phi(0,2) = 1$. In terms of the characteristic function, the cocycle ϕ contains the term $\chi_{0,2}$. By computing the quotients for all pairs, one obtains

$$\phi = \chi_{0,2} + \chi_{1,2} + 2\chi_{1,0} + 2\chi_{2,0}.$$

The same argument was applied to R_∞ to show that the quandle R_∞ is an Alexander extension of R_n by R_∞, for any positive integer n.

Similar techniques give us untwisted cocycles [**4**], with explicit formulas for these 2-cocycles as follows. In this case, the extension is called an *abelian extension*, denoted by $E = E(X, A, \phi)$ for $\phi \in Z^2_{\mathrm{Q}}(X; A)$, and the quandle operation on $E = A \times X$ is defined by $(a_1, x_1) * (a_2, x_2) = (a_1 + \phi(x_1, x_2), x_1 * x_2)$.

- For any positive integers q and m, $E = \mathbb{Z}_{q^{m+1}}[T, T^{-1}]/(T - 1 + q)$ is an abelian extension $E = E(\mathbb{Z}_{q^m}[T, T^{-1}]/(T - 1 + q), \mathbb{Z}_q, \phi)$ of $X = \mathbb{Z}_{q^m}[T, T^{-1}]/(T - 1 + q)$ for some cocycle $\phi \in Z^2_{\mathrm{Q}}(X; \mathbb{Z}_q)$.
- For any positive integer q and m, the quandle $E = \mathbb{Z}_q[T, T^{-1}]/(1 - T)^{m+1}$ is an abelian extension of $X = \mathbb{Z}_q[T, T^{-1}]/(1 - T)^m$ over $\mathbb{Z}_q$: $E = E(X, \mathbb{Z}_q, \phi)$, for some $\phi \in Z^2_{\mathrm{Q}}(X; \mathbb{Z}_q)$.

Furthermore, for untwisted 2-cocycles, an interpretation of the cocycle knot invariant was given [**4**] as an obstruction to extending a given coloring of a knot diagram by a quandle X to a coloring by an abelian extension E. Similar interpretations for twisted case or knotted surface case are unknown.

Ohtsuki [**38**] defined a new cohomology theory for quandles and an extension theory, together with a list of problems in the subject.

Acknowledgments. We gratefully acknowledge the contributions of our collaborators in these projects: Mohammed Elhamdadi, Daniel Jelsovsky, Seiichi Kamada, Louis Kauffman, Laurel Langford, and Marina Nikiforou. We thank the organizers of the 2001 GTC for their hard work and for allowing us the opportunity to present these results. As we were finishing this paper we received a copy of the survey [**28**]. We are pleased to refer the reader to that paper as well.

References

[1] Artin, E., *Zur Isotopie zweidimensionalen Flächen im R_4*, Abh. Math. Sem. Univ. Hamburg **4** (1926), 174–177.

[2] Brieskorn, E., *Automorphic sets and singularities*, Contemporary math., **78** (1988), 45–115.

[3] Brown, K. S., *Cohomology of groups*, Graduate Texts in Mathematics, **87**. Springer-Verlag, New York-Berlin (1982).

[4] Carter, J, S.; Elhamdadi, M.; Nikiforou, M.; Saito. M., *Extensions of quandles and cocycle knot invariants*, Preprint. http://xxx.lanl.gov/math/abs/GT0107021.

[5] Carter, J, S.; Elhamdadi, M.; Saito. M., *Twisted Quandle homology theory and cocycle knot invariants*, Algebraic and Geometric Topology, **2** (2002), 95-135.

[6] Carter, J.S.; Jelsovsky, D.; Kamada, S.; Langford, L.; Saito, M., *State-sum invariants of knotted curves and surfaces from quandle cohomology*, Electron. Res. Announc. Amer. Math. Soc. **5** (1999), 146–156.

[7] Carter, J.S.; Jelsovsky, D.; Kamada, S.; Langford, L.; Saito, M., *Quandle cohomology and state-sum invariants of knotted curves and surfaces*, to appear in Trans. Amer. Math. Soc., available at http://xxx.lanl.gov/abs/math.GT/9903135.

[8] Carter, J.S.; Jelsovsky, D.; Kamada, S.; Saito, M., *Computations of quandle cocycle invariants of knotted curves and surfaces*, Advances in Math. **157** (2001), 36–94.

[9] Carter, J.S.; Jelsovsky, D.; Kamada, S.; Saito, M., *Quandle homology groups, their betti numbers, and virtual knots*, J. of Pure and Applied Algebra **157** (2001), 135–155.

[10] Carter, J.S.; Kamada, S.; Saito, M., *Alexander numbering of knotted surface diagrams*, Proc. Amer. Math. Soc. **128** (2000), 3761–3771.

[11] Carter, J.S.; Kamada, S.; Saito, M., *Geometric interpretations of quandle homology*, J. of Knot Theory and its Ramifications **10** (2001), 345–358.

[12] Carter, J.S.; Kamada, S.; Saito, M., *Diagrammatic computations for quandles and cocycle knot invariants*, to appear in Contemporary Math., Proceedings of a conference on quantum topology, held in San Francisco, 1999, available at http://xxx.lanl.gov/abs/math.GT/0102092.

[13] Carter, J.S.; Kauffman, L.H.; Saito, M., *Structures and diagrammatics of four dimensional topological lattice field theories*, Advances in Math. **146** (1999), 39–100.

[14] Carter, J.S.; Saito, M., *Knotted surfaces and their diagrams*, American Mathematical Society, Mathematical surveys and monographs, vol. 55, 1998.

[15] Crane, L.; Kauffman, L.H.; Yetter, D.N., *State-sum invariants of 4-manifolds*, J. Knot Theory Ramifications, **6** (1997) 177–234.

[16] Dijkgraaf, R.; Witten, E., *Topological gauge theories and group cohomology*, Comm. Math. Phys. **129** (1990), 393–429.

[17] Drinfel'd, V.G., *Quantum groups*, in Proceedings of the International Congress of Mathematicians (Berkeley, Calif., 1986), Amer. Math. Soc., Providence, RI, 1987, pp. 798–820.

[18] Fenn, R.; Rourke, C., *Racks and links in codimension two*, Journal of Knot Theory and Its Ramifications **1** (1992), 343–406.

[19] Fenn, R.; Rourke, C.; Sanderson, B., *James bundles and applications*, preprint, available at http://www.maths.warwick.ac.uk/~bjs/ .

[20] Flower, J., *Cyclic Bordism and Rack Spaces*, Ph.D. Dissertation, Warwick, 1995.

[21] Fox, R.H., *A quick trip through knot theory*, in Topology of 3-Manifolds, Ed. M.K. Fort Jr., Prentice-Hall (1962), 120–167.

[22] Fox, R.H. ; Milnor, J.W., *Singularities of 2-spheres in 4-space*, Manuscript, Circa 1957.

[23] Greene, M. T., *Some Results in Geometric Topology and Geometry*, Ph.D. Dissertation, Warwick, 1997.

[24] Hillman, J.A., *Finite knot modules and the factorization of certain simple knots*, Math. Ann. **257** (1981), 261–274.

[25] Inoue, A., *Quandle homomorphisms of knot quandles to Alexander quandles*, J. Knot Theory Ramifications, **10** (2001), 813–821.

[26] Joyce, D., *A classifying invariant of knots, the knot quandle*, J. Pure Appl. Alg., **23** (1982), 37–65.

[27] Kamada, N.; Kamada, S., *Abstract link diagrams and virtual knots*, J. Knot Theory Ramifications, **9** (2000) 93–106.

[28] Kamada, S., *Knot invariants derived from quandles and racks*, preprint, 2001.

[29] Kauffman, L. H., *Knots and Physics*, World Scientific, Series on knots and everything, vol. 1, 1991.

[30] Kauffman, L. H., *Virtual knot theory*, European J. Combin. **20** (1999), 663–690.

[31] Kawauchi, A., *Three dualities on the integral homology of infinite cyclic coverings of manifolds*, Osaka J. Math. **23** (1986), 633–651.

[32] Kawauchi, A., *The first Alexander modules of surfaces in 4-sphere*, in Algebra and Topology (Taejon, 1990), Proc. KAIST Math. Workshop, 5, KAIST, Taejon, Korea (1990), pp. 81–89.

[33] Litherland, R.A., Letter to Cameron Gordon: Example 12 is the 2-twist spun trefoil.

[34] Litherland, R.A.; Nelson, S., *The Betti numbers of some finite racks*, preprint, available at http://xxx.lanl.gov/abs/math.GT/0106165.

[35] Lopes, P., *Quandles at finite temperatures I*, preprint, available at http://xxx.lanl.gov/abs/math.QA/0105099.

[36] Matveev, S., *Distributive groupoids in knot theory*, (Russian) Mat. Sb. (N.S.) 119(161) (1982), no. 1, 78–88, 160.

[37] Mochizuki, T., *Some calculations of cohomology groups of finite Alexander quandles*, preprint.

[38] Ohtsuki, T. (ed), *Problems on invariants of knots and 3-manifolds*, preprint.

[39] Rourke, C.; Sanderson, B., *There are two 2-twist-spun trefoils*, preprint, available at http://xxx.lanl.gov/abs/math.GT/0006062.

[40] Ruberman, D., *Doubly slice knots and the Casson-Gordon invariants*, Trans. Amer. Math. Soc. **279** (1983), 569–588.

[41] Satoh, S.; Shima, A., *The 2-twist spun trefoil has the triple number four*, preprint.

[42] Sumners, D.W., *Invertible Knot Cobordisms*, in Proceedings of the 1968 Georgia Topology Conference, Topology of Manifolds, ed. Cantrell, J. C, and Edwards, C.H. Jr., Markham, 1970.

[43] Wakui, M., *On Dijkgraaf-Witten invariant for 3-manifolds*, Osaka J. Math. **29** (1992), 675–696.

[44] Zeeman, E.C., *Twisting spun knots*, Trans. Amer. Math. Soc., **115** (1965) 471–495.

University of South Alabama, Mobile, AL 36688
E-mail address: `carter@jaguar1.usouthal.edu`

University of South Florida, Tampa, FL 33620
E-mail address: `saito@math.usf.edu`

Proceedings of Symposia in Pure Mathematics
Volume **71**, 2003

Boundary train tracks of laminar branched surfaces

Tao Li

1. Introduction

Essential laminations are codimension one objects in 3-manifolds that generalize both incompressible surfaces and taut foliations. It has been shown in [**9**] that if a closed orientable 3-manifold contains an essential lamination, then its universal cover is $\mathbb{R}^3$. Many 3-manifolds contain essential laminations. For example, the Property P conjecture has been proved for alternating knots [**2, 13**] and arborescent knots [**17**] by constructing essential laminations in knot manifolds. Most of the constructions of essential laminations were given by constructing a branched surface first and then showing that it carries useful essential laminations. In [**10**], it is shown that if a branched surface satisfies some standard conditions and it contains no sink disk, then it fully carries an essential lamination, and (except for the case of a lamination of T^3 by planes) any essential lamination is carried by such a branched surface. The conditions in [**10**] are easy to check in general, but in many situations we want to put some restrictions on the essential laminations. In particular, if we have a manifold with torus boundary and a properly embedded branched surface (with boundary), a natural question is whether this branched surface carries a lamination whose boundary is a union of circles with slope s. This question is interesting because if it carries such a lamination, then the lamination extends to a lamination in the manifold obtained by Dehn filling along the slope s. The next question is whether this extended lamination is essential in the closed manifold after Dehn filling. In this paper, we prove the following theorem which says that given a branched surface with certain easy-to-check conditions, for any slope realizable by its boundary train track, we always get an essential lamination in the manifold after Dehn filling (along this slope).

THEOREM 2.5. *Let M be an irreducible and orientable 3-manifold whose boundary is an incompressible torus. Suppose B is a laminar branched surface and $\partial M - \partial B$ is a union of bigons. Then, for any rational slope $s \in \mathbb{Q} \cup \infty$ that can be realized by the train track ∂B, if B does not carry a torus that bounds a solid torus in $M(s)$, $M(s)$ contains an essential lamination.*

2000 *Mathematics Subject Classification.* Primary 57M50, 57M25.

Partially supported by NSF grant DMS 0102316.

The branched surfaces (with boundary) constructed in [**12, 13, 14, 15**] all satisfy the hypotheses in Theorem 2.5 (after a simple splitting in a special case), and the interval of slopes computed in [**12, 13, 14, 15**] using affine laminations are in fact the exact intervals of realizable slopes as described in Theorem 2.5. In section 3, we also give a way to calculate realizable slopes by splitting and simplifying train tracks.

As an application, we show that:

THEOREM 4.1. *Let K be a knot in S^3 and R be a minimal genus Seifert surface. If R can be decomposed as a $2n$-Murasugi sum, then $K(s)$ has a taut foliation for any slope $s \in (1-n, n-1)$, where $K(s)$ is the closed 3-manifold after Dehn surgery along the slope s.*

The following is an easy corollary to Theorem 4.1. It says that, except for one case (i.e. the case when the Murasugi sum is a plumbing a trivial twisted band and a nonplanar surface), such a knot has property P. In section 4, we also discuss a construction of essential laminations for this exceptional case using Theorem 2.5.

COROLLARY 4.2. *Let K be a knot in S^3 and R be a minimal genus Seifert surface of K. Suppose R can be decomposed as a $2n$-Murasugi sum. Then, unless the Murasugi sum is a plumbing of a trivial twisted band and a nonplanar surface, K has property P. In particular, if $n \neq 2$, K has property P.*

NOTATION. In this paper, we denote a small regular neighborhood of X by $N(X)$, and denote the interior of X by $int(X)$.

2. Laminar branched surfaces with boundary

In this section, we first explain the relation between essential laminations and branched surfaces without sink disks, which is the main result in [**10**]. Then, we consider branched surfaces in a 3-manifold with torus boundary. We show that if a branched surface does not contain any (half) sink disk and satisfies some standard conditions, then the question of whether this branched surface carries a lamination that extends to an essential lamination in the manifold after Dehn filling totally depends on its boundary train track.

Let B be a branched surface and L be the branch locus of B. We call the closure (under path metric) of each component of $B - L$ a *sector* of B. L is a collection of smooth immersed curves in B. Let Z be the union of double points of L. We associate with every component of $L - Z$ a normal vector (in B) pointing in the direction of the cusp, as shown in Figure 2.1. We call it the *branch direction* of this arc. We call a disk sector of B a *sink disk* if the branch direction of every smooth arc (or curve) in its boundary points into the disk. See Figure 2.1 for an example.

Note that if we "blow an air bubble" into the interior of a sink disk, it will destroy the sink disk by definition, but nothing really changes. So, we need to eliminate these trivial "air bubbles" first in the branched surface.

Let X be an orientable, irreducible and closed 3-manifold and B be an incompressible branched surface in X. Let $N(B)$ be a fibered neighborhood of B and K be a component of $M - int(N(B))$. If K is a 3-ball, since B is incompressible, we can give K a fiber structure $D^2 \times I$ with $D^2 \times \partial I \subset \partial_h N(B)$ and $\partial D^2 \times I \subset \partial_v N(B)$. We call K a *$D^2 \times I$ region*

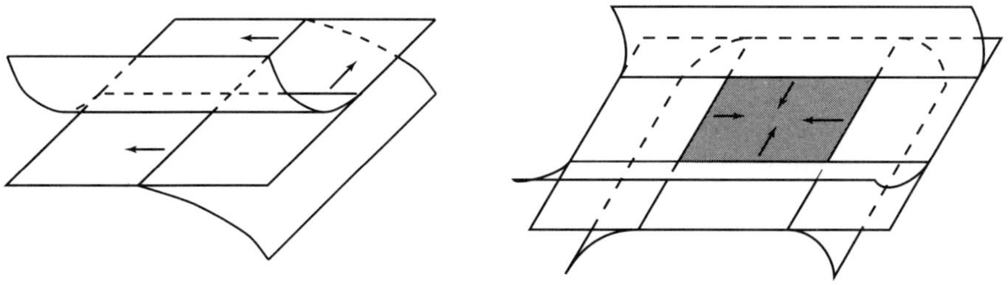

(a) (b)

FIGURE 2.1

DEFINITION 2.1 ([**10**]). Let D_1 and D_2 be the two disk components of the horizontal boundary of a $D^2 \times I$ region K in $M - int(N(B))$. Hence, D_1 and D_2 are also two disk components of $\partial_h N(B)$ and $D_1 \cup D_2 = D^2 \times \partial I \subset \partial K$. Thus, $\pi(\partial D_1) = \pi(\partial D_2)$ is a circle in the branch locus L, where $\pi : N(B) \to B$ is the collapsing map. If π restricted to the interior of $D_1 \cup D_2$ is injective, i.e., the intersection of each I-fiber of $N(B)$ with $int(D_1) \cup int(D_2)$ is either empty or a single point, then we call K a *trivial* $D^2 \times I$ *region*, and we say that $\pi(D_1 \cup D_2)$ forms a *trivial bubble* in B ($\pi(D_1 \cup D_2)$ bounds a 3-ball in $M - B$).

If K is a trivial $D^2 \times I$ region as above, then $B \cup K$ is a fibered neighborhood of another branched surface $\bar{B}$ which can be obtained by collapsing the trivial bubble. This operation of collapsing a trivial bubble does not change properties of B, since any surface or lamination is carried by B if and only if it is carried by $\bar{B}$. Note that not all $D^2 \times I$ regions are trivial. For example, the $D^2 \times I$ region in the simplest Reeb branched surface is not trivial (its top is glued to its bottom). As we pointed out before, because of the combinatorial nature of the definition of sink disk, we would like to eliminate these trivial bubbles first.

DEFINITION 2.2 ([**10**]). Let B be a branched surface in N and suppose we have eliminated all the trivial bubbles. We call B a *laminar branched surface* if B is incompressible, there is no Reeb component and B contains no sink disk.

The following theorem is proved in [**10**], it gives some kind of equivalence relation between essential laminations and branched surfaces without sink disks.

THEOREM 2.3 ([**10**]). *Suppose N is a closed and orientable 3-manifold. Then*

(a) *Every laminar branched surface in N fully carries an essential lamination.*
(b) *Any essential lamination in N that is not a lamination by planes is fully carried by a laminar branched surface.*

Furthermore, if $\lambda \subset N$ is a lamination by planes (hence $M = T^3$), then any branched surface carrying λ must contain a sink disk and hence is not a laminar branched surface.

In this paper, we are interested in obtaining a closed manifold through Dehn surgery, and we would like to see, given a branched surface with boundary, whether this branched surface carries useful laminations. We first define the relative version of sink disk and trivial bubble.

Let M be an irreducible 3-manifold whose boundary is an incompressible torus. Suppose B is a properly embedded branched surface with boundary ($\partial B \subset \partial M$). Let L be the branch locus of B, and D be the closure (in the path metric) of a disk component of $B - L$. We call D a *half sink disk* if $\partial D \cap \partial M \neq \emptyset$ and the branch direction of each arc in $\partial D - \partial M$ points into D. Note that $\partial D \cap \partial M$ may not be connected.

Now, we define the relative version of trivial $D^2 \times I$ regions and trivial bubbles in the same way as before. Let B be an incompressible branched surface properly embedded in a 3-manifold M ($B \cap \partial M = \partial B$), and let $N(B)$ be a fibered neighborhood of B. Suppose K is a component of $M - int(N(B))$ and K admits a fiber structure $D^2 \times I$ such that $D^2 \times \partial I = K \cap \partial_h N(B)$, and $\partial D^2 \times I = K \cap (\partial M \cup \partial_v N(B))$ with the same fiber structure as that of $\partial_v N(B)$ at $(\partial D^2 \times I) \cap \partial_v N(B)$. Let D_1 and D_2 be the two components of $D^2 \times \partial I$. If π restricted to the interior of $D_1 \cup D_2$ is injective, i.e., the intersection of any I-fiber of $N(B)$ with $int(D_1) \cup int(D_2)$ is either empty or a single point, then we call K a *trivial* $D^2 \times I$ *region*, and we say that $\pi(D_1 \cup D_2)$ forms a *trivial bubble* in B. As before, a trivial bubble can be eliminated by pinching B.

DEFINITION 2.4. Let B is a properly embedded branched surface in a 3-manifold M. We say B is a *laminar branched surface* if, after eliminating trivial bubbles, the followings are satisfied.

(1) $\partial_h N(B)$ is essential in the following sense: $\partial_h N(B)$ is incompressible and ∂-incompressible in $M - int(N(B))$, there is no monogon in $M - int(N(B))$ and no component of $\partial_h N(B)$ is a sphere or a disk properly embedded in M.
(2) $M - int(N(B))$ is irreducible and $\partial M - int(N(B))$ is incompressible in $M - int(N(B))$.
(3) B contains no Reeb branched surface (see [**9**] for the definition of Reeb branched surface).
(4) B has no sink disk or half sink disk.

Let T be a torus, then the isotopy class of every nontrivial simple closed curve in T uniquely corresponds to a rational slope in $\mathbb{Q} \cup \infty$. Let τ be a train track in T and $s \in \mathbb{Q} \cup \infty$ be a rational slope. We say that s can be *realized* by τ if τ fully carries a union of simple closed curves with slope s, i.e., we can split τ into a union of simple closed curves with slope s. For any slope $s \in \mathbb{Q} \cup \infty$, we denote by $M(s)$ the manifold after Dehn filling along s.

THEOREM 2.5. *Let M be an irreducible and orientable 3-manifold whose boundary is an incompressible torus. Suppose B is a laminar branched surface and $\partial M - \partial B$ is a union of bigons. Then, for any rational slope $s \in \mathbb{Q} \cup \infty$ that can be realized by the train track ∂B, if B does not carry a torus that bounds a solid torus in $M(s)$, $M(s)$ contains an essential lamination.*

REMARK 2.6. (1) Since ∂M is a torus and $\partial_h N(B)$ is essential as in definition 2.4, by a standard index argument, $\partial M - \partial B$ must be a union of bigons and annuli. If $\partial M - \partial B$ has an annular component, ∂B can realize at most one slope which is the slope of the boundary of this annular component.
(2) Note that if B carries a torus, then this torus must be incompressible in M. Thus, if $int(M)$ admits a hyperbolic structure and B does not carry

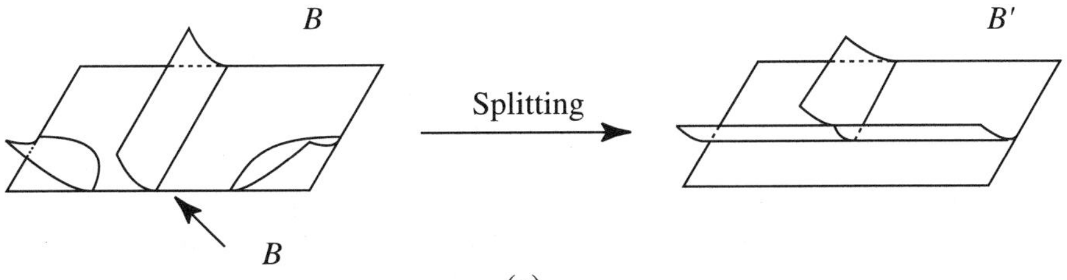

(a)

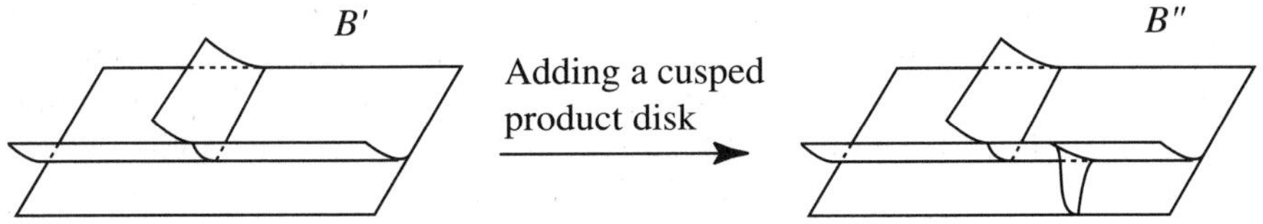

(b)

FIGURE 2.2

a ∂-parallel torus (which is easy to check in practice), then B does not carry any torus at all and the last requirement on B in Theorem 2.5 is satisfied.

PROOF. If s can be realized by ∂B, s can also be realized by the boundary train track of the new branched surface after we collapse a trivial bubble. Thus, we can assume that B has no trivial bubble. Now, we split B in a small neighborhood of ∂B, as shown in Figure 2.2 (a), such that ∂B becomes a union of simple closed curves with slope s. Note that during the splitting, no new component of $\partial_v N(B)$ is created (but two components of $\partial_v N(B)$ may merge and become one component in the new branched surface after splitting). We denote the branched surface after splitting by B'.

Now, we can do Dehn filling along the slope s by adding a disk along each circle in $\partial B'$ and capping off the 2-spheres. We denote the branched surface in $M(s)$ obtained from B' as above by B_s. By Theorem 2.3, it suffices to show that B_s is a laminar branched surface in the closed manifold $M(s)$.

Let D be a product disk in $M - int(N(B'))$ with one vertical boundary arc in ∂M and the other vertical boundary arc in $\partial_v N(B)$, i.e., D is a compressing disk for $M - int(N(B'))$ such that $\partial D \cap \partial_v N(B')$ is a vertical arc in $\partial_v N(B')$ and $\partial D \cap \partial M$ is a vertical arc in an annular component of $\partial M - int(N(B'))$. Then, $\pi(D)$ is a triangle with vertices A, B, C such that $A, B \in \partial B'$ and $C \in L$, where L is the branch locus of B and $\pi : N(B') \to B'$ is the collapsing map. Note that the edge $\overline{AB}$ is an embedded vertical arc of (the closure of) an annular component of $\partial M - \partial B'$ and $int(D)$ is embedded (edges $\overline{AC}$ and $\overline{BC}$ may cross each other on B' from different sides). We call such a disk $\pi(D)$ a *cusped product disk*. If Δ is a cusped product disk, we can slide Δ at $\partial\Delta$ to make $B' \cup \Delta$ a new branched surface, as shown in Figure 2.2 (b). Since the effect of the splitting from B to B' on the complement of B is connecting the components of $M - B$ by a product region near

∂M, adding appropriate cusped product disks to B' (and modifying them to form a branched surface B'' as above) can be consider as the inverse of the splitting from B to B'. More precisely, we can add a set of cusped product disks to B' to form a branched surface B'' such that each component of $M - B''$ is isotopic (in M) to a corresponding component of $M - B$, as illustrated in Figure 2.2. We denote this set of cusped product disks as above by $\mathcal{D} = \{D_1, \ldots, D_k\}$.

Suppose a region P in $M - int(N(B'))$ becomes a $D^2 \times I$ region in $M(s) - int(N(B_s))$ after the Dehn filling. Since $\partial M - \partial B$ is a union of bigons, P is cut by the cusped product disks in $\mathcal{D}$ as above into a union of $D^2 \times I$ regions in $M - int(N(B''))$. In other words, there is a union of $D^2 \times I$ regions $K_1, \ldots, K_n$ in $M - int(N(B))$ that are connected and become the region P after the splitting from B to B'. Since we have assumed that B has no trivial bubbles, by the definition of trivial $D^2 \times I$ region, there must exist an I-fiber in $N(B)$ with both endpoints in $int(D^2) \times \partial I \subset \partial K_1$ (here we identify K_1 with $D^2 \times I$). As the splitting from B to B' happens in a small neighborhood of ∂M, there must be an I-fiber in $N(B')$ with both endpoints in the horizontal boundary of P. Hence, after the Dehn filling, P cannot produce a trivial $D^2 \times I$ region. So, there is no trivial $D^2 \times I$ region in $M(s) - int(N(B_s))$.

Since the splitting from B to B' happens in a small neighborhood of ∂M, as shown in Figure 2.2(a), any sink disk in B_s must come from a half sink disk in B. As B has no sink disk or half sink disk, B_s contains no sink disk.

Next, we show that $\partial_h N(B_s)$ is incompressible in $M(s) - int(N(B_s))$. Since B is incompressible and B' is obtained by splitting B, by **[9]**, B' is also incompressible in M (this can also be proved using the argument below). Hence, there is no compressing disk for $\partial_h N(B')$ in $M - int(N(B'))$. Suppose there is a compressing disk for $\partial_h N(B_s)$, then there is a planar surface Q properly embedded in $M - int(N(B'))$ with one boundary component (which we call the outer boundary) in $\partial_h N(B')$ and other boundary components (which we call the inner boundary components) essential simple closed curves in $\partial M - N(B')$. After the Dehn filling, Q becomes a compressing disk for $\partial_h N(B_s)$. The outer boundary of Q is an essential simple closed curve in $\partial_h N(B_s)$.

As $\partial M - \partial B$ is a union of bigons, the cusped product disks in $\mathcal{D}$ cut Q into a union of disks. Let $D \in \mathcal{D}$ be such a cusped product disk that intersects Q. We can assume D is transverse to Q and $D \cap Q$ contains no simple closed curves (by a standard cutting and pasting argument). Let A, B, C be the three vertices of D with edge $\overline{AB} \subset \partial M$ and $C \in L$. We call $\alpha \subset D$ a ∂-parallel arc if both endpoints of α lie in the same edge. Suppose there is an arc $\alpha \subset D \cap Q$ with both endpoints in the same edge $\overline{AC}$. We can assume α is outermost in D, i.e., α and a subarc of $\overline{AC}$ bound a subdisk Δ in D that contains no other intersection arcs. We cut Q along α and then glue two parallel copies of Δ to it. Note that in Q, α is an arc with both endpoints in the outer boundary of Q. So, the surgery above yields two planar surfaces. As the outer boundary of Q is essential in $\partial_h N(B_s)$, the outer boundary of at least one of the two resulting planar surfaces must be essential in $\partial_h N(B_s)$. Thus, by choosing Q to be a planar surface with the least intersections with disks in $\mathcal{D}$, we can assume that $Q \cap D$ contains no ∂-parallel arc to edges $\overline{AC}$ and $\overline{BC}$. If there is an intersection arc with both endpoints in $\overline{AB}$, we can do a surgery as above and get a planar surface with fewer inner boundary components. Therefore, we can assume that $Q \cap D$ contains no ∂-parallel arc in D for any $D \in \mathcal{D}$.

Let γ be an inner boundary component of Q. So, γ lies in an annulus H of $\partial M - int(N(B'))$. We may choose γ to be outermost in H, i.e., γ and a component of ∂H, which we denote by β, bound an annulus in H that contains no other components of ∂Q. Since $\partial M - \partial B$ is a union of bigons, there must be a cusped product disk D in $\mathcal{D}$ such that $\gamma \cap \partial D \neq \emptyset$. By our construction, $\gamma \cap \partial D$ is a single point. Suppose A, B, C are the three vertices of D as before with edge $\overline{AB} \subset \partial M$, $A \in \beta$ and $C \in L$. Let $\alpha' \subset D \cap Q$ be the intersection arc containing $\gamma \cap \partial D$, and by choosing an appropriate γ, we can assume that the other endpoint of α' is on the edge $\overline{AC}$ (note that $A \in \beta$ as above). By our choice of β, α' must be outermost in D, i.e., α' cuts off a small triangle Δ from D containing $A \in \beta$ as a vertex and Δ contains no other arcs in $Q \cap D$. Now, we can isotope Q in $M - int(N(B'))$ near the annulus H by pushing γ towards β and eventually isotoping γ off ∂M so that γ becomes a circle in $\partial_h N(B')$ parallel to β. After this isotopy, α' becomes a ∂-parallel arc in D with both endpoints in the edge $\overline{AC}$. Hence, we can do surgery on Q as before to get another planar surface with fewer boundary components. Since the outer boundary of Q is an essential curve in $\partial_h N(B_s)$ and β is trivial in $\partial_h N(B_s)$, after this surgery, the outer boundary of the new planar surface is nontrivial in $\partial_h N(B_s)$. Therefore, by choosing Q to have the least intersection with disks in $\mathcal{D}$, Q must have no inner boundary, i.e., Q is a compressing disk for $\partial_h N(B')$ in M, which contradicts that B' is an incompressible branched surface in M. Therefore, $\partial_h N(B_s)$ is incompressible in $M(s)$.

By the same argument as above, it is easy to see that $M(s) - int(N(B_s))$ is irreducible and there is no monogon in $M(s) - B_s$.

Now, by Theorem 2.5, we only need to check that B_s does not contain any Reeb branched surface, i.e., B_s does not carry a sublamination of a Reeb foliation of a solid torus. Suppose B_s contains a Reeb branched surface. Then, there is a torus T embedded in $N(B_s)$ and transverse to the I-fibers of $N(B_s)$ such that T bounds a solid torus V in $M(s)$ and $N(B_s) \cap V$ fully carries a sublamination of Reeb foliation. Let U be the solid torus glued to ∂M during the Dehn filling. By the construction of B_s, $U \cap B_s$ is a union of meridional disks of U. By our hypothesis that B does not carry a torus that bounds a solid torus in $M(s)$, $U \cap T \neq \emptyset$. Since B_s has no disk of contact and since every leaf (except for the boundary torus) in a Reeb foliation is a plane, $V - int(N(B_s))$ must be a union of $D^2 \times I$ regions.

Let K be a $D^2 \times I$ region in $M(s) - int(N(B_s))$ that intersects U, and let Y be a component of $U \cap K$. We denote the core of U by C (U is considered as a small tubular neighborhood of this simple closed curve C) and let η be the component of $C \cap Y$. Since $\partial M - \partial B$ is a union of bigons, there must be a cusped product disk in $\mathcal{D}$ that intersects ∂Y and the intersection of Y with this cusped product disk must be isotopic to η in Y. Since K is a $D^2 \times I$ region, η must be a vertical arc, in other words, η is isotopic to an I-fiber of K. Therefore, $K \cap U$ is a union of vertical cylinders for each $D^2 \times I$ region K, i.e., after some isotopy, every component of $K \cap U$ is a cylinder consisting of I-fibers in the $D^2 \times I$ structure of K. If we extend the lamination in V to a Reeb foliation of the solid torus by adding plane leaves to fill all the $D^2 \times I$ regions in V, as the intersection of C with each $D^2 \times I$ region is a union of vertical arcs, we can assume each arc in $C \cap V$ is transverse to the Reeb foliation. However, this is impossible since there is no compact arc properly embedded in a solid torus and transverse to the Reeb foliation. Thus, B_s contains

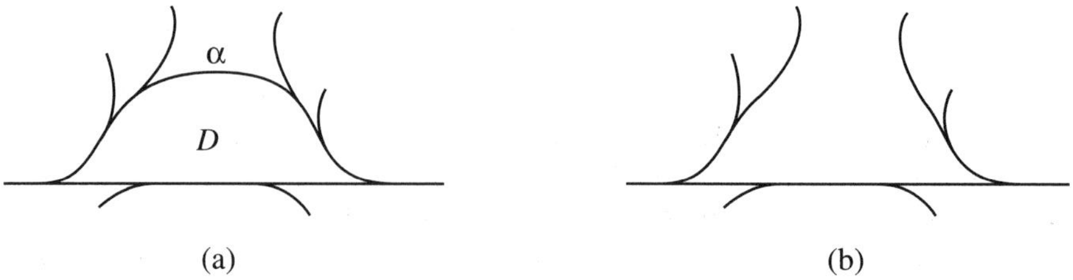

FIGURE 3.1

no Reeb branched surface and hence B_s fully carries an essential lamination in $M(s)$ by [**10**]. □

Theorem 2.5 reduces the question of finding a lamination (carried by branched surface B) that extends to an essential lamination in the closed manifold after Dehn fillings to the question of calculating slopes that can be realized by the boundary train track of B. In the following section, we discuss realizable slopes for some simple train tracks and how to reduce a complicated train track to simple ones. These discussions will be used in section 4.

3. Boundary train tracks

Let T be a torus and τ be a train track in T. The nonmanifold points of τ are called *switches*. Let D be the closure (under the path metric) of a component D_0 of $T - \tau$. We say that D is an *embedded bigon* if D is a bigon and the closure of D_0 in T is the same as D, in other words, there is no arc in τ which bounds D_0 on both sides. Let Z be the union of switches in τ. Then, each endpoint of (the closure of) an arc γ in $\tau - Z$ has a cusp direction pointing into or out of γ similar to the branch direction for branched surfaces. We call (the closure under path metric of) an arc γ in $\tau - Z$ a *sink arc* if the cusp directions of both endpoints of γ point into γ. We call D a *trivial bigon* if D is an embedded bigon and there is no sink arc in at least one boundary edge of D. Note that if D is a trivial bigon and the boundary edge η_1 of D contains no sink arc then there must be an arc α in $\eta_1 - Z$ with cusp directions of both endpoints pointing out of α, as shown in Figure 3.1(a). Moreover, the cusp directions at the switches in $\eta_1 - \alpha$ are coherent in each component, and $\tau - \alpha$ is a sub train track of τ as shown in Figure 3.1(b).

LEMMA 3.1. *Let τ be a train track in a torus T and D be a trivial bigon as above. Let η_1 and η_2 be the two boundary edges of D and suppose η_1 contains no sink arc. Let α be the arc in $\eta_1 - Z$ whose cusp directions at both endpoints point out of α as above. Then, a slope can be realized by τ if and only if it can be realized by the train track $\tau - \alpha$.*

PROOF. Suppose τ fully carries a lamination μ with slope s. Since D is an embedded bigon, we can pinch those arcs in μ that are carried by α to the edge η_2. Thus, $\tau - \alpha$ also fully carries μ.

Suppose μ is a lamination fully carried by $\tau - \alpha$. We can isotope some arcs carried by the edge η_2 across this bigon D to the edge η_1. By taking two parallel copies of μ (when μ consists of finitely many curves) if necessary, we can assume η_2 still carries some arcs in μ after this isotopy, i.e., τ fully carries a lamination that is either μ or two copies of μ. Hence, the slope of μ can also be realized by τ. □

If $T - \tau$ is a union of bigons, we can always split τ (in different ways) and apply Lemma 3.1 to get several simple train tracks without losing realizable slopes. Next, we calculate the sets of realizable slopes for some train tracks that we will use in section 4.

EXAMPLE 3.2. Let α be a simple closed curve with slope ∞ and β be a simple closed curve with slope 0 in a torus T. Suppose $\alpha \cap \beta$ is a single point O. Then, we can cut α (at O) into a compact arc and move the two endpoints of this arc apart a little along β. There are two ways to move the two endpoints apart as shown in Figure 3.2 (a, b). Then, we can deform them into train tracks as shown in Figure 3.2 (c, d, e, f). Then, by assigning a positive weight to each arc in the train track, one can easily calculate the slopes that can be realized by each of the four train tracks:

- Any slope in the interval $(0, 1)$ can be realized by the train track in Figure 3.2 (c).
- Any slope in the interval $(-1, 0)$ can be realized by the train track in Figure 3.2 (d).
- Any slope in the interval $(-\infty, 0)$ can be realized by the train track in Figure 3.2 (e).
- Any slope in the interval $(0, \infty)$ can be realized by the train track in Figure 3.2 (f).

EXAMPLE 3.3. Suppose we have n copies of the arc γ in Figure 3.2 (c) forming a train track as shown in Figure 3.3 (a). Similar to example 3.2, by assigning positive weights to arcs in the train track, it is easy to see that the interval of realizable slopes for the train track in Figure 3.3 (a) is $(0, n)$. Note that this can also be calculated by splitting the train track and applying Lemma 3.1.

Suppose we have n copies of the arc ρ in Figure 3.2 (d) forming a train track as shown in Figure 3.3 (b). As above, the interval of realizable slopes for the train track in Figure 3.3 (b) is $(-n, 0)$.

EXAMPLE 3.4. If we put the arcs γ and ρ in Figure 3.2 (c) and (d) together along the $0/1$ curve, we can form another train track as shown in Figure 3.4 (a). There are 3 ways to split this train track. The first splitting yields the train track in Figure 3.4 (b) and this train track only carries circles with slope 0. The second splitting yields the train track in Figure 3.4 (c), and the third splitting yields the train track in Figure 3.4 (d). Note that the train track in Figure 3.4 (d) is the same train track (on the torus T) as the one in Figure 3.4 (e) which contains a trivial bigon. By Lemma 3.1, the train track in Figure 3.4 (e) has the same interval of realizable slopes as the train track in Figure 3.4 (f), and by example 3.2, this interval is $(0, 1)$. Similarly, the interval of realizable slopes for the train in Figure 3.4 (c) is $(-1, 0)$. Thus, the interval of realizable slopes for the train track in Figure 3.4 (a) is $(-1, 1)$. Note that this interval can also be calculated easily by assigning a positive weight to each arc.

EXAMPLE 3.5. If we put n copies of Figure 3.4 (a) along the $0/1$ curve, as shown in Figure 3.5, we can split each copy in the train track similarly. Hence, the interval of slopes realizable by the train track in Figure 3.5 is $(-n, n)$.

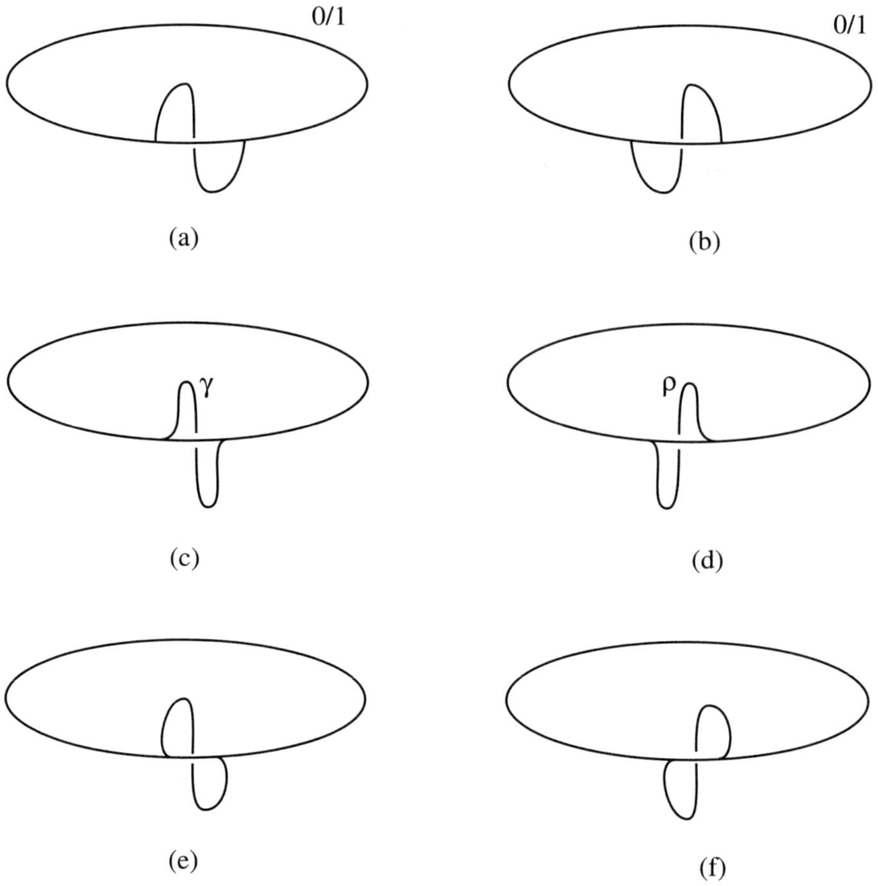

FIGURE 3.2

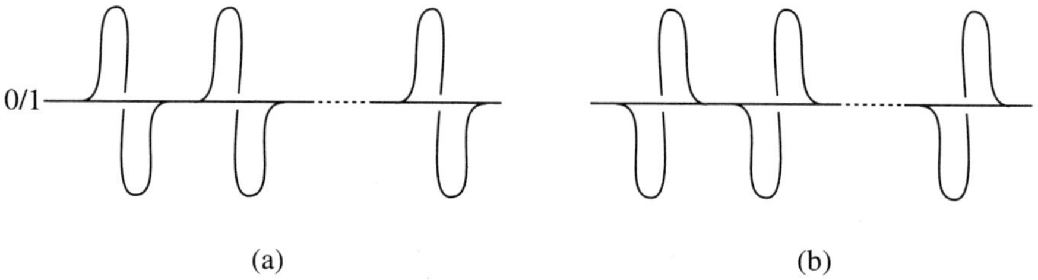

FIGURE 3.3

EXAMPLE 3.6. The interval of realizable slopes for the train track in Figure 3.6 (a) is $(-1, \infty)$, and the interval of realizable slopes for the train track in Figure 3.6 (b) is also $(-1, \infty)$. This can be calculated either by splitting as in example 3.4 or by assigning a positive weight to each arc of the train track.

EXAMPLE 3.7. We consider the train track in Figure 3.5 as the union of a simple closed curve C with slope 0 and $2n$ arcs $\alpha_1, \ldots, \alpha_{2n}$. We say that an α_i is

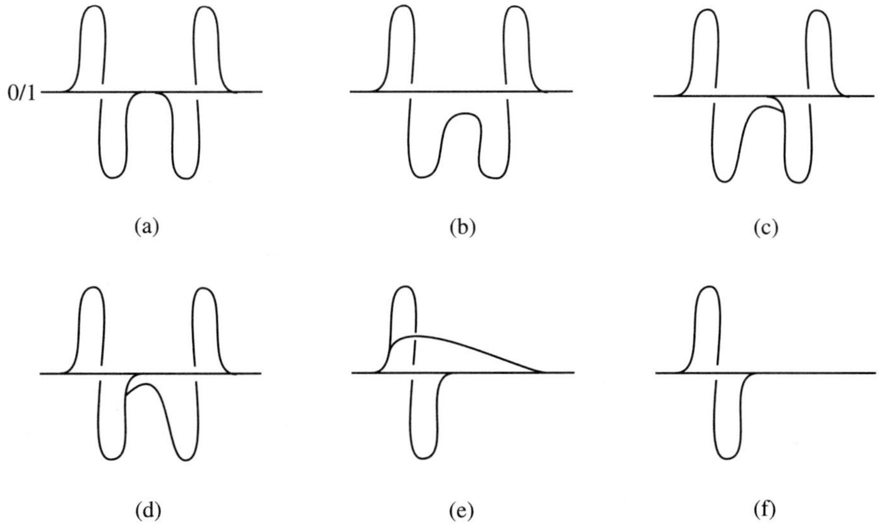

FIGURE 3.4

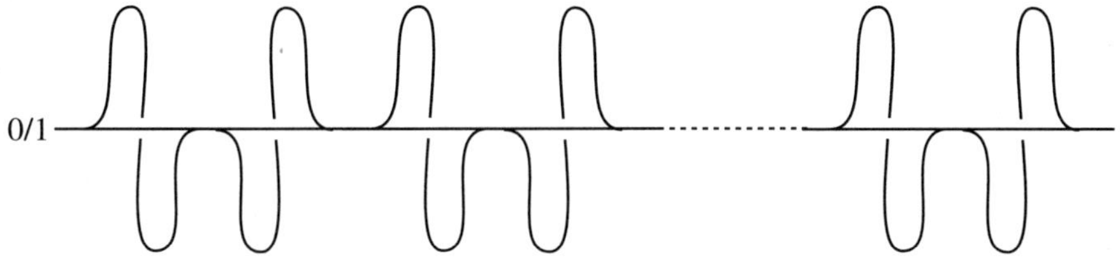

FIGURE 3.5

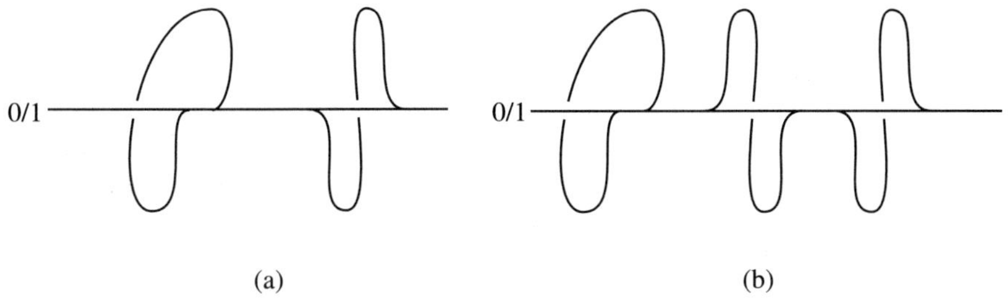

FIGURE 3.6

of positive type if $C \cup \alpha_i$ forms the train track in Figure 3.2 (c) (it can realize any slope in $(0,1)$), and we say α_i is of negative type if $C \cup \alpha_i$ forms the train track in Figure 3.2 (d) (it can realize any slope in $(-1,0)$). Let τ be a train track obtained by erasing k_1 arcs of positive type and k_2 arcs of negative type from the train track in Figure 3.5. Similar to the argument before, the interval of realizable slopes for τ is $(-n+k_2, n-k_1)$.

4. Murasugi sum

Let K be a knot in S^3 and R be a minimal genus Seifert surface. Suppose R can be decomposed into a $2n$-Murasugi sum of compact oriented surfaces R_1 and R_2, i.e., there is a 2-sphere H in S^3 decomposing S^3 into two 3-balls B_1 and B_2 such that: $R = R_1 \cup R_2$, $R_i \subset B_i$ $(i = 1, 2)$, and $R_1 \cap H = R_2 \cap H = D$, where D is a $2n$-gon. See [**3**] for a picture of a Murasugi sum.

When $n = 1$, the Murasugi sum is known as connected sum, and when $n = 2$, the Murasugi sum is known as plumbing. It was proved in [**3, 4**] that R is a minimal genus Seifert surface for K if and only if R_i is a minimal genus Seifert surface for the link ∂R_i $(i = 1, 2)$.

THEOREM 4.1. *Let K be a knot in S^3 and R be a minimal genus Seifert surface for K. If R can be decomposed as a $2n$-Murasugi sum, then $K(s)$ has a taut foliation for any slope $s \in (1-n, n-1)$, where $K(s)$ is the closed 3-manifold obtained from Dehn surgery along the slope s.*

PROOF. Let H be the 2-sphere and $D = R \cap H$ be the $2n$-gon as in the definition of Murasugi sum. Let $N(K)$ be a small tubular neighborhood of K such that $R \cap N(K)$ is an annular neighborhood of $K = \partial R$ and $H \cap N(K)$ consists of $2n$ meridional disks of the solid torus $N(K)$. To simplify notation, we still use R, D and H to denote $R - int(N(K))$, $D - int(N(K))$ and $H - int(N(K))$ respectively. Now, R is a surface properly embedded in $M = S^3 - int(N(K))$ with ∂S a circle of slope $0/1$.

By [**3, 4, 5**], $H - D$ gives rise to a taut sutured manifold decomposition. Hence, we can deform $H \cup R$ to a branched surface B such that $M - int(N(B))$ is a taut sutured manifold whose suture is the union of $\partial_v N(B)$ and $\partial M - int(N(B))$. B can be constructed by sliding the disk $H - int(D)$ away from D along R. The branched locus of B consists of $2n$ disjoint arcs with endpoints in ∂M (they correspond to the $2n$ edges of D). B has 4 sectors corresponding to D, $H - D$, $R_1 - D$ and $R_2 - D$. To simplify notation, we still use D, $H - D$, $R_i - D$ to denote the 4 sectors. Note that the sectors $R_1 - D$ and $R_2 - D$ have disjoint neighborhoods in M. The branch direction of L points into $R_i - D$ $(i = 1, 2)$ and out of D and $H - D$. See Figure 3.2 in [**5**] for a picture.

Since $M - int(N(B))$ is a taut sutured manifold, B is an incompressible branched surface. Moreover, by the construction, B does not carry any closed surface and in particular B does not carry any torus. The boundary train track ∂B in the torus ∂M consists of ∂S and $2n$ arcs as shown in Figure 3.5.

Case 1. Neither $R_1 - D$ or $R_2 - D$ is a disk.

In this case, B contains no sink disk or half sink disk, and hence B is a laminar branched surface. As B does not carry a torus, by Theorem 2.5, for any slope s that is realizable by the train track ∂B, B fully carries a lamination λ whose boundary is a union of circles with slope s, and λ extends to an essential lamination in $M(s)$. Since the train track ∂B is as shown in Figure 3.5, by example 3.5, the slope s can be any rational slope in $(-n, n)$. Furthermore, this essential lamination extends to a taut foliation in $M(s)$ as follows. This basically follows from Gabai's sutured manifold decomposition.

As in the proof of Theorem 2.5, we split B along ∂B in a small neighborhood of ∂M and get a branched surface B' such that $\partial B'$ is a union of circles with slope s and λ is fully carried by B'. For each component Z of $M - int(N(B'))$, there is a union of

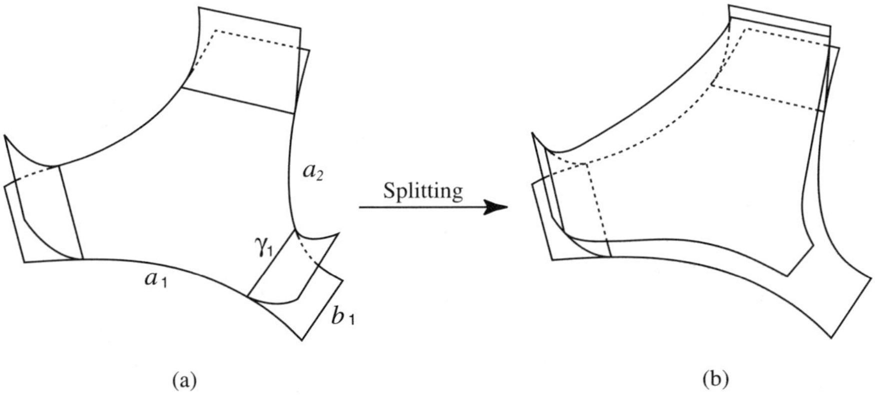

FIGURE 4.1

product disks $\mathcal{D}$ that decompose Z into some components isotopic to corresponding components of $M - int(N(B))$. Since each component of $M - int(N(B))$ is a taut sutured manifold, by Lemma 3.12 in [**6**], $M - int(N(B'))$ is a taut sutured manifold with suture consisting of annuli $\partial_v N(B')$ and $\partial M - int(N(B'))$. Since K is a knot in S^3, by the thin position argument in [**7**], there is a taut sutured manifold hierarchy for $M - int(N(B'))$ which extends λ to a foliation whose boundary is a union of circles. Note that although the slope of the boundary suture in our case is not 0, since the slope cannot be ∞, the combinatorial argument in [**7**] also works and we also get such a taut sutured manifold hierarchy [**11**], which leads to a taut foliation. A similar argument for this can also be found in the proof of Corollary 1.10 in [**13**].

Case 2. $R_1 - D$ is a disk but $R_2 - D$ is not a disk.

In this case, $R_1 - int(D)$ is a half sink disk. What we will do next is to split B to eliminate this half sink disk. Let N_1 be a small neighborhood of $R_1 - D$ in M, $D_1 = N_1 \cap R$ and $\hat{D}_1 = N_1 \cap B$. Thus, D_1 is a $2n$-gon with n edges in ∂M, and $\hat{D}_1$ is a branched surface consisting of the $2n$-gon D_1 and n "tails", as shown in Figure 4.1 (a). Let $a_1, a_2, \ldots, a_n$ be n edges of D_1 that lie in ∂M and $b_1, b_2, \ldots, b_n$ be the other n edges. The branched locus of $\hat{D}_1$ consists of n arcs, which we denote by $\gamma_1, \gamma_2, \ldots, \gamma_n$. Each arc γ_j is an arc properly embedded in D_1 and parallel to b_j with endpoints in α_j and α_{j+1} ($\alpha_{n+1} = \alpha_1$). Then, we split $\hat{D}_1$ starting from γ_1, as shown in Figure 4.1, and we call this splitting a *standard splitting with respect to* γ_1. Let $\hat{B}$ be the branched surface after this splitting. Since $R_2 - D$ is not a disk, after this splitting, there is no half sink disk and hence $\hat{B}$ is a laminar branched surface.

Before this splitting, ∂B consists of a simple closed curve ∂R with slope 0 and $2n$ arcs $\alpha_1, \alpha_2, \ldots, \alpha_{2n}$, as shown in Figure 3.5. As in example 3.7, we say that an α_i is of positive type if $\partial R \cup \alpha_i$ forms the train track in Figure 3.2 (c) (it can realize any slope in $(0, 1)$), and we say α_i is of negative type if $\partial R \cup \alpha_i$ forms the train track in Figure 3.2 (d) (it can realize any slope in $(-1, 0)$). If we split the train track ∂B by sliding an arc α_i of negative type over an arc of positive type as shown in Figure 3.4 (d), the resulting train track (as we pointed out in example 3.4) realize the same interval of slopes as the train track obtained by simply erasing this arc α_i from ∂B.

Now, we calculate the slopes that are realizable by the train track $\partial\hat{B}$. We need to see how the train track changes near $a_1, a_2, \ldots, a_n$ during the splitting. Let α_i and α_j be the two arcs containing the two endpoints of γ_1 respectively in ∂B. Since R is orientable, one arc, say α_i, is of positive type and the other arc α_j is of negative type. As the two endpoints of γ_1 lie in a_1 and a_2 respectively, the effect of this spitting near a_1 and a_2 is sliding α_i and α_j over corresponding adjacent arcs as shown in Figure 3.4 (c, d). Thus, the train track after the splitting near a_1 and a_2 realize the same interval of slopes as the train track obtained by erasing α_i and α_j from ∂B. Moreover, near a_k for $k \neq 1, 2$, the effect of splitting B on the boundary train track is creating a trivial bigon, as shown in Figure 4.1. Thus, by Lemma 3.1, the interval of realizable slope does not change after this splitting near a_k for $k \neq 1, 2$. Therefore, the train track $\partial\hat{B}$ realizes the same interval of slopes as the train track obtained by erasing an arc α_i of positive type and an arc α_j of negative type, and hence by example 3.7, the interval of realizable slopes for $\partial\hat{B}$ is $(-(n-1), n-1)$.

Case 3. Both $R_1 - D$ and $R_2 - D$ are disks.

In this case, both $R_1 - int(D)$ and $R_2 - int(D)$ are half sink disks. Let N_i be a small neighborhood of $R_i - D$ in M, $D_i = N_i \cap R$ and $\hat{D}_i = N_i \cap B$ ($i = 1$ or 2). By choosing N_i small enough, we can assume that $\hat{D}_1$ and $\hat{D}_2$ are disjoint and as shown in Figure 4.1 (a). Now, we split $\hat{D}_1$ and $\hat{D}_2$ as follows.

We keep the same notation as in case 2, and describe ∂B as the union of ∂R and $2n$ arcs $\alpha_1, \alpha_2, \ldots, \alpha_{2n}$. Suppose α_1 is of positive type and we denote the two endpoints of α_1 by x and y (x and y are two switches in the train track ∂B). Let γ_x and γ_y be the two components of the branch locus of B that contain x and y respectively. We can assume that $\gamma_x \subset \hat{D}_1$ and $\gamma_y \subset \hat{D}_2$. Since α_1 is of positive type, the arc α_j (resp. α_k) containing the other endpoint $\partial\gamma_x - x$ (resp. $\partial\gamma_y - y$) is of negative type. Now, similar to case 2, we perform standard splitting on $\hat{D}_1$ with respect to γ_x and standard splitting on $\hat{D}_2$ with respect to γ_y. The effect of the two splittings on ∂B near α_1, α_j and α_k is sliding the three arcs over adjacent arcs. Similar to case 2, the train track after the two splittings above realize the same interval of slopes as the train track obtained by erasing α_1, α_j and α_k from ∂B. Since α_1 is of positive type and α_j and α_k are of negative type, the interval of realizable slope is $(-(n-2), n-1)$ by example 3.7.

Then, we replace α_1 by an arc of negative type and repeat the splitting above to eliminate the two half sink disks. By the same argument, the boundary train track of the resulting branched surface realizes the same interval of slopes as the train track obtained by erasing one arc of negative type and two arcs of positive type from ∂B. Hence, the interval we get is $(-(n-1), n-2)$. By combining the two intervals together, we get that any slope in $(-(n-1), n-1)$ can be realized by an essential lamination carried by B. As in case 1, this essential lamination extends to a taut foliation. □

Corollary 4.2 is a quick application of Theorem 4.1 and some well-known results.

COROLLARY 4.2. *Let K be a knot in S^3 and R be a minimal genus Seifert surface of K. Suppose R can be decomposed as a $2n$-Murasugi sum. Then, unless the Murasugi sum is a plumbing of a trivial twisted band and a nonplanar surface, K has property P. In particular, if $n \neq 2$, K has property P.*

PROOF. If $n = 1$, the Murasugi sum is a connected sum and the Property P for connected sum is well-known. So, we assume $n \neq 1$ and by Theorem 4.1 $K(s)$ has infinite fundamental group for any $s \in (1-n, n-1)$. If $n \neq 2$ (i.e., $n \geq 3$), $[-1,1] \subset (1-n, n-1)$. As $K(s)$ is a homology sphere only if $s = \pm 1/m$ for some integer m, K has property P if $n > 2$.

Now we consider the case that $n = 2$, i.e., the minimal genus Seifert surface is a plumbing of R_1 and R_2. If R_i is not an annulus for both $i = 1, 2$, $R_i - D$ is not a disk and hence the branched surface B as above contains no sink disk or half sink disk. By case 1 in the proof of Theorem 4.1, $K(s)$ contains a taut foliation for any $s \in (-2, 2)$ and the property P for K follows.

If R_1 is an annulus, then a small neighborhood of R_1 can be considered as a tubular neighborhood of a knot k'. If k' is a nontrivial knot, then K is a satellite knot. Property P is known for satellite knots [**8**]. If K is not a satellite knot and R_1 is an annulus, then R_1 must be a trivial twisted band. Furthermore, if R_2 is also an annulus, then R is a plumbing of two trivial twisted bands and hence K is a 2-bridge knot. Property P for 2-bridge knots is also known [**16, 12, 1**]. Hence, unless R is a plumbing of a trivial twisted band and a nonplanar surface, K has property P. □

In the next theorem, we consider the case that R is a plumbing of a trivial twisted band and another surface. This theorem was first proved by Roberts (in a different situation) [**13**] using techniques of affine lamination. We give another proof here by splitting the branched surface and using Theorem 2.5.

THEOREM 4.3. *Let K be a knot in S^3 and R be a minimal genus Seifert surface of K. Suppose R is a plumbing of two surfaces R_1 and R_2, and R_1 is a trivial twisted band. Then, depending on the direction of the twist for R_1, $K(s)$ has a taut foliation for any slope s in $(-\infty, 1)$ or $(-1, \infty)$.*

PROOF. Let $L = \partial R_1$. We can regard R_1 as a surface properly embedded in $M_L = S^3 - int(N(L))$. Then, we cut M_L open along R_1 and get a manifold $T = M_L - int(N(R_1))$. Since R_1 is a trivial twisted band, T is a solid torus. The two sides of R_1 are two annuli in ∂T. Moreover, we can consider B_2 ($R_2 \subset B_2$) as a 3-ball in T whose intersection with ∂T is a square corresponding to the plumbing square. As T is a solid torus, there is a compressing disk E for T that does not intersect $int(B_2)$. So, we can view E as a properly embedded disk in $M - int(N(R))$, where $M = S^3 - int(N(K))$, and by [**6**], E gives rise to a taut sutured manifold decomposition. Hence, we can add E to R forming a branched surface B such that $M - int(N(B))$ is a taut sutured manifold. In fact, we can view E directly from the picture of R_1 (see Figure 11 in [**13**] for a picture). Suppose R_1 is a band with m full twists. Figure 4.2 (a) gives a schematic picture (in the case $m = 3$) of the branched surface $B = R \cup E$ near the annulus R_1 (see Figure 14 in [**13**] for a picture of B near ∂M). In particular, the branch locus of B is a union of disjoint arcs on R_1. The boundary train track of ∂B is as shown in Figure 4.2 (b).

If the number of twists $m > 1$, then B contains half sink disks on R_1 as shown in Figure 4.2(a) (with $m = 3$). Now, we can split B by moving E along R_1 changing B from Figure 4.2(a) to a branched surface as shown in Figure 4.2(c) which does not contain any half sink disk. This splitting changes the boundary train track from Figure 4.2(b) to Figure 4.2(d), and after this splitting we get a laminar branched surface. The boundary train track of Figure 4.2(d) contains many trivial bigons.

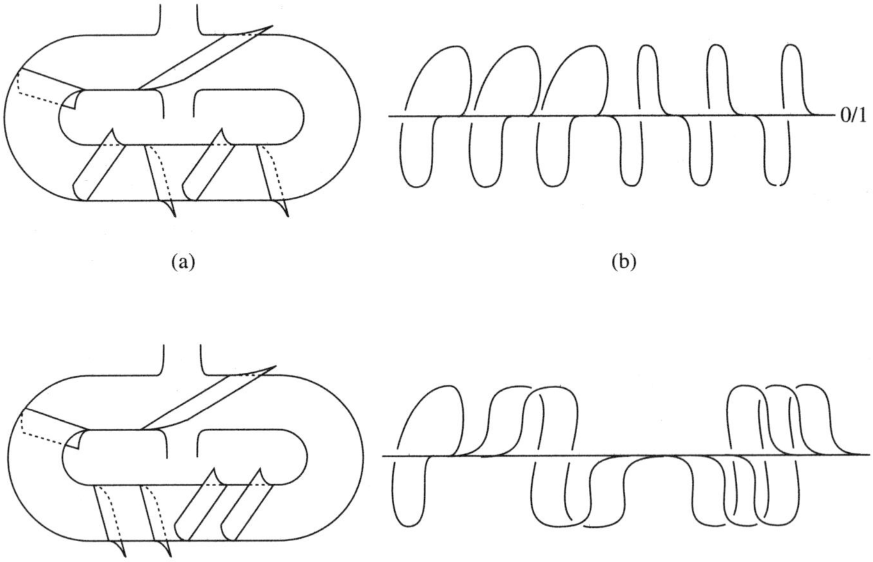

(a) (b)

(c) (d)

FIGURE 4.2

After eliminating these trivial bigons as in Lemma 3.1, we get a train track in Figure 3.6 (b). Therefore, by Theorem 2.5, Lemma 3.1 and example 3.6, we get that, depending on the direction of the twist for R_1, $K(s)$ contains an essential lamination for any slope $s \in (-1, \infty)$ or $(-\infty, 1)$. As in the proof of Theorem 4.1, this lamination extends to a taut foliation.

If R_1 is a Hopf band, i.e., $m = 1$, then B contains no half sink disk and ∂B is the train track as in Figure 3.6 (a). In this case, we do not need any splitting and the theorem follows from the same argument. □

References

[1] Delman, C., *Essential laminations and Dehn surgery on 2-bridge knots.* Topology Appl. **63** (1995), no. 3, 201–221.

[2] Delman, C.; Roberts, R. *Alternating knots satisfy Strong Property P.* Comment. Math. Helv. **74** (1999), no. 3, 376–397.

[3] Gabai, D., *The Murasugi sum is a natural geometric operation.* Low-dimensional topology, 131–143, Contemp. Math., **20**, Amer. Math. Soc., Providence, RI, 1983.

[4] Gabai, D., *The Murasugi sum is a natural geometric operation II.* Combinatorial methods in topology and algebraic geometry, 93–100, Contemp. Math., **44**, Amer. Math. Soc., Providence, RI, 1985

[5] Gabai, D., *Detecting fibred links in S^3.* Comment. Math. Helv. **61** (1986), no. 4, 519–555.

[6] Gabai, D., *Foliations and the topology of 3-manifolds*, J. Differential Geometry. **18** (1983), 445-503.

[7] Gabai, D., *Foliations and the topology of 3-manifolds III.* J. Differential Geometry. **26** (1987), 479-536.

[8] Gabai, D., *Surgery on knots in solid tori.* Topology. **28** (1989), no. 1, 1–6.

[9] Gabai, D.; Oertel, U., *Essential laminations in 3-manifolds*, Ann. of Math. **2** (1989), no. 1, 41–73.

[10] Li, T. *Laminar branched surfaces in 3-manifolds.* Preprint.

[11] Luecke, J. *Personal communication.*
[12] Naimi, R. *Constructing essential laminations in 2-bridge knot surgered 3-manifolds.* Pacific J. Math. **180** (1997), 153-186
[13] Roberts, R., *Constructing taut foliations*, Comment. Math. Helv. **70** (1995), 516–545.
[14] Roberts, R., *Taut foliations in punctured surface bundles, I.* Proc. London Math. Soc. (3) **82** (2001), no. 3, 747–768.
[15] Roberts, R., *Taut foliations in punctured surface bundles, II.* Proc. London Math. Soc. (3) **83** (2001), no. 2, 443–471.
[16] Takahashi, Moto-o, *Two-bridge knots have property P.* Mem. Amer. Math. Soc. **29** (1981), no. 239.
[17] Wu, Ying-Qing *Sutured manifold hierarchies, essential laminations, and Dehn surgery.* J. Differential Geom. **48** (1998), no. 3, 407–437

DEPARTMENT OF MATHEMATICS, OKLAHOMA STATE UNIVERSITY, STILLWATER, OK 74078
E-mail address: tli@math.okstate.edu

Proceedings of Symposia in Pure Mathematics
Volume **71**, 2003

Strongly Irreducible Surface Automorphisms

Saul Schleimer

ABSTRACT. A surface automorphism is *strongly irreducible* if every essential simple closed curve in the surface has nontrivial geometric intersection with its image. We show that a three-manifold admits only finitely many inequivalent surface bundle structures with strongly irreducible monodromy.

1. Introduction

A surface automorphism $h : F \to F$ is *strongly irreducible* if every essential simple closed curve $\gamma \subset F$ has nontrivial geometric intersection with its image, $h(\gamma)$. This paper shows that a three-manifold admits only finitely many inequivalent surface bundle structures with strongly irreducible monodromy. This imposes a serious restriction; for example, any three-manifold which fibres over the circle and has $b_2(M) \geq 2$ admits infinitely many distinct surface bundle structures.

The main step is an elementary proof that all *weakly acylindrical* surfaces inside of an irreducible triangulated manifold are isotopic to fundamental normal surfaces. As weakly acylindrical surfaces are a larger class than the acylindrical surfaces this strengthens a result of Hass [**5**]; an irreducible three-manifold contains only finitely many acylindrical surfaces.

Section 2 gives necessary topological definitions, examples of strongly irreducible automorphisms, and precise statements of the theorems. The required tools of normal surface theory are presented in Section 3. Next, Section 4 defines weakly acylindrical and proves that every such surface is isotopic to a fundamental surface. In the spirit of the Georgia Topology Conference the paper ends by listing several open questions.

Many of the ideas and terminology discussed come from the study of Heegaard splittings as in [**2**] and in my thesis [**11**]. This paper, in particular Theorem 4.2, owes an obvious debt to [**6**] by Jaco and Oertel. I thank Ian Agol for simplifying my original proof of Proposition 2.3 and Dave Bachman for critiquing an early version of this paper.

2. Definitions and Examples

This section lays out the necessary definitions, states the main theorems precisely, and gives examples of families of strongly irreducible surface automorphisms.

Let F be a closed, orientable, genus$(F) > 1$ surface. Let $h : F \to F$ be an automorphism of F. If γ_0 and γ_1 are simple closed curves in F then the *geometric intersection number*, $i(\gamma_0, \gamma_1)$, is the minimum of $|\gamma_0' \cap \gamma_1'|$ taken over all γ_i' isotopic to γ_i.

DEFINITION. The map h is *strongly irreducible* if $i(h(\gamma), \gamma) > 0$ for every essential simple closed curve $\gamma \subset F$.

If h is not strongly irreducible then h is *weakly reducible.*

REMARK 2.1. A *reducible* surface automorphism is one which admits an invariant set of disjoint essential simple closed curves, up to isotopy. Thus reducible maps are also weakly reducible.

REMARK 2.2. There exist weakly reducible pseudo-Anosov maps with arbitrarily high stretch factor — the main ideas required for the construction may be found in [**9**].

EXAMPLE. Let P be a regular $4g$-gon in the hyperbolic plane with angle at the vertices equal to $2\pi/4g$. Here g is assumed to be two or larger. Glue opposite sides of P by isometries to obtain F, an orientable surface of genus g. Let $h : P \to P$ be a counter-clockwise rotation of P about its center, O, through an angle of $2\pi/4g$. Let h' be the induced isometry of F.

PROPOSITION 2.3. The periodic map h' is strongly irreducible.

All of the hyperbolic trigonometry needed in the proof may be found in Chapter 7 of [**1**].

PROOF. Suppose that R is the distance between O and V, where V is a vertex of P. Then $\cosh(R) = \cot^2(2\pi/8g)$. Both O and V give fixed points, $x_O, x_V \in F$, of h'. Suppose that $\gamma \subset F$ is a simple closed geodesic. There is a point of γ which lies within $R/2$ of either x_O or x_V. This last holds because the points of F which are not this close to one of x_O or x_V form a disjoint union of disks.

Suppose there is a point of γ within distance $R/2$ of the point x_O. (The other case is similar.) Pick $\widetilde{\gamma} \subset \mathbb{H}^2$, a lift of γ, which lies within $R/2$ of O. Let L be the distance between $\widetilde{\gamma}$ and O. Let θ be the visual angle which $\widetilde{\gamma}$ occupies, as viewed from O. (When $\widetilde{\gamma}$ meets x_O take $\theta = \pi$.) Then $L \leq R/2$ and $\cosh(L) = 1/\sin(\theta/2)$. Applying a "double-angle" formula to $\cosh(R)$ yields

$$\cosh(R/2) = 1/\sqrt{2}\sin(2\pi/8g).$$

Hyperbolic cosine is an increasing function on the positive reals so

$$1/\sin(\theta/2) \leq 1/\sqrt{2}\sin(2\pi/8g).$$

As sine is increasing in the interval $[0, \pi/2]$ we deduce that $\theta > 2\pi/4g$. Thus the visual angle of $\widetilde{\gamma}$ is greater than $2\pi/4g$ and $h(\widetilde{\gamma}) \cap \widetilde{\gamma} \neq \emptyset$. It follows that $h'(\gamma) \cap \gamma \neq \emptyset$. This gives the desired conclusion as the intersection and geometric intersection numbers agree for geodesics. □

PROPOSITION 2.4. If h is a pseudo-Anosov map then there is an $n \in \mathbb{N}$ such that h^n is strongly irreducible.

Here we only sketch a proof; the numbers in parentheses refer to theorems in Kapovich's book [**8**] which we take as our reference for $\mathcal{PML}(F)$, the space of projectively measured laminations of F. Recall that geometric intersection extends to a continuous function on $\mathcal{ML} \times \mathcal{ML}$ (11.26).

Let $\lambda^{\pm}$ be the stable and unstable laminations for h. Let U, V be small neighborhoods of $\lambda^{\pm}$ (respectively) in $\mathcal{PML}(F)$. Choose U and V so that all $x \in U$, $y \in V$ have $i(x, y) > 0$. This is possible because the geometric intersection between λ^+ and λ^- is non-zero (11.49). In particular, $U \cap V = \emptyset$.

There is an $m \in \mathbb{N}$ such that if $x \in \mathcal{PML}(F) \smallsetminus V$ then $h^m(x) \in U$ (11.47). Now, $n = m + 1$ gives the desired conclusion. To see this, pick an essential simple closed curve $y \subset F$. Denote the corresponding element of $\mathcal{PML}(F)$ again by y. There is a integer k such that $h^k(y) \in V$ but $h^{k+1}(y) \notin V$. (Again, 11.47.) So $i(y, h^n(y)) = i(h^k(y), h^{k+1+m}(y)) > 0$. This completes the proof sketch.

REMARK 2.5. The notions irreducible and strongly irreducible may be generalized by introducing the *curve complex* of F, $\mathcal{C}(F)$, and defining the translation distance, $\tau(h)$, of h's action on $\mathcal{C}(F)$. The deep results of [**10**] give a positive integer $n(g)$ depending only on $g = \text{genus}(F)$ such that: If h is pseudo-Anosov then $\tau(h^{n(g)}) \geq 2$ and so $h^{n(g)}$ is strongly irreducible. This greatly improves upon Proposition 2.4.

We now turn from examples to the main objects of interest: surface bundles over the circle.

DEFINITION. If $h : F \to F$ is a surface automorphism then let M_h be the *mapping torus* of h. So

$$M_h \cong F \times I/(x, 1) \sim (h(x), 0).$$

The mapping torus admits a natural map to the circle, $\pi_h : M_h \to S^1$. The embedded surfaces $F \times \{t\} \subset M$ are called the *fibres* of this map while h is the *monodromy* and $g(F) = \text{genus}(F) \geq 2$ is the *genus* of the bundle. As an example, if h is a strongly irreducible periodic automorphism, then M_h is an atoroidal Seifert fibred space.

Fix M, a closed orientable three-manifold.

DEFINITION. A *surface bundle structure* on M is a pair (h, ϕ), where h is a surface automorphism and ϕ is a homeomorphism from M_h to M.

Let (h, ϕ) and (h', ϕ') be two surface bundle structures on M. Suppose that $\psi : M_h \to M_{h'}$ is a homeomorphism such that $\pi_h = \pi_{h'} \circ \psi$ and ϕ is isotopic to $\phi' \circ \psi$. Then the two bundle structures on M are *equivalent*. Thus (h, ϕ) is equivalent to (h', ϕ') if and only if h' is conjugate to h and the fibres of the two bundle structures are isotopic in M.

Here is a precise statement of the theorem alluded to in the introduction.

THEOREM 4.3. *Suppose that M is a closed, orientable three-manifold. Then M admits only finitely many inequivalent surface bundle structures with strongly irreducible monodromy.*

Closely related is our:

THEOREM 4.4. *Suppose that M is a closed, orientable three-manifold. There is a positive real number $c(M)$ such that if (h, ϕ) is a surface bundle structure on*

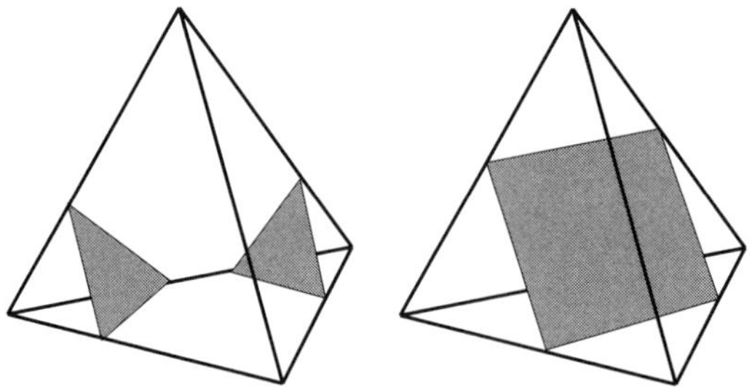

FIGURE 1. Normal triangles and normal quad

M with genus $g > 1$ *then* h^i *is weakly reducible for all integers* i *where* $1 \leq i \leq c(M) \cdot (2g-2)$.

If $c(M) \cdot (2g-2) < 1$ then the theorem is vacuous as no such i exists.

REMARK 2.6. If M is atoroidal then M has only finitely many inequivalent surface bundle structures in each genus. (This simply because there are only finitely many incompressible surfaces, up to isotopy, of each genus. See [**6**].) Thus, in the atoroidal case, Theorem 4.4 implies Theorem 4.3.

REMARK 2.7. Suppose M is a closed, atoroidal three-manifold admitting infinitely many inequivalent surface bundle structures. Then this manifold and Theorem 4.4 provide another proof that the constant $n(g)$ of Remark 2.5 must tend to infinity as g does.

REMARK 2.8. Again, consider the translation distance $\tau(h)$ where h is an automorphism of the closed, orientable, genus at least two, surface F. One obtains an analogue of a theorem of Hartshorn [**4**] regarding Heegaard splittings: If (h, ϕ) is a surface bundle structure on M and G is a two-sided incompressible surface in M then either G is isotopic to a fibre, G is a torus (and h is reducible), or $\tau(h) \leq -\chi(G)$.

As a corollary, deduce that if h has translation distance greater than $-\chi(F)$, where F is the fibre, then F is the unique minimal genus incompressible surface in M, up to isotopy. Weak conclusions about the shape of the Thurston norm ball and on the structure of M's symmetry group follow.

3. Normal surfaces

This section presents the required bare minimum of normal surface theory. For a more complete treatment consult [**6**] or [**7**].

Fix a closed orientable three-manifold M and choose T, a triangulation of M. Suppose $F \subset M$ is a closed embedded surface. The *weight* of F, $w(F)$, is the number of intersections between F and the one-skeleton of T. The surface F is *normal* if F intersects every tetrahedron of T in a disjoint collection of *normal triangles and quadrilaterals*. See Figure 1. In each tetrahedron there are four types of normal triangle and three types of normal quad.

LEMMA 3.1. (Haken [**3**]) *Suppose that* (M, T) *is closed and irreducible. If* $F \subset M$ *is embedded and incompressible then* F *is isotopic to a normal surface* F' *with* $w(F') \leq w(F)$.

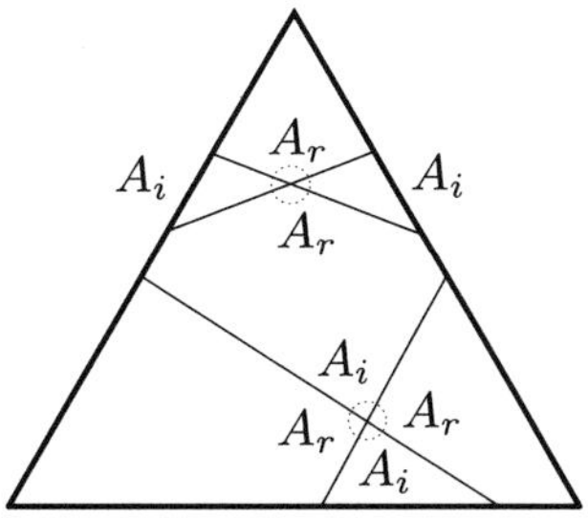

FIGURE 2. Regular and irregular annuli

A *normal isotopy* of M fixes every simplex of T setwise. Two normal surfaces, G and H, are *compatible* if in each tetrahedron G and H have the same types of quad (or one or both have no quads.) In this situation we form the *Haken sum* $F = G + H$ as follows:

Normally isotope H to make G transverse to H, to make $\Gamma = G \cap H$ transverse to the skeleta of T, and to minimize the number of curves in Γ. The components of Γ are the *exchange curves.* For every such $\gamma \subset \Gamma$ let $R(\gamma)$ be the closure (taken in M) of $\eta_M(\gamma)$, an open regular neighborhood of γ. The set $R(\gamma)$ is a solid torus containing γ as a core curve.

Then $(\partial R(\gamma)) \setminus (G \cup H)$ is a union of annuli. Taking closures divide these into two sets, the *regular* and *irregular* annuli, $A_r(\gamma)$ and $A_i(\gamma)$, as indicated by Figure 2. Finally, as in Figure 3, form the surface

$$F = \left((G \cup H) \setminus \bigcup_{\Gamma} \{\eta_M(\gamma)\} \right) \cup \bigcup_{\Gamma} \{A_r(\gamma)\}.$$

Each connected component of $(G \cup H) \setminus \bigcup_{\Gamma} \{\eta_M(\gamma)\}$ is a *patch* of the sum $G + H$ while each closed annulus $A_r(\gamma)$ is a *seam.* Note that F is again a normal surface which, up to normal isotopy, does not depend on the choices made in the above construction.

A normal surface is *fundamental* if it admits no such decomposition. A fundamental result due to Haken is:

LEMMA 3.2. (See [**7**]) *A closed orientable triangulated three-manifold* (M^3, T) *contains only finitely many fundamental normal surfaces, up to normal isotopy.*

REMARK 3.3. The proof of Lemma 3.2 requires only that the manifold M admit a finite triangulation. We restrict ourselves to the closed and orientable case to avoid an unnecessarily technical discussion.

Each "cut-and-paste" operation involved in the Haken sum $F = G + H$ may be recorded by an embedded *exchange band* $(C(\gamma), \partial C(\gamma)) \subset (N(\gamma), A_r(\gamma))$. That is, the band is embedded in $N(\gamma)$ with boundary inside of $A_r(\gamma)$. See Figure 3. Each $C(\gamma)$ is either an annulus or a Mobius band. The exchange bands record enough information to reverse the sum. Note also that each seam is a regular neighborhood (in F) of a boundary component of some $C(\gamma)$.

A sum $F = G + H$ is *reduced* if F cannot be realized as a sum $G' + H'$ where G' and H' are again normal, isotopic to G and H respectively, with $|G' \cap H'| < |G \cap H|$.

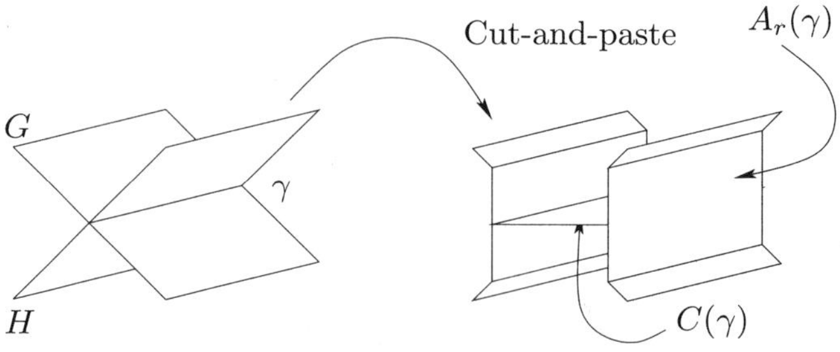

FIGURE 3. The exchange band for γ.

Note that the isotopy between G and G' (H and H') need not be normal. Lemma 3.4 is a key technical result for [**6**] and our Theorem 4.2.

LEMMA 3.4. (Jaco and Oertel [**6**]) *Suppose that (M^3, T) is a closed, orientable, irreducible, triangulated three-manifold. Suppose that $F \subset M$ is an incompressible normal surface which is least weight in its isotopy class. If the sum $F = G + H$ is reduced then no patch of $G + H$ is a disk.*

4. Weakly acylindrical surfaces

Here the *weakly acylindrical* surfaces are defined. Equipped with this definition and Theorem 4.2 we will prove Theorem 4.3, the main goal of the paper.

Suppose that N is a compact, orientable three-manifold with non-empty boundary.

DEFINITION. An embedded annulus $(A, \partial A) \subset (N, \partial N)$ is *essential* in N if A is incompressible and boundary-incompressible.

Now fix M, a closed orientable three-manifold. Let $F \subset M$ be a closed, embedded, incompressible, two-sided surface with genus at least two.

DEFINITION. The surface F is *cylindrical* if $N = M \smallsetminus \eta_M(F)$ admits an essential annulus. If N does not admit an essential annulus then F is *acylindrical.*

DEFINITION. The surface F is *strongly cylindrical* if there exists an embedded annulus $(A, \partial A) \subset (M, F)$ such that $A \smallsetminus \eta_M(F)$ is essential in $N = M \smallsetminus \eta_M(F)$. On the other hand, if no such annulus exists then F is *weakly acylindrical.*

REMARK 4.1. If F is acylindrical then F is weakly acylindrical. Note that if F is a fibre of a surface bundle then F is never acylindrical, as the complement of F is homeomorphic to $F \times I$. However, if the monodromy is strongly irreducible then F is weakly acylindrical. For separating surfaces the notions acylindrical and weakly acylindrical coincide.

THEOREM 4.2. *Suppose that (M, T) is a closed, orientable, irreducible, triangulated three-manifold. Suppose $F \subset M$ is weakly acylindrical. Then F is isotopic to a fundamental normal surface.*

Before proceeding with the proof of Theorem 4.2 we remark that the normal surface techniques used may easily be replaced by the methods of branched surface theory.

PROOF OF THEOREM 4.2. Fix $F \subset M$ as in the hypothesis. Recall that weakly acylindrical includes the property of being incompressible. Applying Lemma 3.1 isotope F to be normal and least weight in its isotopy class. Set $N = M \setminus \eta_M(F)$. Recall also that F is two-sided, and has genus two or greater.

Suppose that F is not fundamental. Pick a reduced decomposition $F = G + H$. This sum admits some exchange band A. Let $A_N = A \cap N$. If A is a Mobius band then A's *double*, defined below, must be compressible or boundary-compressible in $N = M \setminus \eta_M(F)$. If A is an annulus then A_N is itself compressible or boundary-compressible in N. The proof deals with each of these possibilities in turn, showing that all lead to contradiction.

Suppose that A is an exchange Mobius band. As M is orientable, A is one-sided. Let X be a closed regular neighborhood of A_N, taken in N. Let $B = X \cap \partial N$. Then B is an annulus on the boundary of the solid torus X. The *double* of A, $\widetilde{A}$, is the closure of $(\partial X) \setminus B$. Note that $\widetilde{A}$ is an embedded annulus with $(\widetilde{A}, \partial \widetilde{A}) \subset (N, \partial N)$. As F is weakly acylindrical $\widetilde{A}$ must compress or boundary-compress in $N = M \setminus \eta_M(F)$.

Suppose that $\widetilde{A}$ compresses along a disk E. Note that $E \cap X = \partial E$. Compress $\widetilde{A}$ along E to obtain a pair of disks C and D. Then $S = B \cup C \cup D$ is a two-sphere bounding $\overline{\mathbb{RP}^3 \setminus \mathbb{B}^3}$ on the side which meets A. Thus $M \cong \mathbb{RP}^3$, contradicting the fact that F was two-sided and incompressible.

Suppose that $\widetilde{A}$ instead boundary-compresses along a disk E. Let Y be the closure of $\eta_{N \setminus X}(E)$. Boundary-compressing $\widetilde{A}$ along E gives a disk D. That is, D is the closure of $(\partial(X \cup Y)) \setminus \partial N$. Since F is incompressible and two-sided, ∂D bounds a disk $D' \subset \partial N$. Note that D' meets $X \cup Y$ along $\partial D'$ only, as $\partial N \cap (X \cup Y)$ is nonplanar. Now, as F is incompressible, the two-sphere $D \cup D'$ bounds a three-ball, Z, on the side not meeting A. It follows that $X \cup Y \cup Z$ is a solid torus with boundary equal to a component of ∂N. However this is impossible, as every component of ∂N has genus equal to that of F.

Thus the sum $G + H$ has no exchange Mobius bands. Instead, suppose that A is an exchange annulus. As M is orientable A is two-sided. The weakly acylindrical hypothesis forces A_N to be compressible or boundary-compressible in $N = M \setminus \eta_M(F)$.

Suppose that A_N is compressible along a disk E. Compress A along E to obtain disks C and D with $\partial C \cup \partial D = \partial A$. As F is incompressible ∂C bounds a disk, $C' \subset F$. Now $\partial C = \partial C'$ is contained inside of a seam, say $A_r(\gamma)$. Thus C'', the closure of $C' \setminus A_r(\gamma)$, is a disk which is a union of patches and seams of $G + H$. An innermost disk in C'' must be a disk patch, contradicting Lemma 3.4.

Finally, suppose that A_N is boundary-compressible along a disk E. Boundary-compress A_N to obtain a disk D. By incompressibility $\partial D \subset \partial N$ bounds a disk $D' \subset \partial N$. Let Z be the three-ball with boundary $D \cup D'$. If $E \subset Z$ then A_N is compressible (see Figure 4) yielding contradiction as in the proceeding paragraph.

If $E \cap Z = \emptyset$ then the situation is more delicate. Let W be the closure of the component of $N \setminus A_N$ containing E and Z. Then W is a solid torus having E as a meridional disk. Note that $A_N \subset \partial W$.

Let X be the closure of $\eta_N(A_N)$. Let $\widetilde{A}$ be the closure of $\partial X \setminus \partial N$. Take $F' = (\partial N \setminus X) \cup \widetilde{A}$. Then $F' = T \cup F''$ where $T = \partial\left(\overline{W \setminus X}\right)$ is a torus and F is isotopic to F''. (The surface F' is obtained by performing an *irregular exchange* along the annulus A.) To check that F'' is indeed isotopic to F recall that F is

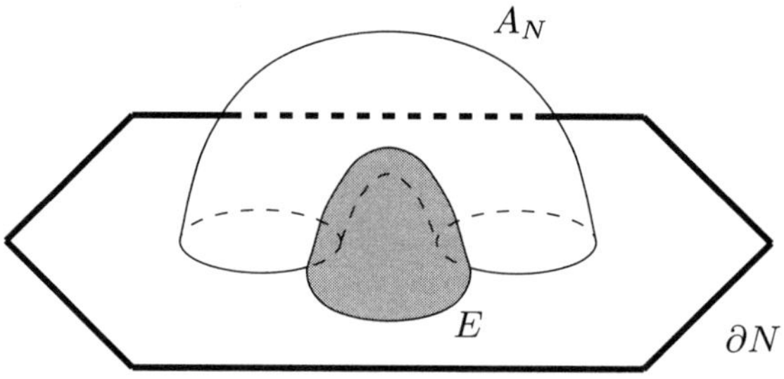

FIGURE 4. A boundary-compressible tube

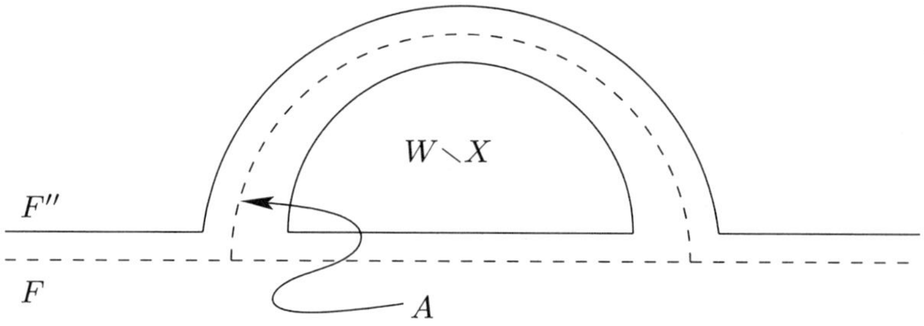

FIGURE 5. An irregular exchange along A

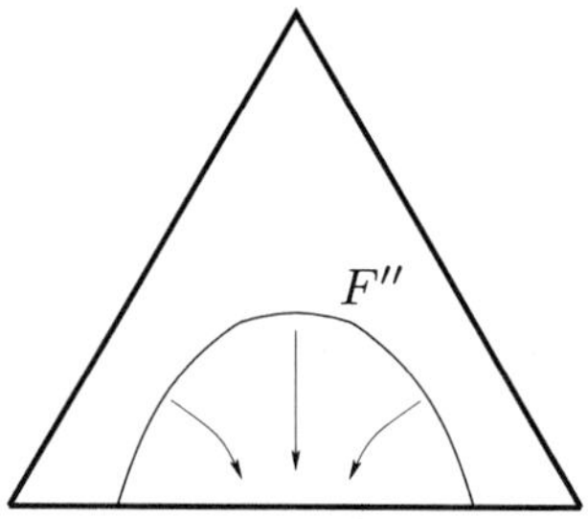

FIGURE 6. The weight-reducing isotopy

two-sided and note that the annulus $\partial N \cap (W \cup X)$ may be isotoped, relative to its boundary, across the solid torus $W \cup X$. See Figure 5 for a schematic picture of the cross-section of W.

Either $w(F'') < w(F)$ or F'' is not normal. (Again, see Figure 2.) In the latter case there is an isotopy of F'' to F''', supported in a small neighborhood of some face in the two-skeleton, which reduces weight by two. See Figure 6. Both possibilities contradict Lemma 3.1 because F is least weight in its isotopy class. □

We now deduce our main theorem as a corollary of Theorem 4.2:

THEOREM 4.3. *Suppose that M is a closed, orientable three-manifold. Then M admits only finitely many inequivalent surface bundle structures with strongly irreducible monodromy.*

PROOF. It follows from Theorem 4.2 and Lemma 3.2 that there are finitely many isotopy classes of weakly acylindrical surfaces in M. As in Remark 4.1, any bundle with strongly irreducible monodromy has a weakly acylindrical fibre. Finally, a given closed embedded surface in M is isotopic to the fibre of at most two bundle structures, up to equivalence. This gives the desired conclusion. □

Now suppose that M is a closed, orientable three-manifold with triangulation T. Let $\{F_i\}$ be the fundamental normal surfaces in T with negative Euler characteristic while $\{T_i\}$ and $\{P_i\}$ are those fundamental surfaces of Euler characteristic zero and positive, respectively. By Lemma 3.2 each of these collections is finite. Let $K = |\{F_i\}|$ be the number of fundamental surfaces with negative Euler characteristic. Let $P = \max\{-\chi(F_i)\}$.

We end this section by sketching a proof of:

THEOREM 4.4. *If (h, ϕ) is a surface bundle structure on M with genus $g > 1$ then h^i is weakly reducible for all integers i where $1 \leq i \leq c(M,T) \cdot (2g-2)$ and $c(M,T) = 1/(3KP)$.*

PROOF. Suppose that $F \subset M$ is a fibre of a surface bundle structure with monodromy h. Isotope F to be normal with respect to the triangulation T and the least weight such. Suppose that F decomposes as a Haken sum. Then, by Theorem 2.2 of [**6**], $F = \sum n_i F_i + \sum m_i T_i$ where the summands with nonzero coefficient are fundamental, incompressible normal surfaces. (Note that no fundamental surface with positive Euler characteristic appears as a summand; this follows directly from Lemma 3.4.)

Recall that $g(F) = \text{genus}(F) \geq 2$. Thus some of the n_i are nonzero. Reindex to obtain $n_1 \geq n_i$ for all i. If F_1 is two-sided then rewrite the Haken sum as $F = nG + H$, with $n = n_1$, $G = F_1$, and H equal to the sum of the remaining terms. If F_1 is one-sided then take $F = nG + H$ where n is the integer part of $n_1/2$, $G = 2F_1$ is the double of F_1, and H is the sum of the remaining terms.

Recall that Euler characteristic is additive under Haken sum. Now, if $n = 0$ then $n_i = 0$ or 1 for all i. It follows that $-\chi(F) \leq KP$ and thus $-\chi(F) \cdot c(M,T) < 1$. In this case the theorem is trivially satisfied.

Assume from now on that n is positive. An easy estimate shows that $n \geq -\chi(F)/(3KP)$. Make $F = nG + H$ a reduced sum by isotoping G and H if necessary. The surface nG is n normally parallel copies of G. Label these, in order, $G_1, G_2, \ldots, G_n$. The surface H is nonempty and intersects G because F is connected. So $H \cap nG$ decomposes into parallel (in H) families of curves, each family of size n. Let $\{\gamma_i\}_{i=1}^n$ be one such family. Each $\gamma_i \subset G_i$ yields an exchange annulus for the sum $F = nG + H$. An argument identical to the proof of Theorem 4.3 shows these annuli to be essential.

Pick $m \leq n$. The exchange annuli will lift to the mth cyclic cover of the surface bundle. Thus h^m is not strongly irreducible. □

5. Questions

Here are several questions, not all of which are necessarily difficult:

- Can a periodic surface automorphism be irreducible but not strongly irreducible?
- Are weakly acylindrical surfaces vertex surfaces? (See [**7**] for the definition of a vertex surface.)

- Give an algorithm to recognize strongly irreducible surface automorphisms or, more generally, compute translation distance.
- How are the strongly irreducible bundle structures on M distributed among the fibred faces of the Thurston norm ball?
- Is there a meaningful stabilization theory for surface bundle structures on M?
- Suppose that h is pseudo-Anosov. What does the translation distance of h imply about the hyperbolic geometry of the mapping torus M_h?
- Can the function $n(g)$, as in Remark 2.5, be given more explicitly? Remark 2.7 only suggests a lower bound.

References

[1] Alan F. Beardon. *The geometry of discrete groups.* Springer-Verlag, New York, 1983.

[2] A. J. Casson and C. McA. Gordon. Reducing Heegaard splittings. *Topology Appl.*, 27(3):275–283, 1987.

[3] Wolfgang Haken. Some results on surfaces in 3-manifolds. In *Studies in Modern Topology*, pages 39–98. Math. Assoc. Amer. (distributed by Prentice-Hall, Englewood Cliffs, N.J.), 1968.

[4] Kevin Hartshorn. *Heegaard splittings: the distance complex and the stablization conjecture.* PhD thesis, U.C. Berkeley, 1999.

[5] Joel Hass. Acylindrical surfaces in 3-manifolds. *Michigan Math. J.*, 42(2):357–365, 1995.

[6] William Jaco and Ulrich Oertel. An algorithm to decide if a 3-manifold is a Haken manifold. *Topology*, 23(2):195–209, 1984.

[7] William Jaco and Jeffrey L. Tollefson. Algorithms for the complete decomposition of a closed 3-manifold. *Illinois J. Math.*, 39(3):358–406, 1995.

[8] Michael Kapovich. *Hyperbolic manifolds and discrete groups.* Birkhäuser Boston Inc., Boston, MA, 2001.

[9] D. D. Long and H. R. Morton. Hyperbolic 3-manifolds and surface automorphisms. *Topology*, 25(4):575–583, 1986.

[10] Howard A. Masur and Yair N. Minsky. Geometry of the complex of curves. I. Hyperbolicity. *Invent. Math.*, 138(1):103–149, 1999.

[11] Saul Schleimer. *Almost normal Heegaard splittings.* PhD thesis, U.C. Berkeley, 2001.

Saul Schleimer, Department of Mathematics, UIC, 851 South Morgan Street, Chicago, Illinois 60607

E-mail address: saul@math.uic.edu

Proceedings of Symposia in Pure Mathematics
Volume **71**, 2003

Problems in Foliations and Laminations of 3–manifolds

Danny Calegari

1. Introduction and basic definitions

1.1. Notation. I will try to use consistent notation throughout, so for instance, $\mathscr{F}, \mathscr{G}$ denote foliations, Λ denotes a lamination, λ, μ, ν denote leaves, etc. For a given object X in a manifold M, $\widetilde{X}$ will denote the pullback of X to the universal cover $\widetilde{M}$.

1.2. Attribution. I have tried to credit questions to their originators. There are certain caveats to this, however:

- I have made no attempt to verify that my sources were themselves original.
- I have occasionally edited or reformulated questions; in so doing, it is possible that I have perverted the intent of the originator.
- I have not systematically checked other problem lists to see if the questions here were posed independently elsewhere.

1.3. Other problem lists. The format and taste of this problem list is greatly influenced by Gabai's well known problem list [**45**]. However I have tried to choose questions which complement, rather than overlap with, Gabai's choice of problems.

1.4. Basic objects. A reference for the basic theory of foliations is [**24**]. An overview of the state of the subject as it stood in 1990 is contained in [**42**]. [**45**] also contains background and numerous examples.

DEFINITION 1.1. A codimension one foliation $\mathscr{F}$ of a 3–manifold M is *taut* if there is a circle γ transverse to $\mathscr{F}$ intersecting every leaf.

A basic reference is [**41**]. Note that a foliation of an atoroidal 3–manifold is taut iff it contains no torus leaf.

For a taut foliation $\mathscr{F}$ of M, $\widetilde{M}$ may be taken to be an open subset of $\mathbb{R}^3$ so that $\widetilde{\mathscr{F}}$ consists of the intersection of $\widetilde{M}$ with the horizontal planes in $\mathbb{R}^3$. These leaves are all disks, and the leaf space L of $\widetilde{\mathscr{F}}$ is a simply–connected 1–manifold. There is a canonical *holonomy representation*

$$\rho_H : \pi_1(M) \to \text{Homeo}(L)$$

Key words and phrases. Foliation, lamination, branched surface, hyperbolic geometry, pseudo-Anosov flow, problem list.

DEFINITION 1.2. A taut foliation is said to be or have

- $\mathbb{R}$–*covered* if the leaf space of $\widetilde{\mathscr{F}}$ is homeomorphic to $\mathbb{R}$.
- *one–sided branching* if the leaf space of $\widetilde{\mathscr{F}}$ branches in one direction.
- *two–sided branching* if the leaf space of $\widetilde{\mathscr{F}}$ branches in both directions.

All possibilities are realized, for foliations of atoroidal 3–manifolds.

DEFINITION 1.3. A codimension one *lamination* Λ in a 3–manifold M is a foliation by two–dimensional leaves of a closed subset of M. A lamination is *essential* if it contains no sphere leaf or torus leaf bounding a solid torus, if complementary regions are irreducible and have incompressible boundary, and admit no compressing monogons.

A taut foliation is an example of an essential lamination. A closed union of leaves of a taut foliation is another example.

DEFINITION 1.4. The complementary regions to an essential lamination fall into two kinds of pieces, which can be chosen uniquely up to isotopy: I–bundles $\Sigma_i \times I$ over noncompact surfaces Σ_i, called *interstitial regions*, and compact pieces called *guts* $\mathfrak{G}_i$ meeting the I–bundles along annuli called *interstitial annuli*. An essential lamination is *genuine* if it has nonempty gut — i.e. if some complementary region is not an I–bundle.

To construct the partition of a complementary region C into guts and interstices, take as a first approximation of the interstices the characteristic I–bundle of $C, \partial C$, and then remove all I–bundles over non–compact components.

For more details on essential and genuine laminations, see [**52**], [**48**] and [**50**]. Note that though foliations can always be co–oriented in some finite cover of the ambient manifold, laminations admit local obstructions to co–orientability.

DEFINITION 1.5. A genuine lamination has *solid torus guts* if all gut regions are neutered ideal polygon bundles over S^1. It is *full* if some complementary region is an ideal polygon bundle over S^1. It is *very full* if *every* complementary region is a bundle region.

There is a procedure to turn *co–oriented* very full genuine laminations into taut foliations, by filling in the complementary regions with *monkey saddles*. These are disks which are asymptotic along the boundary leaves of a complementary region in alternating directions. The complementary regions can be given a bundle structure over S^1 where these saddle leaves are the fibers.

DEFINITION 1.6. A *pseudo–Anosov flow* X on a manifold M is a flow which, away from a finite number of closed orbits γ_i, is *Anosov* — that is, there is a decomposition of TM as a sum $E_s \oplus E_u \oplus TX$ where the time t flow contracts E_u by some factor λ^{-t} and expands E_s by λ^t, for some $\lambda > 0$. Along the singular orbits, the local picture is that of a *semi–branched cover* of an Anosov orbit of degree $n/2$ for $n \geq 3$. The distributions E_s and E_u are integrable, and integrate to give two singular foliations $\mathscr{F}_s, \mathscr{F}_u$ the *stable* and *unstable* integral foliations. If one splits open $\mathscr{F}_s, \mathscr{F}_u$ along the singular leaves and takes the closure, one obtains very full genuine laminations Λ_s, Λ_u.

The basic examples of pseudo–Anosov flows are the suspensions of pseudo–Anosov automorphisms of surfaces. See [**116**].

DEFINITION 1.7. A lamination Λ has *hyperbolic leaves* if the leaves of $\widetilde{\Lambda}$ are (uniformly) quasi–isometric to $\mathbb{H}^2$ with the path metric induced from some metric on M.

A theorem of Candel ([**23**]) implies that for M atoroidal, any essential lamination Λ can be given a metric of *constant* curvature -1. In particular, all such laminations have hyperbolic leaves.

DEFINITION 1.8. Suppose $\mathscr{F}$ is a taut foliation of M with hyperbolic leaves. A *universal circle* for $\mathscr{F}$ is a circle S^1_{univ} with a group action

$$\rho_{\text{univ}} : \pi_1(M) \to \text{Homeo}(S^1_{\text{univ}})$$

satisfying the following properties:

(1) For all leaves λ of $\widetilde{\mathscr{F}}$ there is a *monotone* map

$$\phi_\lambda : S^1_{\text{univ}} \to S^1_\infty(\lambda)$$

where ϕ_λ varies continuously as a function of λ. Here a map between circles is *monotone* if the preimage of every point is contractible.

(2) For each leaf λ of $\widetilde{\mathscr{F}}$ and each $\alpha \in \pi_1(M)$ the following diagram commutes:

$$\begin{array}{ccc} S^1_{\text{univ}} & \xrightarrow{\rho_{\text{univ}}(\alpha)} & S^1_{\text{univ}} \\ \phi_\lambda \downarrow & & \downarrow \phi_{\alpha(\lambda)} \\ S^1_\infty(\lambda) & \xrightarrow{\alpha} & S^1_\infty(\alpha(\lambda)) \end{array}$$

(3) For any leaf λ, the *associated gaps* are the maximal connected open intervals in S^1_{univ} mapped to points by ϕ_λ. The complement of the gaps in S^1_{univ} is the *core* associated to λ. Then for any pair μ, λ of incomparable leaves, the core associated to λ is contained in a single gap associated to μ, and vice versa.

Universal circles are introduced in [**118**]; examples and discussion are found in [**15**], [**18**] and [**14**]. Since the map between the universal circle and $S^1_\infty(\lambda)$ is monotone for each λ, it easily follows that the Euler class of the representation ρ_{univ} is equal to the Euler class of $T\mathscr{F}$.

DEFINITION 1.9. A *branched surface* $\mathscr{B}$ is a certain kind of 2–complex in a 3–manifold with a well–defined C^1 tangent space everywhere. The place where the branched surface is not a surface is called the *branch locus*; it is a 1–complex, and the branch locus is not a 1–manifold only at isolated points, where two lines of the branch locus cross each other transversely.

Along the nonsingular part of the branch locus of a branched surface, there is a natural transverse orientation, which points in the direction in which the two branches come together. Where two arcs of the branch locus cross, there is one complementary quadrant where both co–orientations are inward, and one where both co–orientations are outward.

DEFINITION 1.10. A *sink disk* for a branched surface is an embedded disk $D \subset \mathscr{B}$ with boundary contained in the branch locus, such that the transverse orientation points inward all along ∂D.

Sink disks are introduced by Li in [**80**]. He shows, amongst other things, that branched surfaces without sink disks fully carry essential laminations, and that every essential lamination (with some very simple exceptions) is carried by a branched surface without sink disks.

1.5. 3–manifold topology. Throughout this paper, numerous reference is made, implicitly or explicitly, to elements of the theory of 3–manifolds. A basic reference for this theory is [**70**]. A reference for the theory of geometrization is [**114**].

1.6. Acknowledgements. Many thanks to Charles Delman, Tao Li, Curt McMullen, Saul Schleimer and participants in the problem session on foliations and laminations at the 2001 Georgia International Topology Conference. Ian Agol and Nathan Dunfield should be singled out for special thanks, for their considerable contributions to this problem list.

2. Existence questions

QUESTION 2.1. *Which hyperbolic 3–manifolds admit taut foliations? Give an effective procedure to decide if a hyperbolic 3–manifold admits a taut foliation. For a useful property* p *of a 3–manifold, give an explicit construction of infinitely many 3–manifold with property* p *with/without a taut foliation. Same question for essential laminations.*

REMARK. (1) It is known by [**103**] that there exist infinitely many hyperbolic 3–manifolds without taut foliations. These manifolds are so special that their fundamental groups admit no nontrivial actions on (potentially non–Hausdorff) 1–manifolds. Here an action on a non–Hausdorff 1–manifold is trivial if it has a global fixed point *on the maximal Hausdorff quotient*. The proof is somewhat *ad hoc* and relies on the special structure of the fundamental groups of these manifolds (they are Dehn fillings of certain punctured torus bundles with negative trace). It should be remarked that Hatcher initially speculated that such manifolds would be good candidates to be taut foliation free.

The method of proof is not entirely satisfactory for two reasons. Firstly, the *certificate* that the manifolds in question admit no taut foliation is very long, and it seems unlikely that it could be further compressed within a similar approach. (See [**22**] for a different method which provides a much shorter certificate that the Weeks manifold admits no taut foliation). Secondly, the criterion by which the existence of taut foliations is ruled out is not *a priori* sharp. For instance, if $f : M \to N$ has positive degree and N admits a taut foliation $\mathscr{F}$, then $\pi_1(M)$ acts on the leaf space of $\widetilde{\mathscr{F}}$ without a global fixed point; on the other hand, it is unclear whether such an M admits a taut foliation in general. In particular, it is important to understand when a 3–manifold admitting a nontrivial action on a non–Hausdorff 1–manifold admits a taut foliation.

The methods of [**103**] are extended by Fenley in [**38**] to show that infinitely many of these hyperbolic manifolds actually admit no essential laminations, by showing that their fundamental groups admit no nontrivial actions on order trees. It suffers from the same shortcomings as [**103**], and is therefore perhaps more of theoretical and philosophical than of practical interest.

(2) Work of Agol [**2**] gives strong restrictions on the topology of the guts of a tight genuine lamination on a hyperbolic manifold M of small volume. If a taut foliation $\mathscr{F}$ is not monotone equivalent to a minimal foliation, some minimal set is a genuine lamination. Thus it might be interesting to give examples of 3–manifolds without taut foliations of large volume and/or injectivity radius.

(3) Agol and Li [**3**], combining work of Li [**80**] and Brittenham [**10**], give an algorithm to decide whether a given 3–manifold admits an essential lamination. For, if a 3–manifold admits a nowhere dense essential lamination, Brittenham shows it admits an essential lamination in normal form with respect to any triangulation, and therefore is carried by one of a finite constructible set of branched surfaces. With more work, using technology of Gabai [**46**], one can show that every essential nowhere dense lamination is *fully carried* by one of a finite constructible set of branched surfaces. Li shows that every essential lamination can, after blowing up, be carried by a branched surface without a sink disk (except for the special case of a foliation of T^3 by planes). So the algorithm proceeds by taking each of the finite constructed set of branched surfaces, and splits it open repeatedly in all possible ways. After some finite time, either every branched surface reveals an obstruction to being fully split open, or some branched surface can be split open to a branched surface with no sink disk. How computationally efficient is this algorithm? Is there any case in which it might actually be implemented on a computer, e.g. for the Seifert–Weber dodecahedral space?

Note that the method of Agol and Li decides whether a given 3–manifold admits a taut foliation. For, one can check that one of the finite constructible set of branched surfaces which fully carries an essential lamination has I–bundle complementary region.

(4) Properties p for which this question is interesting might include Hakenness, or more generally, laminar properites. Obstructions to the existence of a taut foliation which are homological in nature probably do not obstruct the existence of genuine laminations. In particular, the local non co–orientability is a strong advantage. Since taut foliations can be perturbed to symplectically semi–fillable contact structures by Eliashberg and Thurston [**32**], one might be able to use contact or symplectic 4–manifold techniques (e.g. gauge theory) to rule out the existence of taut foliations on certain manifolds which contain genuine laminations. Conversely it would be very interesting to give an example of a hyperbolic 3–manifold with a semi–fillable contact structure but no taut foliation.

(5) For a given manifold M and a hyperbolic knot $K \subset M$, the set of surgeries on K giving rise to a tautly foliated or laminar manifold, if not everything, is probably quite complicated in some cases. There are probably arithmetic criteria (e.g. some persistent lamination might be co–orientable or not after certain surgeries and therefore fillable or otherwise), inequalities (e.g. branched surfaces supporting foliations realizing certain slopes), bounded cohomology (e.g. Milnor–Wood), etc. There are existence results

of Gabai–Mosher [**93**] which say that if K is a hyperbolic knot, there is a degeneracy slope δ associated to K such that the stable and unstable laminations coming from a pseudo–Anosov flow on $M - K$ stay essential for surgery on all slopes α with $\Delta(\delta, \alpha) > 1$. Bootstrap techniques to determine degeneracy slopes are exploited in [**22**] to give a partial table of which small volume hyperbolic manifolds are laminar.

QUESTION 2.2. *Is there an effective algorithmic procedure to produce and recognize a hyperbolic knot of depth n for any given n? What about $\geq n$?*

REMARK. (1) Satellite knots of arbitrary depth were produced by Cantwell and Conlon. A single explicit example of a hyperbolic knot of depth ≥ 2 is constructed by Kobayashi in [**74**]. Brittenham has a class of examples which are conjecturally of depth ≥ 2. But what about higher depth? See also question 10.3.
(2) Algorithms exist to detect whether a knot is of depth 0 ([**69**]) or of depth 1 but they are not very useful in practice. Part of the problem is that finding Thurston norm minimizing surfaces is NP complete by [**5**]. On the other hand, perhaps there is a fast algorithm to *check* that a foliation is of minimal depth. A "good" algorithm producing a knot of depth n would also produce a certificate for this fact. The paper [**107**] may be relevant here.
(3) Lackenby's technique of taut ideal triangulations [**77**] is useful in practice for finding and especially *certifying* minimal genus representatives for relative homology classes in open manifolds with torus boundary.

QUESTION 2.3. *Given a collection $\mathscr{C}$ of topological or geometric types of surface, what 3–manifolds admit a taut foliation $\mathscr{F}$ whose leaves are all homeomorphic or coarsely quasi–isometric to an element of the collection $\mathscr{C}$?*

REMARK. (1) Every topological surface is a leaf of a foliation (even a taut foliation) of some 3–manifold M, by a construction of Cantwell and Conlon [**25**].
(2) The Reeb stability theorem implies that a foliation of a 3–manifold with a spherical leaf is a bundle (possibly over an orbifold). Moreover, taut foliations of hyperbolic 3–manifolds do not contain torus or Klein bottle leaves.
(3) By a theorem of Candel [**23**], if $\mathscr{F}$ is a taut foliation of an atoroidal 3–manifold M, there is a metric on M with respect to which every leaf has constant curvature -1. In particular, the universal covers of all the leaves are isometric to $\mathbb{H}^2$. Moreover, if there is a leaf (of $\mathscr{F}$, not $\widetilde{\mathscr{F}}$) with amenable growth, then its closure supports an invariant transverse measure, and therefore contains leaves of polynomial growth, by Plante's theorem. In particular, if M is a rational homology sphere, every leaf of an orientable/co–orientable taut foliation $\mathscr{F}$ has *uniformly* exponential growth.
(4) In great generality this question is probably not so interesting, but in some special cases it is probably quite interesting. For instance, if $\mathscr{C}$ consists just of a plane, M must be T^3 and the foliation must be monotone equivalent to one by planes of irrational slope. If $\mathscr{C}$ consists of just planes

and cylinders and $\mathscr{F}$ is $\mathbb{R}$–covered, either M is Solv, or there are cylinders whose nontrivial loops are homotopic to each other. If one adds the geometric constraint that the shortest loops in the cylinders are of bounded length, if the ambient manifold is atoroidal and therefore has word–hyperbolic fundamental group, these loops must be homotopic by an annulus of controlled diameter etc.

(5) The character of this problem would be very different if one asked for foliations with the property that *almost all* leaves were homeomorphic or coarsely quasi–isometric to an element of $\mathscr{C}$. If μ is a harmonic measure on a foliated space, then any leaf λ in the support of μ is quasi–homogeneous; that is, any local geometric feature of λ on any scale recurs with definite density in any sufficiently large subset of λ. This is more or less a consequence of the fact that the support of μ is roughly the subset where leafwise random walks are recurrent. In this vein, there is a very satisfying theorem of Ghys that for Λ *any* (abstract) compact Riemann surface lamination, and μ any harmonic measure, then μ–almost all leaves are homeomorphic to one of the following six possibilities:
   - The plane
   - The Loch–Ness monster
   - The cylinder
   - Jacob's ladder
   - S^2 – Cantor set
   - S^2– Cantor set with countably many handles accumulating to every point in the removed set.

   Ghys' theorem is proved in [**55**].

QUESTION 2.4. *Let X be a vector field on a 3–manifold. When is there a foliation $\mathscr{F}$ of M transverse to X?*

REMARK. (1) If M is Seifert–fibered and X is a vector field tangent to the circle fibers, there are many restrictions on the existence of $\mathscr{F}$. The *Milnor–Wood* inequality says that the Euler class of a foliated circle bundle is at most as large in absolute value as the Euler characteristic of the base orbifold [**88**],[**123**]. This result can be phrased in terms of bounded cohomology, as saying that the norm of the Euler class is equal to 1 [**60**],[**117**]. Moreover, results of Brittenham and Thurston [**9**][**111**] imply that any taut foliation of a Seifert–fibered manifold is either transverse to the circle fibers or contains "vertical" leaves — leaves foliated by circles. Complete necessary and sufficient conditions for the existence of a foliation transverse to the foliation by circles of a Seifert fibered 3–manifold were established by Naimi [**94**] building on results of Jankins and Neumann [**73**].

(2) By analogy with the circle bundle case, one can ask whether there exists a *slithering* of M over S^1 for which the slithering map Z is realized by time 1 flow along X. Define an equivalence relation $\sim$ on M which identifies pairs of points which are an integer distance apart on a flow line of X (parameterized by the flow speed). Then one can think of $M/\sim$ as a circle bundle over the non–commutative space M/X, the leaf space of X. One would like to prove an analogue of the Milnor–Wood inequality for some appropriate fundamental classes for $M/\sim$ and M/X. See [**29**]. One

can vary the question by allowing (nonconstant) reparameterizations of X.

(3) More generally, one can study a homology theory closely related to the *Godbillon homology* of X, whose i–chains are equivalence classes of singular maps of simplices to M, where two such maps $\sigma_1, \sigma_2 : \Delta \to M$ are equivalent if there is a homotopy $S : \Delta \times I \to M$ where $S(\cdot, 0) = \sigma_1(\cdot)$, $S(\cdot, 1) = \sigma_2(\cdot)$ and $S(p, I)$ is contained in a flowline of X. What is the cohomology of the *bounded* cochains on this complex?

3. Rigidity and Moduli

QUESTION 3.1. *Let M be atoroidal. Is there a natural refinement of the polyhedral structure of the unit ball of the Thurston norm to a polyhedron $\mathscr{P}_{\mathscr{F}}$ whose faces parameterize the set of taut of foliations of M in the following sense:*

(1) *Each open face c of $\mathscr{P}_{\mathscr{F}}$ corresponds to a pseudo–Anosov flow X_c or equivalence classes of pseudo–Anosov flows. The taut foliations "parameterized" by $\mathscr{P}_{\mathscr{F}}$ can be isotoped to be transverse (or* almost transverse*) to X_c, or some sublamination is monotone equivalent into the stable or unstable singular foliation of X_c.*

(2) *Every geometric limit of taut foliations $\mathscr{F}_i$ associated to a cell c should be associated to some cell of the closure $\bar{c}$.*

(3) *There should be a natural polyhedral map from $\mathscr{P}_{\mathscr{F}}$ to (some polyhedral subdivision of) the unit ball of the Thurston norm.*

REMARK. (1) Basic references for the Thurston norm are [**115**] and [**96**]. For depth 0 (surface bundles), fibered homology classes are a union of top dimensional faces, by [**115**]. For depth 1, a refinement of the polyhedral structure is necessary, but is accomplished by Cantwell and Conlon in ([**26**]). Their procedure applies to depth 1 foliations of sutured manifolds, and therefore inductively, should give a polyhedral parameterization of all finite depth foliations on a fixed manifold, up to some appropriate equivalence relation.

(2) For a given fixed manifold M, Gabai has shown in [**46**] there is a triangulation τ such that every taut foliation of M can be put into normal form with respect to τ. There is a natural polyhedral structure $\mathscr{P}_\tau$ on the space of all normal surfaces carried by τ. $\mathscr{P}_{\mathscr{F}}$ should be intermediate between the unit ball for the Thurston norm and $\mathscr{P}_\tau$.

Note for certain classes of foliations ($\mathbb{R}$–covered, one–sided branching, minimal foliations etc.) there are *existence* results for transverse pseudo–Anosov flows and, in many cases, stability results which say that these flows are unique up to isotopy for small deformations of the foliations ([**15**],[**18**], [**20**]). It seems plausible that a deformation of normal foliations which preserves the normal disk types in each simplex should be mild enough to ensure stability of a transverse flow. Note that for *finite depth foliations* one has the theorem of Lackenby that a taut sutured manifold hierarchy can be placed in a position resembling normal form, relative to *any* triangulation ([**76**]).

(3) Kronheimer and Mrowka ([**75**]) found a characterization of the unit ball of the dual Thurston norm by means of differential geometry. Is there

some way to detect the depth of a taut foliation or the codimension of the face it corresponds to by geometric properties of an approximating tight contact structure?

(4) By Gabai and Mosher, there is a pseudo–Anosov flow almost transverse to any finite depth foliation of a hyperbolic 3–manifold [**93**]. For depth > 1, there are choices involved in the construction of the flow; any natural equivalence relation on pseudo–Anosov flows would have to take this ambiguity into account. Mosher in particular has developed a sophisticated picture of the relationship between pseudo–Anosov and, more generally, pA flows on 3–manifolds and the Thurston norm, for instance in the papers [**90**],[**91**],[**92**].

(5) Thurston's theorem that euler classes of taut foliations are contained in the unit ball of the dual Thurston norm follows from the existence of a *universal circle* for a taut foliation, and from the Milnor–Wood inequality. One potential source of polyhedra $\mathscr{P}_{\mathscr{F}}$ might be unit balls of L_1 norms on certain natural quotients of the vector space $\Omega^1(M)$ of 1–forms on M. Even an infinite dimensional polyhedron parameterizing taut foliations would be very interesting.

QUESTION 3.2 (McMullen). *Generalize the* Teichmüller polynomial *from the fibered faces of the Thurston norm ball to the other faces (of some possibly generalized polyhedron, perhaps the polyhedron sought in question 3.1).*

REMARK. (1) A reference for the Teichmüller polynomial associated to a fibered face of the Thurston norm ball is [**86**]. The fibered face itself is the Newton polygon of the polynomial associated to that face; amongst other information, the polynomial encodes the (topological) entropy of the (topologically unique) transverse pseudo–Anosov flow associated to the face. If the singular stable and unstable foliations of the pseudo–Anosov flow are co–orientable, similar information is encoded in Fried's dynamical zeta function [**39**].

(2) The entropy of the pseudo–Anosov flow arising from a surface bundle over a circle is intimately related to the entropy of the automorphism of the fiber. Loosely speaking, the Teichmüller polynomial is a kind of "characteristic polynomial" for the action of the monodromy on the space of invariant transverse measures for an invariant train track. [**83**] gives a nice exposition of these ideas.

(3) For a depth n foliation where $n > 0$, one has invariant laminations for the end–periodic automorphisms of the fibers of greatest depth (see e.g. [**33**]), and one can study the entropy of these automorphisms. Global genuine laminations which can be collapsed to pseudo–Anosov flows are built up from these laminations at lowest depth inductively. A transverse pseudo–Anosov flow X for a fibration or a transversely measured foliation $\mathscr{F}$ is like a Teichmüller geodesic for the family of hyperbolic metrics on the leaves of $\mathscr{F}$ (see [**86**]). The action of $\pi_1(M)$ on this Teichmüller geodesic is conjugate to the action on the transverse measures in the appropriate projective class for the singular foliations associated to X. For more general (not transversely measured) foliations with transverse pseudo–Anosov flows, there might exist refined polynomials with values in $\mathbb{Z}[GL(n,\mathbb{R})]$ for some appropriate representation of $GL(n,\mathbb{R})$ in Homeo($\mathbb{R}$). For instance, there

are many interesting slitherings of hyperbolic 3–manifolds with transverse $\widetilde{SL(2,\mathbb{R})}$ structure, associated to branched covers over unit tangent bundles of hypebolic orbifolds, where the branch loci are timelike with respect to the usual Lorentz structure on $PSL(2,\mathbb{R})$. See [**117**] or questions 8.4 and 8.7.

4. Minimal surfaces

QUESTION 4.1. *Suppose $\mathscr{F}$ is a taut foliation of M. Characterize the space of metrics on M for which $\mathscr{F}$ can be isotoped to consist of minimal surfaces.*

REMARK. (1) For general results on minimal surfaces relevant to this question, see [**67**], [**99**], [**106**].

(2) By Sullivan [**110**] a C^2 foliation of a 3–manifold can be made minimal with respect to *some* Riemannian metric on the ambient manifold iff it is taut. This is equivalent to the existence of a nonsingular divergence free *normal vector field* — i.e. one which is perpendicular to $T\mathscr{F}$ everywhere. Given any transverse nonsingular vector field X for which $\mathscr{L}_X\omega = 0$ for some nonsingular 3–form ω, one can find a metric on M for which X is normal and ω is the volume form.

(3) A minimal surface in a hyperbolic manifold has intrinsic curvature ≤ -1 everywhere. By Gauss–Bonnet, this puts an upper bound on the area of a compact minimal surface. Suppose $\mathscr{F}$ is a taut foliation in a hyperbolic 3–manifold enjoying some comparable bound on the leafwise intrinsic curvature. If $\mathscr{F}$ admits a nontrivial invariant transverse measure μ (not necessarily of full support), one obtains nontrivial constraints on the growth rate and topology of leaves in the support of μ.

(4) Ben Andrews has constructed an area–minimizing flow on surfaces with principle curvatures < 1 to a minimal surface with the same properties. Thurston has observed that elliptic PDE flows generally fail to preserve integrability of plane fields, since after a short time, such flows tend to make structures real analytic; but there are many C^∞ foliations, even taut foliations, which admit no real analytic structure, and even manifolds which admit no (nontrivial) transversely real analytic foliation at all. Even so, it might be interesting to understand the behaviour of the leaves of $\mathscr{F}$ under Andrews' flow.

(5) A minimal surface in $\mathbb{E}^3$ or $\mathbb{H}^3$ has a natural real analytic structure, which one can think of as a kind of *tangential rigidity* condition. Zeghib shows in [**125**] that such tangential rigidity for foliations implies a transverse Lipschitz regularity; that is, any minimal foliation of $\mathbb{E}^3$ or $\mathbb{H}^3$ should be transversely Lipschitz. This already shows that certain taut foliations of hyperbolic 3–manifolds cannot be made minimal with respect to the hyperbolic metric, and perhaps can be used to show no foliation of a hyperbolic 3–manifold by minimal surfaces can occur. Notice in codimension 2 a minimal foliation would actually have to be geodesic; Zeghib showed that there are no nontrivial geodesic foliations of any codimension of any closed hyperbolic manifold ([**124**]). On the other hand, if $\mathscr{F}$ is proper and all leaves have the continuous extension property, one can inductively

minimize leaves of fixed depth. The closure of the union will be an essential (but not genuine) lamination, monotone equivalent to the original foliation; but there seems to be no reason to expect it to be a foliation.

(6) One weakened version of the question is as follows: is there an *a priori* upper bound on the infimum of the mean curvature of a taut foliation $\mathscr{F}$ in a hyperbolic 3–manifold M, independent of $\mathscr{F}$ and M?

QUESTION 4.2. *Given a collection of taut foliations $\mathscr{F}_i$ of M, what are the obstructions to finding a metric on M for which the $\mathscr{F}_i$ (after an isotopy) are simultaneously minimal?*

REMARK. (1) Some kind of "stably minimal" property holds at the PL level. For foliations $\mathscr{F}_i$ whose tangent fields are sufficiently close as 2–plane fields, we can find a triangulation τ of M with respect to which the $\mathscr{F}_i$ can all be made simultaneously normal, in such a way that the co–orientations on the one skeleton of τ coming from the co–orientations on $\mathscr{F}_i$ agree. For sufficiently close $\mathscr{F}_i$, we can find a triangulation such that the 1–skeleton with this orientation is recurrent; such a 1–skeleton admits a combinatorial weight $w : \tau^1 \to \mathbb{Z}^+$ such that for each vertex, the sum of the weights of the incoming edges is equal to the sum of the weights of the outgoing edges. With respect to such a weight, these foliations are "minimal" polyhedral surfaces. For similar ideas, see [**67**] and [**47**].

(2) Maybe this question become easier if we can replace the $\mathscr{F}_i$ by monotone equivalent laminations Λ_i. This is especially plausible if the $\mathscr{F}_i$ are finite depth — the leaves at each depth could be constructed inductively, and one could ignore the problem of filling up the remaining gaps with the leaves of greatest depth.

5. Reeb components

Every closed 3–manifold admits a foliation with at most one Reeb component. One way to see this is to use an open–book decomposition. A more profound construction due to Gabai shows the core of the Reeb component can be essentially any knot ([**41**]). Every open 3–manifold admits a foliation with possibly infinitely many Reeb components. This follows from Thurston's local construction of a foliation from an anti–orientation on a triangulation ([**113**]), and Moise's theorem that 3–manifolds can be triangulated.

QUESTION 5.1. *How many Reeb components must a foliation of an open 3–manifold contain?*

REMARK. (1) There are two questions here, a topological one and a geometric one. The topological question is perhaps not so interesting. To make the geometric question precise, some kind of quantification is necessary. First, the 3–manifold should have uniformly bounded geometry, and then one should uniformly bound the local geometry of the foliation, and either require all Reeb components to have bounded diameter, or weight them by the length of a core circle.

Here it would be useful to distinguish between answers *none, finitely many, infinitely many*, and if the third alternative, find some increasing function $f : \mathbb{R} \to \mathbb{R}$ such that the ball of radius r must contain at least $O(f(r))$ such components.

(2) Perhaps one should modify this question to measure generalized Reeb components. For example, the Whitehead manifold can be obtained as an increasing union of foliated "plugs" centered on a single Reeb component.

QUESTION 5.2. *What generalizations of the notion of taut foliation make sense on an open 3–manifold?*

REMARK. This question has different flavors in the topological and the geometric world.

(1) For M the interior of a compact manifold with boundary, there are well–understood generalizations of a taut foliation or a taut sutured structure; see [**41**].
(2) If M is an infinite cover of some closed N, and N admits a vector field X which is volume–preserving for some volume form on N, a foliation $\mathscr{F}$ of M transverse to the cover $\widehat{X}$ probably has some useful properties. Maybe one should ask for $\mathscr{F}$ to be a Lipschitz section of $\widehat{X}$, in the sense that the transverse flow is uniformly bounded away from $T\mathscr{F}$. In this case, leaves of $\mathscr{F}$ are approximate minimal surfaces for some equivariant metric.
(3) One could weaken the transversality condition on $\widehat{X}$ to an asypmtotic or a statistical condition. For example: we could ask for every flowline of $\widehat{X}$ to be eventually transverse to $\mathscr{F}$, Or: assume $\mathscr{F}$ is co–oriented; then one could ask for every $\epsilon > 0$ there should exist a t such that for each flowline γ of $\widehat{X}$, every segment of γ of length t is transverse to $\mathscr{F}$ in the positive direction.
(4) Rather than looking at a vector field X, one could ask for a single loop γ in N such that every lift $\widehat{\gamma}$ of γ is transverse to $\mathscr{F}$ in the positive direction, and every leaf of $\mathscr{F}$ intersects some lift. Again, this could be made into an asymptotic or statistical condition.

6. Sublaminations and superlaminations

6.1. Essential laminations.

QUESTION 6.1. *Characterize those essential laminations which contain genuine sublaminations.*

REMARK. Again, *characterize* can take several flavors. For example:

(1) Geometric conditions. For instance, if $\pi_1(M)$ is Gromov–hyperbolic, a lamination whose universal cover contains a quasi–geodesically embedded leaf contains a genuine sublamination.
(2) Topological conditions. For instance, if $\mathscr{F}$ is $\mathbb{R}$–covered or has one–sided branching, it cannot contain a genuine sublamination.
(3) Algorithmic conditions. Given a lamination, described somehow in finite terms, give an algorithm to determine whether it contains a genuine sublamination.

Given a branched surface $\mathscr{B}$, can the algorithm of Agol and Li can be modified to tell whether it carries (possibly not fully) a genuine lamination? (see [**107**] for related questions). The set of branched surfaces in a 3–manifold which fully carry an essential lamination can be constructed, so the problem reduces to showing, for two branched surfaces $\mathscr{B}, \mathscr{B}'$ whether some splitting of $\mathscr{B}'$ is carried by $\mathscr{B}$.

6.2. Genuine laminations.

QUESTION 6.2. *Suppose* Λ *is a* full genuine lamination; *i.e. it has some complementary region which is an ideal polygon bundle over a circle. Suppose* M *is hyperbolic. Is the core circle of this region isotopic to a geodesic? Does it have a noncoalescable insulator family?*

REMARK. (1) Noncoalescable insulator families are introduced in [**44**] where they are used to establish (virtual) topological rigidity of hyperbolic manifolds.

(2) Maybe bundle regions are exactly the wrong kind of region for this analogy. Does it work better if we ask whether cores of interstitial annuli of non–bundle regions have insulator families? After all, these regions look more like pared manifolds, where the annuli are like the pared locus, and therefore maybe have short geodesic representatives. Note that such interstitial annuli are unknotted in the universal cover, and therefore, if $\pi_1(M)$ is residually finite, in some finite cover they lift to circles isotopic to geodesics. See e.g. [**40**]

(3) Given some decomposition of $M - \Lambda$ into guts $\mathfrak{G}_i$ and interstitial I–bundles, the *filling lemma* from [**22**] says that Λ can be *filled* to a lamination Λ' with gut regions isotopic to the gut regions of Λ, and for which every interstitial I–bundle is a product $I \times S^1 \times \mathbb{R}^+$. Thus if γ is the core of a gut region which is a polygon bundle over S^1, it is also the core of a complementary region (of a possibly different lamination) of an ideal polygon bundle over S^1. Perhaps such ideal polygon bundles are *artificial*; the more robust examples would be complementary regions to laminations with every leaf dense — i.e. minimal laminations.

QUESTION 6.3 (Thurston). *Suppose* Λ *is a genuine lamination. When can* Λ *be "filled in" to a very full lamination* Λ'*? Does it help for* M *to be hyperbolic?*

REMARK. (1) If Λ' is a co–oriented very full lamination, it can be filled in with monkey saddles to give a taut foliation. This implies for instance that closed leaves in any $\Lambda \subset \Lambda'$ should be Thurston norm minimizing, and $\pi_1(M)$ should admit an action on a universal circle with no global fixed point. On the other hand, if we do not require Λ' to be co–oriented, there are no known homological obstructions. There are examples of homologically trivial closed surfaces which are leaves of very full laminations. ([**4**])

(2) If Λ is an essential lamination of M, then the universal cover $(\widetilde{M}, \widetilde{\Lambda})$ is topologically a product $(\mathbb{R}^2, \lambda) \times \mathbb{R}$ where λ is a lamination of $\mathbb{R}^2$ ([**49**]).
Since λ is proper, there is a natural circular ordering on the ends of the leaves of λ, but this depends strongly on the (non–unique) choice of λ. If one can choose an equivariant representation $(\mathbb{R}^2, \lambda) \times \mathbb{R}$ of $(\widetilde{M}, \widetilde{\Lambda})$, one obtains a representation of $\pi_1(M)$ in $\text{Homeo}(S^1)$ with infinite image. It follows by [**8**] that $\pi_1(M)$ is isomorphic to a subgroup of $\text{Homeo}(S^1)$. This happens for instance when Λ is tight and has solid torus guts. See [**22**].

6.3. Loosesse laminations.

Definition 6.4. A lamination Λ is *loosesse* if it satisfies all the properties of essentiality except possibly the *no compressing monogons* condition. This concept was introduced by Gabai.

Question 6.5 (Gabai). *Are loosesse laminations good for anything? Are leaves of the universal cover of a loosesse lamination properly embedded? If M contains a loosesse lamination, does $\widetilde{M} = \mathbb{R}^3$?*

Remark. Brittenham has observed that for *branched surfaces*, the "no compressing monogons" condition can be substantially weakened to "no monogon bundles over S^1".

If M contains a loosesse lamination Λ for which $\widetilde{M} = \mathbb{R}^3$, is there an analogue of [**49**] to the effect that $(\widetilde{M}, \widetilde{\Lambda})$ is topologically $(\mathbb{R}^2, \lambda) \times \mathbb{R}$ for some (not necessarily proper) lamination λ of $\mathbb{R}^2$?

Question 6.6. *Give an example of a lamination in an atoroidal manifold — perhaps loosesse — which can never be realized by minimal surfaces for any metric, but which certifies some useful topological property of M.*

Remark. It is unknown whether every essential lamination can be realized by minimal surfaces for some metric; in some sense they seem to function as though they can be. But laminations containing compressing monogons can *never* be minimal surfaces for any metric.

7. Branched surfaces and triangulations

Question 7.1 (Agol). *Characterize branched surfaces embedded in 3–manifolds which can be non–trivially split to a homeomorphic copy of themselves.*

Remark.
(1) If $\mathscr{B}$ is such a branched surface, one can set $\mathscr{B}_0 = \mathscr{B}$ and $\mathscr{B}_i$ the result of performing such a splitting on $\mathscr{B}_{i-1}$. If one fixes a triangulation τ transverse to $\mathscr{B}$, this procedure gives rise to a partial lamination $\mathscr{L}$ carried by $\mathscr{B}$ ("partial" because the injectivity radius of the subsurface which is being split open might be bounded above). If $\mathscr{B}$ carries a lamination, is $\mathscr{L}$ contained in some lamination Λ carried by $\mathscr{B}$?
(2) If $\mathscr{B}$ supports a nontrivial transverse measure μ, it supports a nontrivial measure projectively invariant under splitting (this follows from the fact that the set of nontrivial projective measures is a convex set, and the splitting operation induces a projective transformation of such a set, which necessarily has a fixed point). One can also look at nontrivial transverse measures twisted by some fixed holonomy representation of the fundamental group of $\mathscr{B}$. Is there some underlying geometric meaning of such measures? A basic example is a branched surface carrying the stable or unstable laminations of a pseudo–Anosov flow transverse to a surface bundle (such a branched surface can be derived from a sequence of splittings and collapses of an supporting train track).

Question 7.2 (Agol). *Develop a theory of hierarchies for branched surfaces.*

Remark. A branched surface $\mathscr{B}$ would be decomposed sequentially along train tracks into simple pieces. The idea would be to try to prove a gluing theorem about how essential laminations carried by the simple pieces of the hierarchy glue together to give essential laminations in the branched surface.

Something like this is done by Li in [**80**] for branched surfaces without sink disks.

Recall the notion of a *taut ideal triangulation* as introduced by Lackenby in [**77**], as a combinatorial refinement of certain ideal triangulations of hyperbolic 3–manifolds with torus cusps. Such objects give rise to branched surfaces which are very simple — their worst singularities look like train tracks $\times I$.

QUESTION 7.3 (Dunfield). *Which boundary slopes are realized by essential laminations carried by a fixed taut ideal triangulation? Give an algorithm.*

REMARK. Though these branched surfaces do not have sink disks, they have many *half sink–disks*, so not all of Li's constructions for manifolds with boundary can be made to work. These are good candidates for the "automatic laminations" in question 7.8.

QUESTION 7.4 (Schleimer). *When does a Haken sum operation make sense for a pair of laminations in normal form with respect to a fixed triangulation?*

REMARK. For transversely measured laminations, this question is less interesting. Although, see Hatcher [**68**].

A prerequisite must be that the laminations intersect each tetrahedron in the same kind of quadrilateral pieces. This is not too far from the condition that they are simultaneously carried by a fixed branched surface. So a better question might be — what is the structure of the set of laminations carried by a given branched surface (up to monotone equivalence, say)? Is there a natural topology for which it is star–like? Or contractible?

7.1. Normal surfaces. The theory of normal surfaces was introduced by Haken in [**64**], although similar ideas were introduced in a more limited scope by Kneser. Any foliation can be normalized relative to some triangulation [**7**].

QUESTION 7.5. *Let M be a 3–manifold, and Λ an essential lamination. Let C be a cycle representing the fundamental class of M. Is there a cycle C' with the same Gromov norm as C, and another essential lamination Λ' which is normal with respect to C'?*

REMARK.

(1) If Λ is an incompressible surface, C can be normalized by [**2**]. In fact, Agol points out in his paper that the same method works if Λ is merely tight. Presumably, the cycle C' and the lamination Λ' will be derived from C and Λ by some process of straightening the simplices of C relative to Λ, and compressing and evacuating ([**46**]) Λ with respect to its intersection with C.

A positive answer would extend Agol's volume bounds to manifolds with essential (not necessarily tight) laminations.

(2) Does it help to assume all the coefficients of C are positive? In this case, for any singular simplex $\sigma : \Delta^3 \to M$ in the support of C, and any leaf λ of Λ intersecting $\sigma(\Delta^3)$, we can *develop* (not uniquely) C along any path $\gamma \subset \lambda$ which does not pass through any edges or vertices. For, once γ comes to a face τ of $\sigma(\Delta^3)$, there is another singular simplex $\sigma' : \Delta^3 \to M$ whose boundary absorbs some of the mass of τ; then Λ can be developed along γ into σ', and so on inductively. This gives a particular order in which to try to compress and evacuate the leaves of Λ relative to C.

(3) If C is a virtually embedded cycle — i.e. it comes from a triangulation in some finite cover $\widehat{M}$, Λ can be lifted to $\widehat{\Lambda} \subset \widehat{M}$ and evacuated there. Can it be evacuated equivariantly?

DEFINITION 7.6. A *taut local orientation* is a choice of ordering for the vertices of each tetrahedron in a triangulation with the following properties:

(1) The star of each vertex is ordered compatibly with a local foliation in normal form
(2) Every oriented loop is homotopically essential

Such objects are dual to branched surfaces with certain properties; in particular, certain examples give rise to branched surfaces which carry nothing, but fully carry essential laminations in finite covers. See [**13**] or [**19**].

QUESTION 7.7. *Suppose $\mathscr{B}$ is a branched surface in M which is dual to a taut local orientation. Is there a finite cover of M in which the pullback of $\mathscr{B}$ fully carries a lamination? What about an amenable cover?*

Say an *automatic lamination* is given by the following data:

- A branched surface $\mathscr{B}$ with some fixed cell decomposition, so that there is a notion of combinatorial paths supported by this surface which can be enumerated.
- A lamination Λ fully carried by $\mathscr{B}$.
- A finite state automaton A which recognizes combinatorial paths in $\mathscr{B}$ which represent geodesic paths in leaves of Λ.

REMARK. Note that if Λ is essential and M is atoroidal, leaves of Λ are uniformly δ–negatively curved, so such an automaton exists relative to an oracle which recognizes combinatorial paths in $\mathscr{B}$ which are carried by some leaf of Λ.

QUESTION 7.8. *Do branched surfaces without sink disks carry automatic laminations?*

REMARK.
(1) This seems possible, insofar as Li's construction of a lamination carried by such a branched surface is local and explicit, and describes a procedure for recursively splitting open the branched surface. One subtle issue might be the choice of an extension over foliated circle bundles over punctured surfaces of positive genus: the existence of such an extension follows from the fact that every homeomorphism of S^1 is a commutator of length at most 2. But it is not clear whether a recursive description of such a commutator can be given "automatically".
(2) Every time a combinatorial path passes over a source branch, there are two possible routes it can take. The sequence of splittings (positive, negative) so obtained is like a continued fraction expansion for an irrational number. The automaton A can only remember a bounded length of this sequence; this constrains the possible transverse structures supported by an automatic lamination. In particular, any automatic lamination should admit a natural transverse bilipshitz structure.
(3) This is related to question 8.7 on the existence of *automatic left–orderings* on finitely presented groups.

QUESTION 7.9. *Give a useful definition of* thin position *for an embedded graph* $\Gamma \subset M$ *with respect to a taut foliation* $\mathscr{F}$. *If* Γ *is the* 1*–skeleton* τ^1 *for a triangulation, can one find an isotopy such that the leaves of* $\mathscr{F}$ *are made up of polyhedral disks of bounded index?*

REMARK. For $\mathscr{F}$ a foliation by closed surfaces, one can make the leaves either normal or almost normal. If $\mathscr{F}$ is finite depth, one can normalize the depth 0 leaves. The depth 1 leaves are asymptotic to the depth 0 leaves, so they may be assumed to be normal away from a compact piece; some kind of relative minimal sweepout argument might produce an isotopy so that these leaves are almost normal. Continuing inductively, it seems that one can bound the index of the disks appearing in a leaf by a function of their depth. This argument should be made more precise.

8. Leaf spaces and transverse structures

The most significant nonexistence results in this area are the main theorems of [**103**] and [**38**] which give infinitely many examples of hyperbolic 3–manifolds M for which there is no nontrivial action of $\pi_1(M)$ on a simply–connected 1–manifold, respectively order tree.

QUESTION 8.1 (Thurston). *Suppose* M *is irreducible. Suppose further that* $\pi_1(M)$ *admits a nontrivial action on* $\mathbb{R}$. *When does* M *admit a taut foliation with a transverse* $(\pi_1(M), \mathbb{R})$ *structure?*

REMARK. (1) Given a representation $\rho : \pi_1(M) \to \text{Homeo}(\mathbb{R})$, let

$$(E_\rho, \mathscr{F}_\rho) \to M$$

be the associated foliated bundle. A section $s : M \to E_\rho$ transverse to $\mathscr{F}_\rho$ pulls back $\mathscr{F}_\rho$ to a foliation of M with a transverse $(\pi_1(M), \mathbb{R})$ structure.

(2) Given an action of a group Γ on a simply connected 1–manifold L, when is there an immersion $i : L \to \mathbb{R}$ such that the action of Γ descends to an action on $\mathbb{R}$? One should allow some flexibility in addressing this question, so that for instance one should be allowed to substitute monotone equivalent actions. Does it make a difference to weaken the quality of i so that it is not necessarily an immersion, but merely (partial) order preserving? Under what circumstances is there a *unique* maximally ordered Γ–invariant order–preserving quotient $L \to L_o$? What if one imposes extra conditions on Γ, for instance that it should be amenable (especially if one allows monotone equivalent actions)?

(3) One can obviously generalize this question to actions on order trees.

QUESTION 8.2. *Suppose* $\mathscr{F}$ *is an* $\mathbb{R}$*–covered foliation of an atoroidal* 3*–manifold* M. *Is the holonomy representation* ρ_H *of* $\pi_1(M)$ *on* $\mathbb{R}$ *conjugate to a group of coarse* 1*–quasi–isometries? i.e. is there a positive valued function* $C : \pi_1(M) \to \mathbb{R}$ *such that*

$$|d(\rho_H(\alpha)(p), \rho_H(\alpha)(q)) - d(p,q)| \leq C(\alpha)$$

for all $\alpha \in \pi_1(M)$ *and* $p, q \in \mathbb{R}$*?*

REMARK. (1) If the value of C can be chosen independent of α, it follows that any two leaves of $\widetilde{\mathscr{F}}$ are contained within bounded neighborhoods of each other. Such foliations are said to be *uniform*. All known examples of

$\mathbb{R}$–covered but nonuniform foliations of atoroidal 3–manifolds satisfy the condition above.

(2) If M is allowed to be toroidal, the action of $\pi_1(M)$ on L can fail to satisfy the condition above. It is compatible with all known examples that for toroidal manifolds, the action of $\pi_1(M)$ should always be conjugate to a group of coarse quasi–isometries.

See [**117**] and [**12**] for more details.

QUESTION 8.3. *Suppose M is atoroidal and admits a taut foliation. Must it admit an $\mathbb{R}$–covered foliation?*

REMARK. (1) There are many examples in [**22**] of small volume hyperbolic manifolds which admit essential laminations, but which do not admit $\mathbb{R}$–covered foliations. In fact, their fundamental groups admit no nontrivial representation in Homeo($\mathbb{R}$).

(2) If one is prepared to pass to a finite cover, one probably cannot find an example based on the nonexistence of group actions on certain 1–manifolds. For, with suitable homological hypotheses (which can be achieved by passing to a finite cover), the fundamental group of a tautly foliated manifold acts on $\mathbb{R}$ without a global fixed point. See [**22**] for details. On the other hand, [**103**] gives many examples of fundamental groups of hyperbolic 3–manifolds which admit no actions on $\mathbb{R}$ without a global fixed point, but which admit taut foliations and therefore have faithful circle actions. The Euler class of these circle actions are torsion. In general, for any 3–manifold M and for any 2–plane field ξ on M there is a 3–manifold N together with a map $f : N \to M$ that induces an isomorphism on homology (even over $\pi_1(M)$) such that f is transverse to τ, and such that the induced 2–plane field $f^*\xi$ on N is homotopic to a taut foliation. For, Thurston's construction [**113**] lets us construct a foliation $\mathscr{F}$ on M in a neighborhood of the 2–skeleton of an anti–oriented triangulation transverse to τ. By drilling out spiralling paths, one can arrange for the complementary regions to the foliated region to be solid tori S_i. Then N can be obtained from M by drilling out these solid tori S_i and filling in circle bundles B_i over punctured surfaces. Any homeomorphism of S^1 can be realized as a commutator of bounded length, so the foliations of $\mathscr{F}|_{\partial S_i}$ can be extended to foliations of B_i transverse to the circle fibers. This gives a demonstrably taut foliation in the appropriate homology class on a manifold with the required properties. This construction is due to Thurston [**119**], and gives many examples of circle actions of fundamental groups of hyperbolic manifolds whose Euler class is torsion but non–zero.

(3) Brittenham, Naimi and Roberts ([**11**]) have produced examples of graph manifolds which admit taut foliations but do not admit $\mathbb{R}$–covered foliations.

(4) A related question from [**45**] is: if M admits an essential lamination, does it admit a tight essential lamination? (an essential lamination is *tight* if the leaf space of the universal cover is Hausdorff). Quasigeodesic laminations in hyperbolic manifolds are tight, so a positive answer to question 10.4 would give the most positive answer to this question.

QUESTION 8.4 (Agol). *For a fixed manifold M, describe the structure of the set of all essential laminations with a transverse* $\widetilde{SL(2,\mathbb{R})}$ *structure.*

REMARK. (1) For a fixed triangulation τ, the set of normal foliations supported by τ with a $\widetilde{SL(2,\mathbb{R})}$ structure can be algorithmically described, and has the natural structure of a semi–algebraic set. By Gabai [**46**], there is a fixed triangulation τ of M so that every essential lamination of M can be made normal with respect to τ; in fact, there is a constructive procedure to find a finite set of branched surfaces which fully carry every nowhere dense essential lamination in M.

(2) One can study more generally the space of actions of $\pi_1(M)$ on $SL(2,\mathbb{R})$–trees, which one could ask to be either partially ordered or not. In this context, it is probably useful to differentiate between *complete* and *incomplete* trees. Any properly embedded interval I in an $SL(2,\mathbb{R})$–tree, has a well–defined *coarse length*, a non–negative integer $l(I)$ which is the number of times I wraps around S^1 under projection. An $SL(2,\mathbb{R})$–tree is *complete* if every properly embedded tight ray r has infinite coarse length. In particular, if some element α stabilizes some point $p \in r$, it stabilizes infinitely many points on r which exit the noncompact end of r. This problem is probably a good warm–up to understanding the space of actions of $\pi_1(M)$ on arbitrary order trees or 1–manifolds.

(3) Suppose Γ acts on an $\mathbb{R}$ order tree T in such a way that for every nontrivial γ, and for every properly embedded copy l of $\mathbb{R}$ in T invariant under γ, the set of fixed points of γ is isolated, and is either empty or exits both noncompact ends of l, and the translation direction of γ on the complementary intervals of l alternate. Suppose furthermore that there is *no* sequence γ_i with consecutive fixed points p_i, q_i contained in a fixed segment I for which $p_i, q_i \to r$ for some $r \in I$. Does T carry the natural structure of an $SL(2,\mathbb{R})$–tree?

QUESTION 8.5. *Suppose M admits a minimal taut foliation. What is the best analytic (transverse) quality of a taut foliation it admits? Can we find a minimal foliation such that the holonomy groupoid is of type* III_λ *for some algebraic* λ*? What about if one asks for a foliation monotone equivalent to the first? Homotopic?*

REMARK. There are well–known topological obstructions to smoothing transverse structures. [**100**] gives some important examples of foliations which can be made transversely C^1 but not C^2. At the level of Haefliger structures, the obstructions are related to homology of various groups of homeomorphisms of $\mathbb{R}^n$ ([**62**],[**63**],[**120**]).

A pseudogroup $\mathscr{G}$ of transformations of a measure space (X,μ) is said to be of *type* III_λ if for each $g \in \mathscr{G}$, the Radon–Nikodym derivative $d(g_*\mu)/d\mu$ is well–defined on a set of full measure and takes values in $\langle\lambda\rangle$, the abelian group of powers of λ. Such measures and actions arise frequently from automatic structures; the theory of Patterson–Sullivan measures is also relevant.

An obvious obstruction is that $\pi_1(M)$ must act on *some* 1–manifold with this quality. Question 7.8 and question 8.10 are relevant here. See [**29**] for an abstract discussion of transverse measure theory for foliated manifolds.

QUESTION 8.6. *Is there a universal constant c such that a hyperbolic 3–manifold M whose fundamental group $\pi_1(M)$ can be ordered out to radius c can be left–ordered? Or weaker, is there an effective method to compute such a $c(M)$ for a given M?*

REMARK. (1) A 3–manifold group $\pi_1(M)$ is left orderable iff it admits a faithful representation in $\text{Homeo}^+(\mathbb{R})$. If M admits an $\mathbb{R}$–covered foliation, the representation $\pi_1(M) \to \text{Homeo}(L)$ may not be faithful, but the kernel is a surface group, and is itself left–orderable, so the representation is monotone equivalent to a faithful one.

(2) The set of finitely presented groups which are not left–orderable is recursively enumerable; on the other hand, the set of finitely presented groups which *are* left–orderable is not. Does hyperbolic geometry, or more generally word–hyperbolic geometry make a difference here? A word–hyperbolic group looks more or less free on a large scale; free groups are certainly left–orderable. A positive answer to this question would give an algorithm to detect left–orderability (of hyperbolic 3–manifold groups). In fact, any M which admits a taut foliation, a pseudo–Anosov flow, or certain kinds of genuine laminations, has a faithful representation $\pi_1(M) \to \text{Homeo}(S^1)$ coming from a universal circle; if M is a rational homology sphere, this representation lifts to $\widetilde{\text{Homeo}(S^1)}$ when restricted to the commutator subgroup, a finite index subgroup. Such periodic representations might be easier to detect or rule out than arbitrary ones. In particular, in **[22]** it is proved that the Weeks manifold does not admit a pseudo–Anosov flow, or a tight essential lamination. On the other hand, there are examples of 3–manifolds which act faithfully on S^1 but not on $\mathbb{R}$. See **[22]** for more details.

QUESTION 8.7. *Let* T *be some class of abstract computers; e.g. finite state automata, Turing machines, Turing machines relative to some oracle O, etc. A* T–order *on a group G is a left–invariant order such that there is a machine $T \in \mathsf{T}$ which recognizes the positive cone $G^+ \subset G$. What kinds of* T*–orders are possible for fundamental groups G of hyperbolic 3–manifolds?*

REMARK. (1) A special case consists of the class T of finite state automata. One can ask whether there exists an automatic order on G; i.e. whether there is an FSA which can detect *geodesic* words with values in the positive cone. Of course, G must already be an automatic group for there to be any chance at all. It turns out that this class of groups is very small. There are automatic orders on $\mathbb{Z}^n$ for all positive integers n, but there is no automatic order on F_2, the free group on two generators. A proof is as follows. Let $F_2 = \langle a, b \rangle$ be a generating set. A left–invariant order on F_2 is more or less equivalent to a faithful action of F_2 on $\mathbb{R}$ by orientation–preserving homeomorphisms. We may also assume this action is minimal. Let 0 be a point with trivial stabilizer, so that the positive cone are the elements g with $g(0) > 0$.

Suppose w acts as a positive translation on a w–invariant interval I, with lowest point $p < 0$ and highest point $p' > 0$. Since the action is minimal, there is $q \in I$ and a word v such that $v(q) = p'$. Let x_1 be a word with $x_1(0) = p_1^+$, where $p_1^+ > p$ is very close to p. Then

$$w^{-n}vw^n x_1(0) > 0$$

for $n > N_1$, and

$$w^{-n}vw^n x_1(0) < 0$$

for $n \leq N_1$ for some very large N_1. We can choose x_i with $x_i(0) \to p$ for which there are corresponding $N_i \to \infty$. If A is a fixed finite state automaton, eventually N_i exceeds the memory capacity of A, and A is incapable of distinguishing $w^{-N_i}vw^{N_i}x_i$ from $w^{-N_i-1}vw^{N_i+1}x_i$.

Conversely, if every w fixes at most 1 point on $\mathbb{R}$, the image of the group is solvable, and therefore not faithful.

Since every hyperbolic 3–manifold group contains a quasigeodesically embedded F_2, this implies that no hyperbolic 3–manifold group admits an automatic order.

(2) If a finitely generated automatic group G is left–orderable, given a lexicographic ordering on representatives of G there is a well–defined *lexicographically first* left–order. What is the algorithmic quality of this order? Note that given an automatic group G, there is an algorithm which either certifies that G is not left–orderable, or recursively enumerates the words in the lexicographically first left–order.

(3) The general question of whether or not a finitely presented group is left–orderable is undecidable. But even if a group is left–orderable, it might be much harder to describe an explicit left–order than merely to certify orderability.

8.1. Universal circles.

QUESTION 8.8. *Let $\Lambda^\pm$ be a pair of laminations of S^1 which are transverse to each other and have finite area complementary domains. Suppose Γ is a group of automorphisms of S^1 which preserves $\Lambda^\pm$ and acts minimally on the leaves of either lamination. When is Γ commensurable with $\pi_1(M)$ for M a hyperbolic 3–manifold?*

REMARK. (1) If $\Gamma = \pi_1(M)$ for some M, must M admit a pseudo–Anosov flow X for which the associated action on the universal circle of X is conjugate to the given action of $\pi_1(M)$ on S^1? Actions of 3–manifold groups on circles associated to pseudo–Anosov flows are constructed in [**22**].

(2) What properties should a lamination Λ or collection of laminations of S^1 satisfy to ensure good geometric control over the subgroup of $\text{Homeo}(S^1)$ preserving Λ? For instance, let $\Lambda^\pm$ be as above and Γ a group of automorphisms of $\Lambda^\pm$: is Γ word hyperbolic? Is there a natural topology on the space of such pairs $\Lambda^\pm$ for which a large group Γ of automorphisms exists? What do non–trivial convergent sequences (of commensurability classes of hyperbolic 3–manifolds) look like in such a topology?

(3) Given a pair of laminations $\Lambda^\pm$ of S^1 as above, the natural quotient of S^1 by the leaf relations of $\Lambda^\pm$ is topologically S^2. If Γ is a group of automorphisms of $\Lambda^\pm$, we get a representation $\Gamma \to \text{Homeo}(S^2)$ and a flat bundle over a $B\Gamma$. The topological group $\text{Homeo}(S^2)$ is homotopy equivalent to $O(3,\mathbb{R})$ by Smale; are the cohomology classes pulled back via the associated classifying map $B\Gamma \to BO(3,\mathbb{R})$ bounded in norm in terms of combinatorial data that can be read off from $\Lambda^\pm$?

QUESTION 8.9. *What possibilities are there for universal circles S^1_{univ} for a fixed manifold? For a fixed foliation? For what taut foliations is there a* unique minimal *universal circle?*

REMARK. (1) A universal circle S^1_{univ} for a taut foliation $\mathscr{F}$ formally gives rise to a nontrivial invariant lamination of S^1_{univ} for each direction in which $\widetilde{\mathscr{F}}$ branches ([**20**]). Given an action of a group G on a circle S^1 preserving a lamination Λ, one can modify the action in quite drastic ways: one can take an orbit class of complementary region, and "flip" the lamination along every edge of every region in the orbit. If T is the dual planar tree to Λ, this is equivalent to re–embedding T in the plane in such a way that the circular order is reversed at every vertex in an orbit class (see [**22**]). If there is more than one orbit class of complementary region, this operation is not a topological conjugacy, and the circle actions in question are quite distinct. One can almost certainly use this operation to produce examples of minimal circle actions of hyperbolic 3–manifold groups which are not universal circles for any foliations.

(2) There is a unique minimal universal circle for $\mathscr{F}$ an $\mathbb{R}$–covered foliation, or one with one–sided branching. These minimal universal circles are rigid under certain kinds of deformations of the foliation, in many cases.

(3) The best answer would do two things: firstly it would give a good description of the "representation space" of $\pi_1(M)$ in $\mathrm{Homeo}(S^1)$ and secondly it would explicitly characterize the subspace of representations of geometric origin.

It is probably unreasonable to expect an algebraic structure on the representation space, but a *polyhedral structure* might be more promising, by analogy with question 3.1. A sequence of isotopies of a fixed pseudo–Anosov flow degenerating to a flow corresponding to a higher codimension face should give rise to a monotone relation between the universal circles of the leaf space of the flows.

(4) Universal circles for certain kinds of genuine laminations and for pseudo–Anosov flows are constructed in [**22**]. Perhaps every essential lamination gives rise to a universal circle.

QUESTION 8.10. *What is the best analytic quality for the action of $\pi_1(M)$ on a universal circle S^1_{univ}?*

REMARK. (1) It is probably unreasonable to expect the action to be C^2 if M is hyperbolic. On the other hand, *any* minimal action of a finitely generated group on S^1 is conjugate to a Lipschitz action, with respect to a harmonic measure. For an introduction to the rich panorama of subgroups of homeomorphisms of S^1 stretching between Lipschitz and C^2, see [**122**].

(2) The following related question was suggested by Barry Mazur. Let G be a finitely presented group, and let X, Y be compact smooth manifolds with $\dim(X) < \dim(Y)$. Suppose $\rho_X : G \to \mathrm{Homeo}(X)$ and $\rho_Y : G \to \mathrm{Homeo}(Y)$ are minimal actions such that there is a surjection $f : X \to Y$ which intertwines ρ_X and ρ_Y. When are ρ_X, ρ_Y actions by diffeomorphisms for some appropriate smooth structures on X, Y, and how can one find construct such a smooth structure on X from a smooth structure on Y and from f? This question is most interesting when X is

a circle or an interval, but there are probably many interesting examples of Y. What seems plausible is that high dimensional cohomology of G pulled back from the action on Y represents an obstruction to stiffening the analytic quality of the action on X.

9. Classical 3–manifold theory

QUESTION 9.1. *Is there a* universal *transverse surgery description of tautly foliated manifolds, in the sense that there is a fixed M such that for every tautly foliated manifold $N, \mathscr{F}$ there is a link $L \subset N$ transverse to $\mathscr{F}$ so that M is obtained from N by surgery on L?*

REMARK. (1) The idea is to get an analogue of Lickorish's theorem (that every closed 3–manifold is surgery on a link in S^3) in the foliated context. One should also have in mind Thurston's observation (see [**1**]) that the process of repeatedly drilling out the shortest geodesics in a sequence of hyperbolic manifolds tends to stabilize to a "universal" sequence. Question 10.5 is probably relevant to an approach along these lines. One should concentrate on taut foliations for this question, so that there are enough transverse knots to "see" the whole topology of N. One idea might be to start with a triangulation τ in normal form with respect to $\mathscr{F}$, and look at the interaction of the associated Heegaard decomposition with $\mathscr{F}$.

(2) We can modify the question to ask whether, given two tautly foliated manifolds $M_1, \mathscr{F}_1$ and $M_2, \mathscr{F}_2$, there are links L_i transverse to $\mathscr{F}_i$ such that $M_1 \backslash L_1 = M_2 \backslash L_2$.

QUESTION 9.2. *Give a collection of fundamental operations on foliations and an explicit family of* base foliations *such that every tautly foliated manifold $M, \mathscr{F}$ is obtained from one of the base family by repeated application of fundamental operations.*

REMARK. (1) It is unclear if even the following naive list of operations
- branched covers and pushforward
- cut–and–shear
- torus connect sum
- tangential surgery
- surface bundle plumbing along transverse train–tracks
- monotone deformations
- cut–and–paste furrows along loops with linear holonomy

and base foliations
- foliations of circle bundles
- transversely measured foliations
- finite depth foliations and those representing irrational rays in H_2
- foliations derived from alternating knots by Delman–Roberts methods

would fail to suffice. The idea is really to develop some more robust invariants of foliations which vary either not at all, or in some well–understood way under basic operations, and at the same time to give a procedure for recognizing *orphans* with respect to a given list of operations.

(2) A model theorem along these lines is the recent result of Noah Goodman and, independently, Giroux, improving work of Gabai, which says that the set of fibered links in a homology 3–sphere M are related by a sequence of elementary stabilizations and destabilizations, where the stabilization operation is plumbing with a Hopf band.

9.1. Persistence questions.

QUESTION 9.3. *What is the most general class of knots to which the techniques of Delman–Roberts (in constructing persistent laminations) can be extended?*

REMARK. The main reference is [**30**].

The idea of this question is that the lamination should be constructed in an essentially algorithmic way from a *projection* of the knot. A related question asks what information essential laminations can give about projections of knots; for instance, when can foliated techniques be used to certify that a knot projection has minimal number of crossings? Is the problem of minimal crossing number for knot projection NP–complete?

QUESTION 9.4 (Delman). *Suppose K is a non–torus alternating knot. Then essential laminations can be constructed which realize every (nontrivial) boundary slope. Can essential laminations be constructed with an even sided bundle complementary region containing K, so that every nontrivial surgery can be filled in with a monkey saddle?*

9.2. Distance of Heegaard splittings.

QUESTION 9.5. *It is known* [**66**] *that if a 3–manifold M contains an essential surface of genus g, the* distance *of any Heegaard splitting of M has distance at most $2g$. Does the "distance filtration" put any useful structure on the essential laminations supported by a given M? i.e. if M admits Heegaard splittings of distance at least $2g$, what can one say about the essential laminations Λ contained in M?*

REMARK. A preliminary technical issue would be to decide what a *tight* intersection of Λ with the splitting surface Σ should be, and then try to tighten any given Λ with respect to Σ, perhaps by means of an infinite process, by analogy with Brittenham's principle (see [**10**] and [**46**]).

10. Hyperbolic geometry

10.1. Asymptotic geometry of leaves.

QUESTION 10.1. *Suppose $\mathscr{F}$ is a taut foliation of a hyperbolic 3–manifold M with two–sided branching. Must there be a leaf λ of $\widetilde{\mathscr{F}}$ whose complement contains an open halfspace of $\mathbb{H}^3$ on either side? We call such a leaf* asymptotically separated.

REMARK. This question is equivalent to the condition that the limit set of some leaf of $\widetilde{\mathscr{F}}$ is not all of S^2_∞. A conjecture of Fenley says that $\lambda_\infty = S^2_\infty$ for some leaf λ of $\widetilde{\mathscr{F}}$ iff $\mathscr{F}$ is $\mathbb{R}$–covered.

QUESTION 10.2. *Do leaves of $\widetilde{\Lambda}$ for Λ an essential lamination have the* continuous extension property? *More generally, what is the relationship between the action* *of* $\pi_1(M)$ *on various ideal boundaries of* $\widetilde{M}$ *arising from the foliated structure (e.g. universal circles) and the ideal boundaries arising from the geometry of* $\widetilde{M}$.

REMARK. (1) The seminal result in this area is the theorem of Cannon and Thurston [**27**] that leaves of surface bundles over S^1 extend continuously to sphere filling curves. The method of Cannon and Thurston has been greatly extended and improved by Fenley to deal with many (most?) finite depth foliations, and certain other taut foliations [**34**],[**35**],[**36**].

(2) A *measurable* extension property is much easier to establish, and follows from exponential growth of hyperbolic spaces, and the fact that a leaf of an essential lamination in the universal cover is uniformly properly embedded in its ϵ neighborhood for some ϵ. See [**17**]. Note that even in cases where the continuous extension property is known leafwise (e.g. finite depth foliations) it is not known for all or any universal circles (although Fenley has some unpublished partial results on this question).

QUESTION 10.3. *Suppose* $\mathscr{F}$ *is a finite depth foliation of a hyperbolic 3–manifold. What is the relationship (if any) between the Hausdorff dimension of the limit set of a leaf* λ *of* $\widetilde{\mathscr{F}}$ *and the depth of* $\mathscr{F}$ *or* λ*?*

REMARK. (1) Conjecturally, a foliation is $\mathbb{R}$–covered iff the limit set of any (and therefore every) leaf is all of S^2_∞. A proper foliation of depth > 0 contains a quasigeodesic leaf; in particular, depth 0 is distinguished from any other finite depth by the property that the limit set of every leaf has Hausdorff dimension 2. More generally, if $M' \subset M$ is the union of the leaves of highest depth, the components of $\widetilde{M}'_\infty$ and λ_∞ for leaves λ of highest depth should agree.

(2) For depth one leaves, the components of M' are the interiors of compact hyperbolic manifolds with quasigeodesic boundary. For such manifolds, elements of π_1 can be enumerated by a finite state automaton. Does this mean that the Hausdorff dimension of the limit set is a special value of an L–function?

(3) One can numerically estimate the Hausdorff dimension of the limit set of a noncompact leaf, using a harmonic measure for the universal circle to determine which directions at infinity (in the intrinsic geometry of the leaf) are most distorted in the ambient geometry. This should give a fast numerical algorithm to tell whether or not a foliation of a hyperbolic manifold is of depth one. On the other hand, when M has cusps, there are numerical problems with this approach, since it is slow to enumerate elements of $\pi_1(M)$ which exit a cusp. This is especially troubling, since the case one would be most interested in is foliations of knot or link complements. It is an interesting problem to find therefore a fast numerical algorithm to estimate the Hausdorff dimension of the limit set of a leaf in a foliation of a cusped manifold. Obviously, the foliation itself must be described in some computationally effective manner. Foliations carried by Lackenby's taut ideal triangulations [**77**] are good candidates for study.

10.2. Realization questions.

QUESTION 10.4 (Thurston). *Suppose M an atoroidal 3–manifold admits an essential lamination. Does it admit a (necessarily genuine) lamination with quasi–geodesic leaves?*

REMARK. (1) Any atoroidal manifold with an essential lamination contains a genuine lamination Λ, and therefore has word–hyperbolic fundamental group ([**20**],[**50**]). If such a lamination is not quasi–geodesic, perhaps one can find another (more quasi–geodesic?) lamination transverse to it which is an "eigenlamination" for the extrinsic distortion of the geometry of $\widetilde{\Lambda}$. As a caveat, one should note that in general there is no naive notion of a *universal circle* for a genuine lamination, so some new ideas are necessary. Note that faithful circle actions are associated to certain essential laminations in [**22**], but these do not give monotone relations between circles at infinity of leaves, as a real universal circle should. Rather they function as a kind of circular parameterization of the *leaf space* of the laminations. Moreover, even such naive circular parameterizations do not exist for certain non–tight genuine laminations.

(2) A lamination with quasi–geodesic leaves is tight, so a first step would be to try to replace a non–tight lamination with a tight one. Associated to a cataclysm (i.e. a collection $\{\mu_i\}$ of incomparable non–separated leaves of $\widetilde{\Lambda}$ which are the limit of a comparable monotone sequence) in a non–tight lamination Λ, there is a geodesic lamination of $\mathbb{H}^2$ whose complementary regions parameterize the μ_i.

(3) One reason to be interested in the existence of such a lamination is that it would be an important technical advantage in Thurston's program to geometrize laminar 3–manifolds. Given a choice of conformal structure on the leaves of Λ, one can try to generalize the *skinning map* as a map from $\text{Teich}(\bigcup_C \partial C)$ to itself, for each complementary region C. In the case of a taut foliation, it might proceed as follows. A universal circle gives a monotone relation between the circles at infinity of the leaves of $\widetilde{\mathscr{F}}$. In fact, for λ_i a maximal collection of incomparable leaves above or below μ, for any leaf μ of $\widetilde{\mathscr{F}}$, there is a monotone relation between $S^1_\infty(\mu)$ and a "cactus" made from the the union $\bigcup_i S^1_\infty(\lambda_i)$. For each leaf λ of $\widetilde{\mathscr{F}}$, we should choose a pair of points

$$\tau^{\pm}(\lambda) \in \text{Teich}(D^2) \times \text{Teich}(\overline{D^2})$$

a product of universal Teichmüller spaces as an initial guess. Then monotone relations between circles at infinity give a new marking $\sigma(\tau^+)$ obtained by sewing together the markings $\tau^+(\lambda_i)$ by the monotone relation. There are numerous problems with such a program; for instance, there are many possibilities for leaves λ_i; moreover, there is no natural decomposition of a laminated manifold along a collection of leaves into a compact "fundamental domain", especially if $\mathscr{F}$ is minimal. Secondly, the domain obtained by gluing together copies of $\tau^+(\lambda_i)$ will typically not be a quasidisk, but will be pinched along an infinite lamination. One can try to remedy this problem by letting $\sigma(\tau^+)$ be equal, not to the union of the $\tau^+(\lambda_i)$ sewn together, but by τ^+ bent in the direction of this pinching lamination. But more importantly, no foliation of a hyperbolic manifold can have all leaves quasi–geodesically embedded; in particular, for all but the most trivial cases, there should be no fixed point for the skinning map in $(\mathrm{Teich}(D^2) \times \mathrm{Teich}(\overline{D^2}))^L$, but rather in some natural compactification, analogous to the Thurston boundary of Teichmüller space.

For a quasigeodesic lamination, at least a fixed point for the skinning map should exist in some universal Teichmüller space, and proving geometrization in this case might be considerably easier.

10.3. Interactions of foliations with hyperbolic structures.

QUESTION 10.5. *What do short geodesics look like with respect to taut foliations? Is there a universal ϵ such that for every hyperbolic manifold M, every taut foliation $\mathscr{F}$ of M, and every geodesic γ with $|\gamma| < \epsilon$, γ is either isotopic into a leaf of $\mathscr{F}$ or isotopic to be transverse to $\mathscr{F}$? What about homotopic?*

REMARK. (1) For a fixed taut foliation $\mathscr{F}$, leaves of $\widetilde{\mathscr{F}}$ are uniformly properly embedded in their ϵ–neighborhoods, for some uniform ϵ. If $\mathscr{F}$ does not have 2–sided branching, *every* loop is homotopic to either a transverse or a tangential loop; this question asks whether branching of $\widetilde{\mathscr{F}}$ takes place on a scale uniformly bounded below over all taut foliations.

(2) For M a hyperbolic manifold with a short geodesic γ, it is an empirical observation of Rubinstein, Jaco and Sedgewick that a 1–efficient triangulation (see [**72**]) has a "layered" solid torus corresponding to γ, which has a very standard combinatorial form. We can try to normalize $\mathscr{F}$ relative to such a triangulation, possibly evacuating a subfoliation in the process ([**46**]). Can one use the explicit combinatorial structure of the layered solid torus to control the interaction of γ with $\mathscr{F}$?

QUESTION 10.6. *Is there a uniform bound on the Godbillon–Vey invariants of the taut foliations of a hyperbolic manifold in terms of its volume?*

REMARK. Some estimate for foliations on fixed M can be obtained from the theorem of Gabai [**46**] that there is a fixed triangulation τ of M such that every taut foliation of M can be made normal with respect to τ. On the other hand, there are no known *a priori* bounds on how many simplices such a τ must have in terms of the number of simplices in a minimal triangulation, or what relation τ has to the geometry of M. For instance, there are geodesic triangulations which cannot be made "virtually transverse" to certain taut foliations; i.e. there is no finite cover in which the pullback of the triangulation can be normalized relative to the pullback of the foliation.

QUESTION 10.7. *Suppose $\mathscr{F}$ is a taut foliation of a hyperbolic 3–manifold M. Let*

$$\pi : \widetilde{M} \to L$$

be the projection to the leaf space of $\widetilde{\mathscr{F}}$.

1. *For γ a random walk in $\widetilde{M}$ (which is isometric to $\mathbb{H}^3$), what is the typical behaviour of $\pi(\gamma)$?*
2. *Does a random walk in $\widetilde{M}$ converge to a definite end of L?*
3. *What if we replace "random walk" with "random geodesic" in the previous question?*
4. *Is the pushforward of asymptotic behaviour well–defined? That is, is it true that for a set of geodesics γ of full measure, for all γ' a bounded*

distance from some γ the pushforward of γ and γ' exit the same end of L?

(5) *Suppose $\mathscr{F}$ has one–sided branching. Does a random walk always exit L in the unbranching direction?*

(6) *Suppose there is a positive probability of a random walk exiting a proper positive or negative end of L. Does this imply $\mathscr{F}$ is $\mathbb{R}$–covered? What about if there is a positive probability of a random walk being recurrent?*

REMARK. (1) For the basic theory of random walks and Brownian motion, see e.g. [**108**].

(2) For $\mathscr{F}$ transversely measured, L is naturally isomorphic to $\mathbb{R}$. Then the pushforward of a random walk or geodesic on $\widetilde{M}$ under π is a random walk on $\mathbb{R}$, and in particular the pushforward is recurrent.

(3) The question about foliations with one–sided branching has a positive answer if the limit set of some (and therefore every) leaf of $\widetilde{\mathscr{F}}$ is not all of S^2_∞. For in this case, the limit set of any leaf is a dendrite of measure 0, and the complement is entirely on the nonbranching side of the leaf.

(4) One has to be very careful in defining the behaviour of a random walk on L; coarsely equivalent models for random walks on $\widetilde{M}$ (e.g. combinatorial, piecewise geodesic, etc.) might give rise to very different behaviour in L. For instance, for many $\mathscr{F}$ with 2–sided branching, we can find a pair of proper lines γ_1, γ_2 in $\widetilde{M}$ which are asymptotic (i.e. the distance between them converges to 0) but for which γ_1 exits an end of L and γ_2 is contained in a leaf. On the other hand, if γ_1, γ_2 are a bounded distance apart, and γ_1 exits to an *incomparable end* of L (i.e. the limit of a sequence of leaves $x_1 > y_1 < x_2 > y_2 < x_3 > y_3 < \ldots$ where the x_i are all incomparable and the y_i are all incomparable) then γ_2 exits to the *same* incomparable end. Moreover, if $\mathscr{F}$ arises from a slithering or branched slithering, if γ_1 exits *any* end, γ_2 a bounded distance away exits the same end. If $\pi(\gamma)$ exits an end of L, with probability one does it exit an incomparable end? (Yes in many cases)

(5) If $\mathscr{F}$ is finite depth, we can "trap" the behaviour of a random walk by compact leaves. If $\mathscr{F}$ is not a surface bundle over a circle, the random walk has a definite probability of making an incomparable detour in a fixed time interval. The tree of incomparable detours branches with ramification bounded below; so a random walk will exit an incomparable end with probability 1 in this case. Probably the same result is true if $\widetilde{\mathscr{F}}$ contains any leaf which is asymptotically separated. Conjecturally, every leaf of every $\mathscr{F}$ with 2–sided branching should be asymptotically separated. However, part of the motivation for studying this question would be to address this conjecture, rather than the other way around.

QUESTION 10.8. *Suppose Λ is an essential lamination of a hyperbolic manifold M. Is Λ isotopic to a lamination whose curvature is bounded below everywhere by -2?*

REMARK. A complete properly embedded minimal surface in $\mathbb{H}^3$ has curvature bounded below by -2 everywhere. Also see question 4.1.

11. Foliated Teichmüller theory

11.1. Deformations, mapping–class groups. For $\mathscr{F}$ a taut foliation of an atoroidal 3–manifold, any conformal class of leafwise metric on $\mathscr{F}$ can be "uniformized" to a leafwise hyperbolic metric. Co–deformations of such metrics are parameterized by leafwise quadratic holomorphic differentials, and presumably analogues of many other elements of one–dimensional complex analysis can be found.

QUESTION 11.1. *What kind of nontrivial "mapping class elements" are possible for taut foliations?*

REMARK. (1) If $\mathscr{F}$ is a foliation of an atoroidal manifold M arising from a slithering over S^1, there is a canonical (up to isotopy) transverse pseudo–Anosov flow X associated to the slithering. This flow is not in general leaf preserving at all times, but the time one flow takes $\mathscr{F}$ to itself, and qualifies as a "mapping class element" of infinite order. Slitherings, and slightly more generally, uniform foliations are the only foliations for which the time one flow of a transverse vector field takes leaves to leaves; are all such mapping class elements isotopic to the canonical automorphism?

(2) For arbitrary taut $\mathscr{F}$ and atoroidal M, the mapping class group of M is finite, by results of [**20**] and [**51**]. If ϕ is a mapping class of $M, \mathscr{F}$ of infinite order, for some finite power ϕ will be isotopic to the identity. Lift ϕ to $\widetilde{\phi} : \widetilde{M} \to \widetilde{M}$. It follows that there is a universal t such that for every leaf λ of $\widetilde{\mathscr{F}}$, the Hausdorff distance between λ and $\widetilde{\phi}(\lambda)$ is $\le t$. In particular, the space of leaves between λ and $\widetilde{\phi}(\lambda)$ is isotopic to a product, and so if $\mathscr{F}$ branches, every leaf corresponding to a non–Hausdorff point of L is fixed by $\widetilde{\phi}$. For example, if $\mathscr{F}$ is finite depth, the only leaves which can move are those of highest depth. If $\mathscr{F}$ is $\mathbb{R}$–covered but not uniform, there is a monotone equivalent foliation $\mathscr{G}$ to which an element homotopic to ϕ descends, which fixes $\widetilde{\mathscr{G}}$ leafwise.

11.2. Transverse Teichmüller flows.

QUESTION 11.2. *A foliation is taut iff it admits a volume–preserving transverse flow. Pseudo–Anosov flows are good candidates for "best" such transverse flows, when they exist, which is frequently. Is there an analytic construction of pseudo–Anosov flows, by analogy with Bers' proof of Thurston's classification of surface automorphisms?*

REMARK. Fix a conformal sturcture on $\mathscr{F}$, and let C be a transverse cone field. Amongst unit speed vector fields X transverse to $\mathscr{F}$, there are quasiconformally extremal ones, by a standard compactness argument. It is not *a priori* clear that there should be an extremal vector field amongst *all* transverse unit speed vector fields. Moreover, one would want a "best" flow to be metric independent. If $\mathscr{F}$ arises from a slithering, it makes sense to parameterize a transverse vector field so that the time 1 flow is a unit of the slithering; the dilatation of the time 1 map is well–defined with respect to a leafwise conformal structure. Of course, one must simultaneously optimize over all transverse flows *and all choices of leafwise conformal structure.*

For more general foliations, one needs to decide how to naturally parameterize the leafwise flow in order to measure (infinitesimal) dilatation. One plausible choice is to measure the result of flow with (transverse) speed equal to the magnitude of a harmonic transverse measure.

QUESTION 11.3. *Suppose M is atoroidal and $\mathscr{F}$ arises from a slithering over S^1. Let X be pseudo–Anosov transverse to $\mathscr{F}$, such that the time 1 flow Z takes $\mathscr{F}$ to itself. Lift to $\widetilde{M}$ and let $\lambda, Z^n(\lambda)$ be leaves of $\widetilde{\mathscr{F}}$, both uniformized as $\mathbb{H}^2$ by Candel's theorem. Can Z be approximated by mapping class elements between compact surfaces? That is, are there integers n_i, a sequence Σ_i of hyperbolic surfaces and $\phi_i : \Sigma_i \to \Sigma_i$ Teichmüller representatives for the isotopy class $[\phi_i]$ such that the composition*

$$\widetilde{\phi_i^{-1} Z^{n_i}} : \mathbb{H}^2 \to \mathbb{H}^2$$

is a k_i–quasi–isometry, where $k_i \to 1$?

REMARK. (1) Implicitly in this question we are choosing an identification of $\mathbb{H}^2$ with λ and $Z^n(\lambda)$. We could do this by choosing some $\alpha \in \pi_1(M)$ for which $\alpha Z^n(\lambda)$ is very close to λ and compose it with a nearest point map, or we could minimize k_i over all isometries of $\mathbb{H}^2$ to itself. Either method gives the same answer.

(2) One reason to be interested in the existence of such ϕ_i is that it might help to uniformize M. We know that the mapping cylinders of the ϕ_i are hyperbolic, and therefore estimates as above would give us approximate hyperbolic structures on larger and larger subsets of $\widetilde{M}$. What is lacking is an *a priori* comparison between the geometry of a singular solv metric on the universal cover of a 3–manifold which fibers over S^1 and the geometry of the hyperbolic metric, depending only on a bound on the degree of branching on the singular locus, and a bound c on the supremum of the distance from a point in $\widetilde{M}$ to the singular locus. Note that without such a bound, no *a priori* geometric comparison is possible.

11.3. Algebraic and analytic geometry of foliations.

QUESTION 11.4. *If $\mathscr{F}$ is a taut foliation, one can let γ_i be a collection of transverse circles to $\mathscr{F}$ intersecting every leaf and study the space of functions $\mathscr{O}(\sum n_i\gamma_i)$ which are leafwise holomorphic, with poles of order at most n_i along γ_i. (Here the notation $\sum n_i\gamma_i$ is meant to suggest a divisor on a Riemann surface). How do these function spaces change as a function of γ_i? What is the effect of certain "topological" operations on the γ_i; e.g. crossing changes, cabling etc.*

REMARK. Ghys used the fact that $\mathscr{O}(D)$ is large for certain D to show that taut foliations admit leafwise embeddings in complex projective space of sufficiently high dimensions.

12. Coarse foliations

QUESTION 12.1. *Suppose $\rho : \pi_1(M) \to \mathbb{R}$ is a 1–cochain with bounded coboundary; i.e. there is a uniform C so that*

$$|\rho(\alpha) + \rho(\beta) - \rho(\alpha\beta)| < C$$

for all $\alpha, \beta \in \pi_1(M)$. Consider $\pi_1(M)$ as a metric space by thinking of it as the vertices of some Cayley graph. Let $L_\rho = \rho^{-1}(I) \subset \pi_1(M)$.

(1) *Are the coarse connected components of L_ρ coarsely simply connected?*

(2) *If the answer is "yes", is there a sense in which L_ρ and its translates by $\pi_1(M)$ can be thought of as coarse minimal planes in $\pi_1(M)$?*

REMARK. (1) If L_ρ is coarsely connected for some I, we say ρ is *weakly uniform*, and if it is both coarsely connected and coarsely simply connected, it is *uniform*. One should think of L_ρ as a coarse kernel for ρ; under this analogy, coarsely connected corresponds to finitely generated, and coarsely simply connected corresponds to finitely presented.

As one might guess from the Scott core theorem, for $\pi_1(M)$ a 3–manifold group, ρ is weakly uniform iff it is uniform ([**21**]). This may be thought of as a kind of "coarse Stallings fibration theorem", or as a coarsening of Novikov's theorem. In [**21**] it is also shown that such $\pi_1(M)$ either contain $\mathbb{Z} \oplus \mathbb{Z}$, or are word hyperbolic.

(2) The potential usefulness of this question comes from the fact that 1–cochains with bounded coboundary are very common on hyperbolic 3–manifolds. In fact, if the geometrization conjecture is true, *every* 3–manifold with infinite π_1 admits infinitely many independent such ρ. This is in stark contrast to the fact that many 3–manifolds with infinite π_1 — even hyperbolic 3–manifolds — do not admit taut foliations, by [**103**].

QUESTION 12.2. *Does every hyperbolic 3–manifold admit a* taut cone field*? That is, a cone field C which is recurrent and supports only homotopically essential loops.*

REMARK. (1) Taut cone fields are introduced in [**19**]. They exist on small Seifert fibered spaces with infinite π_1 which do not admit essential laminations. Examples of taut cone fields are cone fields supported by pseudo–Anosov flows. However, not every hyperbolic 3–manifold admits a pseudo–Anosov flow. For instance, the Weeks manifold does not admit such a flow ([**22**]).

(2) Do cone fields stay taut after most surgeries (i.e. those outside finitely many slopes) on a supported loop? Is there a class of examples where the cone field stays taut for all nontrivial surgeries? See also question 9.4.

(3) By Sullivan [**109**], there is a closed 2–form positive on the de Rham currents supported by a taut cone field. So, for instance, if M is a rational homology sphere, any complete surface (possibly immersed and noncompact) transverse to the cone field enjoys a uniform linear isoperimetric inequality.

12.1. Coarse invariants; deformation.

QUESTION 12.3. *What deformations of a foliation or lamination should be thought of as "inessential"? For instance — monotone equivalence, cut–and–shear along a surface or transverse lamination, isotopy of branch locus in a branched cover, isomorphic universal circles etc.*

REMARK. This question is prompted by the phenomenon that certain geometric objects associated to foliations — e.g. transverse pseudo–Anosov flows — are invariant under quite drastic deformations of the foliation.

An added bonus of a "good" deformation–equivalence class for foliations might be a finite–dimensional parameter space, perhaps analogous to the unit ball of the Thurston norm for finite–depth foliations. It seems that an advantage of the theory of tight contact structures over the theory of taut foliations is a better and more robust equivalence relation. E.g. finiteness up to isotopy of tight contact structures on an atoroidal 3–manifold [**56**].

13. Numerical invariants

13.1. Godbillon–Vey. The Godbillon–Vey invariant was introduced in [**57**]. It is a invariant of the foliated cobordism class of a sufficiently smooth (C^2 is sufficient) codimension one foliation of a 3–manifold. A basic reference for this invariant is [**121**].

QUESTION 13.1. *Suppose $\mathscr{F}$ is a minimal taut C^2 foliation of an atoroidal 3–manifold M with*

$$\mathfrak{gv}(\mathscr{F})[M] \neq 0$$

Is there a choice of 1–form α with $T\mathscr{F} = \ker(\alpha)$ for which the Godbillon–Vey form ω (where $d\alpha = \alpha \wedge \omega$) has the same sign? i.e. either $\omega \wedge d\omega \geq 0$ everywhere or ≤ 0 everywhere. Say that such a foliation has monotone wobble.

REMARK. (1) Some condition on tori is essential. If T is a torus leaf separating M into two components $M_1 \coprod M_2$, then $\int_{M_i} \omega \wedge d\omega$ is independent of the choice of α, for each i. So we could splice together two foliated manifolds with torus boundary, one with positive Godbillon–Vey invariant, the other with negative invariant in such a way that the sum was nonzero. More generally, if M_1, M_2 are two manifolds foliated by $\mathscr{F}_1, \mathscr{F}_2$, and γ_1, γ_2 are transverse circles, the effect of *torus connect sum* of M_1, M_2 along the γ_i gives a foliated manifold whose Godbillon–Vey invariant is the sum of the invariants for M_1 and M_2. A form α for which ω is contact must have $\|\omega \wedge d\omega\|$ potentially very large in a neighborhood of the separating torus $\partial N(\gamma_1) = \partial N(\gamma_2)$.

(2) Minimality is also essential, or else regions of positive and negative twisting can be separated by some exceptional minimal set or even closed leaf. On the other hand, for a minimal foliation, positive twisting can be propagated along neighborhoods of a leafwise path to regions of negative twisting, by analogy with the process of deforming a transitive confoliation to a contact structure.

(3) If $\omega \wedge d\omega \geq 0$ can be achieved everywhere, when can it be achieved so that the zero locus $\omega \wedge d\omega = 0$ is a collection of knots, and M, ω is locally modelled on a branched cover of a contact manifold over a Legendrian link?

(4) If $\omega \wedge d\omega > 0$ everywhere — that is, if ω is actually a contact structure, then in particular $T\mathscr{F}$ admits a nowhere zero section, and the Euler class of $T\mathscr{F}$ is zero. This implies that $\mathscr{F}$ has no closed leaves.

QUESTION 13.2. *For $\mathscr{F}$ as in the previous question, suppose there is a choice of α for which ω is a contact form. Is the contact structure defined by ω necessarily tight?*

13.2. Foliated Gromov norms. For $M, \mathscr{F}$ a foliated manifold, the *foliated Gromov norm* on H_i is the infimum of the L_1 norm on chains representing homology classes, where we restrict attention to chains which are *transverse* to $\mathscr{F}$ (in the sense that the induced foliation of each simplex should be affine). A reference for this material is [**16**]. Notice that to define the Godbillon–Vey invariant of $\mathscr{F}$, the transverse structure must be at least Lipschitz + quadratic variation (see [**122**]); by contrast, the foliated Gromov norm depends only on the *homeomorphism* type of the foliation.

QUESTION 13.3. *Calculate the norm of the fundamental class of a hyperbolic 3–manifold for some taut foliation $\mathscr{F}$ with two–sided branching.*

REMARK. If $\mathscr{F}$ is $\mathbb{R}$–covered or has one–sided branching, the foliated Gromov norm agrees with the usual Gromov norm.

If some leaf of $\widetilde{\mathscr{F}}$ is *asymptotically separated* (see question 10.1) then the value of the norm on this fundamental class is *strictly* greater than the value of the usual Gromov norm on this class.

QUESTION 13.4. *Let $\mathscr{F}, \mathscr{G}$ be taut foliations on a hyperbolic manifold M. Are there examples where there is a finite cover of M such that a sequence of isotopies of the lift of $\mathscr{G}$ converges geometrically to the lift of $\mathscr{F}$, but no such sequence of isotopies exists in M?*

REMARK. If this can be done, the foliated Gromov norm of $\mathscr{F}$ is at least as large as that of $\mathscr{G}$ for every homology class.

QUESTION 13.5. *Suppose $\mathscr{F}$ is a foliation (possibly $\mathbb{R}$–covered) of a hyperbolic 3–manifold. Define a foliated Gromov norm using* cubical *chains. Is the value of the foliated norm on the fundamental class always* strictly *greater than the value of the usual (cubical) Gromov norm?*

REMARK. Cubical chains cannot necessarily be straightened rel. vertices to be transverse to a given foliation. Suppose $\mathscr{F}$ is a transversely measured foliation. Lift this transverse measure to a function $f : \widetilde{\mathscr{F}} \to \mathbb{R}$. Then for a random regular cube C of side length t, the ordering of the vertices of C induced by f converges to a random distribution as $t \to \infty$.

QUESTION 13.6. *What kinds of local order structure are there on a family of deformations of a (taut) foliation? Can one use such structures to define co–ordinates on the "space of deformations" of a taut foliation?*

REMARK.

(1) For instance, local order structure could be provided by numerical invariants, such as the Godbillon–Vey invariant, or foliated Gromov norms.
(2) A family $\mathscr{F}_t$ obtained by cut and shear along some particular submanifold, where the re–gluings f_t vary monotonically in some sense should be considered a monotone deformation. E.g. in local co–ordinates, the differences $f_t - f_s$ should be non–negative functions for $t > s$.

13.3. Invariants for laminations.

QUESTION 13.7. *Is there some notion of a Godbillon–Vey invariant for a lamination?*

REMARK. One can look at eigendirections of pinching for holonomy and construct a leafwise vector field; a Godbillon–Vey type invariant should be a topological measure of how much this vector field twists in the positive direction. Even an invariant taking values in $+, -, 0$ would be interesting, where the invariant would be

+ if for some metric, the transverse twisting ("helical wobble" is Thurston's term) is always positive, – if for some metric it is always negative, and 0 if neither can be achieved. Is this even well–defined? Maybe one would need to restrict attention to the behavior at infinity — e.g. ask for a metric such that for some choice of partition into guts and interstices, the helical wobble over the interstices is always positive.

14. Immersed objects

QUESTION 14.1. *Is there a geometric notion for a 3–manifold analogous to LERFness for foliations? What properties could a manifold have so that immersed essential laminations are virtually embedded?*

REMARK. It might be better to restrict attention to foliations dual to one–cochains with bounded coboundary, since these have a more "algebraic" description.

QUESTION 14.2. *Let $\mathscr{F}$ be a taut foliation of M. Can leaves of $\mathscr{F}$ be approximated by compact essential surfaces? That is, given a leaf λ of $\mathscr{F}$ and a point $p \in \lambda$, is there a sequence of immersed incompressible surfaces $\phi_i : \Sigma_i \to M$ and points $p_i \in \Sigma_i$ such that the images under ϕ_i of the balls of radius r_i about p_i where $r_i \to \infty$ converge on compact sets to p, λ?*

REMARK.

(1) If one does not ask for incompressibility, it is easy to find such approximating surfaces; for, if $D \subset \lambda$ is a big embedded subsurface, we can just take two parallel copies of D tubed together by a collection of thin annuli, one for each boundary component of D.
(2) If $\pi_1(M)$ is LERF, each of the surfaces Σ_i lifts to an embedded surface in some finite cover. An interesting additional constraint is to ask for these lifts to be Thurston norm minimizing.
(3) Suppose we can choose the Σ_i which converge *uniformly* to some leaf λ in the sense that for every ϵ there is an i such that $(\phi_i)_* T\Sigma_i$ makes an angle of less than ϵ at every point with $T\lambda$. Then the de Rham cycles $\phi_i([\Sigma_i])$ converge to a de Rham cycle supported by $\mathscr{F}$; since $\mathscr{F}$ is taut, this implies $H_1(M;\mathbb{R})$ is nontrivial. On the other hand, the support of the limiting cycle need not be all of $\mathscr{F}$. This can happen, for instance, if the Σ_i are surfaces of a fibration of M over S^1 representing homology classes projectively limiting to a class contained in a high codimension face of the Thurston norm. However, if λ is dense, every transverse loop to $\mathscr{F}$ is eventually transverse to some Σ_i, and therefore pairs positively with the projective homology class represented by Σ_i. It follows that every transverse loop to $\mathscr{F}$ is nontrivial in homology, and therefore $\mathscr{F}$ is a perturbation of a surface bundle over S^1.

14.1. Total foliations; families of foliations. A *total* foliation on an n manifold is a collection of n foliations $\mathscr{F}_i$ for $i = 1 \dots n$ which are mutually transverse everywhere and in sufficiently small charts, are topologically conjugate to the total foliation of $\mathbb{R}^n$ by translates of the co–ordinate planes.

QUESTION 14.3. *What 3–manifolds admit total taut foliations?*

REMARK. Every 3–manifold admits a total foliation, by a local construction due to Hartdorp [**65**]. Suspensions of Anosov automorphisms of tori admit total taut foliations. Some examples arise in the theory of slitherings ([**117**]). Anosov flows give rise to some good examples. It is probably better in general for M a hyperbolic manifold to look for taut foliations of M with *genuine* transverse pseudo–Anosov flows; these exist in very many examples.

QUESTION 14.4. *Are there any interesting examples of total genuine laminations?*

REMARK.
(1) One can generalize total foliations to a $\mathscr{G}$–structure where $\mathscr{G}$ is the pseudogroup of homeomorphisms between subsets of $\mathbb{R}^3$ which permute translates of the co–ordinate planes, but are not required to take horizontal to horizontal, etc. For M a manifold with a $\mathscr{G}$–structure, there is a representation $\pi_1(M) \to \mathsf{S}_4$ whose kernel defines a finite cover with a total foliation (in the usual sense). For *total laminations*, there are local obstructions, much as there are local obstructions to co–orientability. It is reasonable to use this more flexible definition of total lamination, since it encompasses many known examples (e.g. pseudo–Anosov flows transverse to foliations). Is it useful to think of a total genuine lamination as some kind of intrinsically three–dimensional analogue of a pseudo–Anosov flow?
(2) The immersed meridional surface in a cubulation of strictly negative curvature is an example of a total genuine lamination. The complementary regions to a total genuine lamination should be cubes, products of negatively curved polygons with intervals, and polyhedra satisfying the condition of Andre'ev's theorem (see [**71**]) for all right–angles. It should be straightforward to find a direct proof that manifolds with total genuine laminations admit CAT 0 metrics.
(3) In [**81**] examples are constructed of hyperbolic manifolds which do not admit nonpositive cubulations. It would be interesting to know whether these manifolds can admit total genuine laminations.

14.2. Amenability.

QUESTION 14.5. *What is the weakest useful 2–dimensional object that might be present in every atoroidal 3–manifold? For instance, does every hyperbolic 3–manifold M contain an immersed quasigeodesic surface Σ of amenable growth?*

REMARK.
(1) Of course, the virtual infinite positive Betti number conjecture would imply that every hyperbolic 3–manifold contains an immersed quasigeodesic *compact* surface. The point here is to think of as weak a condition as possible, so that there are (hopefully?) many examples to study.
(2) The closure of the image of such an immersed surface Σ supports a de Rham cycle by the usual method of Goodman–Plante [**58**]. If M is a rational homology sphere, then for any loop γ in M, large subsets of Σ must intersect γ positively and negatively with approximately equal frequency. Does this imply ergodicity of the geodesic flow?

15. Miscellaneous

QUESTION 15.1. *What possibilities are there for (co–oriented) laminations in a 3–manifold whose transverse spaces are* well–ordered*?*

Is there a (useful) theory of branched surfaces with ordinal–valued weights?

REMARK. Probably the condition that laminations be geometrically realized in a 3–manifold implies that the order types will be *countable*. There is an *evacuation* procedure to replace a well–ordered lamination by a simpler one: given Λ, we can replace Λ by Λ_∞, the sublamination whose leaves are labelled by limit ordinals in a given chart. This will be a proper sublamination. To get an interesting theory, one would want to study laminations for which this procedure does not strictly decrease some notion of complexity — so that the order types should be at least uncountable.

For branched surfaces with no realization issues, more exotic and interesting ordinals could be used; perhaps one can use this to give "geometric" interpretations of interesting theorems in logic.

QUESTION 15.2. *Is there a good notion of* taut foliated cobordism*? Are there numerical invariants of the equivalence classes this induces on taut foliations which are finer than the Godbillon–Vey invariant?*

References

[1] C. Adams, *Isometric cusps in hyperbolic 3–manifolds*, Mich. Math. J. **46** 3 (1999), 515–531

[2] I. Agol, *Lower bounds on volumes of hyperbolic Haken 3–manifolds*, preprint, math.GT/9906182

[3] I. Agol and T. Li, *An algorithm to detect laminar 3–manifolds*, preprint, math.GT/0201310

[4] I. Agol and W. Thurston, personal communication

[5] I. Agol, J. Hass and W. Thurston, *The computational complexity of knot genus and spanning area*, preprint, math.GT/0205057

[6] L. Ahlfors and L. Bers, *Riemann's mapping theorem for variable metrics*, Ann. Math. **72** (1960), 413–429

[7] M–T. Benameur, *Triangulations and the stability theorem for foliations*, Pacific J. Math. **179** (1997), 221–239

[8] S. Boyer, D. Rolfsen and B. Wiest, *Orderable 3–manifold groups*, preprint

[9] M. Brittenham, *Essential laminations in Seifert–fibered spaces*, Topology, **32** 1 (1993), 61–85

[10] M. Brittenham, *Essential laminations and Haken normal form*, Pacific J. Math. **168** 2 (1995), 217–234

[11] M. Brittenham, R. Naimi and R. Roberts, *Graph manifolds and taut foliations*, J. Diff. Geom. **45** 3 (1997), 446–470

[12] D. Calegari, *$\mathbb{R}$–covered foliations of hyperbolic 3–manifolds*, Geom. Top. **3** (1999), 137–153

[13] D. Calegari, *Foliations transverse to triangulations of 3–manifolds*, Comm. Anal. Geom. **8** 1 (2000), 133–158

[14] D. Calegari, *Foliations and the geometry of 3–manifolds*, Dissertation for a PhD at UC Berkeley, (2000)

[15] D. Calegari, *The geometry of $\mathbb{R}$–covered foliations*, Geom. Top. **4** (2000), 457–515

[16] D. Calegari, *The Gromov norm and foliations*, Geom. and Func. Anal **10** (2000), 1423–1447

[17] D. Calegari, *Almost continuous extension for taut foliations*, Math. Res. Lett. **8** (2001), no. 5–6, 637–640

[18] D. Calegari, *Foliations with one–sided branching*, Geom. Ded. to appear

[19] D. Calegari, *Useful branched surfaces which carry nothing*, preprint, math.GT/0010146

[20] D. Calegari, *Promoting essential laminations*, preprint

[21] D. Calegari, *Bounded cochains on 3–manifolds*, preprint, math.GT/0111270

[22] D. Calegari and N. Dunfield, *Laminations and groups of homeomorphisms of S^1*, preprint, math.GT/0203192

[23] A. Candel, *Uniformization of surface laminations*, Ann. Sci. École Norm. Sup (4) **26** no.4, (1993), 489–516

[24] A. Candel and L. Conlon, *Foliations I*, AMS Graduate studies in mathematics **23**

[25] J. Cantwell and L. Conlon, *Every surface is a leaf*, Topology **26** (1987), 265–285

[26] J. Cantwell and L. Conlon, *Foliation cones*, Geom. Top. Monographs 2 Proceedings of the Kirbyfest, 35–86

[27] J. Cannon and W. Thurston, *Group invariant Peano curves*, preprint

[28] A. Casson and D. Jungreis, *Convergence groups and Seifert–fibered 3–manifolds*, Invent. Math. **118** 3 (1994), 441–456

[29] A. Connes, *Non–Commutative geometry*, Academic Press (1994)

[30] C. Delman and R. Roberts, *Alternating knots satisfy strong property P*, Comm. Math. Helv. **74** 3 (1999), 376–397

[31] A. Denjoy, *Sur les courbes définies par les équations différentiables à la surface du tore*, J. Math. Pures Appl. **11** (1932), 333–375

[32] Y. Eliashberg and W. Thurston, *Confoliations*, AMS

[33] S. Fenley, *Depth 1 foliations and end–periodic automorphisms*, PhD thesis at Princeton University, (1992)

[34] S. Fenley, *Asymptotic properties of depth one foliations in hyperbolic 3–manifolds*, J. Diff. Geom. **36** 2 (1992), 269–313

[35] S. Fenley, *Limit sets of foliations in hyperbolic 3–manifolds*, Topology **37** no.4 (1998), 875–894

[36] S. Fenley, *Foliations with good geometry*, Jour. AMS **12** 3 (1999), 619–676

[37] S. Fenley and L. Mosher, *Quasigeodesic flows in hyperbolic 3–manifolds*, Topology, **40** (2001), no. 3, 530–537

[38] S. Fenley, *Laminar free hyperbolic 3–manifolds*, preprint.

[39] D. Fried, *Flow equivalence, hyperbolic systems and a new zeta function for flows*, Comment. Math. Helv. **57** no. 2 (1982), 237–259

[40] L. Funar and S. Gadgil, *Topological geodesics and virtual rigidity*, Alg. Geom. Top. **1** (2001) 369–380

[41] D. Gabai, *Foliations and the topology of 3–manifolds*, J. Diff. Geom. **18** no. 3, (1983), 445–503

[42] D. Gabai, *Foliations and 3–manifolds*, Proceedings of the ICM Vol. 1 (Kyoto, 1990) 609–619

[43] D. Gabai, *Convergence groups are Fuchsian groups*, J. Diff. Geom. **36** 2 (1992), 269–313

[44] D. Gabai, *On the geometrical and topological rigidity of hyperbolic 3–manifolds*, Bull. AMS **31** 2 (1994), 228–232

[45] D. Gabai, *Problems in foliations and laminations*, in Geometric Topology (ed. W. Kazez) proceedings of the 1993 Georgia International Topology Conference; vol. 2 (1997), 1–33

[46] D. Gabai, *Essential laminations and Kneser normal form*, J. Diff. Geom. **53** no. 3, (1999), 517–574

[47] D. Gabai, *Combinatorial volume–preserving flows and taut foliations*, Comm. Math. Helv. **75** 1 (2000), 109–124

[48] D. Gabai and W. Kazez, *Homotopy, isotopy and genuine laminations of 3–manifolds*, in Geometric Topology (ed. W. Kazez) proceedings of the 1993 Georgia International Topology Conference; vol. 1 (1997), 65–77

[49] D. Gabai and W. Kazez, *Order trees and laminations of the plane*, Math. Res. Lett. **4** (1997), no. 4, 603–616

[50] D. Gabai and W. Kazez, *Group negative curvature for 3–manifolds with genuine laminations*, Geom. Top. 2 (1998), 65–77

[51] D. Gabai and W. Kazez, *The finiteness of the mapping class group for atoroidal 3–manifolds with genuine laminations*, J. Diff. Geom. **50** (1998), no. 1, 123–127

[52] D. Gabai and U. Oertel, *Essential laminations in 3–manifolds*, Ann. Math. (2) **130** no. 2, (1989), 41–73

[53] L. Garnett, *Foliations, the ergodic theorem and Brownian motion*, J. Func. Anal. **51** (1983), 285–311

[54] E. Ghys, *Flots d'Anosov sur les 3–variétés fibrées en cercles*, Erg. Theory Dynam. Systems **4** no. 1, (1984), 67–80

[55] E. Ghys, *Topologie des feuilles génériques*, Ann. Math. **141** (1995), 387–422

[56] E. Giroux and K. Honda, *Finiteness of tight contact structures on 3–manifolds*, research announcement

[57] C. Godbillon and J. Vey, *Un invariant des feuilletages de codimension* 1, Acad. des Sciences. Paris Comptes Rendus. ser. A. **273** (1971), 92–95

[58] S. Goodman and J. Plante, *Holonomy and averaging in foliated sets*, J. Diff. Geom. **14** 3 (1979), 401–407

[59] P. Griffiths and J. Harris, *Principles of algebraic geometry*, John Wiley and sons (1978)

[60] M. Gromov, *Volume and bounded cohomology*, IHES Publ. Math. **56** (1982), 5–99

[61] M. Gromov, *Partial differential relations*, Springer–Verlag (1986)
[62] A. Haefliger, *Structures feuilletées et cohomologie à valeur dans un faisceau de groupoides*, Comm. Math. Helv. **32** (1957/58), 248–329
[63] A. Haefliger, *Variétés feuilletées*, Ann. Scuola Norm. Sup. Pisa **16** (1962), 367–397
[64] W. Haken, *Theorie der Normalflächen*, Acta Math. **105** (1961), 245–375
[65] D. Hardorp, *All compact orientable three dimensional manifolds admit total foliations*, Mem. AMS **26** (1980)
[66] K. Hartshorn, *Heegaard splittings: The distance complex and the stabilization conjecture*, Ph.D. thesis, UC Berkeley (1999)
[67] J. Hass, *Minimal surfaces in foliated manifolds*, Comm. Math. Helv. **61** (1986), 1–32
[68] A. Hatcher, *Measured lamination spaces for 3–manifolds*, preprint.
[69] G. Hemion, *On the classification of homeomorphisms of 2–manifolds and the classification of 3–manifolds*, Acta Math. **142** 1–2 (1979), 123–155
[70] J. Hempel, *3–manifolds*, No. 86 Ann. Math. Stud. Princeton University Press (1976)
[71] C. Hodgson and I. Rivin, *A characterization of compact convex polyhedra in hyperbolic 3–space*, Invent. Math. **111** 1 (1993), 77–111
[72] W. Jaco and H. Rubinstein, *0–Efficient triangulations of 3–manifolds*, preprint
[73] M. Jankins and W. Neumann, *Rotation numbers of products of circle homeomorphisms*, Math. Ann. **271** 3 (1985), 381–400
[74] T. Kobayashi, *Example of hyperbolic knot which do not admit depth 1 foliation*, Kobe. J. Math. **13** 2 (1996), 209–221
[75] P. Kronheimer and T. Mrowka, *Scalar curvature and the Thurston norm*, Math. Res. Lett. **4** 6 (1997), 931–937
[76] M. Lackenby, *Exceptional surgery curves in triangulated 3–manifolds*, preprint; math.GT/9907093
[77] M. Lackenby, *Taut ideal triangulations of 3–manifolds*, Geom. Top. **4** (2000), 369–395
[78] O. Lehto, *Univalent functions and Teichmüller space*, GTM 109 Springer–Verlag (1987)
[79] T. Li, *Immersed surfaces, Dehn surgery and essential laminations*, PhD thesis at Caltech (2000)
[80] T. Li, *Laminar branched surfaces in 3–manifolds*, Geom. Top. **6** (2002), 153–194
[81] T. Li, *Boundary curves of surfaces with 4–plane property*, preprint
[82] R. Lickorish, *A foliation for 3–manifolds*, Ann. Math. **82** 2 (1965), 414–420
[83] S. Matsumoto, *Topological entropy and Thurston's norm of atoroidal surface bundles over the circle*, J. Fac. Sci. Univ. Tokyo Sect. 1A Math. **34** no. 3, (1987), 763–778
[84] S. Matsumoto, *Some remarks on foliated S^1 bundles*, Invent. Math. **90** no. 2, (1987), 343–358
[85] C. McMullen, *Renormalization and 3–manifolds which fiber over the circle*, No. 142 Ann. Math. Stud. Princeton University Press (1996)
[86] C. McMullen, *Polynomial invariants for fibered 3–manifolds*, Ann. Scient. École Norm. Sup. **33** (2000), 519–560
[87] G. Meigniez, *Bouts d'un groupe opérant sur la droite: 2. applications á la topologie des feuilletages*, Tóhoku Math. J. **43** (1991), 473–500
[88] J. Milnor, *On the existence of a connection with curvature zero*, Comm. Math. Helv. **32** (1958), 215–223
[89] E. Moise, *Geometric topology in dimensions 2 and 3*, Springer–Verlag GTM **47** (1977)
[90] L. Mosher, *Surfaces and branched surfaces transverse to pseudo–Anosov flows on 3–manifolds*, Jour. Diff. Geom. **34** (1991), 1–36
[91] L. Mosher, *Dynamical systems and the homology norm of a 3–manifold, I: Efficient intersection of surfaces and flows*, Duke Math. Jour. **65** no. 3 (1992), 449–500
[92] L. Mosher, *Dynamical systems and the homology norm of a 3–manifold, II*, Invent. Math. **107** (1992), 243–281
[93] L. Mosher, *Laminations and flows transverse to finite depth foliations. Part i: branched surfaces and dynamics*, preprint
[94] R. Naimi, *Foliations transverse to fibers of Seifert manifolds*, Comment. Math. Helv. **69** no. 1 (1994), 155–162
[95] S. Novikov, *Topology of foliations*, Trud. Mosc. Math. Ob. **14** (1965), 268–304
[96] U. Oertel, *Homology branched surfaces: Thurston's norm on $H_2(M^3)$*, Low dimensional topology and Kleinian groups; LMS No. 112 (1986), 253–272
[97] C. Palmeira, *Open manifolds foliated by planes*, Ann. Math. **107** (1978), 109–131

[98] H. Pittie, *Characteristic classes of foliations*, Research Notes in Mathematics **10**, Pitman Publishing (1976)
[99] J. Pitts, *Existence and regularity of minimal surfaces on Riemannian manifolds*, Princeton University Press Mathematical Notes **27** (1981)
[100] D. Pixton, *Nonsmoothable, nonstable group actions*, Trans. AMS **229** (1977), 259–268
[101] H. Poincaré, *Sur les courbes définies par les équations différentielles*, J. de Mathématiques **1** 4, (1885)
[102] G. Reeb, *Sur certaines proprietes topologiques des varietes feuilletees*, Act. Sci. Indust. **1183** (1952), 91–158
[103] R. Roberts, J. Shareshian and M. Stein, *Infinitely many hyperbolic 3–manifolds which contain no Reebless foliation*, preprint.
[104] D. Rolfsen, *Knots and Links*, Publish or Perish (1976)
[105] H. Rosenberg, *Foliations by planes*, Topology **7** (1968), 131–138
[106] H. Rubinstein, *Minimal surfaces in geometric 3–manifolds*, preprint
[107] S. Shields, *Branched surfaces and the simplest foliations of 3–manifolds*, Pac. J. Math. **177** 2 (1997), 305–327
[108] D. Stroock, *Probability theory — an analytic view*, Cambridge Univ. Press (1993)
[109] D. Sullivan, *Cycles for the dynamical study of foliated manifolds and complex manifolds*, Invent. Math. **36** (1976), 225–255
[110] D. Sullivan, *A homological characterization of foliations consisting of minimal surfaces*, Comm. Math. Helv. **54** (1979), 218–223
[111] W. Thurston, *Foliations of 3–manifolds that are circle bundles*, PhD thesis at UC Berkeley (1972)
[112] W. Thurston, *Noncobordant foliations of S^3*, Bulletin of the AMS **78** no. 4, (1972), 511–514
[113] W. Thurston, *A local construction of foliations for 3–manifolds*, Proc. Symp. Pure Math. XXVII (part I) (1973), 315–319
[114] W. Thurston, *The geometry and topology of three–manifolds*, Lecture notes from Princeton University 1978–1980
[115] W. Thurston, *A norm for the homology of 3–manifolds*, Mem. AMS **59** (1986) no. 339, pp. i–vi and 99–130
[116] W. Thurston, *On the geometry and dynamics of diffeomorphisms of surfaces*, Bull. AMS. **19** 2 (1988), 417–431
[117] W. Thurston, *Three–manifolds, Foliations and Circles, I*, eprint math.GT/9712268
[118] W. Thurston, *Three–manifolds, Foliations and Circles, II*, unfinished manuscript
[119] W. Thurston, personal communication
[120] T. Tsuboi, *Homology of diffeomorphism groups, and foliated structures*, translated from the Japanese in Sugaku Expositions **3** no. 2, (1990), 145–181
[121] T. Tsuboi, *Characterization of Godbillon–Vey classes*, translated from the Japanese in Sugaku Expositions **8** no. 2, (1995), 165–182
[122] T. Tsuboi, *Homological and dynamical study on certain groups of Lipschitz homeomorphisms of the circle*, J. Math. Soc. Jap. **47** 1 (1995), 1–30
[123] J. Wood, *Foliations on 3–manifolds*, Ann. Math. **89** (1969), 336–358
[124] A. Zeghib, *Sur les feuilletages géodésiques continus des variétés hyperboliques*, Invent. Math. **114** no.1 (1993), 193–206
[125] A. Zeghib, *Lipschitz regularity in some geometric problems* preprint

Department of Mathematics, California Institute of Technology, Pasadena CA 91125

E-mail address: `dannyc@its.caltech.edu`

Proceedings of Symposia in Pure Mathematics
Volume **71**, 2003

Problems in Low Dimensional Contact Topology

John B. Etnyre and Lenhard L. Ng

During the 2001 Georgia International Topology Conference, two problem sessions were held concerning contact geometry. The first was run by Emmanuel Giroux and focused on problems in three dimensional contact topology; the second was run by John Etnyre and focused on Legendrian knots and contact homology. This article collects problems from the sessions, and adds some background. Sections 1 through 5 deal with three dimensional contact geometry, and Sections 6 through 10 deal with Legendrian knots.

We would like to thank the organizers of the Georgia Conference for running the problem sessions and encouraging us to write this problem list. We would also like to thank Ko Honda and Will Kazez, who took exceptional notes during the problem sessions, and Josh Sabloff, who compiled a list of some of the problems discussed here.

I. Three Dimensional Contact Geometry

In this section the reader is assumed to be familiar with the basic notions in contact geometry, see [**31**], and 3-manifold topology, see [**49**].

1. Existence and types of contact structures

QUESTION 1. *Which 3-manifolds have tight contact structures, and what types of tight contact structures do they admit?*

This question breaks down into several subquestions, but first note that a tight contact structure may be:

(1) weakly fillable (or weakly semi-fillable),
(2) strongly fillable (or strongly semi-fillable),
(3) holomorphically fillable (or holomorphically semi-fillable),
(4) universally tight, or
(5) virtually overtwisted.

Recall that a holomorphically fillable contact structure is also a strongly fillable contact structure, and strongly fillable implies weakly fillable. Moreover, a weakly

2000 *Mathematics Subject Classification.* Primary 57R17; secondary 53D35, 57M27.
Key words and phrases. Contact structure, tight, Legendrian knot, convex surface.
JE supported in part by NSF Grant # DMS-9705949.
LN supported by the American Institute of Mathematics and NSF Grant # DMS-9729992.

fillable contact structure is tight [**19, 48**]. It is known that there are tight contact structures that are not weakly fillable [**34**] and weakly fillable contact structures that are not strongly fillable (the first such examples were found by Eliashberg [**22**] on the 3-torus, see also [**15**]). However, on homology spheres we know that weakly fillable structures must also be strongly fillable [**69**]. For a more thorough discussion of the notions of fillability and their relations, see [**28**].

Tight		
$\cup \not\,$		
Weakly symplectically semi-fillable	$\supsetneq$	Strongly symplectically semi-fillable
$\cup ?$		$\cup ?$
Weakly symplectically fillable	$\supsetneq$	Strongly symplectically fillable
		$\cup ?$
		Holomorphically fillable

If a contact structure is fillable, it is interesting to consider the number and type of fillings. In particular:

QUESTION 2. *Do there exist only finitely many (strong or weak) fillings* (X^4, ω) *of a fixed* (M^3, ξ)*, up to blowup and blowdown?*

Actually, it is not too hard to show there are always infinitely many strong fillings if we allow the fundamental group of X to vary [**74**]. So it is most interesting to consider this question when X is required to be simply connected.

The main technique used to show a contact structure is not fillable is Seiberg-Witten theory. This technique was initiated by Lisca [**58**] and only applies to manifolds with positive curvature. It would be exciting to find other methods for showing that a contact structure is not fillable.

QUESTION 3. *How can one determine if a contact structure is fillable?*

It is obvious that universally tight and virtually overtwisted contact structures form disjoint sets, but it is not clear if all tight contact structures must fall into one of these sets. For any 3-manifold with a residually finite fundamental group, it is known (and not hard to prove) that any tight contact structure is either universally tight or virtually overtwisted. Recall that a group G is residually finite if every nontrivial element of G is in the complement of some finite index normal subgroup of G. Of course, it is conjectured that all 3-manifolds have residually finite fundamental groups.

There are no known connections between a tight structure being fillable (in any sense) and being universally tight or virtually overtwisted.

QUESTION 4. *Are universally tight contact structures fillable?*

It is known that the converse is not true. An easier question might be the following.

QUESTION 5. *If* ξ *is a universally tight contact structure on* M*, is there some finite cover of* M *in which the pulled-back contact structure is fillable?*

We recall that the Poincaré homology sphere with the nonstandard orientation $-P$ does not admit a tight contact structure [**32**]. Moreover, any tight contact structure on the connected sum $M\#N$ can be decomposed into a tight contact structure on M and a tight contact structure on N [**8**]. Thus any 3-manifold of the form $M\#(-P)$ does not admit any tight contact structure, so when discussing existence of tight contact structures, we will always consider irreducible 3-manifolds.

If one believes the conjectured picture of 3-manifolds coming from Thurston's Geometrization conjecture, then the main question breaks into two:

QUESTION 6. *Do all hyperbolic manifolds admit a tight contact structure? What type of tight structures do they support?*

QUESTION 7. *Which Seifert fibered spaces admit tight contact structure? Which are fillable? Which are universally tight or virtually overtwisted?*

Eliashberg and Thurston [**26**] have shown that a Reebless foliation may be perturbed into a tight (in fact, weakly semi-fillable and universally tight) contact structure. Moreover, Gabai [**40**] can construct such a foliation on any irreducible 3-manifold with $b_1 \geq 1$. Thus, when considering the above questions, one can focus on rational homology spheres, or consider strong fillability or virtually overtwisted contact structures.

Gompf [**46**] has constructed strongly fillable contact structures on many Seifert fibered spaces. Given the above results, here is our main concern: Let $\Sigma(\frac{\alpha_1}{\beta_1}, \frac{\alpha_2}{\beta_2}, \frac{\alpha_3}{\beta_3})$ be a Seifert fibered space over S^2 with three singular fibers with invariants $\frac{\alpha_1}{\beta_1}$, $\frac{\alpha_2}{\beta_2}$, $\frac{\alpha_3}{\beta_3}$; then there is a strongly fillable contact structure on $\Sigma(\frac{\alpha_1}{\beta_1}, \frac{\alpha_2}{\beta_2}, \frac{\alpha_3}{\beta_3})$ if $e_0 \neq -1$, where $e_0 = \sum_{i=1}^{3} \lfloor -\frac{\alpha_i}{\beta_i} \rfloor$ and $\lfloor x \rfloor$ is the greatest integer less than or equal to x. Moreover, Gompf can also construct fillable contact structures on many Seifert fibered spaces with $e_0 = -1$; see [**46**].

The only manifold currently known not to admit a tight contact structure is $-P$, which is a Seifert manifold over S^2 with invariants $-\frac{1}{2}, \frac{1}{3}, \frac{1}{5}$.

There are several other manifolds that do not admit weakly fillable contact structures. In [**58, 59**], Lisca has shown that the Seifert manifolds with invariants $(-\frac{1}{2}, \frac{1}{3}, \frac{1}{5}), (-\frac{1}{2}, \frac{1}{3}, \frac{1}{4}), (-\frac{1}{2}, \frac{1}{3}, \frac{1}{3})$ do not admit weakly fillable contact structures. It would be very interesting to know:

QUESTION 8. *Do the Seifert manifolds with invariants*

$$(-\frac{1}{2}, \frac{1}{3}, \frac{1}{5}), (-\frac{1}{2}, \frac{1}{3}, \frac{1}{4}), (-\frac{1}{2}, \frac{1}{3}, \frac{1}{3})$$

admit tight contact structures at all?

It is believed, at least by some, that these manifolds do admit tight contact structures. This result would be particularly interesting, because then these would be the first known manifolds admitting tight contact structures but not admitting any fillable structures.

As for hyperbolic manifolds, very little is known about the existence of tight contact structures. There are no known constructions of contact structures in terms of hyperbolic structures. It was once conjectured that all hyperbolic manifolds admit taut foliations. If this were true, we could perturb them into tight contact structures, but it has recently been shown that there are hyperbolic 3-manifolds without taut foliations [**6, 70**]. It is still possible, however, that all hyperbolic manifolds admit a tight contact structure.

Another important question is:

QUESTION 9. *Do all Haken manifolds admit a tight contact structure?*

Any manifold with $b_1 \geq 1$ is known to be Haken, and from the work of Eliashberg-Thurston [**26**] and Gabai [**40**], it is known that these manifolds admit semi-fillable, universally tight contact structures; see also [**54**] for a proof that avoids foliation theory. Thus our main interest here again is rational homology spheres.

2. Uniqueness and classification

Of course the main problem here is:

PROBLEM 10. *Classify tight contact structures on all 3-manifolds.*

Currently contact structures are classified on the following manifolds:

(1) $S^3, S^2 \times S^1$ [**20**]
(2) Lens spaces [**30, 43, 50**]
(3) Torus bundles [**43, 51**]
(4) Circle bundles [**44, 51**].

The classification is also known on some Seifert fibered spaces. For example, $\#\mathit{Tight}(\Sigma(2,3,5)) = 0$ or 1, depending on orientation, where $\mathit{Tight}(M)$ denotes the set of tight contact structures up to isotopy. See [**41**] for further results on Seifert fibered spaces.

A specific problem whose resolution would have great importance for the classification of contact structures is:

PROBLEM 11. *Classify tight contact structures on* $\Sigma\times[0,1]$, *where* Σ *is a surface of genus* $g > 1$.

In [**55**], Honda, Kazez, and Matić classify tight contact structures on $\Sigma \times [0,1]$, when the relative Euler class of the contact structure is extremal, $e(\xi)(\Sigma) = \pm(2g-2)$, and some mild conditions hold on the dividing curves. It seems much more difficult to understand the situation when the Euler class is not extremal.

This problem has relevance to the classification of tight contact structures on:

(1) surface bundles over circles,
(2) 3-manifolds with Haken decompositions, and
(3) general 3-manifolds.

The application of Problem 11 to surface bundles is obvious: given a tight contact structure on a surface bundle, split it along a convex fiber; the result is $\Sigma\times[0,1]$, and the dividing curves (or characteristic foliation) on the boundary components are related by the monodromy of the bundle. Now if one could normalize the dividing curves on a surface fiber, then one could apply the solution to Problem 11 to obtain an upper bound on the number of tight contact structures on the surface bundle. To get a lower bound, one could use Legendrian surgery and "state transition," see [**52**].

A *Haken decomposition* of a 3-manifold is a way of successively cutting the manifold along incompressible surfaces until one arrives at a disjoint union of three-balls. Haken decompositions have been very useful in studying manifolds that support them [**40, 75**]. Honda, Kazez, and Matić [**54**] have used these decompositions to construct tight contact structures on manifolds with $b_1 \geq 1$. The hardest step in this process is understanding the first cut. This would be greatly facilitated by answering Problem 11; it might then be possible to generalize the results of [**54**].

To study a general 3-manifold, one can use *Heegaard decompositions.* Any 3-manifold can be obtained by gluing two handlebodies together; such a representation of a 3-manifold is called a Heegaard decomposition. These decompositions have been very useful in studying 3-manifold topology (e.g., branched covers, Casson invariants, ...). So to understand tight contact structures on a 3-manifold, one might first want to consider:

PROBLEM 12. *Classify tight contact structures on handlebodies.*

For handlebodies of genus 1, this has been done [**43, 50**]. Using techniques from [**50**], it is not hard to give an upper bound on the number of tight contact structures on a handlebody with fixed dividing curves on its boundary, but an actual classification seems difficult. If one could solve this problem, then to understand tight structures on an arbitrary 3-manifold, one would have to understand how to glue together tight structures on the handlebodies. Honda has developed a method called "state transition" for gluing tight contact structures together. In trying to implement this method, a solution to Problem 11 would be helpful.

Short of a complete classification, we can ask more specific questions. For example, we can start with the following motivational question.

QUESTION 13. *Which manifolds admit only finitely many tight contact structures?*

In [**9, 53**], it has been shown that any closed oriented irreducible toroidal 3-manifold admits infinitely many universally tight contact structures. Moreover, it has recently been announced, by Colin, Giroux, and Honda, that any closed oriented atoroidal 3-manifold admits only finitely many tight contact structures (for this statement, it is important to keep in mind that 0 is a finite number!). In light of these results, the above question, as stated, has been answered; however, there are, of course, many interesting refinements of this question. Foremost among them is:

QUESTION 14. *Do all 3-manifolds admit only finitely many virtually overtwisted contact structures?*

In [**9, 53**], the infinity of universally tight contact structures were constructed by first constructing one special universally tight contact structure, in which the incompressible torus T has a neighborhood contactomorphic to $(T \times (-1,1), \xi_1)$, where

$$\xi_n = \ker(\sin(2\pi nz)dx + \cos(2\pi nz)dy),$$

x, y are coordinates on T, and z is the coordinate on $(-1,1)$. Then one obtains infinitely many tight contact structures by replacing $(T \times (-1,1), \xi_1)$ with $(T \times (-1,1), \xi_n)$. Intuitively, one is making the contact structure twist more near T. When doing this, one must prove that all the contact structures obtained are universally tight, and that they are different. To prove tightness, it is important that we begin with a special tight contact structure. To conclude that there are infinitely many different tight contact structures, one needs to know that the "twisting along T is finite." This "twisting" is measured in [**9, 11**] using an invariant called the torsion of T (see below), and in [**53**] using the twisting of various Legendrian knots transverse to T. In any event, it is not clear that a similar construction can be performed on virtually overtwisted contact structures.

The *torsion* of a torus T in a contact manifold (M, ξ) is the largest n for which $(T \times (-1,1), \xi_n)$ can be contact embedded in M so that $T \times \{0\}$ is isotopic to T;

if no such ξ_n can be embedded, the torsion is defined to be 0. We would like to answer the following questions.

QUESTION 15. *Are there only finitely many contact structures with minimal torsion?*

QUESTION 16. *Is torsion always finite in a tight contact structure?*

For example, if we drop the hypothesis that the torus is normal in Colin's theorem [**9**], does an arbitrary tight ξ have finite torsion? Note that for the infiniteness theorems of Colin and Honda-Kazez-Matić, which rely, more or less, on torsion being finite, the contact structures constructed are very special. We can also ask:

QUESTION 17. *How are torsion and fillability related?*

In many specific examples (e.g., the tight contact structures on T^3), we know that only the tight contact structures with minimal torsion are strongly fillable, while the rest are only weakly fillable [**22**]. Is this always the case? Specifically, we ask:

QUESTION 18. *Can large torsion sometimes imply no filling?*

When studying 3–manifolds one usually decomposes them in various ways. First, since any orientable 3–manifold has a unique prime decomposition, one usually restricts to irreducible manifolds. One may also do this when considering tight contact structures since for a connected sum $M_1 \# M_2$, we know $Tight(M_1 \# M_2) = Tight(M_1) \times Tight(M_2)$; see [**8**]. Next, a 3–manifold is usually decomposed along incompressible tori. Specifically, given an irreducible orientable 3–manifold M, there is a collection of incompressible tori $T_1, \ldots, T_k$ such that the components of $M \setminus \cup T_i$ are atoroidal or Seifert fibered; moreover, this collection is unique if it is minimal with respect to this property. These tori give the Johannson-Shalen-Jaco (JSJ) decomposition of M.

QUESTION 19. *Is there a contact JSJ decomposition?*

More explicitly, suppose that $T_1, \ldots, T_k$ is a minimal system of tori given by the JSJ decomposition. Can we classify contact structures on M if we have a classification on the pieces of $M \setminus (\cup T_i)$?

A related question is:

QUESTION 20. *Suppose two embeddings $(T^2 \times I, \xi_i) \subset (M, \xi)$ are smoothly isotopic and maximal. Are they necessarily contact isotopic?*

3. Convex contact structures and open book decompositions

A vector field v on a contact manifold (M, ξ) is called a *contact vector field* if the flow of v preserves ξ. The contact structure ξ is called *convex* if there is a gradient like contact vector field on M. Recently Giroux has proved (using ideas from [**42**]) that any contact structure on a closed 3-manifold is convex. This gives us another tool to study contact structures.

QUESTION 21. *How are properties like fillability and virtual overtwistedness reflected in convex vector fields associated to a given contact structure?*

For example:

QUESTION 22. *Does the dynamics of the flow of a contact vector field say anything about the contact structure?*

Giroux [**45**] has provided an interesting approach to Question 21. Recall that an *open book decomposition* of a closed 3-manifold M is a fibration of the complement of a link L, called the *binding* of the open book. If F is the fiber of the fibration of $M \setminus L$, then $M \setminus L$ is obtained from $F \times [0,1]$ by gluing $F \times \{1\}$ to $F \times \{0\}$ by a diffeomorphism $\phi : F \to F$ called the *monodromy*. Following Thurston and Winkelnkemper [**76**], one can find a 1-form λ on F such that $d\lambda$ is a volume form on F. Then $\alpha = dt + t \cdot \lambda + (1-t) \cdot \phi^* \lambda$, $t \in [0,1]$, is a contact form on $F \times [0,1]$ that descends to a contact form on $M \setminus L$, and it is not hard to extend α over M. The contact structure $\xi = \ker \alpha$ is said to be *adapted* to the open book decomposition. Giroux has shown how to construct a contact vector field on (M, ξ) that is transverse to the fibers of the open book. Thus from an open book decomposition, one gets an adapted convex contact structure. Moreover, given a convex contact structure, Giroux can construct an open book decomposition to which the contact structure is adapted. Hence there is a one-to-one correspondence between convex contact structures and open book decompositions.

Generalizing work of Loi and Piergallini [**62**] (see also [**2, 63, 66**]), Giroux can show that a contact structure is Stein fillable if and only if it is adapted to an open book decomposition whose monodromy is a product of positive Dehn twists. Thus we can refine Question 21 to:

QUESTION 23. *What conditions on the link or monodromy map of an open book decomposition imply that the adapted contact structure is tight? (Weakly or strongly) symplectically fillable? Universally tight? Virtually overtwisted?*

Fix M and a contact structure ξ. Let $\mathcal{B}(\xi)$ be the set of all open book decompositions to which ξ is adapted. Let $g(\xi) = \min\{\text{genus}(F)\}$, where the minimum is taken over all fibers F of fibrations in $\mathcal{B}(\xi)$.

QUESTION 24. *Can $g(\xi)$ be effectively computed for any class of contact structures? If so, is it an effective invariant of the contact structure? Is it related to any properties of ξ?*

QUESTION 25. *What is the minimum of $g(\xi)$ over all tight structures (or over all universally tight structures, etc.) on M?*

We also ask:

QUESTION 26. *Can the contact homology or symplectic field theory* [**25**] *of a contact structure be computed in terms of an adapted open book decomposition?*

Since a contact structure can be reconstructed from an adapted open book decomposition, it is clear that contact homology can be computed from the open book data. We would like to see an effective algorithm that takes dynamical information about the monodromy (and/or topological information about the binding) of the open book decomposition and outputs the contact homology.

Open book decompositions have been used to study the topology of 3-manifolds for quite some time now. So we end this section with a very open ended problem.

PROBLEM 27. *Find purely topological applications of Giroux's correspondence between convex contact structures and open book decompositions.*

Though answering this problem might seem overly optimistic, Giroux and Goodman [**45**] have been able to answer some longstanding questions about fibered links using this machinery.

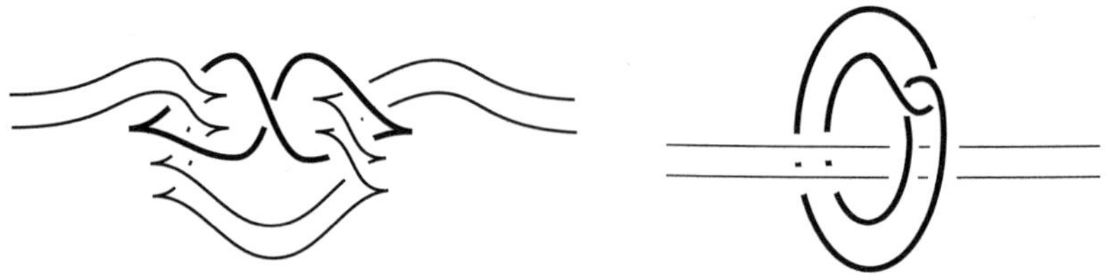

FIGURE 1. Legendrian surgery on the lighter curves on the left can yield a virtually overtwisted contact structure. The darker curve bounds an immersed disk, as shown on the right, with respect to which its twisting number is 0.

4. Contactomorphisms

There is very little known concerning the group of contactomorphisms of a given contact structure.

PROBLEM 28. *Given a contact structure ξ on M, understand the group Diff(M, ξ) of contactomorphisms of ξ.*

Several interesting subproblems are:

PROBLEM 29. *Compute $\pi_0(\mathit{Diff}(M, \xi))$.*

PROBLEM 30. *Study the inclusion map from Diff(M, ξ) to Diff$^+(M)$, the group of orientation-preserving diffeomorphisms.*

For S^3 with the standard tight contact structure, this map is an isomorphism on the π_0 level. What can one say about higher homotopy groups? The inclusion is not surjective on π_0 for tight contact structures on T^3, since a preferred direction of "twisting" is picked out by a contact structure. We also know that for (T^3, ξ_n), the kernel of the inclusion on π_0 contains at least n elements.

PROBLEM 31. *Compute Diff(M, ξ) when ξ is overtwisted.*

Eliashberg's work in [**17**] should be very helpful here, but there are technical hypotheses in most of the theorems in [**17**] which make it difficult to address this problem directly. In [**16**], it has been shown that π_0 for an overtwisted contact structure contains at least two elements. In order to address Problem 31 using [**17**], one might first want to consider the following.

QUESTION 32. *Does there exist an overtwisted contact structure with two overtwisted disks which are not contact isotopic?*

Note that Eliashberg's theorem does not state that homotopy classes of plane fields are in one-to-one correspondence with overtwisted contact structures; this is only true at the π_0 level.

5. Miscellaneous

Gompf [**46**] has shown that if a contact structure is obtained from Legendrian surgery on a Legendrian knot as in Figure 1, and the immersed disk on the left of Figure 1 lifts to an embedded disk in a finite cover, then the contact structure is virtually overtwisted. The reason is that there is a Hopf-like unknot bounding an immersed overtwisted disk that can be unwound in a cover.

QUESTION 33. *Can we always find such an immersed disk in a virtually overtwisted contact structure?*

If the virtually overtwisted phenomenon could be sufficiently well understood, one could try to address the following question.

QUESTION 34. *Given a virtually overtwisted contact structure, is there some procedure to find an associated universally tight structure?*

Said another way, are all contact structures obtained from universally tight contact structures via some surgery-type construction?

When we perform a Lutz twist on a contact structure, a tube of overtwisted disks is produced. This raises the natural question:

QUESTION 35. *Can an overtwisted disk always be completed to an overtwisted tube?*

This is unlikely, given the fact that $-P$ does not support a tight contact structure. However, perhaps the putative overtwisted tube is only immersed, while each overtwisted disk is actually embedded.

QUESTION 36. *If a contact structure has an overtwisted tube, is there an un-Lutz-twist which yields a tight structure?*

II. Legendrian Knots

Please refer to the article by Etnyre [**31**], elsewhere in this volume, for introductory definitions concerning Legendrian and transverse knots. Unless otherwise stated, the ambient contact manifold is assumed to be (S^3, ξ_{std}). Sections 6 through 8 deal with the problem of classifying Legendrian and transverse knots; Sections 9 and 10 address applications of these knots to contact geometry and topology.

6. Examples

If $\mathcal{K}$ is a topological knot type in any contact 3-manifold, let $\mathcal{L}(\mathcal{K})$ denote the set of Legendrian knots of type $\mathcal{K}$. The main problem in Legendrian knot theory is the following.

PROBLEM 37. *Given $\mathcal{K}$, classify the Legendrian knots in $\mathcal{L}(\mathcal{K})$ up to Legendrian isotopy.*

In the case (S^3, ξ_{std}), recall that there are two "classical invariants" of Legendrian knots, tb and r. A knot type is called *Legendrian simple* if two Legendrian knots of this type with equal classical invariants must be Legendrian isotopic. For Legendrian simple knots, classification is concluded by calculating precisely which values of (tb, r) can be achieved.

All of the prime knot types which we can presently classify are Legendrian simple:

(1) the unknot in any tight contact structure [**24**]
(2) torus knots in any tight contact structure [**33**]
(3) the figure eight knot in $(S^3, \xi_{\mathrm{std}})$ [**33**].

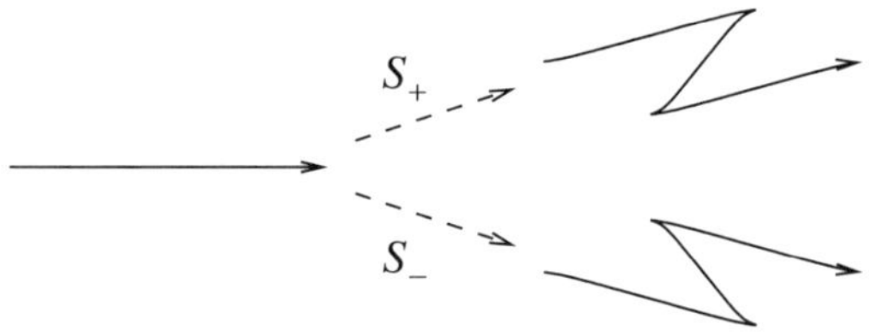

FIGURE 2. Stabilization of a Legendrian knot replaces a segment of the knot's front (yz) projection with a zigzag as shown. There are two different stabilizations $S_\pm$, which change r by ± 1.

We can also classify Legendrian knots in a non-prime knot type if we know the classification for the prime summands [**35**]; see Section 7. Thus, in our discussions below, we will usually restrict attention to prime knot types. In addition, for all of the above knot types, any Legendrian knot which does not maximize tb in its knot type can be obtained from a maximal-tb knot through the operation of stabilization, which adds zigzags to the front (yz) projection of the knot; see Figure 2. (It is straightforward to show that the stabilizations $S_\pm$ are well-defined operations up to Legendrian isotopy.) The classification results of [**33, 35**], and probably future classifications, rely on the powerful theory of convex surfaces developed by Giroux; see Etnyre's article [**31**].

Not all knot types are Legendrian simple, as we will see in Section 8. For instance, the 5_2 knot shown in Figure 4 (a few pages hence) is not Legendrian simple. (Here we use the usual knot terminology from, e.g., [**71**].)

We digress for a moment to discuss the situation of transverse knots. Call a knot type *transversely simple* if all transverse knots in this knot type are completely determined by their self-linking number. Using characteristic foliation techniques, Eliashberg [**21**] showed that the unknot is transversely simple; similarly, Etnyre [**29**] and Etnyre-Honda [**33**] have shown that torus knots and the figure eight knot are transversely simple. Through a completely different approach, using braid foliations, Birman and Wrinkle [**5**] have proved that all so-called exchange reducible knots are transversely simple; by work of Menasco [**64**], these include all iterated torus knots. Birman and Menasco [**4**] have adapted difficult results in braid theory to construct families of knots which are not transversely simple.

Further results about transverse knots seem likely to arise from their connection with Legendrian knots. Given an oriented Legendrian knot K, we obtain two transverse knots K_+ and K_- by pushing the knot off of itself in a direction tangent to the contact field so that the orientation induced on K_+ and K_- as transverse knots agrees/disagrees with the orientation on K. (Note there is some discrepancy in the literature as to the sign convention. The one given here seems natural from many perspectives, but the other choice is also frequently used, e.g. in [**3**].) Conversely, any transverse knot is transverse isotopic to a knot thus obtained. Now if we denote by $S_\pm$ the operations of positive and negative stabilization on Legendrian knots, then two transverse knots $(K_1)_+, (K_2)_+$ are transverse isotopic if and only if the Legendrian knots K_1, K_2 are stably isotopic, i.e., $S_-^n K_1, S_-^n K_2$ are isotopic for some $n \geq 0$; see, e.g., [**27, 33**]. It follows that classifying transverse knots up to transverse isotopy is equivalent to classifying Legendrian knots up to stable isotopy.

In trying to attack the main Legendrian problem, we would like to mention some particularly important subproblems.

PROBLEM 38. *Classify Legendrian knots in $\mathcal{L}(\mathcal{K})$, where*

(1) *$\mathcal{K}$ is a fibered knot;*
(2) *$\mathcal{K}$ is a hyperbolic knot;*
(3) *$\mathcal{K}$ is not Legendrian simple;*
(4) *$\mathcal{K}$ is any knot type in any overtwisted contact structure.*

Here are some remarks about these subproblems.

Problem 38 (1): It is currently possible (though not likely) that prime fibered knots are always Legendrian simple (for non-prime knots see below). We note that the only fibered knots of genus 1 are the figure eight knot and the right- and left-handed trefoil knots, and Legendrian knots in all of these knot types have been classified. The next step is to look at genus 2 fibered knots. It is probable that the techniques in [**33**] can be used in this situation (at least sometimes), but it seems that some new ideas will be needed as the genus of the knot grows.

Problem 38 (2): We currently know very little about the Legendrian classification of hyperbolic knots; so far, we only understand Legendrian figure eight knots. As mentioned in Section 1, it would be very helpful to understand interactions between hyperbolic and contact geometry, and some clues to this interaction could come from understanding some more Legendrian knots in hyperbolic knot types. In particular, it would be great to have a Legendrian classification in a non-fibered hyperbolic knot type. However, it might be easier to try knot types that are both hyperbolic and fibered. W. Kazez has suggested looking at the knot 8_{20}; this is a fibered hyperbolic knot whose holonomy seems to be interesting.

Problem 38 (3): There are currently no prime knot types $\mathcal{K}$ that are not Legendrian simple for which we can classify $\mathcal{L}(\mathcal{K})$. The most obvious one to consider is the 5_2 knot type.

CONJECTURE 39. *There are exactly two Legendrian knots realizing the* 5_2 *knot type with* $\text{tb} = 1$ *and* $r = 0$. *All other Legendrian knots realizing* 5_2 *are stabilizations of these, and are determined by their* tb *and* r.

Problem 38 (4): It is a surprising fact that much progress has been made towards classifying Legendrian knots in tight contact structures, but little is known about Legendrian knots in overtwisted contact structures. Call a Legendrian knot K in an overtwisted contact structure *loose* if the contact structure on the complement of L is overtwisted. It seems that most Legendrian knots in overtwisted contact structures are loose, but there are also non-loose Legendrian knots [**16**]. If two loose Legendrian knots have the same knot type and tb and r, then they are Legendrian isotopic [**16, 24**]. Thus it would be interesting to know:

QUESTION 40. *For what contact manifolds and knot types $\mathcal{K}$ are all Legendrian knots in $\mathcal{L}(\mathcal{K})$ loose?*

For example, let ξ' be the contact structure on S^3 obtained from the tight contact structure by applying one Lutz twist.

CONJECTURE 41. *If ξ is any overtwisted contact structure on S^3 besides ξ', then all Legendrian unknots are loose.*

CONJECTURE 42. *Any Legendrian unknot in ξ' with* tb $\neq 1$ *or* $r \neq 0$ *is loose, and there are exactly two Legendrian unknots with* tb $= 1$ *and* $r = 0$, *one loose, one not.*

As a final remark, we note that any Legendrian knot that violates the Bennequin inequality is automatically loose.

7. Structure Results

A prototypical structure result for Legendrian knots concerns connected sums. Given two Legendrian knots $L_1 \in \mathcal{L}(\mathcal{K}_1)$ and $L_2 \in \mathcal{L}(\mathcal{K}_2)$, we can form their connected sum $L_1 \# L_2$ in a standard way (see, e.g., [**12**]). In [**35**] it is shown that connected sum of Legendrian knots gives a bijection

$$\left(\frac{\mathcal{L}(\mathcal{K}_1) \times \mathcal{L}(\mathcal{K}_2)}{\sim}\right) \longrightarrow \mathcal{L}(\mathcal{K}_1 \# \mathcal{K}_2),$$

where the equivalence relation $\sim$ is of two types:

(1) $(L_1, S_\pm(L_2)) \sim (S_\pm(L_1), L_2)$,
(2) $(L_1, L_2) \sim (L_2, L_1)$ if $\mathcal{K}_1 = \mathcal{K}_2$.

Thus we can classify $\mathcal{L}(\mathcal{K}_1 \# \mathcal{K}_2)$ in terms of $\mathcal{L}(\mathcal{K}_1)$ and $\mathcal{L}(\mathcal{K}_2)$. Using this result and the classification of Legendrian negative torus knots, one may easily find non-Legendrian-simple fibered knot types $\mathcal{K}$ for which we can completely classify $\mathcal{L}(\mathcal{K})$.

Virtually nothing is known about the structure of $\mathcal{L}(\mathcal{K})$ for general knot types $\mathcal{K}$. One important notion here is *destabilizing* a Legendrian knot, i.e., isotoping it to become a stabilization of another Legendrian knot, and then removing the stabilization; this raises tb by 1.

QUESTION 43. *Do all Legendrian knots destabilize until they reach the maximal possible* tb *for their knot type?*

It seems unlikely that the answer to this question is yes in general, but it is yes for all knot types for which the question has been answered [**33**]. Question 43 is important since the method used in all current classification results is to show Legendrian knots destabilize to maximal tb knots, and then to classify these. This scheme would have real difficulties if the answer to the following question were yes.

QUESTION 44. *Is there a knot type with Legendrian representatives that do not destabilize but have arbitrarily negative* tb*?*

Give a knot type $\mathcal{K}$, let $\mathcal{K}_{(p,q)}$ be the (p,q)-cable of $\mathcal{K}$; that is, a representative of $\mathcal{K}_{(p,q)}$ sits on the boundary of a tubular neighborhood of a representative of $\mathcal{K}$ as a (p,q) curve. The techniques in [**33**] should allow one to address the following.

PROBLEM 45. *Classify Legendrian knots in $\mathcal{L}(\mathcal{K}_{(p,q)})$ in terms of $\mathcal{L}(\mathcal{K})$.*

In addition to connected sum and cabling, there are many other operations one can perform on knots.

QUESTION 46. *What can one say about Legendrian knots that are in a knot type obtained from*

(1) *Whitehead doubling or some other satellite construction?*
(2) *Murasugi sums?*
(3) *your other favorite topological abuse of knots?*

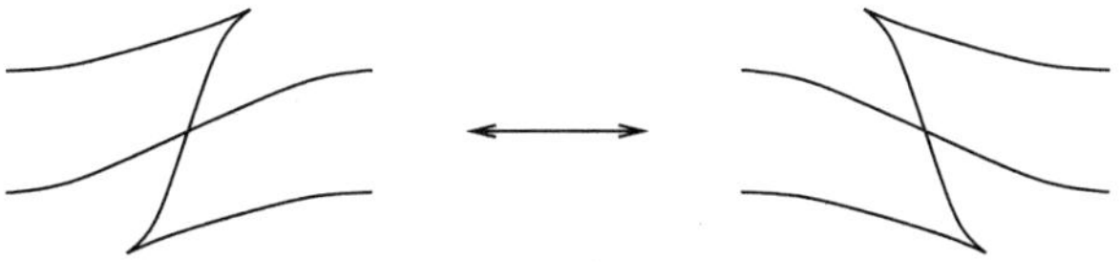

FIGURE 3. Basic move preserving tb and topological class, but not necessarily Legendrian isotopy class.

For background on Legendrian satellites, see, e.g., [**68**]. It is plausible that Whitehead doubling two non-isotopic Legendrian knots leads to non-isotopic Legendrian knots; if the original knots have the same tb, then the doubles have the same tb and r. Examples similar to the non-isotopic knots of Figure 4 can be constructed by doubling unknots with the same tb but different r.

Here are some other interesting structural questions.

QUESTION 47. *How are the Legendrian knots in a knot type affected by crossing changes of the knot type?*

For example, is there a formula for how the maximal value of tb changes under crossing changes?

QUESTION 48. *If K_1 and K_2 are topologically isotopic with the same* tb, *is there a set of local moves which, along with Legendrian isotopy, will send K_1 to K_2?*

A basic move suggested by L. Ng is given in Figure 3. This move seems to suffice in many examples. However, there are examples in which this is not sufficient, as one can show using the characteristic algebra (see Section 8).

A result of Fuchs and Tabachnikov [**39**] states that two topologically isotopic Legendrian knots K_1 and K_2, with identical classical invariants, must have $S_+^{n_1}S_-^{n_2}K_1$ and $S_+^{n_1}S_-^{n_2}K_2$ Legendrian isotopic for some $n_1, n_2 \geq 0$. (Note that we need both positive and negative stabilizations.) By considering Legendrian knots that are topologically connected sums of negative torus knots, one sees that the number of necessary stabilizations can be arbitrarily large [**35**], but one can ask:

QUESTION 49. *Are there bounds on the number of stabilizations necessary to make two Legendrian knots isotopic, in terms of the topology of the knot type?*

8. Legendrian invariants and contact homology

The first example of a knot type which is not Legendrian simple was produced independently by Chekanov [**12**] and Eliashberg and Hofer [**23**]. As mentioned before, the knot type is 5_2; the two non-isotopic Legendrian 5_2 knots with equal classical invariants are shown in Figure 4.

The method used to distinguish between the two 5_2 knots is a new, powerful invariant of Legendrian knots, motivated by the development of contact homology [**23**], which applies Gromov's holomorphic-curve techniques to contact manifolds. This invariant takes the form of a *differential graded algebra* (DGA), a free non-commutative unital algebra with a grading and differential; the homology of the algebra is precisely the contact homology of S^3 relative to the Legendrian knot (see [**36**]).

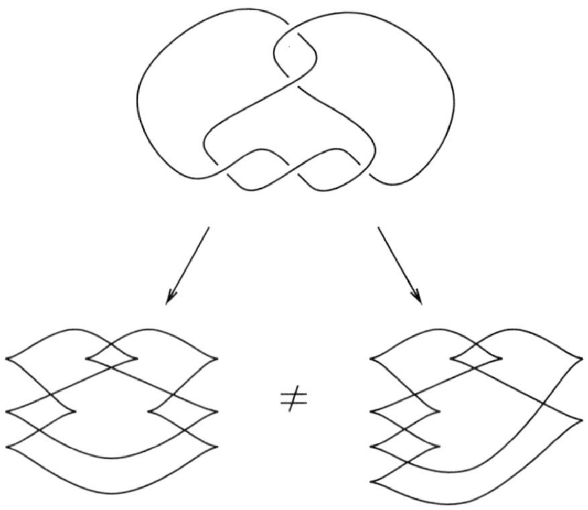

FIGURE 4. Example, due to Chekanov and Eliashberg-Hofer, of two Legendrian knots, both of topological type 5_2 and with $tb = 1$ and $r = 0$, which are not Legendrian isotopic.

Although contact homology is difficult to compute in general, this relative contact homology has a simple combinatorial formulation due to Chekanov [**12**]; in addition, it can be treated completely combinatorially, without reference to the general theory of contact homology. Chekanov's definition allows us to write down the DGA invariant explicitly, given the knot diagram of a Legendrian knot. Two Legendrian knots are then isotopic only if they have equivalent DGAs, under an algebraic equivalence called "stable tame isomorphism." The formulation of the DGA has subsequently been refined in [**36, 65, 67**].

In practice, because of the difficulty of manipulating the DGA directly, applications have used simpler invariants derived from the DGA. The easiest are Poincaré-type polynomials which calculate the graded dimension of linearizations of the DGA. Chekanov used these polynomials to show that the knots in Figure 4 are not Legendrian isotopic. Along the same lines, it can be shown [**12, 27**] that there exist knot types and tb and r for which there are arbitrarily many non-isotopic Legendrian representatives.

A more involved DGA-derived invariant is the characteristic algebra [**67**]. This can be used to answer (in the negative) the Legendrian mirror question of Fuchs and Tabachnikov [**39**], which asks whether a Legendrian knot with $r = 0$ must necessarily be Legendrian isotopic to its mirror, its image under the contactomorphism $(x, y, z) \mapsto (x, -y, -z)$. Further applications of the characteristic algebra are given in [**67**], including knots distinguished without using any sort of grading. It seems possible that the characteristic algebra is also linked to topology.

CONJECTURE 50 (Ng). *The abelianized characteristic algebra is a topological invariant. More precisely, it depends only on $\mathcal{K}$ and* tb.

QUESTION 51. *Fully understand the space of DGAs modulo stable tame isomorphism. In particular, are there other useful invariants which can be derived from the DGA?*

Along these lines, it would be particularly nice to answer:

QUESTION 52. *If two DGAs have the same homology, are they stable tame isomorphic?*

Chekanov and Pushkar [**13**] have discovered another non-classical invariant of Legendrian knots, using admissible decompositions of fronts. No applications of this invariant are currently known, besides examples in which the Poincaré-polynomial technique can already be used. In fact, the following seems possible; see [**38**] for a result in this direction.

CONJECTURE 53. *The Chekanov-Pushkar invariant can be deduced from Poincaré polynomials for the corresponding DGA.*

Analogues of the DGA theory can be derived for contact manifolds other than (S^3, ξ_{std}).

PROBLEM 54. *Construct combinatorial relative contact homology theories for Legendrian knots in other contact manifolds.*

This has been done for the solid torus $S^1 \times \mathbb{R}^2$ with the standard tight contact structure by Ng and Traynor, and, with more difficulty, for circle bundles over closed orientable surfaces with a contact structure transverse to the fibers [**73**]. We note that for $S^1 \times \mathbb{R}^2$, Traynor [**77**] has defined another non-classical invariant for a particular class of links, using generating functions and algebraic topology; mysteriously, for these links, all current calculations indicate that this invariant coincides with a link-graded Poincaré polynomial derived from the DGA.

We now return to (S^3, ξ_{std}). One major remaining difficulty with the DGA invariant is that it vanishes for stabilized Legendrian knots.

PROBLEM 55. *Are there useful contact-homology-type invariants for stabilizations?*

This problem is of particular importance in trying to understand transverse knots; see Section 6.

One possible approach to understanding invariants for stabilizations is to use satellite constructions, which in many cases produce knots with nontrivial DGAs. The simplest possible satellite, the Whitehead double, yields no interesting invariants for stabilizations [**68**], but more complicated satellites may encode useful information.

Another possible approach is to apply the symplectic field theory of Eliashberg-Givental-Hofer [**25**], which generalizes contact homology, to Legendrian knots.

PROBLEM 56. *Combinatorially understand the full symplectic field theory for Legendrian knots.*

The algebraic structure behind absolute symplectic field theory is somewhat well understood [**25**], but this is far from the case for the relative symplectic field theory needed for the Legendrian case. Here, the algebraic formalism has resisted several attempts at a reasonable definition.

Given a Legendrian knot K in (M, ξ), there is a natural $(\mathrm{tb}(K) - 1)$ surgery on K, called *Legendrian surgery*, which produces another contact manifold; see [**18, 46, 78**] for background.

PROBLEM 57. *Let K be a Legendrian knot inside (S^3, ξ_{std}), and let (M, ξ) be the Stein fillable contact structure obtained by performing Legendrian surgery on K. Is there a way to calculate the contact homology of (M, ξ) from the DGA invariant associated to K?*

Though finding such a relation would be great, it seems unlikely that one can determine the contact homology of (M, ξ) merely from the DGA of K; however, one can almost certainly compute the contact homology of (M, ξ) from the information which would define the symplectic field theory of K. Thus it is probably more realistic to try to solve Problem 56 first before attacking this problem. As a possibly simpler first step, we ask the following.

QUESTION 58. *If (M, ξ) in Problem 57 is obtained from a stabilized Legendrian knot, does the contact homology of (M, ξ) have a simple form? (Is is almost trivial? Does it equal the base field?)*

9. Relation to contact geometry

We now turn from intrinsic properties of Legendrian knots to applications of Legendrian knots to contact geometry. Bennequin [**3**] famously established the existence of an exotic contact $\mathbb{R}^3$ by studying transverse unknots, and Legendrian knots have played a similar role in distinguishing between contact structures on the 3-torus [**56**] and homology spheres [**1**]. In each case, one investigates what sorts of Legendrian (or transverse) knots can be carried by the contact structure.

PROBLEM 59. *Distinguish other contact structures using Legendrian knots.*

A more precise version of this problem might be given by the following question.

QUESTION 60. *Can tight contact structures ξ_1 and ξ_2 on the same manifold M always be distinguished by finding a Legendrian knot of a certain* tb *(or twisting number) and r which exists in one, but not in the other?*

For example, the tight contact manifolds $(T^3 = \mathbb{R}^3/\mathbb{Z}^3, \xi_n)$, $n \in \mathbb{Z}^+$, given by the 1-forms $\alpha_n = \sin(2\pi nz)dx + \cos(2\pi nz)dy$, are distinguished by the maximal twisting number among curves isotopic to $x = y = const$ [**56**]. More generally, tight contact structures on T^2 bundles over S^1 are determined by their homotopy class of plane field and by information about Legendrian knots [**51**].

It would be interesting to see if the Lisca-Matić examples [**60**] of contact structures can also be distinguished in this way. These are tight structures on the Brieskorn sphere $\Sigma(p, q, pqn-1)$, obtained by surgery on distinct Legendrian knots. Via Seiberg-Witten theory, the contact structures are distinguished by the Chern classes of Stein 4-manifolds which bound $\Sigma(p, q, pqn-1)$. Could one use Legendrian knot theory instead of Seiberg-Witten theory to distinguish these examples?

We next address the effect of Legendrian surgery on contact structures. It is known [**18**] that Legendrian surgery on a knot in a fillable contact structure produces another fillable structure.

QUESTION 61. *Does Legendrian surgery preserve tightness?*

If we allow manifolds with boundary, the answer is no [**52**] (see also [**10**]). Nothing is known for the closed case.

QUESTION 62. *Consider two Legendrian knots K_1, K_2 with the same* tb *and* r *which are not isotopic. Let (M, ξ_1) and (M, ξ_2) be the contact manifolds obtained by performing Legendrian surgery on K_1 and K_2, respectively. Are ξ_1 and ξ_2 necessarily isotopic?*

One might like to begin working on this problem by considering the maximal tb examples of the 5_2 knot.

Here are some final questions.

QUESTION 63. *Let K be a maximal* tb *Legendrian knot in (S^3, ξ_{std}). Is $(S^3 \setminus K, \xi_{std}|_{S^3 \setminus K})$ universally tight?*

QUESTION 64. *If K has non-maximal* tb *but is not a stabilization (cf. Question 43), is $(S^3 \setminus K, \xi_{std}|_{S^3 \setminus K})$ universally tight?*

Though it is possible for the answers of these questions to be yes, that would be quite amazing. More realistically, we can ask:

QUESTION 65. *If K is a maximal* tb *Legendrian knot in (S^3, ξ_{std}), is the $\mathbb{Z}$-cover of $S^3 \setminus K$ tight?*

Understanding these questions could be helpful in addressing Question 61, as well as providing a connection to the topology of the knot.

10. Relation to topology

Legendrian knot theory has deep connections with topology as well as with contact geometry, principally through the information encoded in the Thurston-Bennequin invariant of a Legendrian knot in (S^3, ξ_{std}). For a knot type $\mathcal{K}$, let $\overline{\text{tb}}(\mathcal{K})$ denote the maximal tb over all Legendrian knots of type $\mathcal{K}$. This number is finite; see the next paragraph.

Bennequin's proof of the existence of an exotic contact structure on S^3 reduces, in modern terminology, to establishing that the unknot in (S^3, ξ_{std}) satisfies $\overline{\text{tb}} = -1$. His method extends to an upper bound on $\overline{\text{tb}}(\mathcal{K})$ for all $\mathcal{K}$, in terms of the genus of $\mathcal{K}$. Since then, upper bounds for $\overline{\text{tb}}$ have been obtained in terms of the slice genus, HOMFLY polynomial, and Kauffman polynomial; see [**37**] for a survey of results. In particular, Rudolph [**72**] (q.v. [**57, 60**]) has shown that if $\overline{\text{tb}}(\mathcal{K}) \geq 0$, then $\mathcal{K}$ is not slice.

In this vein, consider the smooth knot concordance group $\mathcal{C}$; this is a direct sum of countably many $\mathbb{Z}$, $\mathbb{Z}/2\mathbb{Z}$, and possibly $\mathbb{Z}/4\mathbb{Z}$ factors. It is currently unknown whether there are any, or even countably many, $\mathbb{Z}/4\mathbb{Z}$ factors. One can see (at least some of) the $\mathbb{Z}/2\mathbb{Z}$ factors from the fact that amphichiral knots have order 2. The result of Rudolph allows us to see some of the $\mathbb{Z}$ factors as well, through Legendrian knot theory; if $\overline{\text{tb}}(\mathcal{K}) \geq 0$, then $\mathcal{K}$ is of infinite order in $\mathcal{C}$.

QUESTION 66. *Can we use Legendrian knots to improve or reprove other results concerning slice knots?*

See, e.g., [**7, 61**]. One might be able to prove results such as the following.

CONJECTURE 67 (Etnyre). *A knot is either of order 2, or concordant to a knot for which either it or its mirror has $\overline{\text{tb}} \geq 0$.*

We can consider topological properties besides sliceness, as well.

PROBLEM 68. *Relate Legendrian knot invariants to the Alexander module.*

Questions 63, 64, and 65 are relevant to this problem. We also have the following related problem; see [**14**] for a start.

PROBLEM 69. *Find relationships between Legendrian knot invariants and the Alexander norm or Thurston norm.*

It would also be illuminating to consider:

PROBLEM 70. *Understand more deeply the relation between $\mathcal{L}(\mathcal{K})$ and $\overline{\mathrm{tb}}(\mathcal{K})$ on the one hand, and the HOMFLY and Kauffman polynomials on the other.*

According to Fuchs-Tabachnikov, there are no finite-type invariants of Legendrian knots besides tb, r, and standard topological finite-type invariants. Can one extract non-finite-type topological knot invariants by looking at $\mathcal{L}(\mathcal{K})$? In particular:

QUESTION 71. *Is $\overline{\mathrm{tb}}$ a non-finite-type invariant?*

One possible motivation for applying Legendrian knot theory to topological knot theory is that Legendrian knots are very sensitive to mirroring; e.g., $\mathcal{K}$ and its mirror can have vastly different $\overline{\mathrm{tb}}$.

Finally, showing that a Legendrian knot is determined by its complement is somewhat easy, using Eliashberg's classification of tight contact structures on (S^3, ξ_{std}). However, the topological knot complement problem requires the ingenuity of Gordon and Luecke [**47**].

PROBLEM 72. *Use Legendrian knot theory to reprove the topological knot complement result.*

References

[1] S. Akbulut and R. Matveev, A note on contact structures, *Pacific J. Math.* **182** (1998), 201–204.

[2] S. Akbulut and B. Ozbagci, Lefschetz fibrations on compact Stein surfaces, *Geom. Topol.* **5** (2001), 319–334.

[3] D. Bennequin, Entrelacements et équations de Pfaff, *Astérisque* **107–108** (1983), 87–162.

[4] J. Birman and W. Menasco, Stabilization in the braid groups, preprint, math.GT/0203227.

[5] J. Birman and N. Wrinkle, On transversally simple knots, *J. Diff. Geom.* **55** (2000), 325–354.

[6] D. Calegari and N. Dunfield, Criteria for non existence of taut foliations, in preparation.

[7] T. Cochran and K. Orr, Not all links are concordant to boundary links. *Ann. of Math. (2)* **138** (1993), 519–554.

[8] V. Colin, Chirurgies d'indice un et isotopies de sphères dans les variétés de contact tendues, *C. R. Acad. Sci. Paris Sér. I Math.* **324** (1997), 659–663.

[9] V. Colin, Une infinité de structures de contact tendues sur les variétés toroïdales, *Comment. Math. Helv.* **76** (2001), 353–372.

[10] V. Colin, Chirurgies de Dehn admissibles dans les variétés de contact tendues, *Ann. Inst. Fourier (Grenoble)* **51** (2001), 1419–1435.

[11] V. Colin, Sur la torsion des structures de contact tendues, *Ann. Sci. École Norm. Sup. (4)* **34** (2001), 267–286.

[12] Y. Chekanov, Differential algebra of Legendrian links, *Invent. Math.*, to appear.

[13] Y. Chekanov and P. Pushkar, Arnold's four cusp conjecture and invariants of Legendrian knots, in preparation.

[14] O. Dasbach and B. Mangum, On McMullen's and other inequalities for the Thurston norm of link complements, *Algebr. Geom. Topol.* **1** (2001), 321–347.
[15] F. Ding and H. Geiges, Symplectic fillability of tight contact structures on torus bundles, *Algebr. Geom. Topol.* **1** (2001), 153–172.
[16] K. Dymara, Legendrian knots in overtwisted contact structures on S^3, *Ann. Global Anal. Geom.* **19** (2001), 293–305.
[17] Y. Eliashberg, Classification of overtwisted contact structures on 3-manifolds, *Invent. Math.* **98** (1989), 623–637.
[18] Y. Eliashberg, Topological characterization of Stein manifolds of dimension > 2, *Internat. J. Math.* **1** (1990), 29–46.
[19] Y. Eliashberg, Filling by holomorphic discs and its applications, *Geometry of Low-Dimensional Manifolds*, Vol. II, Donaldson and Thomas, eds. (Cambridge University Press, Cambridge, 1990).
[20] Y. Eliashberg, Contact 3-manifolds twenty years since J. Martinet's work, *Ann. Inst. Fourier (Grenoble)* **42** (1992), 165–192.
[21] Y. Eliashberg, Legendrian and transversal knots in tight contact 3-manifolds, *Topological methods in modern mathematics (Stony Brook, NY, 1991)*, 171–193 (Publish or Perish, Houston, 1993).
[22] Y. Eliashberg, Unique holomorphically fillable contact structure on the 3-torus, *Internat. Math. Res. Notices* **1996**, 77–82.
[23] Y. Eliashberg, Invariants in contact topology, *Doc. Math. J. DMV Extra Volume ICM 1998* (electronic), 327–338.
[24] Y. Eliashberg and M. Fraser, Classification of topologically trivial Legendrian knots, in *Geometry, topology, and dynamics (Montreal, PQ, 1995)*, CRM Proc. Lecture Notes, 15 (Amer. Math. Soc., Providence, 1998).
[25] Y. Eliashberg, A. Givental, and H. Hofer, Introduction to symplectic field theory, *GAFA 2000 (Tel Aviv, 1999)*, *Geom. Funct. Anal.* **2000**, Special Volume, Part II, 560–673.
[26] Y. Eliashberg and W. Thurston, *Confoliations* (Amer. Math. Soc., Providence, 1998).
[27] J. Epstein, D. Fuchs, and M. Meyer, Chekanov-Eliashberg invariants and transverse approximations of Legendrian knots, *Pacific J. Math.* **201** (2001), 89–106.
[28] J. Etnyre, Symplectic convexity in low-dimensional topology, *Top. Appl.* **88** (1998), 3–25.
[29] J. Etnyre, Transversal torus knots, *Geom. Topol.* **3** (1999), 253–268.
[30] J. Etnyre, Tight contact structures on lens spaces, *Commun. Contemp. Math.* **2** (2000), 559–577.
[31] J. Etnyre, Introductory lectures on contact geometry, elsewhere in these proceedings.
[32] J. Etnyre and K. Honda, On the non-existence of tight contact structures, *Ann. of Math. (2)* **153** (2001), 749–766.
[33] J. Etnyre and K. Honda, Knots and contact geometry I: torus knots and figure eight, *J. Sympl. Geom.* **1** (2001), 63–120.
[34] J. Etnyre and K. Honda, Tight contact structures with no symplectic fillings, *Invent. Math.* **148** (2002), 609–626.
[35] J. Etnyre and K. Honda, Konts and contact geometry II: connected sums, preprint, math.SG/0205310.
[36] J. Etnyre, L. Ng, and J. Sabloff, Coherent orientations and invariants of Legendrian knots, *J. Sympl. Geom.*, to appear; also available at math.SG/0101145.
[37] E. Ferrand, On Legendrian knots and polynomial invariants, *Proc. Amer. Math. Soc.* **130** (2002), 1169–1176.
[38] D. Fuchs, Chekanov-Eliashberg invariants of Legendrian knots: existence of augmentations, preprint.
[39] D. Fuchs and S. Tabachnikov, Invariants of Legendrian and transverse knots in the standard contact space, *Topology* **36** (1997), 1025–1054.
[40] D. Gabai, Foliations and the topology of 3-manifolds, *J. Diff. Geom.* **18** (1983), 115–503.
[41] P. Ghiggini, S. Schönenburger, On the classification of tight contact structures, elsewhere in these proceedings.
[42] E. Giroux, Convexité en topologie de contact, *Comment. Math. Helv.* **66** (1991), 637–677.
[43] E. Giroux, Structures de contact en dimension trois et bifurcations des feuilletages de surfaces, *Invent. Math.* **141** (2000), 615–689.

[44] E. Giroux, Structures de contact sur les variétés fibrées en cercles au-dessus d'une surface, *Comment. Math. Helv.* **76** (2001), 218–262.
[45] E. Giroux, Talks at Stanford, December 2000, and at the University of Georgia, May 2001.
[46] R. Gompf, Handlebody construction of Stein surfaces, *Ann. of Math. (2)* **148** (1998), 619–693.
[47] C. McA. Gordon and J. Luecke, Knots are determined by their complements. *J. Amer. Math. Soc.* **2** (1989), 371–415.
[48] M. Gromov, Pseudoholomorphic curves in symplectic manifolds, *Invent. Math.* **82** (1985), 307–347.
[49] J. Hempel, *3-Manifolds*, Ann. of Math. Stud. 86 (Princeton University Press, Princeton, 1976).
[50] K. Honda, On the classification of tight contact structures I, *Geom. Topol.* **4** (2000), 309–368.
[51] K. Honda, On the classification of tight contact structures II, *J. Diff. Geom.* **55** (2000), 83–143.
[52] K. Honda, Gluing tight contact structures, *Duke Math. J.*, to appear; also available as math.GT/0102029.
[53] K. Honda, W. Kazez, and G. Matić, Convex decomposition theory, *Int. Math. Res. Not.* **2002**, 55–88.
[54] K. Honda, W. Kazez, and G. Matić, On the Gabai-Eliashberg-Thurston theorem, preprint, math.GT/0111123.
[55] K. Honda, W. Kazez, and G. Matić, Tight contact structures on fibered hyperbolic 3-manifolds, preprint, math.GT/0110104.
[56] Y. Kanda, The classification of tight contact structures on the 3-torus, *Comm. Anal. Geom.* **5** (1997), 413–438.
[57] P. Kronheimer and T. Mrowka, Monopoles and contact structures, *Invent. Math.* **129** (1997), 509–525.
[58] P. Lisca, Symplectic fillings and positive scalar curvature, *Geom. Topol.* **2** (1998), 103–116.
[59] P. Lisca, On symplectic fillings of 3-manifolds, Proceedings of 6th Gökova Geometry-Topology Conference, *Turkish J. Math.* **23** (1999), 151–159.
[60] P. Lisca and G. Matić, Tight contact structures and Seiberg-Witten invariants, *Invent. Math.* **129** (1997), 509–525.
[61] C. Livingston and S. Naik, Obstructing four-torsion in the classical knot concordance group, *J. Diff. Geom.* **51** (1999), 1–12.
[62] A. Loi and R. Piergallini, Compact Stein surfaces with boundary as branched covers of B^4, *Invent. Math.* **143** (2001), 325–348.
[63] R. Matveyev, talk at Stanford, December 2000.
[64] W. Menasco, On iterated torus knots and transversal knots, *Geom. Topol.* **5** (2001), 651–682.
[65] K. Mishachev, Relative homotopy splitting of differential algebra of Legendrian link, preprint.
[66] A. Mori, A note on Thurston-Winkelnkemper's construction of contact forms on 3-manifolds, *Osaka J. Math.*, **39** (2002), 1–11.
[67] L. Ng, Computable Legendrian invariants, *Topology*, to appear; also available as math.GT/0011265.
[68] L. Ng, The Legendrian satellite construction, preprint, math.GT/0112105.
[69] H. Ohta and K. Ono, Simple singularities and topology of symplectically filling 4-manifold, *Comment. Math. Helv.* **74** (1999), 575–590.
[70] R. Roberts, J. Shareshian, and M. Stein, research announcement.
[71] D. Rolfsen, *Knots and links* (Publish or Perish Press, Berkeley, 1976).
[72] L. Rudolph, An obstruction to sliceness via contact geometry and "classical" gauge theory, *Invent. Math.* **119** (1995), 155–163.
[73] J. Sabloff, Invariants of Legendrian knots in circle bundles, in preparation.
[74] I. Smith, Torus fibrations on symplectic four-manifolds, *Turkish J. Math.* **25** (2001), 69–95.
[75] W. Thurston, Hyperbolic structures on 3-manifolds, III: Deformations of 3-manifolds with incompressible boundary, *Ann. of Math. (2)*, to appear; also available as math.GT/9801058.
[76] W. Thurston and H. Winkelnkemper, On the existence of contact forms, *Proc. Amer. Math. Soc.* **52** (1975), 345–347.
[77] L. Traynor, Generating function polynomials for legendrian links, *Geom. Topol.* **5** (2001), 719–760.

[78] A. Weinstein, Contact surgery and symplectic handlebodies, *Hokkaido Math. J.* **20** (1991), 241–251.

UNIVERSITY OF PENNSYLVANIA, PHILADELPHIA, PA 19104
E-mail address: `etnyre@math.upenn.edu`
URL: `http://www.math.upenn.edu/~etnyre`

AMERICAN INSTITUTE OF MATHEMATICS, PALO ALTO, CA 94306
E-mail address: `ng@alum.mit.edu`
URL: `http://alum.mit.edu/www/ng`

Titles in This Series

71 **Gordana Matić and Clint McCrory, Editors,** Topology and geometry of manifolds (University of Georgia, Athens, Georgia, 2001)

70 **Michael D. Fried and Yasutaka Ihara, Editors,** Arithmetic fundamental groups and noncommutative algebra (Mathematical Sciences Research Institute, Berkeley, California, 1999)

69 **Anatole Katok, Rafael de la Llave, Yakov Pesin, and Howard Weiss, Editors,** Smooth ergodic theory and its applications (University of Washington, Seattle, 1999)

68 **Robert S. Doran and V. S. Varadarajan, Editors,** The mathematical legacy of Harish-Chandra: A celebration of representation theory and harmonic analysis (Baltimore, Maryland, 1998)

67 **Wayne Raskind and Charles Weibel, Editors,** Algebraic K-theory (University of Washington, Seattle, 1997)

66 **Robert S. Doran, Ze-Li Dou, and George T. Gilbert, Editors,** Automorphic forms, automorphic representations, and arithmetic (Texas Christian University, Fort Worth, 1996)

65 **M. Giaquinta, J. Shatah, and S. R. S. Varadhan, Editors,** Differential equations: La Pietra 1996 (Villa La Pietra, Florence, Italy, 1996)

64 **G. Ferreyra, R. Gardner, H. Hermes, and H. Sussmann, Editors,** Differential geometry and control (University of Colorado, Boulder, 1997)

63 **Alejandro Adem, Jon Carlson, Stewart Priddy, and Peter Webb, Editors,** Group representations: Cohomology, group actions and topology (University of Washington, Seattle, 1996)

62 **János Kollár, Robert Lazarsfeld, and David R. Morrison, Editors,** Algebraic geometry—Santa Cruz 1995 (University of California, Santa Cruz, July 1995)

61 **T. N. Bailey and A. W. Knapp, Editors,** Representation theory and automorphic forms (International Centre for Mathematical Sciences, Edinburgh, Scotland, March 1996)

60 **David Jerison, I. M. Singer, and Daniel W. Stroock, Editors,** The legacy of Norbert Wiener: A centennial symposium (Massachusetts Institute of Technology, Cambridge, October 1994)

59 **William Arveson, Thomas Branson, and Irving Segal, Editors,** Quantization, nonlinear partial differential equations, and operator algebra (Massachusetts Institute of Technology, Cambridge, June 1994)

58 **Bill Jacob and Alex Rosenberg, Editors,** K-theory and algebraic geometry: Connections with quadratic forms and division algebras (University of California, Santa Barbara, July 1992)

57 **Michael C. Cranston and Mark A. Pinsky, Editors,** Stochastic analysis (Cornell University, Ithaca, July 1993)

56 **William J. Haboush and Brian J. Parshall, Editors,** Algebraic groups and their generalizations (Pennsylvania State University, University Park, July 1991)

55 **Uwe Jannsen, Steven L. Kleiman, and Jean-Pierre Serre, Editors,** Motives (University of Washington, Seattle, July/August 1991)

54 **Robert Greene and S. T. Yau, Editors,** Differential geometry (University of California, Los Angeles, July 1990)

53 **James A. Carlson, C. Herbert Clemens, and David R. Morrison, Editors,** Complex geometry and Lie theory (Sundance, Utah, May 1989)

For a complete list of titles in this series, visit the AMS Bookstore at **www.ams.org/bookstore/**.

ISBN 0-8218-3507-6

9 780821 835074

PSPUM/71